Seismogenesis and Earthquake Forecasting: The Frank Evison Volume II

Edited by
Martha K. Savage
David A. Rhoades
Euan G. C. Smith
Matthew C. Gerstenberger
David Vere-Jones

Previously published in *Pure and Applied Geophysics* (PAGEOPH), Volume 167, Nos. 8–9, 2010

Birkhäuser

Editors

Martha K. Savage
Institute of Geophysics
Victoria University of Wellington
Wellington
New Zealand
martha.savage@vuw.ac.nz

David A. Rhoades
GNS Science
Lower Hutt
New Zealand
d.rhoades@gns.cri.nz

Euan G.C. Smith
Institute of Geophysics
Victoria University of Wellington
Wellington
New Zealand
Euan.Smith@vuw.ac.nz

Matthew C. Gerstenberger
GNS Science
Lower Hutt
New Zealand
M.gerstenberger@gns.cri.nz

David Vere-Jones
Statistics Research Associates
PO Box 12649
Wellington
New Zealand
dvj@paradise.net.nz

ISBN 978-3-0346-0499-4   e-ISBN 978-3-0346-0500-7
DOI 10.1007/978-3-0346-0500-7

Library of Congress Control Number: 2010929575

© Springer Basel AG 2010

This work is subject to copyright. All rights are reserved, whether the whole or part of the material is concerned, specifically the rights of translation, reprinting, re-use of illustrations, recitation, broadcasting, reproduction on microfilms or in other ways, and storage in data banks. For any kind of use, permission of the copyright owner must be obtained.

Cover illustration: Composed of Fig. 1 from "Forecasting the locations of future large earthquakes: an analysis and verification" by A. R. Scherbakov, D. L. Turcotte, J. B. Rundle, K. F. Tiampo and J. R. Holliday (Evison Vol. I) and of Fig. 12 from "New Zealand Earthquake Forecast Testing Centre" by M. C. Gerstenberger and D. A. Rhoades (Evison Vol. II).

Printed on acid-free paper

Springer Basel AG is part of Springer Science+Business Media

www.birkhauser-science.com

Contents

1   Introduction
     M.K. Savage, D.A. Rhoades, E.G.C. Smith, M.C. Gerstenberger, D. Vere-Jones

5   First Results of the Regional Earthquake Likelihood Models Experiment
     D. Schorlemmer, J.D. Zechar, M.J. Werner, E.H. Field, D.D. Jackson, T.H. Jordan, and The RELM Working Group

23   New Zealand Earthquake Forecast Testing Centre
     M.C. Gerstenberger, D.A. Rhoades

39   The Area Skill Score Statistic for Evaluating Earthquake Predictability Experiments
     J.D. Zechar, T.H. Jordan

53   Space–Time Earthquake Prediction: The Error Diagrams
     G. Molchan

65   Identifying Seismicity Levels via Poisson Hidden Markov Models
     K. Orfanogiannaki, D. Karlis, G.A. Papadopoulos

79   Distribution of Seismicity Before the Larger Earthquakes in Italy in the Time Interval 1994–2004
     S. Gentili

105 Predicting the Human Losses Implied by Predictions of Earthquakes: Southern Sumatra and Central Chile
     M. Wyss

113 Space- and Time-Dependent Probabilities for Earthquake Fault Systems from Numerical Simulations: Feasibility Study and First Results
     J. Van Aalsburg, J.B. Rundle, L.B. Grant, P.B. Rundle, G. Yakovlev, D.L. Turcotte, A. Donnellan, K.F. Tiampo, J. Fernandez

125 Spatial Separation of Large Earthquakes, Aftershocks, and Background Seismicity: Analysis of Interseismic and Coseismic Seismicity Patterns in Southern California
     E. Hauksson

145 Earthquake Source Zones in Northeast India: Seismic Tomography, Fractal Dimension and b Value Mapping
     P.M. Bhattacharya, J.R. Kayal, S. Baruah, S.S. Arefiev

159 Seismic Hazard Evaluation in Western Turkey as Revealed by Stress Transfer and Time-dependent Probability Calculations
     P.M. Paradisopoulou, E.E. Papadimitriou, V.G. Karakostas, T. Taymaz, A. Kilias, S. Yolsal

195 Correlation of Static Stress Changes and Earthquake Occurrence in the North Aegean Region
     D.A. Rhoades, E.E. Papadimitriou, V.G. Karakostas, R. Console, M. Murru

213 Aftershock Sequences Modeled with 3-D Stress Heterogeneity and Rate-State Seismicity Equations: Implications for Crustal Stress Estimation
     D.E. Smith, J.H. Dieterich

233 Earthquake Recurrence in Simulated Fault Systems
     J.H. Dieterich, K.B. Richards-Dinger

251 Continuous Observation of Groundwater and Crustal Deformation for Forecasting Tonankai and Nankai Earthquakes in Japan
     S. Itaba, N. Koizumi, N. Matsumoto, R. Ohtani

261 Anomalies of Seismic Activity and Transient Crustal Deformations Preceding the 2005 M 7.0 Earthquake West of Fukuoka
     Y. Ogata

Introduction

Martha K. Savage,[1] David A. Rhoades,[2] Euan G. C. Smith,[1] Matthew C. Gerstenberger,[2] and David Vere-Jones[1]

Professor Frank Evison, OBE, FRSNZ, (20.3.1922–25.01.2005). Inaugural Professor of Geophysics, Victoria University of Wellington, 1967–1988. Inaugural Director, Institute of Geophysics, Victoria University of Wellington, 1971–1988. Professor Emeritus, Victoria University of Wellington, 1988–2005. Photo credit: Robert Cross, VUW Image Services

This special issue is an augmented collection of papers originating from the Evison Symposium on Seismogenesis and Earthquake Forecasting held in Wellington, New Zealand, in February 2008. There are two volumes in the issue. The first volume is published in Pure and Applied Geophysics, Vol. 167, Nos. 6/7, 2010, and, in addition to the research papers, includes a biography of Frank Evison and a list of his publications. Here we describe the papers within Volume II, and thank again all reviewers who contributed with papers in either volume.

Certain papers in this volume continue the statistical seismology theme from Volume I, and others relate more directly to the physics of source processes. The first volume contained some papers of methods to be used in earthquake predictability studies through the Collaboratory for the Study of Earthquake Predictability (CSEP) program. This volume includes discussions of the CSEP program itself and its early results from programs designed to carry out the testing. Schorlemmer et al. discuss the first results of a CSEP analysis program called "Regional Earthquake Likelihood Models" or RELM, being carried out for forecasts of earthquakes that might cause damage in California. Although the study is only halfway through its initial five-year program, preliminary results suggest that most submitted models are better than Poissonian models, and that one model, the Helmstetter et al. main-shock model, is to date out-predicting the others. Gerstenberger and Rhoades describe the New Zealand version of the CSEP initiative, the New Zealand Earthquake Forecast Testing Centre. The five-year testing period started in 2008, and new models are encouraged to be submitted.

Zechar and Jordan are working towards new methods to use in mutually comparing different models in earthquake predictability experiments. Their paper here evaluates a measure called the "area skill score" that they have previously suggested to extend the range of models that can participate in such experiments. They present statistical properties of the area skill score, and describe and illustrate a

---

[1] Institute of Geophysics, Victoria University of Wellington, Box 600, Wellington, New Zealand. E-mail: martha.savage@vuw.ac.nz

[2] GNS Science, Lower Hutt, New Zealand.

preliminary procedure for comparing earthquake prediction strategies based on alarm functions.

The paper by MOLCHAN expands the use of error diagrams to characterise whether a potential predicting variable is useful in terms of the rates of prediction failure versus successes. Initial use in the time dimension was expanded into the spatial dimension. This paper supplements understanding of the spatial dimension analysis by determining the structure of the error diagram for space–time prediction and by analysing the properties of two-dimensional error diagrams.

ORFANOGIANNAKI et al. examine the usefulness of identifying changes in seismicity levels via a type of modelling termed "Poisson hidden Markov Models", applied to an earthquake catalogue in an area of Greece. These models assume that a system has two or more states, and that the probability to change from one state to another is unknown. Their models reproduce seismicity clusters in the catalogue and quantify the dependence of the earthquakes on each other at any particular time. They identify previously unrecognised foreshock occurrences and expect that such recognition may assist future warning of impending earthquakes.

GENTILI develops a new algorithm to seek quiescence before large earthquakes in Italy during the ten-year period from 1994–2004, finding that two-thirds of the earthquakes with magnitude larger than five are preceded by quiescence.

WYSS views prediction in a different light. Instead of predicting when an earthquake will occur, he predicts the human losses that are likely to occur if earthquakes that have been predicted in fact arrive. Here he focuses on earthquakes in southern Sumatra and central Chile, determining that, if tsunami effects are not considered, then fatalities are likely to be less than 1,000 in southern Sumatra, but larger than 1,000 in central Chile.

Other papers include knowledge of fault structure as well as seismicity to determine hazards. VAN AALSBURG et al. develop numerical simulations of the known fault systems in California to determine probability density functions for earthquake occurrences over time periods on the order of thirty years. They incorporate the fault geometry and paleoseismic data from past earthquake occurrences, and they include the probability that faults might interact and cluster in time. They present forecasts for the probabilities of earthquakes in several magnitude ranges occurring within the next thirty years along the San Andreas Fault system.

HAUKSSON uses a catalogue of highly accurate relative earthquake locations in southern California to determine the spatial relations between main shocks, aftershocks and background seismicity. He finds that large earthquakes that slip several meters or more occur on mapped faults; that aftershocks typically occur within 2 km of these faults, and that most background seismicity occurs within about 10 km of the faults, with a rate of occurrence decaying as a function of distance from the fault. The background earthquakes are interpreted as occurring on a network of small faults accommodating damage from the interaction of the main fault with irregularities in geometry.

BHATTACHARYA et al. examine the characteristics of earthquake source zones in northeast India by comparing seismic tomography results with maps of fractal dimensions and b values determined from the regional earthquake catalogue. They find a strong correlation between b value and fractal dimension, and they also determine that several regions of high velocity are located in regions of high seismic activity with high fractal dimensions close to 2.0, indicating that most of the earthquake-associated fractures are approaching a two-dimensional space. High b values are also observed along some active faults.

The contribution an earthquake makes to triggering other earthquakes due to stress increases on nearby faults (Coulomb stress modeling) has become a lively topic for debate. Two papers examine this contribution and yet come to opposite conclusions using very similar techniques on nearly the same data set. They examine the interaction of faults in the Aegean region by comparing the real earthquake distributions with those expected from the evolving stress changes due to tectonic loading and to each earthquake occurrence. PARADISOPOULOU et al. determine that including both the long-term and tectonic strain as well as the near-field stress changes occurring due to past earthquakes is necessary. In contrast, RHOADES et al. find that time-invariant models based on constant tectonic loading are as good as or better than time–varying earthquake likelihood models determined from the evolving stress field.

One approach to understanding earthquakes is to create synthetic earthquake catalogues using certain assumptions regarding the physics of the process to see what features of real catalogues can be explained by variations in physical properties. SMITH and DIETERICH take this approach and model aftershock sequences using 3-D stress heterogeneity in the form of Coulomb static stress change analysis and rate-state seismicity equations calculated in regions of geometrically complex faults. Their synthetic models match several features of real catalogues such as earthquake clustering and Omori decay, and the presence of earthquakes in regions where simpler Coulomb stress modelling predicts "stress shadows" with few earthquakes.

DIETERICH and RICHARDS-DINGER also use simulated fault systems. They examine the effects of earthquake nucleation and fault system geometry on earthquake occurrence. They again find strong clustering both spatially and temporally, corresponding to foreshocks, aftershocks and occasionally large-earthquake pairs. They determine that fault system geometry acts as the primary control of earthquake recurrence statistics. They propose using fault system earthquake simulators to define the empirical probability density distributions for use in regional assessments of earthquake probabilities.

Other geophysical measurements may also be related to earthquake occurrence, and such relationships are examined in some of the papers herein. ITABA et al. compare groundwater and crustal deformation to seismicity recorded on newly installed stations to test previously observed preseismic changes in Shikoku and the Kii Peninsula prior to earthquakes in Tonankai and Nankai, Japan. They find strain changes due to slow slip events on the plate boundary, but do not find significant changes in groundwater at that time.

We conclude the volume with another comparison of seismicity and GPS. OGATA compares anomalies of seismic activity with transient crustal deformations preceding the 2005 M 7.0 earthquake west of Fukuoka, Japan, concluding that aseismic slip triggered changes in seismicity rates as well as in GPS recordings during the ten years leading up to the earthquake.

We thank the following colleagues who have reviewed papers for these two volumes: M. Bebbington, A. Christophersen, R. Console, J. Cousins, R. Davies, K. Felzer, C. Frohlich, B. Fry, J. Hardebeck, D. Harte, A. Helmstetter, M. Imoto, T. Iwata, Y. Kagan, A.M. Lombardi, B. Lund, W. Marzocchi, G. Molchan, R. C. Nicholson, P. M. Paradisopoulou, R. Robinson, J. Rundle, D. Schorlemmer, D. Shanker, R. Shcherbakov, W. Smith, B. Stephenson, K. Tiampo, S. Toda, J. Townend, T. Van Stiphout, F. Wenzel, M. Werner, J. Woessner, M. Wyss, I. Zaliapin, A. Zavyalov, J. Zechar, and J. Zhuang.

First Results of the Regional Earthquake Likelihood Models Experiment

Danijel Schorlemmer,[1] J. Douglas Zechar,[1,2] Maximilian J. Werner,[3] Edward H. Field,[4] David D. Jackson,[5] Thomas H. Jordan,[1] and The RELM Working Group

Abstract—The ability to successfully predict the future behavior of a system is a strong indication that the system is well understood. Certainly many details of the earthquake system remain obscure, but several hypotheses related to earthquake occurrence and seismic hazard have been proffered, and predicting earthquake behavior is a worthy goal and demanded by society. Along these lines, one of the primary objectives of the Regional Earthquake Likelihood Models (RELM) working group was to formalize earthquake occurrence hypotheses in the form of prospective earthquake rate forecasts in California. RELM members, working in small research groups, developed more than a dozen 5-year forecasts; they also outlined a performance evaluation method and provided a conceptual description of a Testing Center in which to perform predictability experiments. Subsequently, researchers working within the Collaboratory for the Study of Earthquake Predictability (CSEP) have begun implementing Testing Centers in different locations worldwide, and the RELM predictability experiment—a truly prospective earthquake prediction effort—is underway within the U.S. branch of CSEP. The experiment, designed to compare time-invariant 5-year earthquake rate forecasts, is now approximately halfway to its completion. In this paper, we describe the models under evaluation and present, for the first time, preliminary results of this unique experiment. While these results are preliminary—the forecasts were meant for an application of 5 years—we find interesting results: most of the models are consistent with the observation and one model forecasts the distribution of earthquakes best. We discuss the observed sample of target earthquakes in the context of historical seismicity within the testing region, highlight potential pitfalls of the current tests, and suggest plans for future revisions to experiments such as this one.

The members of the RELM Working Group are listed in the Acknowledgments section.

[1] Department of Earth Sciences, Southern California Earthquake Center, University of Southern California, 3651 Trousdale Parkway, Los Angeles, CA 90089-0740, USA. E-mail: ds@usc.edu
[2] Lamont-Doherty Earth Observatory, Columbia University, P.O. Box 1000, Palisades, NY 10964, USA.
[3] Swiss Seismological Service, ETH Zurich, Sonneggstrasse 5, 8092 Zurich, Switzerland.
[4] United States Geological Survey, 525 S. Wilson Avenue, Pasadena, CA 91106, USA.
[5] Department of Earth and Space Sciences, University of California Los Angeles, Los Angeles, CA 90095, USA.

Key words: Statistical seismology, earthquake predictability, earthquake statistics, earthquake forecasting and testing, seismic hazard.

1. Introduction

The Regional Earthquake Likelihood Model (RELM) working group formed in 2000 and was supported by the Southern California Earthquake Center (SCEC) and the United States Geological Survey (USGS). The group's main purpose was to improve seismic hazard assessment and to increase understanding of earthquake generation processes. Seismic hazard analysis requires two fundamental components: an earthquake forecast that describes the probabilities of earthquake occurrence in a spatio-temporal volume; and a ground-motion model that transforms each forecasted event into a site-specific estimate of ground-shaking. RELM participants focused on the former component and developed several earthquake forecast models (Bird and Liu, 2007; Console *et al.*, 2007; Ebel *et al.*, 2007; Gerstenberger *et al.*, 2007; Helmstetter *et al.*, 2007; Holliday *et al.*, 2007; Kagan *et al.*, 2007; Petersen *et al.*, 2007; Rhoades, 2007; Shen *et al.*, 2007; Ward, 2007; Wiemer and Schorlemmer, 2007). These models span a broad range of input data and methods: most are based on past seismicity, however some incorporate geodetic data and/or geological insights. See Field (2007) and the special volume of Seismological Research Letters for more details on the RELM project.

In addition to developing forecast models, RELM also explored comparative testing strategies and established a plan for conducting these tests.

The group developed a suite of likelihood tests (SCHORLEMMER et al., 2007) to be implemented within a Testing Center, a facility in which earthquake forecast models are installed as software codes and in which all necessary tests are conducted in an automated and fully prospective fashion (SCHORLEMMER and GERSTENBERGER, 2007). By the end of the 5-year project, 19 earthquake forecasts were submitted for prospective testing in the period of 1 January 2006, 00:00–1 January 2011, 00:00. These forecasts were not installed as software codes in the Testing Center because the RELM group decided to use simple forecast tables; nevertheless, the processing is fully automated and does not require human interaction. All other models in the Testing Center, including the RELM 1-day models, are installed as codes.

Following the conclusion of the RELM project, the Collaboratory for the Study of Earthquake Predictability (CSEP) was formed as a venue to expand upon the RELM experiment and to establish and maintain a Testing Center (JORDAN, 2006). CSEP is built upon a global partnership to promote rigorous earthquake predictability experiments in various tectonic environments. In addition to establishing new testing regions, CSEP is developing new testing methods, introducing new kinds of earthquake forecast models, and improving upon the testing rules suggested by the RELM working group. The U.S. branch of CSEP inherited all RELM earthquake forecasts, as well as the task of testing them according to the rules outlined by SCHORLEMMER et al. (2007) in a Testing Center designed according to SCHORLEMMER and GERSTENBERGER (2007).

All models developed by RELM participants forecast earthquakes in a testing area that covers the state of California and all regions within about one degree of its borders. This test region was chosen to include any earthquake that might cause shaking within the state of California (SCHORLEMMER and GERSTENBERGER, 2007). The RELM working group proposed two major classes of forecasts: 1 day and 5 years (SCHORLEMMER and GERSTENBERGER, 2007). In contrast to daily or yearly periodicity in weather, earthquakes do not follow obvious seasonal or cyclical patterns that could be used to scientifically justify the chosen durations. Rather, the classes are end-user-oriented: The 5-year class is relevant for seismic hazard calculations, while the 1-day class allows a closer look at aftershock hazard forecasts and potential short-term precursor detection. Daily forecasts can make use of all seismicity up to and including the previous day to adapt to new earthquakes and to re-calibrate the model, whereas the 5-year forecasts are fixed at the beginning of the experiment and never updated. Because of this fundamental difference in the setup, models were either submitted for the 1-day class or the 5-year class. Forecasts submitted to the 5-year class were taken to be time-invariant. We briefly describe the models below; a detailed summary of the models is given by FIELD (2007) while the full descriptions of each model can be found in the individual articles in the special volume of Seismological Research Letters (see Table 1).

One of the main goals of RELM was to test models comparatively; to compare models, a significant standardization of the forecasts was necessary. Therefore, all testing rules, the testing period, the testing area, and the earthquake catalog and its processing were defined by SCHORLEMMER and GERSTENBERGER (2007) and agreed upon by the members of the RELM working group. This standardization also required that all RELM models provide grid-based forecasts: earthquake rates specified in latitude/longitude/magnitude bins, and characterized by Poisson uncertainty. Models that declare alarms or forecast fault ruptures were not considered, as no testing method was developed or specified for these kinds of forecasts.

In this paper we describe the different model classes and present the results from the first 2.5 years of testing the time-invariant 5-year RELM forecasts. Because the forecasts were specified as being time-invariant, all forecast rates were halved for the results presented here. We emphasize, however, that these results are preliminary because the forecasts were specified as 5-year forecasts. As more earthquakes occur, the results will likely change. Nevertheless, the results indicate which models are consistent with the observations to date and which models have so far performed best in comparative testing.

Table 1

RELM models being evaluated within the Testing Center

Model	Testing class	Forecasted number of earthquakes	Fraction of area covered by forecast (%)	Reference
EBEL-ET-AL.MAINSHOCK	5-year mainshock	8.6703 (8.6705)	47.37	EBEL et al. (2007)
EBEL-ET-AL.MAINSHOCK.CORRECTED	5-year mainshock	9.2431 (9.2433)	51.74	EBEL et al. (2007)
HELMSTETTER-ET-AL.MAINSHOCK	5-year mainshock	10.5760	100.00	HELMSTETTER et al. (2007)
HOLLIDAY-ET-AL.PI	5-year mainshock	14.4205 (15.0164)	8.29	HOLLIDAY et al. (2007)
KAGAN-ET-AL.MAINSHOCK	5-year mainshock	5.9998 (5.9998)	44.39	KAGAN et al. (2007)
SHEN-ET-AL.MAINSHOCK	5-year mainshock	5.2369 (5.2369)	44.39	SHEN et al. (2007)
WARD.COMBO81	5-year mainshock	9.4812 (16.0582)	26.72	WARD (2007)
WARD.GEODETIC81	5-year mainshock	12.1498 (27.9849)	26.72	WARD (2007)
WARD.GEODETIC85	5-year mainshock	6.9972 (16.1169)	26.72	WARD (2007)
WARD.GEOLOGIC81	5-year mainshock	8.3332 (9.0760)	26.72	WARD (2007)
WARD.SEISMIC81	5-year mainshock	7.9605 (11.1136)	26.72	WARD (2007)
WARD.SIMULATION	5-year mainshock	3.7261 (4.1027)	26.72	WARD (2007)
WIEMER-SCHORLEMMER.ALM	5-year mainshock	11.8693	100.00	WIEMER and SCHORLEMMER (2007)
BIRD-LIU.NEOKINEMA	5-year mainshock+aftershock	27.9514	100.00	BIRD and LIU (2007)
EBEL-ET-AL.AFTERSHOCK	5-year mainshock+aftershock	36.4017 (36.4026)	47.37	EBEL et al. (2007)
EBEL-ET-AL.AFTERSHOCK.CORRECTED	5-year mainshock+aftershock	37.5664 (37.5674)	51.74	EBEL et al. (2007)
HELMSTETTER-ET-AL.AFTERSHOCK	5-year mainshock+aftershock	17.7012	100.00	HELMSTETTER et al. (2007)
KAGAN-ET-AL.AFTERSHOCK	5-year mainshock+aftershock	7.9910 (7.9910)	44.39	KAGAN et al. (2007)
SHEN-ET-AL.AFTERSHOCK	5-year mainshock+aftershock	7.3236 (7.3236)	44.39	SHEN et al. (2007)

All models were submitted before 1 January 2006, except for the EBEL-ET-AL.MAINSHOCK.CORRECTED model and the EBEL-ET-AL.AFTERSHOCK.CORRECTED model, which were submitted 12 November 2006. The forecasted number of earthquakes reported here is the number forecasted in all unmasked cells, followed parenthetically by the number forecasted in all cells (see Masking subsection in the text). The fraction of the area covered by forecast is the portion of the study region for which the model makes an unmasked forecast

2. Models

2.1. 5-Year Models

The forecasts submitted to the 5-year class represent a broad spectrum of models, each of which is built on its own set of scientific hypotheses pertaining to the occurrence of earthquakes. Most of the models use past seismicity as the primary data set for model calibration and parameter value estimation, and they then extrapolate historical seismicity rates into the future. However, some models make use of geological, geodetic, and/or tectonic data.

Large earthquakes are followed by dozens to hundreds of earthquakes in their immediate wake. If a very large event were to occur in California tomorrow, its triggered earthquakes would likely dominate the statistics of the entire 5-year period. Because mainshocks and dependent aftershocks cannot be identified by some physical measurement, a compromise was made to accommodate models which forecast independent mainshocks only. Two forecast subclasses were created: one for forecasts of mainshocks only (*mainshock* models) and one for forecasts of all earthquakes (*mainshock+aftershock* models). SCHORLEMMER and GERSTENBERGER (2007) and SCHORLEMMER et al. (2007) provide details on the declustering procedure that is used at the testing center to create catalogs of mainshocks against which the *mainshock* models are tested. Both classes forecast rates of earthquakes with magnitude greater than or equal to 4.95 with a binning of 0.1 magnitude units (resulting in magnitude bins of [4.95, 5.05), [5.05, 5.15), etc., with a final bin starting at magnitude 8.95 with no upper limit) and a spatial binning of $0.1° \times 0.1°$ with the cell boundaries aligned to the full degrees. The observed magnitude is taken to be the magnitude reported in the Advanced National Seismic System (ANSS) catalog, disregarding the magnitude scale.

2.2. Mainshock Models

Twelve *mainshock* models were submitted to RELM; these were formally registered and published on the RELM website (http://relm.cseptesting.org, see also Table 1 and Figs. 1 and 2). Of these, many were generated by smoothing past seismicity under different assumptions. The models EBEL-ET-AL.MAINSHOCK and EBEL-ET-AL.MAINSHOCK.CORRECTED (see below for the explanation of the double entry), developed by EBEL et al. (2007), average the 5-year rate of $M \geq 5$ earthquakes in 3° by 3° cells from a declustered catalog from 1932 until 2004 and use a Gutenberg-Richter distribution for computing rates per magnitude. The model KAGAN-ET-AL.MAINSHOCK (KAGAN et al., 2007) smooths past earthquakes using a longer catalog dating back to 1800 and it accounts for the spatial extent of large earthquake ruptures.

Rates are calculated using a tapered Gutenberg-Richter distribution with corner magnitude 8. HELMSTETTER et al. (2007) extend this approach to their HELMSTETTER-ET-AL.MAINSHOCK model by including past $M \geq 2$ events since 1984 in the smoothing, by optimizing the smoothing, and by accounting for the spatial variability of the completeness magnitude. The model WARD.SEISMIC81 (WARD, 2007) is also based on smoothing past earthquakes, in this case going back to 1850.

WIEMER and SCHORLEMMER (2007) estimated the a and b values of the Gutenberg-Richter distribution in each latitude/longitude cell to test the hypothesis that spatial variations in these values designate stationary

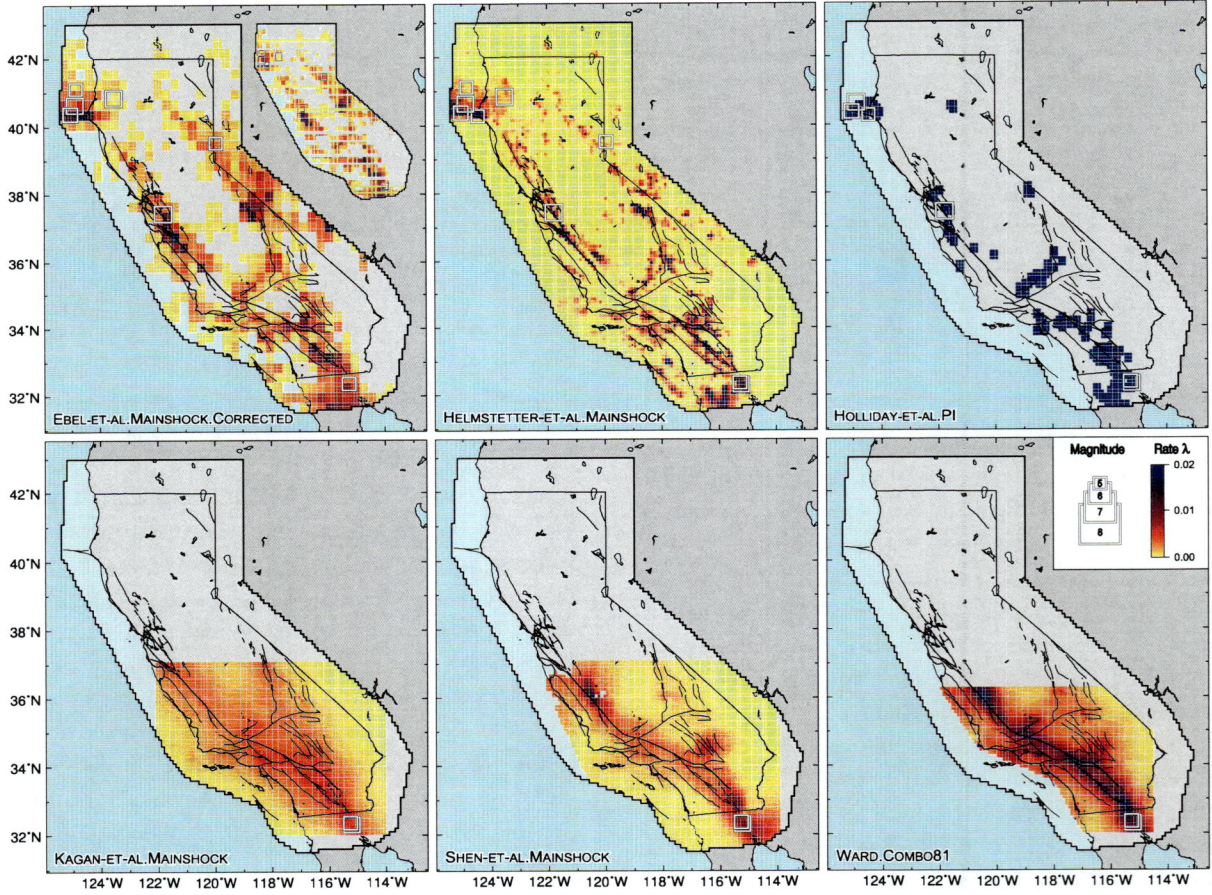

Figure 1
Forecast maps of *5-year mainshock* models. Colors indicate the forecast rate of all events with $M \geq 4.95$ (unmasked areas only), reducing the latitude/longitude/magnitude forecasts to latitude/longitude forecasts by summing over the magnitude bins. The observed target earthquakes are shown as *white squares*; only those earthquakes occurring in unmasked cells are shown for each model. Models from left to right: (*first row*) EBEL-ET-AL.MAINSHOCK.CORRECTED with EBEL-ET-AL.MAINSHOCK as inset, HELMSTETTER-ET-AL.MAINSHOCK, and HOLLIDAY-ET-AL.PI. (*second row*) KAGAN-ET-AL.MAINSHOCK, SHEN-ET-AL.MAINSHOCK, and WARD.COMBO81

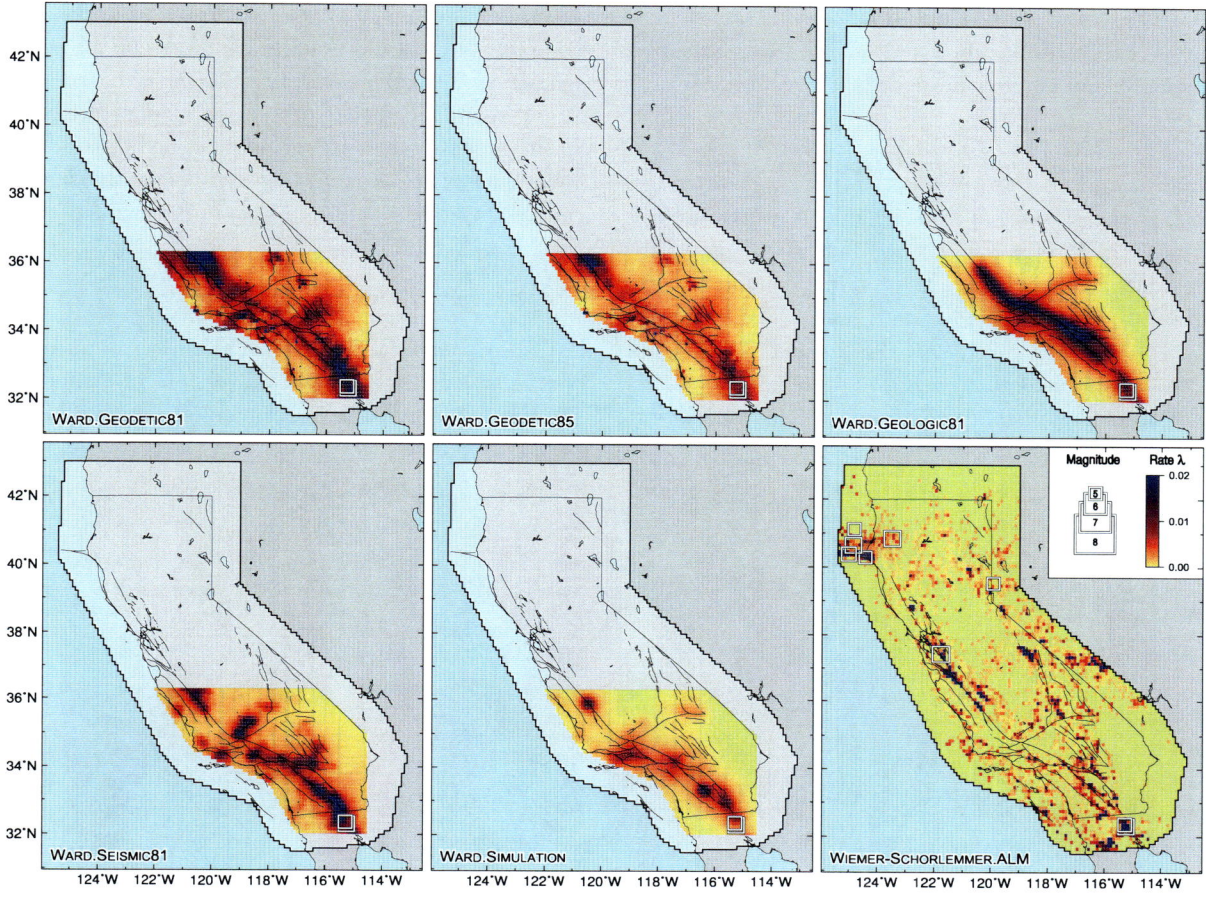

Figure 2

Forecast maps of *5-year mainshock* models. Colors indicate the forecast rate of all events with $M \geq 4.95$ (unmasked areas only), reducing the latitude/longitude/magnitude forecast to latitude/longitude forecasts by summing over the magnitude bins. The observed target earthquakes are shown as *white squares*; only those earthquakes occurring in unmasked cells are shown for each model. Models from left to right: (*first row*) WARD.GEODETIC81, WARD.GEODETIC85, and WARD.GEOLOGIC81. (*second row*) WARD.SEISMIC81, WARD.SIMULATION, and WIEMER-SCHORLEMMER.ALM

asperities that govern the relative frequency of large and small earthquakes (the WIEMER-SCHORLEMMER. ALM model). The model HOLLIDAY-ET-AL.PI, submitted by HOLLIDAY et al. (2007), is based on the assumption that regions of strongly fluctuating seismicity will be the regions of future large earthquakes.

Some models include data other than past earthquake observations. Three models are based solely on geodetic data. In one, SHEN-ET-AL.MAINSHOCK, SHEN et al. (2007) assumed that the earthquake rate is proportional to the horizontal maximum shear strain rate. The magnitude rates are obtained from a spatially-invariant tapered Gutenberg-Richter distribution with corner magnitude 8.02. A second model,

WARD.GEODETIC81 by WARD (2007), uses a larger data set and a different technique to map strain rates to seismicity rates. The sole difference between this and the third model, WARD.GEODETIC85 by WARD (2007), is the maximum magnitude in the truncated Gutenberg-Richter distribution (8.1 and 8.5, respectively).

WARD (2007) also provided a mainshock model based solely on geological data (WARD.GEOLOGIC81). The model is constructed by mapping fault slip rates into a smoothed geological moment rate density and then into seismicity rate, again assuming a spatially invariant truncated Gutenberg-Richter distribution. The model WARD.SIMULATION is based on simulations of velocity-weakening friction on a fixed fault network representing California. The model WARD.COMBO81 presents the average of the seismic, geodetic, and geological models by WARD (2007).

2.3. Mainshock+Aftershock Models

Six *mainshock+aftershock* models were submitted to RELM (see Table 1 and Fig. 3). Of these, all but one are modifications of corresponding *mainshock* forecasts: EBEL *et al.* (2007), KAGAN *et al.* (2007), HELMSTETTER *et al.* (2007) and SHEN *et al.* (2007) calibrated their *mainshock+aftershock* forecast to a complete catalog while their *mainshock*

forecasts were calibrated based on a declustered catalog of past seismicity. The model BIRD-LIU.NEOKINEMA by BIRD and LIU (2007) is based on a local kinematic model of surface velocities derived from geodetic, tectonic, geological, and stress-direction data. The velocities are mapped into seismic moment rate and then into long-term seismicity rate.

2.4. Corrected Forecast Groups

Two additional 5-year model classes were introduced to account for corrected versions of the models by EBEL *et al.* (2007). In their initial submission, the

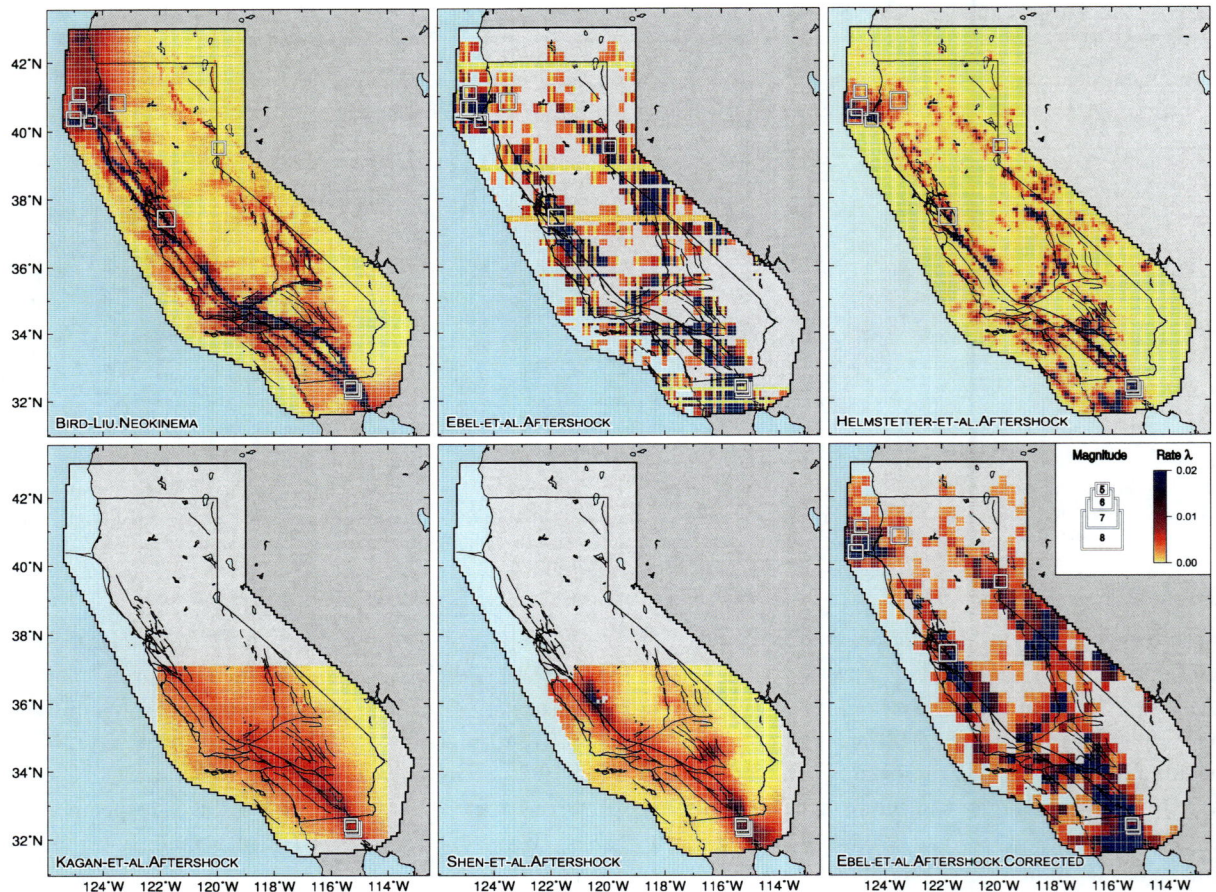

Figure 3
Forecast maps of all *5-year mainshock+aftershock* models. Colors indicate the forecast rate of all events with $M \geq 4.95$ (unmasked areas only), reducing the latitude/longitude/magnitude forecasts to latitude/longitude forecasts by summing over the magnitude bins. The observed target earthquakes are shown as *white squares*; only those earthquakes occurring in unmasked cells are shown for each model.. Models from left to right: (*first row*) BIRD-LIU.NEOKINEMA, EBEL-ET-AL.AFTERSHOCK, and HELMSTETTER-ET-AL.AFTERSHOCK. (*second row*) KAGAN-ET-AL.AFTERSHOCK, SHEN-ET-AL.AFTERSHOCK, and EBEL-ET-AL.AFTERSHOCK.CORRECTED. The EBEL-ET-AL.AFTERSHOCK.CORRECTED model was submitted on 12 November 2006 and is therefore tested against a smaller set of earthquakes

forecasts were erroneous at some locations; they were replaced by a corrected version on 12 November 2006. Because of the logic of truly prospective testing, the *mainshock* class and the *mainshock+aftershock* class were expanded into two groups each. The first group includes all initial RELM submissions and compares them to observations from 1 January 2006 forward, while the second group (denoted by a "corrected" suffix) covers all initial submissions and the corrected version of the model by EBEL et al. (2007). Because the corrected versions were submitted later, testing for this group started at the submission date of the corrected versions.

For any further model addition or correction, a new group will be introduced. Such a group would consist of all existing models and the new submissions, and the starting date for testing would be the submission date of the new contributions.

3. Testing Center

The Testing Center is a multi-computer system running the CSEP Testing Center software. It is divided into four main components: the development system, the integration system, the operational system, and the web presentation system (ZECHAR et al., 2009). The development system is used for software development of the Testing Center software and for model development and installation. After Testing Center software and respective models successfully run on the development system, their functionality is tested on the integration system. Each day this system checks out all necessary software codes and performs unit and acceptance tests for all software programs. This step is introduced to mimic the operational system and to detect possible problems before codes are transferred to the operational system. The operational system has the same setup as the integration system, however the codes are only updated every three months according to the release schedule of new versions of the Testing Center software. On the operational system, all tests are performed according to different scheduling depending on the model groups. All results are copied to the web presentation system from which they can be retrieved.

The design of the Testing Center followed the four main goals as outlined by SCHORLEMMER and GERSTENBERGER (2007):

Transparency. All computer codes are managed in a version control repository and are freely available. Thus, all changes to the codes are documented and a web-based collaboration system allows everyone to monitor the software development. The Testing Center codes are published under the open-source General Public License, and the majority of the models which were submitted as codes are open-source codes and can be used by other researchers. The RELM 5-year models were submitted as simple forecast files which are also freely available on the RELM website (http://relm.cseptesting.org). The Testing Center also catalogs all data files used for generating and testing forecasts. Any of these files is freely available.

Controlled Environment. The Testing Center ensures truly prospective tests of all submitted models with the same data. Any model submission gets time-stamped and will only be tested for periods after the submission date. Such an environment is needed for continuous testing of short-term models like the RELM *1-day* model class. Because modelers cannot modify their models after submission, no conscious or unconscious bias of a modeler is introduced into the forecasts.

Comparability. One of the major purposes of the Testing Center is the comparative testing of models. Models are tested for consistency with the observation and against each other (given the observation) to assess their comparative performance.

Reproducibility. Full reproducibility of any result is perhaps the most important feature of the Testing Center. Each data set used for computing a test is stored in the system. Thus, any forecast and any input data set can be reproduced and the tests can be recomputed at any time. Each test computation also stores the system configuration for full reproducibility.

3.1. Tests for Evaluating the Earthquake Forecasts

SCHORLEMMER et al. (2007) proposed a suite of statistical tests to evaluate probabilistic earthquake forecasts. Similar tests were discussed by JACKSON (1996) and used by KAGAN and JACKSON (1994, 1995) for the evaluation of long-term forecasts of large earthquakes. In the language of statistical hypothesis testing, the tests fall into the class of significance tests: Assuming a null hypothesis (a given forecast model), the distribution of an observable test statistic is simulated; if the observed test statistic (e.g., the number of earthquakes) falls into the upper or lower tail of the distribution, the null hypothesis is rejected. The predictive distributions are constructed from model-dependent Monte Carlo simulations and hence are not assumed to be asymptotically normal. DALEY and VERE-JONES (2004) and HARTE and VERE-JONES (2005) explored performance evaluations based on the entropy score and the information gain.

Three tests are used to evaluate the RELM forecasts: the first two—the L(ikelihood)-Test and the N(umber)-Test—measure the consistency of the forecasts with the observations, while the third—the likelihood R(atio)-Test—measures the relative performance of one model against another. Each of these tests compares forecast rates with observed rates, and although they make slightly different measurements, these tests are not independent metrics.

For the RELM models, the forecast in each bin is the expected Poisson earthquake rate (the mean seismicity rate), which is usually a very small floating point number (e.g., 10^{-4}). To evaluate the likelihood of the model forecast given an observation (which is an integer, usually 0 or 1), the discrete Poisson distribution with mean equal to the forecast is used. For simplicity, the forecasts are stated in terms such that all observations in bins are independent, allowing probabilities to factorize.

3.2. The Number- or N-Test

The N(umber)-Test measures the consistency of the total forecasted rate with the total number of observed earthquakes, summed over all bins. The results of the N-Test indicate whether a forecast has predicted too many earthquakes, too few earthquakes, or a number of earthquakes that is considered to be consistent with the observed number. For example, consider a model which predicted $\lambda = 28.4$ earthquakes in the total space-time-magnitude testing region, and assume that, like the RELM models we consider, the forecast is characterized by Poisson uncertainty. If $\omega = 30$ events were observed during the experiment, the model obtains a quantile score of $\delta = \text{Poi}(\omega = 30 | \lambda = 28.4) = 0.66$ (here Poi stands for the Poisson cumulative distribution function). A model may be rejected if δ is very small (e.g., less than 0.025) or very large (e.g., greater than 0.975), which would indicate that the observed number of earthquakes falls into the far upper or far lower end of the forecast distribution, respectively. This indicates that the number of observed earthquakes is unlikely given the model forecast and, hence, the forecast is inconsistent with the observation. The N-Test disregards the spatial and magnitude distributions of the forecast and the observations, emphasizing each forecast's rate model.

3.3. The Likelihood- or L-Test

The L(ikelihood)-Test measures the consistency of a forecast with the observed rate and distribution of earthquakes. In each latitude-longitude-magnitude bin, the log-likelihood of an observation, given the forecast, is computed (again assuming the Poisson distribution). The log-likelihoods are then summed over all bins. To understand whether this sum—the observed log-likelihood—is consistent with what would be expected if the model were correct, many synthetic catalogs consistent with the model forecast are simulated, and their log-likelihoods calculated. This process produces a distribution of log-likelihoods, assuming that the model of interest is the "true" model. The statistic γ measures the proportion of simulated log-likelihoods less than the observed log-likelihood. If γ is low (e.g., less than 0.05), then the observed log-likelihood is much smaller than what would be expected given the model's veracity. The observation may therefore be considered inconsistent with the model. If γ is very high, the observed likelihood is considerably higher than expected, given the model forecast's veracity. In this case, however, it may be that a model predicted the distribution of earthquakes well but smoothed its forecast too much, and therefore high γ values are not considered grounds for model rejection. For example, consider the case when earthquakes occur only in a model's most highly-ranked bins—those bins with the highest forecast rates. If the model is smooth, simulations consistent with the model would produce more diffuse seismicity than that observed, yielding simulated catalogs with events in bins with lower forecast rates, and thus a very high γ. Considering this effect, the L-Test is one-sided.

3.4. The Likelihood-Ratio- or R-Test

The likelihood R(atio)-Test consists of a pairwise-comparison between forecasts (e.g., forecasts i and j). The observed log-likelihood is calculated for each model forecast, and the difference—the observed likelihood ratio—indicates which model better fits the observations. To understand whether this difference is significant, a null hypothesis that model i is correct is adopted and synthetic catalogs consistent with this model are produced. The likelihood ratio is calculated for each simulated catalog. If the fraction α^{ij} of simulated likelihood ratios less than the observed likelihood ratio is very small (e.g., less than 0.05), the observed likelihood ratio is deemed significantly small enough to reject model i. So that no single forecast is given an advantage, this procedure is applied symmetrically. That is, synthetic catalogs are also simulated assuming model j to be true, and these simulations are used to estimate α^{ji}. Comparing each model with all other models results in a table of α values.

3.5. Masking

Several models are based on data that are not available throughout the entire testing area, and some researchers felt their model was not applicable everywhere in the testing area. For a forecast to cover fully the testing area, a model needs an additional "background" model to fill the gaps. RELM requested that all submitted models cover the entire testing area, although modelers were permitted to mask the area in which they were unable to create their forecast according to their scientific ideas. Thus, the area of the genuine forecast can be identified during testing, although it is also possible to evaluate a model over the entire testing area if a background model is chosen. Currently, only the unmasked areas are tested in the Testing Center; that is, a forecast is only evaluated over bins which are unmasked. For the R-Test, only bins which are unmasked in both forecasts are considered.

3.6. Uncertainties in Observations

The earthquake catalog data used to test forecasts contain measurement uncertainties. To account for these uncertainties in the tests, SCHORLEMMER et al. (2007) proposed generating "modified" catalogs. Each event's location and magnitude is modified using an error distribution suggested by the catalog compilers. Additionally, in the case of mainshock catalogs, declustering according to REASENBERG (1985) is applied using parameters that are sampled as described by SCHORLEMMER and GERSTENBERGER (2007). For each observed catalog, 1000 modified catalogs are generated, and these modified catalogs help to estimate the uncertainty of the test results resulting from the uncertainties of earthquake data.

4. Results

In this section we report preliminary summary results for the first half of the ongoing 5-year RELM experiment in California. Detailed results are available at http://us.cseptesting.org, where they are archived and regularly updated. We remind the reader that these results are preliminary, as they are based on only the first half of the 5-year experiment in progress.

4.1. Observed Earthquakes

Twelve earthquakes with magnitude greater than or equal to 4.95 were reported in the ANSS catalog in the RELM testing region during the first half of the ongoing 5-year experiment. Table 2 lists the properties of these target events. Among the details in Table 2 is the estimated independence probability for each earthquake, computed by a Monte-Carlo application (SCHORLEMMER and GERSTENBERGER, 2007) of the REASENBERG (1985) declustering algorithm. For example, the first target earthquake has an independence probability, P_I, of 21%, indicating that the

Table 2

Observed target earthquakes of magnitude $M_{ANSS} \geq 4.95$ in the testing area

No.	Origin Time (UTC)	Latitude	Longitude	M_{ANSS}	P_I	Mainshock
1	24 May 2006, 4:20	32.31	−115.23	5.37	0.21	Yes
2	19 Jul. 2006, 11:41	40.28	−124.43	5.00	1.00	Yes
3	26 Feb. 2007, 12:19	40.64	−124.87	5.40	1.00	Yes
4	9 May 2007, 7:50	40.37	−125.02	5.20	1.00	Yes
5	25 Jun. 2007, 2:32	41.12	−124.82	5.00	1.00	Yes
6	31 Oct. 2007, 3:04	37.43	−121.77	5.45	1.00	Yes
7	9 Feb. 2008, 7:12	32.36	−115.28	5.10	0.04	Yes
8	11 Feb. 2008, 18:29	32.33	−115.26	5.10	0.96	No
9	12 Feb. 2008, 4:32	32.45	−115.32	4.97	0.02	No
10	19 Feb. 2008, 22:41	32.43	−115.31	5.01	0.26	No
11	26 Apr. 2008, 06:40	39.52	−119.93	5.00	1.00	Yes
12	30 Apr. 2008, 3:03	40.84	−123.50	5.40	1.00	Yes

P_I denotes the independence probability as derived from Monte Carlo declustering simulations. The final column indicates whether the event is considered a mainshock by the REASENBERG (1985) declustering method with standard California parameters and is used to evaluate forecasts in the *5-year mainshock* group

declustering algorithm identified this earthquake as belonging to a cluster in 79% of the declustering iterations, each using a different, Monte Carlo-sampled set of algorithm parameters from a range of plausible values. The independence probabilities were used during evaluation of the *mainshock* and *mainshock.corrected* forecast group models; as mentioned in the previous section, the tests estimate the effect of observation uncertainties by generating modified catalogs, and the independence probability determines in what percentage of the modified catalogs a given earthquake appears.

For the *5-year mainshock* forecast class, only a subset of the events in Table 2 are considered. This subset is determined by applying the REASENBERG (1985) declustering algorithm to the original observed catalog, using standard California parameters. Those events that are not declustered are considered mainshocks and are used to evaluate the *5-year mainshock* forecasts.

An investigation of historical seismicity rates in the RELM testing region indicates that the observed sample of 12 earthquakes (with nine of them mainshocks) in a 2.5-year period is relatively small, but not significantly so. We analyzed the rate of all $M \geq 4.95$ earthquakes from 1 January 1932 to 30 June 2004 using the ANSS catalog. To compare with the experimental observation, we divided this time period into 29 non-overlapping periods of 2 years and 6 months duration; the rates in each period are shown in Fig. 4a. On average, 15.45 earthquakes (with 10.59 of them being mainshocks) were observed during each 2.5-year period, with a sample standard deviation of 9.99. As suggested by JACKSON and KAGAN (1999) (see also (VERE-JONES, 1970; KAGAN, 1973)), we found that the number of earthquakes in each period is better fit by a negative binomial distribution than a Poisson distribution—that is, the best-fit negative binomial distribution obtains a lower Akaike Information Criterion (AIC) value (AKAIKE, 1974) (206.4) than the best-fit Poisson distribution (278.2). The best-fitting negative binomial distribution also provides a marginally better fit to the mainshock rate distribution: the negative binomial model obtains an AIC value of 167.3, whereas the Poisson model obtains an AIC of 168.5. The seismicity rate data and the best fits are shown in Fig. 4b. We find the best-fit negative binomial distribution is described by parameter values $(\tau, \nu) = (2.83, 0.15)$; under this model, the probability to obtain fewer than 12 earthquakes is 41.01%. Accordingly, under the best-fit model for mainshock rates, the probability to obtain fewer than nine mainshocks is 32.91%. Despite our finding that the negative binomial distribution better fits historical rates of seismicity, RELM forecasts were formulated as having Poisson uncertainty, and therefore the tests applied to the models are based on Poisson statistics.

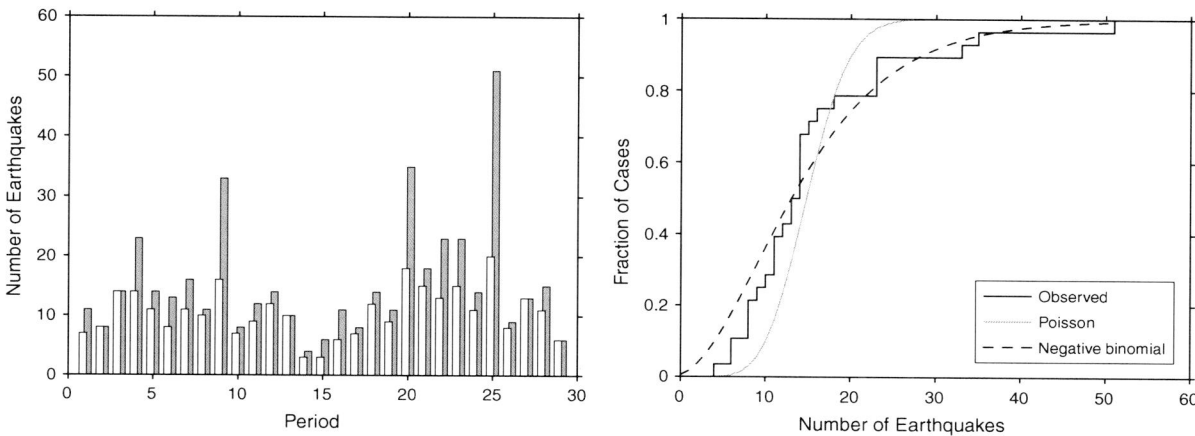

Figure 4
Earthquake rates in California from 1 January 1932 to 30 June 2004. (*left*) *Bar graph* showing the number of earthquakes in 29 non-overlapping periods of 2 years and 6 months duration. *White and gray bars* indicate the number of earthquakes in the declustered catalog, thus mainshocks only, and complete catalog, respectively. (*right*) Cumulative distribution function of the earthquakes rates in the complete catalog from the left frame. The *solid black line* indicates the observation, the *solid gray line* indicates the Poissonian distribution of rate $\lambda = 15.45$, the *dashed black line* indicates the best-fit negative binomial distribution

4.2. Mainshock Models

The summary results for the *mainshock* forecast class are given in Tables 3, 4, and 5. Table 3 lists the quantile scores for the L- and N-Tests. The RELM working group decided a priori to use a significance value of 5%; in the case of the two-sided N-Test, this corresponds to critical values of 2.5% and 97.5%; bold values in the tables indicate that the corresponding forecast is inconsistent with the observed target earthquake catalog. Recall that the γ quantile score, associated with the L-Test, describes how well a forecast matches the observed distribution of earthquakes. A very low γ score is means for rejecting a model, while a very high γ score is suspect, but not grounds for rejection. On the other hand, an extremely low or extremely high δ quantile score—characterizing the overall rate of earthquakes but not including any spatial information—yields rejection. From Table 3 we see that the observations during the first half of the RELM experiment are inconsistent—at the a priori significance level—with the HOLLIDAY-ET-AL.PI, WARD.COMBO81, WARD.GEODETIC81, WARD.GEOLOGIC81, and WARD.SEISMIC81 forecasts. All of these models have overpredicted in the first half of the experiment as indicated by their small δ values. (See also Fig. 5 for a visual comparison of

Table 3

L-Test and N-Test results for the mainshock forecast class

Model	γ	δ
EBEL-ET-AL.MAINSHOCK	0.149	0.503
HELMSTETTER-ET-AL.MAINSHOCK	0.723	0.391
HOLLIDAY-ET-AL.PI	0.992	**[0.011]**
KAGAN-ET-AL.MAINSHOCK	0.974	0.063
SHEN-ET-AL.MAINSHOCK	0.969	0.107
WARD.COMBO81	0.998	**[0.004]**
WARD.GEODETIC81	1.000	**[0.000]**
WARD.GEODETIC85	0.987	0.030
WARD.GEOLOGIC81	0.998	**[0.011]**
WARD.SEISMIC81	0.993	**[0.014]**
WARD.SIMULATION	0.725	0.282
WIEMER-SCHORLEMMER.ALM	0.637	0.256

The statistics γ and δ measure the proportion of simulated likelihoods/numbers less than the observed likelihood/number. Bold values indicate that the observed target earthquake catalog is inconsistent with the corresponding forecast

predicted and observed number of earthquakes per model.)

Table 4 shows the contribution of each earthquake to the resulting likelihoods per model and highlights for each earthquake the model with the highest forecast rate in the respective bin—in other words, which model best forecast the earthquake. The WIEMER-SCHORLEMMER.ALM model provides the highest forecast rate for four earthquakes, and the

Table 4

Result details for the mainshock forecast class

Model		Earthquake								
		1 $M5.37$	2 $M5.00$	3 $M5.40$	4 $M5.20$	5 $M5.00$	6 $M5.45$	7 $M5.10$	11 $M5.00$	12 $M5.40$
EBEL-ET-AL.MAINSHOCK	λ L	$9.55 \cdot 10^{-8}$ -16.16	$3.56 \cdot 10^{-3}$ -5.64	$3.39 \cdot 10^{-4}$ -7.99	n/a	$1.15 \cdot 10^{-4}$ -9.07	$7.64 \cdot 10^{-7}$ -14.08	$9.55 \cdot 10^{-8}$ -16.16	$5.74 \cdot 10^{-4}$ -7.46	$9.55 \cdot 10^{-8}$ -16.16
HELMSTETTER-ET-AL.MAINSHOCK	λ L	$4.59 \cdot 10^{-3}$ -5.39	$6.45 \cdot 10^{-3}$ -5.05	$2.92 \cdot 10^{-4}$ -8.14	$4.14 \cdot 10^{-3}$ -5.49	$2.06 \cdot 10^{-4}$ -8.49	$9.86 \cdot 10^{-4}$ -6.92	$8.50 \cdot 10^{-3}$ -4.78	$8.20 \cdot 10^{-5}$ -9.41	$1.44 \cdot 10^{-4}$ -8.85
HOLLIDAY-ET-AL.PI	λ L	$1.85 \cdot 10^{-3}$ -6.29	$4.66 \cdot 10^{-3}$ -5.37	$1.85 \cdot 10^{-3}$ -6.29	$2.94 \cdot 10^{-3}$ -5.83	n/a	$1.47 \cdot 10^{-3}$ -6.52	$3.70 \cdot 10^{-3}$ -5.60	n/a	n/a
KAGAN-ET-AL.MAINSHOCK	λ L	$3.57 \cdot 10^{-4}$ -7.94	n/a	n/a	n/a	n/a	n/a	$7.12 \cdot 10^{-4}$ -7.25	n/a	n/a
SHEN-ET-AL.MAINSHOCK	λ L	$7.21 \cdot 10^{-4}$ -7.24	n/a	n/a	n/a	n/a	n/a	$1.44 \cdot 10^{-3}$ -6.54	n/a	n/a
WARD.COMBO81	λ L	$1.12 \cdot 10^{-3}$ -6.80	n/a	n/a	n/a	n/a	n/a	$2.08 \cdot 10^{-3}$ -6.18	n/a	n/a
WARD.GEODETIC81	λ L	$1.33 \cdot 10^{-3}$ -6.62	n/a	n/a	n/a	n/a	n/a	$2.48 \cdot 10^{-3}$ -6.00	n/a	n/a
WARD.GEODETIC85	λ L	$7.67 \cdot 10^{-4}$ -7.17	n/a	n/a	n/a	n/a	n/a	$1.43 \cdot 10^{-3}$ -6.55	n/a	n/a
WARD.GEOLOGIC81	λ L	$9.76 \cdot 10^{-4}$ -6.93	n/a	n/a	n/a	n/a	n/a	$1.82 \cdot 10^{-3}$ -6.31	n/a	n/a
WARD.SEISMIC81	λ L	$1.04 \cdot 10^{-3}$ -6.87	n/a	n/a	n/a	n/a	n/a	$1.94 \cdot 10^{-3}$ -6.25	n/a	n/a
WARD.SIMULATION	λ L	$1.87 \cdot 10^{-4}$ -8.59	n/a	n/a	n/a	n/a	n/a	$1.63 \cdot 10^{-5}$ -11.02	n/a	n/a
WIEMER-SCHORLEMMER.ALM	λ L	$5.47 \cdot 10^{-3}$ -5.21	$5.17 \cdot 10^{-3}$ -5.27	$3.45 \cdot 10^{-4}$ -7.97	$2.48 \cdot 10^{-3}$ -6.00	$1.47 \cdot 10^{-8}$ -18.03	$1.64 \cdot 10^{-3}$ -6.41	$1.01 \cdot 10^{-2}$ -4.61	$2.54 \cdot 10^{-5}$ -10.58	$4.32 \cdot 10^{-4}$ -7.75

Contributions of each target earthquake to the log-likelihoods, L, and the forecast rate, λ, of each model for the corresponding bins are shown. For each earthquake, the model with the highest and lowest forecast for the respective bin is highlighted in light gray and dark gray, respectively. Some models do not provide a forecast for the entire space-magnitude testing area and some earthquakes fall into these masked bins, indicated by n/a. Earthquake numbers correspond to those listed in Table 2

Table 5

R-Test results for the mainshock forecast class

	Model	1	2	3	4	5	6	7
1	EBEL-ET-AL.MAINSHOCK	–	[0.000]	[0.000]	[0.000]	[0.000]	[0.000]	[0.000]
2	HELMSTETTER-ET-AL.MAINSHOCK	0.943	–	0.412	0.189	0.703	0.544	0.480
3	KAGAN-ET-AL.MAINSHOCK	0.965	[0.000]	–	[0.010]	0.326	0.369	[0.000]
4	SHEN-ET-AL.MAINSHOCK	0.944	[0.007]	0.783	–	0.964	0.586	[0.000]
5	WARD.GEODETIC85	0.916	[0.000]	0.110	[0.001]	–	0.156	[0.000]
6	WARD.SIMULATION	0.939	[0.000]	[0.001]	[0.001]	[0.002]	–	[0.000]
7	WIEMER-SCHORLEMMER.ALM	0.547	[0.000]	0.130	0.123	0.799	0.614	–

All models which are consistent with the observation in the L- and N-Tests are compared and their corresponding α-values are shown. If printed in bold, the row model (labeled to the left) should be rejected in favor of the column model (labeled at the top). The results show that all models can be rejected in favor of the HELMSTETTER-ET-AL.MAINSHOCK model

HELMSTETTER-ET-AL.MAINSHOCK model has the highest forecast rate for three earthquakes. The EBEL-ET-AL.MAINSHOCK and HOLLIDAY-ET-AL.PI models provide the highest forecast rate for one earthquake each.

The R-Test results for the *mainshock* forecast class are shown in Table 5 and provide a comparative evaluation of the forecasts. This table lists the α quantile scores for each pairwise comparison; for simplicity, we exclude the pairwise comparisons that would include the models shown to be inconsistent by the L- and/or N-Tests. Scores indicating that the corresponding model can be rejected are shown in bold. In this case, such a score indicates that the row model (labeled to the left) should be rejected in favor of the column model (labeled at the top). For example, the α value in the first row and second column indicates that the EBEL-ET-AL.MAINSHOCK forecast should be rejected in favor of the

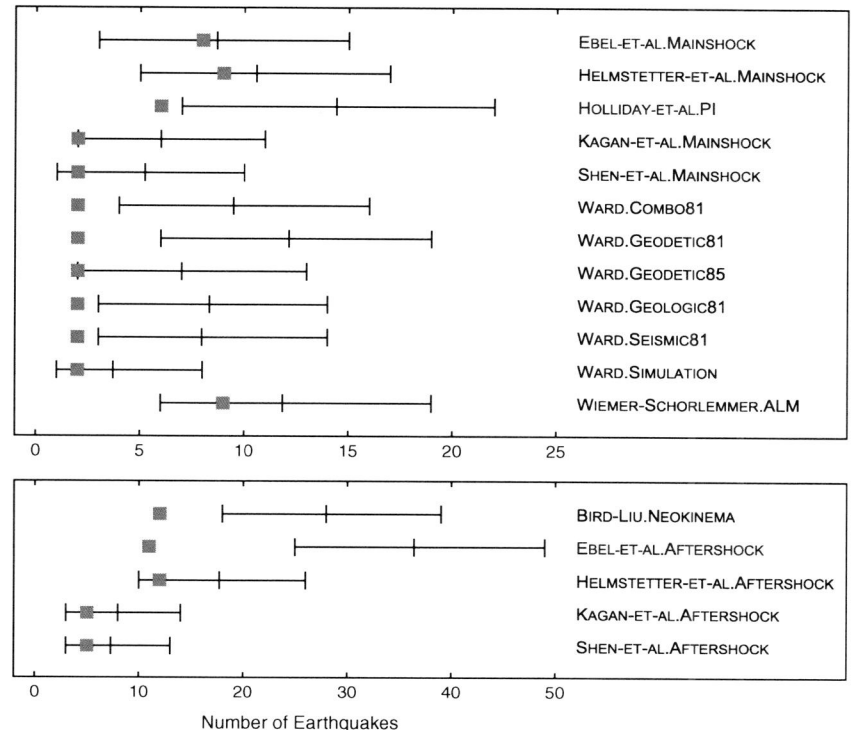

Figure 5
Visual comparison of predicted and observed number of earthquakes per model in the *mainshock* and *mainshock+aftershock* forecast classes. For each model, the *bar* indicates the range of observed earthquake rates that would be consistent with the model, given a Poissonian distribution. The *gray squares* indicate observations per model considering the coverage of the model. If the *gray square* overlaps with the *bar*, the model is consistent with the observation

HELMSTETTER-ET-AL.MAINSHOCK forecast. From this table, we find that only the HELMSTETTER-ET-AL.MAINSHOCK forecast is not rejected (because all other rows contain at least one bold value). Moreover, all models are rejected in favor of the HELMSTETTER-ET-AL.MAINSHOCK forecast (all scores in the second column are bold).

4.3. Mainshock Corrected

As mentioned in the Models section, the *mainshock.corrected* forecast group contains all the same forecasts as the *mainshock* forecast class with one exception: the EBEL-ET-AL.MAINSHOCK.CORRECTED forecast is added and implicitly replaces the EBEL-ET-AL.MAINSHOCK forecast. For consistency, the experiment for this forecast group began on 12 November 2006, so it contains only earthquakes 3–11 from Table 2. The summary results for this forecast group are shown in Tables 6 and 7. In this forecast group, the L- and N-Test results indicate that the observed earthquake distribution is consistent with all forecast models except the WARD.COMBO81 and WARD.GEODETIC81 models, which overpredicted the number of events (Table 6). The R-Test results are similar to the results for the *mainshock* forecast class and indicate that only the HELMSTETTER-ET-AL.MAINSHOCK forecast is not rejected in any pairwise comparison (Table 7).

4.4. Mainshock+Aftershock Models

The summary results for the *mainshock+aftershock* forecast class are shown in Tables 8, 9, and 10. N-Test results show that the BIRD-LIU.NEOKINEMA model and the EBEL-ET-AL.AFTERSHOCK model have each predicted too many earthquakes in the experiment to date (see also Fig. 5). The R-Test results

Table 6

L-Test and N-Test results for the mainshock.corrected forecast group

Model	γ	δ
EBEL-ET-AL.MAINSHOCK	0.085	0.661
EBEL-ET-AL.MAINSHOCK.CORRECTED	0.769	0.300
HELMSTETTER-ET-AL.MAINSHOCK	0.434	0.613
HOLLIDAY-ET-AL.PI	0.984	0.042
KAGAN-ET-AL.MAINSHOCK	0.968	0.098
SHEN-ET-AL.MAINSHOCK	0.969	0.145
WARD.COMBO81	0.998	**[0.015]**
WARD.GEODETIC81	0.997	**[0.003]**
WARD.GEODETIC85	0.984	0.058
WARD.GEOLOGIC81	0.992	0.028
WARD.SEISMIC81	0.990	0.034
WARD.SIMULATION	0.708	0.301
WIEMER-SCHORLEMMER.ALM	0.335	0.488

The statistics γ and δ measure the proportion of simulated likelihoods/numbers less than the observed likelihood/number. Bold values indicate that the observed target earthquake catalog is inconsistent with the corresponding forecast

show that only the HELMSTETTER-ET-AL.AFTERSHOCK forecast is not rejected in any pairwise comparison.

4.5. Mainshock+Aftershock Corrected

As with the *mainshock* and *mainshock.corrected* forecast groups, the *mainshock+aftershock.corrected* forecast group was added to the *mainshock+aftershock* forecast class. The EBEL-ET-AL.AFTERSHOCK.CORRECTED forecast is added and implicitly replaces the EBEL-ET-AL.AFTERSHOCK forecast. For consistency,

Table 8

L-Test and N-Test results for the mainshock+aftershock forecast class

Model	γ	δ
BIRD-LIU.NEOKINEMA	1.000	**[0.001]**
EBEL-ET-AL.AFTERSHOCK	1.000	**[0.000]**
HELMSTETTER-ET-AL.AFTERSHOCK	0.949	0.104
KAGAN-ET-AL.AFTERSHOCK	0.895	0.193
SHEN-ET-AL.AFTERSHOCK	0.896	0.262

The statistics γ and δ measure the proportion of simulated likelihoods/numbers less than the observed likelihood/number. Bold values indicate that the observed target earthquake catalog is inconsistent with the corresponding forecast

the experiment for this forecast group began on 12 November 2006. The summary results for this forecast group are shown in Tables 11 and 12.

As in the *mainshock+aftershock* forecast group, the N-Test results show that the EBEL-ET-AL.AFTERSHOCK model has predicted too many earthquakes in the experiment to date, as has the EBEL-ET-AL.AFTERSHOCK.CORRECTED model. The R-Test results show that only the HELMSTETTER-ET-AL.AFTERSHOCK forecast is not rejected in any pairwise comparison.

5. Discussion

The science of earthquake predictability is an active field with many unsolved problems, including the question of best practices for formulating and

Table 7

R-Test results for the mainshock.corrected forecast group

	Model	1	2	3	4	5	6	7	8	9	10	11
1	EBEL-ET-AL.MAINSHOCK	–	**[0.000]**	**[0.000]**	**[0.000]**	**[0.000]**	**[0.000]**	**[0.000]**	**[0.000]**	**[0.000]**	**[0.000]**	**[0.000]**
2	EBEL-ET-AL.MAINSHOCK.CORRECTED	0.840	–	**[0.003]**	0.406	0.089	0.034	0.278	0.270	0.385	0.445	0.085
3	HELMSTETTER-ET-AL.MAINSHOCK	0.926	0.351	–	0.509	0.339	0.185	0.573	0.536	0.681	0.579	0.630
4	HOLLIDAY-ET-AL.PI	0.489	**[0.004]**	**[0.001]**	–	**[0.003]**	**[0.003]**	**[0.003]**	**[0.006]**	**[0.003]**	0.035	**[0.000]**
5	KAGAN-ET-AL.MAINSHOCK	0.886	0.333	**[0.012]**	0.527	–	0.045	0.453	0.409	0.477	0.478	**[0.007]**
6	SHEN-ET-AL.MAINSHOCK	0.869	0.440	**[0.025]**	0.529	0.676	–	0.974	0.576	0.711	0.654	**[0.010]**
7	WARD.GEODETIC85	0.788	0.135	**[0.002]**	0.631	0.123	**[0.004]**	–	0.225	0.283	0.245	**[0.001]**
8	WARD.GEOLOGIC81	0.701	0.087	**[0.002]**	0.636	0.050	**[0.013]**	0.086	–	0.125	0.190	**[0.004]**
9	WARD.SEISMIC81	0.722	0.104	**[0.005]**	0.732	0.080	**[0.022]**	0.165	0.210	–	0.247	**[0.002]**
10	WARD.SIMULATION	0.761	**[0.001]**	**[0.000]**	**[0.010]**	**[0.004]**	**[0.001]**	**[0.009]**	**[0.009]**	**[0.005]**	–	**[0.000]**
11	WIEMER-SCHORLEMMER.ALM	0.473	**[0.000]**	**[0.000]**	0.286	0.134	0.138	0.600	0.539	0.679	0.651	–

All models are compared and their corresponding α values are shown. If printed in bold, the row model (labeled to the left) should be rejected in favor of the column model (labeled at the top). The results show that all models can be rejected in favor of model HELMSTETTER-ET-AL.MAINSHOCK

Table 9

Result details for the mainshock+aftershock forecast class

Model		Earthquake										
		1 $M5.37$	2 $M5.00$	3 $M5.40$	4 $M5.20$	5 $M5.00$	6 $M5.45$	7/8 $M5.10/5.10$	9/10 $M4.97/5.01$	11 $M5.00$	12 $M5.40$
Bird-Liu.Neokinema	λ	$2.08 \cdot 10^{-3}$	$3.57 \cdot 10^{-3}$	$7.66 \cdot 10^{-4}$	$5.71 \cdot 10^{-3}$	$1.82 \cdot 10^{-3}$	$7.86 \cdot 10^{-4}$	$4.01 \cdot 10^{-3}$	$2.32 \cdot 10^{-3}$	$8.76 \cdot 10^{-5}$	$2.23 \cdot 10^{-4}$
	L	-6.18	-5.64	-7.18	-5.17	-6.31	-7.15	-11.74	-12.83	-9.34	-8.41
Ebel-et-al.Aftershock	λ	$3.43 \cdot 10^{-6}$	$1.70 \cdot 10^{-2}$	$1.45 \cdot 10^{-3}$	n/a	$5.48 \cdot 10^{-4}$	$2.74 \cdot 10^{-5}$	$3.43 \cdot 10^{-6}$	$2.74 \cdot 10^{-3}$	$2.74 \cdot 10^{-3}$	$3.43 \cdot 10^{-6}$
	L	-12.58	-4.09	-6.54		-7.51	-10.50	-25.86	-12.50	-5.90	-12.58
Helmstetter-et-al.Aftershock	λ	$7.71 \cdot 10^{-3}$	$1.14 \cdot 10^{-2}$	$4.90 \cdot 10^{-4}$	$7.13 \cdot 10^{-3}$	$3.63 \cdot 10^{-4}$	$1.63 \cdot 10^{-3}$	$1.48 \cdot 10^{-2}$	$4.03 \cdot 10^{-3}$	$1.45 \cdot 10^{-4}$	$2.41 \cdot 10^{-4}$
	L	-4.87	-4.49	-7.62	-4.95	-7.92	-6.42	-9.13	-11.73	-8.84	-8.33
Kagan-et-al.Aftershock	λ	$4.76 \cdot 10^{-4}$	n/a	n/a	n/a	n/a	n/a	$9.50 \cdot 10^{-4}$	$1.45 \cdot 10^{-5}$	n/a	n/a
	L	-7.65						-14.61	-13.77		
Shen-et-al.Aftershock	λ	$1.01 \cdot 10^{-3}$	n/a	n/a	n/a	n/a	n/a	$2.02 \cdot 10^{-3}$	$2.61 \cdot 10^{-3}$	n/a	n/a
	L	-6.90						-13.11	-12.59		

Contributions of each target earthquake to the log-likelihoods, L, and the forecast rates, λ, of each model for the respective bins. For each earthquake, the model with the highest and lowest forecast for the respective bin is highlighted in light gray and dark gray, respectively. Earthquakes 7 and 8 as well as 9 and 10 occurred in the same bin and are therefore combined in this table. Some models do not provide a forecast for the entire space-magnitude testing area and some earthquakes fall into these masked bins, indicated by n/a. Earthquake numbers correspond to those listed in Table 2

Table 10

R-Test results for the mainshock+aftershock forecast class

	Model	1	2	3
1	Helmstetter-et-al.Aftershock	–	0.372	0.091
2	Kagan-et-al.Aftershock	[0.000]	–	[0.000]
3	Shen-et-al.Aftershock	[0.001]	0.902	–

All models which are consistent with the observation in the L- and N-Tests are compared and their corresponding α values are shown. If printed in bold, the row model (labeled to the left) should be rejected in favor of the column model (labeled at the top). The results show that all models can be rejected in favor of model Helmstetter-et-al.Aftershock

Table 11

L-Test and N-Test results for the mainshock+aftershock.corrected forecast class

Model	γ	δ
Bird-Liu.Neokinema	0.984	0.027
Ebel-et-al.Aftershock	0.994	[0.000]
Ebel-et-al.Aftershock.Corrected	1.000	[0.000]
Helmstetter-et-al.Aftershock	0.692	0.394
Kagan-et-al.Aftershock	0.783	0.402
Shen-et-al.Aftershock	0.706	0.479

The statistics γ and δ measure the proportion of simulated likelihoods/numbers less than the observed likelihood/number. Bold values indicate that the observed target earthquake catalog is inconsistent with the corresponding forecast

Table 12

R-Test results for the mainshock+aftershock.corrected forecast group

	Model	1	2	3	4
1	Bird-liu.Neokinema	–	[0.000]	0.034	[0.002]
2	Helmstetter-et-al.Aftershock	0.067	–	0.433	0.159
3	Kagan-et-al.Aftershock	0.083	[0.001]	–	[0.004]
4	Shen-et-al.Aftershock	0.377	[0.005]	0.928	–

All models which are consistent with the observation in the L- and N-Tests are compared and their corresponding α values are shown. If printed in bold, the row model (labeled to the left) should be rejected in favor of the column model (labeled at the top). The results show that all models can be rejected in favor of model Helmstetter-et-al.Aftershock

evaluating earthquake forecasts. The RELM effort, as one of the first large-scale, prospective, and cooperative predictability experiments, can provide lessons along these lines. RELM experiment participants decided to specify their forecasts as the expected rate of earthquakes in latitude/longitude/magnitude bins, and they decided that the forecasts should be interpreted as having Poisson uncertainty. As we showed in the Observed Earthquakes subsection (and as shown by Jackson and Kagan, 1999), seismicity rates are better fit by a negative binomial distribution than a Poisson distribution; therefore it may be worthwhile for future forecasts to specify an additional parameter per bin (or per forecast) that allows for negative binomial uncertainty. Preferably, a forecast should specify a discrete probability distribution in each bin. This approach would not require the agreement of all participants on one particular distribution to be used for testing and it would also allow for propagating uncertainties of input data into the forecast (Werner and Sornette, 2008). The tests and forecast format that RELM decided to use are relatively simple yet powerful. Nevertheless, they are not without flaws; for example the assumption that observations in each space-time-magnitude bin are independent may sometimes be violated, particularly in the wake of a large earthquake.

Some of these issues will be addressed by considering alternative forecast formats, e.g., by allowing models to specify the likelihood distribution to be used. Moreover, CSEP is incorporating modifications to the current tests and other tests, e.g., alarm-based tests that do not require a specific rate or uncertainty model (MOLCHAN, 1990; MOLCHAN and KAGAN, 1992; KAGAN, 2007; MOLCHAN and KEILIS-BOROK, 2008; ZECHAR and JORDAN, 2008).

The stability of RELM test results—including those presented here—is not easy to understand comprehensively. We made efforts to address stability of the L-Test by exploring a hypothetical predictability experiment. For a given forecast, we determined the bin with the lowest forecast rate, and we generated a modified catalog by adding to the observed catalog one additional event placed in this bin. This additional event represents the most unexpected occurrence according to the model, and we were curious to see if this one event could cause a forecast to be rejected if it otherwise was not rejected. We applied the L-Test to each forecast and the corresponding modified catalog and compared the resulting γ statistic with the observed γ reported in the tables throughout the Results section. We find that there is no simple relationship: some forecasts were rejected while others were not, and rejection depended on the peakedness of a forecast. For example, if a forecast has a very high ratio between its highest and lowest forecast values (i.e., it is very peaked), the most unexpected event has a much stronger effect on the L-Test result than otherwise. In other words, stability of test results is model-dependent, and this issue should be considered carefully in future experiments.

Another aspect of result stability is the duration of the experiment. Five years will most likely not be long enough for a comprehensive and final test result, as it can be questioned how representative the seismicity of these particular five years is. One effect of this problem can be seen in the results of the *mainshock* and *mainshock.corrected* forecast groups.

While in the former group five models are rejected based on N-Test results, only two are rejected in the latter group. The exclusion of about 11 months from testing changes the L-Test considerably. However, the results of the R-Test suggest in both cases that the HELMSTETTER-ET-AL.MAINSHOCK cannot be rejected by any other model.

The fact that some forecasts masked a significant portion of the entire testing area led to the problem that eight of the twelve *mainshock* forecasts were tested against only two earthquakes. Four of these eight were rejected due to overpredicting the number of events. Although only two earthquakes occurred in the unmasked area, this low number indicates that the models are not consistent with the observation as the models expected far more events.

Although the RELM project ended in 2005, efforts to develop testing methods, implement these methods into Testing Center software systems, and expand the scope of experiments to other seismically active regions are ongoing, as is the experiment considered in this study. CSEP, the successor of RELM, took over the entire operation and development and is becoming a global reference project for earthquake predictability research.

Standardization can be considered one of the most important achievements of the RELM project and the Testing Center. The substantial consensus of RELM participants on the tests, rules, and processes is more than just a nucleus for other efforts. The Testing Center software is currently deployed to facilities in New Zealand, Europe, and Japan, and the rules set in California are adopted throughout all new Testing Centers. The next major step will become the unification of all efforts into a global testing program which was made possible only through the successful standardization.

Acknowledgments

This research was supported by the Southern California Earthquake Center (SCEC). SCEC is funded by NSF Cooperative Agreement EAR-0106924 and USGS Cooperative Agreement 02HQAG0008. Tests were performed within the W. M. Keck Collaboratory for the Study of Earthquake Predictability (CSEP)

Testing Center at SCEC, which was made possible by the generous financial support of the W. M. Keck Foundation. We thank the following for their contributions to RELM working group model development: J. A. Baglivo, P. Bird, K. W. Campbell, T. Cao, F. Catalli, D. W. Chambers, C. Chen, R. Console, A. Donnellan, J. E. Ebel, G. Falcone, A. D. Frankel, M. C. Gerstenberger, A. Helmstetter, J. R. Holliday, L. M. Jones, A. L. Kafka, Y. Y. Kagan, Z. Liu, M. Murru, M. D. Petersen, D. A. Rhoades, Y. Rong, J. B. Rundle, Z. Shen, K. F. Tiampo, D. L. Turcotte, S. N. Ward, and S. Wiemer. For stimulating discussion and various contributions to the RELM working group, we thank the following: M. Bebbington, D. Bowman, W. Ellsworth, K. Felzer, S. Hough, D. Sornette, R. Stein, M. Stirling, D. Vere-Jones, and J. Woessner. We thank F. Euchner, N. Gupta, V. Gupta, P. J. Maechling, and J. Yu for computational assistance. We also thank the open source community for the Linux operating system and the many programs used to create the Testing Center. We especially thank M. Liukis for her enthusiastic computational support and Testing Center development. Maps were created using the Generic Mapping Tools (Wessel and Smith, 1998). We thank the editor D. A. Rhoades, J. E. Ebel, and an anonymous reviewer for thoughtful reviews that enhanced the paper. M. J. Werner was supported by the EXTREMES project of the Competence Center Environment and Sustainability of ETH. The SCEC contribution number for this paper is 1230.

Open Access This article is distributed under the terms of the Creative Commons Attribution Noncommercial License which permits any noncommercial use, distribution, and reproduction in any medium, provided the original author(s) and source are credited.

References

Akaike, H. (1974) *A New Look at the Statistical Model Identification*, IEEE Trans. Automatic Control *19*, 716–723.

Bird, P. and Liu, Z. (2007), *Seismic hazard inferred from tectonics: California*, Seismol. Res. Lett. *78*, 37–48, doi:10.1785/gssrl.78.1.37.

Console, R., Murru, M., Catalli, F., and Falcone, G. (2007), *Real time forecasts through an earthquake clustering model constrained by the rate-and-state constitutive law: Comparison with a purely stochastic ETAS model*, Seismol. Res. Lett. *78*, 49–56, doi:10.1785/gssrl.78.1.49.

Daley, D. J. and Vere-Jones, D. (2004), *Scoring probability forecasts for point processes: The entropy score and information gain*, J. Appl. Probab. *41A*, 297–312.

Ebel, J. E., Chambers, D. W., Kafka, A. L., and Baglivo, J. A. (2007), *Non-Poissonian earthquake clustering and the hidden Markov model as bases for earthquake forecasting in California*, Seismol. Res. Lett. *78*, 57–65, doi:10.1785/gssrl.78.1.57.

Field, E. H. (2007), *Overview of the working group for the development of regional earthquake likelihood models (RELM)*, Seismol. Res. Lett. *78*, 7–16, doi:10.1785/gssrl.78.1.7.

Gerstenberger, M. C., Jones, L. M., and Wiemer, S. (2007), *Short-term aftershock probabilities: Case studies in California*, Seismol. Res. Lett. *78*, 66–77, doi:10.1785/gssrl.78.1.66.

Harte, D. and Vere-Jones, D. (2005), *The entropy score and its uses in earthquake forecasting*, Pure Appl. Geophys. *162*, 1229–1253, doi:10.1007/s00024-004-2667-2.

Helmstetter, A., Kagan, Y. Y., and Jackson, D. D. (2007), *High-resolution time-independent grid-based forecast for $M \geq 5$ earthquakes in California*, Seismol. Res. Lett. *78*, 78–86, doi: 10.1785/gssrl.78.1.78.

Holliday, J. R., Chen, C., Tiampo, K. F., Rundle, J. B., Turcotte, D. L., and Donnellan, A. (2007), *A RELM Earthquake forecast based on Pattern Informatics*, Seismol. Res. Lett. *78*, 87–93, doi: 10.1785/gssrl.78.1.87.

Jackson, D. D. (1996), *Hypothesis testing and earthquake prediction*, Proc. Natl. Acad. Sci. USA *93*, 3772–3775.

Jackson, D. D. and Kagan, Y. Y. (1999), *Testable earthquake forecasts for 1999*, Seismol. Res. Lett. *70*, 393–403.

Jordan, T. (2006), *Earthquake predictability, brick by brick*, Seismol. Res. Lett. *77*, 3–6.

Kagan, Y. Y. (1973), *Statistical methods in the study of seismic processes*, Bull. Int. Stat. Inst. *45*, 437–453.

Kagan, Y. Y. (2007), *On earthquake predictability measurement: information score and error diagram*, Pure Appl. Geophys. *164*, 1947–1962, doi:10.1007/s00024-007-0260-1.

Kagan, Y. Y. and Jackson, D. D. (1994), *Long-term probabilistic forecasting of earthquakes*, J. Geophys. Res. *99*, 13685–13700.

Kagan, Y. Y. and Jackson, D. D. (1995), *New seismic gap hypothesis: Five years after*, J. Geophys. Res. *100*, 3943–3959.

Kagan, Y. Y., Jackson, D. D., and Rong, Y. (2007), *A testable five-year forecast of moderate and large earthquakes in southern California based on smoothed seismicity*, Seismol. Res. Lett. *78*, 94–98, doi:10.1785/gssrl.78.1.94.

Molchan, G. M. (1990), *Strategies in strong earthquake prediction*, Phys. Earth Planet. Inter. *61*, 84–98, doi:10.1016/0031-9201(90)90097-H.

Molchan, G. M. and Kagan, Y. Y. (1992), *Earthquake prediction and its optimization*, J. Geophys. Res. *97*, 4823–4838.

Molchan, G. M. and Keilis-Borok, V. (2008), *Earthquake prediction: probabilistic aspect*, Geophys. J. Int. *173*, 1012–1017, doi:10.1111/j.1365-246X.2008.03785.x.

Petersen, M. D., Cao, T., Campbell, K. W., and Frankel, A. D. (2007), *Time-independent and time-dependent seismic hazard assessment for the State of California: Uniform California Earthquake Rupture Forecast Model 1.0*, Seismol. Res. Lett. *78*, 99–109, doi:10.1785/gssrl.78.1.99.

Reasenberg, P. (1985), *Second-order moment of central California seismicity, 1969–1982*, J. Geophys. Res. *90*, 5479–5495.

Rhoades, D. A. (2007), *Application of the EEPAS model to forecasting earthquakes of moderate magnitude in southern California*, Seismol. Res. Lett. *78*, 110–115, doi:10.1785/gssrl.78.1.110.

Schorlemmer, D. and Gerstenberger, M. C. (2007), *RELM Testing Center*, Seismol. Res. Lett. *78*, 30–36, doi:10.1785/gssrl.78.1.30.

Schorlemmer, D., Gerstenberger, M. C., Wiemer, S., Jackson, D. D., and Rhoades, D. A. (2007), *Earthquake Likelihood Model Testing*, Seismol. Res. Lett. *78*, 17–29, doi:10.1785/gssrl.78.1.17.

Shen, Z., Jackson, D. D., and Kagan, Y. Y. (2007), *Implications of geodetic strain rate for future earthquakes, with a five-year forecast of M 5 earthquakes in southern California*, Seismol. Res. Lett. *78*, 116–120, doi:10.1785/gssrl.78.1.116.

Vere-Jones, D. (1970), *Stochastic models for earthquake occurrence*, J. Roy. Stat. Soc. Series B (Methodological) *32*, 1–62.

Ward, S. N. (2007), *Methods for evaluating earthquake potential and likelihood in and around California*, Seismol. Res. Lett. *78*, 121–133, doi:10.1785/gssrl.78.1.121.

Werner, M. J., and Sornette D. (2008), *Magnitude uncertainties impact seismic rate estimates, forecasts, and predictability experiments*, J. Geophys. Res. *113*, B08302, doi:10.1029/2007JB005427.

Wessel, P. and Smith, W. (1998), *New, improved version of Generic Mapping Tools released*, EOS Trans. AGU *79*, 579.

Wiemer, S. and Schorlemmer, D. (2007), *ALM: An Asperity-based Likelihood Model for California*, Seismol. Res. Lett. *78*, 134–140, doi:10.1785/gssrl.78.1.134.

Zechar, J. D. and Jordan, T. (2008), *Testing alarm-based earthquake predictions*, Geophys. J. Int. *172*, 715–724, doi:10.1111/j.1365-246X.2007.03676.x.

Zechar, J. D., Schorlemmer, D., Liukis, M., Yu, J., Euchner, F., Maechling, P. J., and Jordan, T. H. (2009), *The Collaboratory for the Study of Earthquake Predictability perspective on computational earthquake science*, Concurrency and Computation: Practice and Experience, doi:10.1002/cpe.1519.

(Received August 14, 2008, revised February 21, 2009, accepted March 10, 2009, Published online May 11, 2010)

New Zealand Earthquake Forecast Testing Centre

MATTHEW C. GERSTENBERGER[1] and DAVID A. RHOADES[1]

Abstract—The New Zealand Earthquake Forecast Testing Centre is being established as one of several similar regional testing centres under the umbrella of the Collaboratory for the Study of Earthquake Predictability (CSEP). The Centre aims to encourage the development of testable models of time-varying earthquake occurrence in the New Zealand region, and to conduct verifiable prospective tests of their performance over a period of five or more years. The test region, data-collection region and requirements for testing are described herein. Models must specify in advance the expected number of earthquakes with epicentral depths $h \leq 40$ km in bins of time, magnitude and location within the test region. Short-term models will be tested using 24-h time bins at magnitude $M \geq 4$. Intermediate-term models and long-term models will be tested at $M \geq 5$ using 3-month, 6-month and 5-year bins, respectively. The tests applied will be the same as at other CSEP testing centres: the so-called N test of the total number of earthquakes expected over the test period; the L test of the likelihood of the earthquake catalogue under the model; and the R test of the ratio of the likelihoods under alternative models. Four long-term, three intermediate-term and two short-term models have been installed to date in the testing centre, with tests of these models commencing on the New Zealand earthquake catalogue from the beginning of 2008. Submission of models is open to researchers worldwide. New models can be submitted at any time. The New Zealand testing centre makes extensive use of software produced by the CSEP testing centre in California. It is envisaged that, in time, the scope of the testing centre will be expanded to include new testing methods and differently-specified models, nonetheless that the New Zealand testing centre will develop in parallel with other regional testing centres through the CSEP international collaborative process.

Key words: Earthquake forecasting, statistical seismology, New Zealand.

1. Introduction

Learning how to forecast earthquakes is one of the most important problems in seismology. It is important for two reasons. From a scientific perspective, our ability to forecast earthquakes is a measure of our understanding of how earthquakes are generated. From a practical perspective, foreknowledge of an increased hazard of earthquake occurrence in a particular location would be useful for decision-making on the timing of mitigation measures, such as protection and upgrading of building stocks and lifeline networks.

After some years of relative neglect, earthquake forecasting is again becoming a target of geophysicists worldwide. It is now widely recognised that, in order to make progress in this field, there is a need both to develop testable earthquake forecasting models and to conduct verifiable tests of their practical forecasting performance. Internationally, efforts to develop models, agree on testing procedures, and establish testing centres to undertake the performance tests, are gaining momentum (JORDAN, 2006; FIELD, 2007; SCHORLEMMER and GERSTENBERGER, 2007). Broadly speaking, the requirements for a model to be testable are that it must be well-defined, i.e., the forecasts are derived in an unequivocal way from the available data, and capable of generating synoptic estimates of the time-varying rate of earthquake occurrence for any source location and magnitude level within a substantial region of surveillance. Models meeting these requirements are called Regional Earthquake Likelihood Models (RELMs).

A major objective of this study is to establish an earthquake forecast testing centre in the New Zealand region. This includes the specification of the detailed requirements for models to be tested in this centre, including the spatial extent of the test region, the magnitude levels and time periods that will be used, and the grid cells within which forecasts will be made and evaluated. Decisions on such specifications

---

[1] GNS Science, 30-368, Lower Hutt, New Zealand. E-mail: M.gerstenberger@gns.cri.nz

depend on the quality and extent of the New Zealand earthquake catalogue, and the data requirements for models that are presently envisaged for installation in the testing centre. Also to be borne in mind is the maintenance of consistent practices with other similar testing centres, especially the California testing centre of the Collaboratory for the Study of Earthquake Predictability (CSEP). There are many benefits to be derived from maintaining such consistency across the testing centres in the area of software development costs, which are considerable, especially in light of the level of automation that is needed.

A second objective is to install certain existing models into the testing centre. The authors of this report include developers of some of the existing models, namely the STEP—"Short-Term Earthquake Probability" (GERSTENBERGER, 2003; GERSTENBERGER et al., 2005), and EEPAS—"Every Earthquake a Precursor According to Scale" (RHOADES and EVISON, 2005, 2006; RHOADES, 2007)—models. Another existing model is the New Zealand National Seismic Hazard model—NZNSHM—(STIRLING et al., 2002), which is already widely used for underpinning earthquake engineering design codes, as well as for many other practical purposes. Although this model is in principle static, rather than time-varying, it is an important reference model to compare models of time-varying earthquake occurrence against. The Epidemic-Type Aftershock Sequence (ETAS) model (OGATA, 1989, 1998) is probably the most widely used short-term earthquake clustering model, and it is desirable to have one or more versions of the space–time ETAS model in the testing centre. Details of all the installed models are given below.

2. Purpose of the Centre

The New Zealand Earthquake Forecast Testing Centre is being established with the following purposes in mind.

The Centre will encourage modellers to develop testable time-dependent seismicity forecasting models for New Zealand. Many studies carry out retrospective analyses of seismicity, but the results and ideas emanating from such studies need to be verified by tests against future seismicity, and in order for that to occur they must first be incorporated into testable models.

The Centre will establish a testing framework appropriate to New Zealand. There are similarities and differences between the New Zealand region and other regions where testing centres are being established, notably, at present, California and parts of Europe. The differences relate to the style of seismicity, and the extent of coverage and history of the earthquake catalogue. New Zealand is a continent that straddles the boundary between the Pacific and (Indo-)Australian plates. The interaction between ocean and continental plates has produced a complex plate boundary (ANDERSON and WEBB, 1994; STERN et al., 2006). In the north, the Pacific plate subducts under the Australian plate in the Hikurangi subduction zone. The subduction is accompanied by island arc and rhyolitic volcanism in the central north Island, where there is an incipient backarc basin. In the southwest the Australian plate (Tasman Sea) subducts under the Pacific plate in the Fiordland subduction zone. Between the subduction zones there is a transpressive continental collision zone.

The Centre will re-evaluate the RELM/CSEP likelihood-based testing procedure. This is a long-term goal. Initially the New Zealand centre is being set up with the same testing procedures as other CSEP testing centres. It is envisaged that re-evaluation of the present procedures will take place through a collaborative process, and that when changed procedures are agreed to, they will be made available to all regional testing centres using common software.

The Centre will investigate other testing methodologies including ground-motion-based testing. The first generation of testing is for regional earthquake likelihood models which estimate the expected number of earthquakes in any given window of time, magnitude and location. The expected number of exceedances of a given level of ground motion at any location in a given window of time is also a quantity of interest, and indeed is the primary quantity of interest in the national seismic hazard model. A long-term goal is to extend the testing to ground-motion models.

The Centre will test multiple forecast models developed for New Zealand in a 5+ year prospective test. Robust tests require a large number of earthquakes. To obtain a large-enough number of significant earthquakes to test the models against, a period of at least 5 years will be necessary. The number of test earthquakes can be increased by lowering the magnitude threshold for targeted events, but in practice any magnitude less than about 5 has minor impact on the ground-motion hazard. It is therefore substantially more important to forecast the larger earthquakes than small-to-moderate-sized events. Also, since it is not clear that the earthquake process is entirely self-similar, an ability to forecast small earthquakes is not equivalent to an ability to forecast large ones. Therefore it may be unhelpful to lower the magnitude threshold too much. In any case, testing of forecasting models is likely to remain an important ongoing activity.

The Centre will test the impact of individual assumptions within models. The effect of individual assumptions on model performance is not always easy to determine from retrospective studies. For example, a more complex model will always fit existing data better than a simpler one, although this does not mean that it will perform better against future earthquakes. Also, the performance of models on a discrete test grid of time, magnitude and location cells is not the same as its performance measured on continuous scales of time, magnitude and location. For model development, it is often more computationally efficient to measure performance on continuous scales. The impact of individual assumptions is not necessarily the same when assessed on a discrete grid. It is desirable to make the testing centre software available to researchers developing models, so that they can anticipate the effects of the test grid on model performance, and if necessary, adjust their models accordingly before submitting them for testing against future earthquakes.

The Centre will maintain a strong relationship with CSEP. A strong international research community with an interest in evaluating the predictability of earthquakes is now developing within the CSEP framework. It is important that the New Zealand centre can benefit from, and contribute to, the combined knowledge of this research community, as well as the specific software products developed by the CSEP community to facilitate testing.

3. Test Region and Grid Specifications

Following extensive consultations between initial participants and potential participants, including informal meetings of the Wellington-based statistical seismology group, specifications for the test region, and the spatial and magnitude grids were drawn up. Important considerations were that while the quality and completeness of the earthquake catalogue is generally good for earthquake locations inside or close to the edges of the New Zealand Seismograph Network, i.e., for onshore locations, this quality and completeness can be expected to deteriorate quite rapidly for offshore locations.

Boundaries of the test region are shown in Fig. 1, and vertices of the polygon defining the test region are listed in Appendix 1. The test region covers the New Zealand land area plus a region extending about 50 km offshore. Figure 1 also shows the data-collection region, and the vertices of the polygon defining this region are listed in Appendix 2. The data-collection region extends about 50 km in all directions beyond the edge of the test region.

The location grid consists of cells of area $0.1°$ squared centred on 1/10th degree coordinates of latitude and longitude which have their centres within the test region, e.g. $(-41.5 \pm 0.05, 174.5 \pm 0.05)$.

Figure 2 is a map of shallow earthquake epicentres in the New Zealand region. By comparing Figs. 1 and 2, it can be seen that many earthquakes occur outside the test and data-collection regions. However, these regions were chosen for reasons of catalogue completeness and quality as mentioned above and discussed in more detail below.

4. Catalogue Completeness Issues

Catalogue completeness is an important issue to consider when specifying the test and data-collection regions. Broadly speaking we can have regard to the following approximate eras of the New Zealand earthquake catalogue when assessing the change of catalogue completeness with time. (a) Pre 1964; (b) 1964 to 1986; (c) 1987 to 1999; and (d) 2000 on. Although the changes to the seismograph network have taken place gradually over periods of time rather than

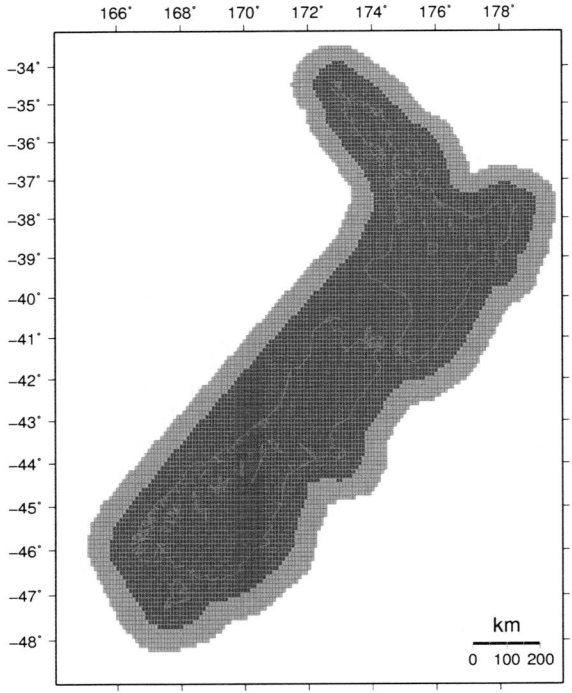

Figure 1
Map showing test region (*darkly shaded*) and buffer zone (*lightly shaded*). The whole shaded region is the data collection region

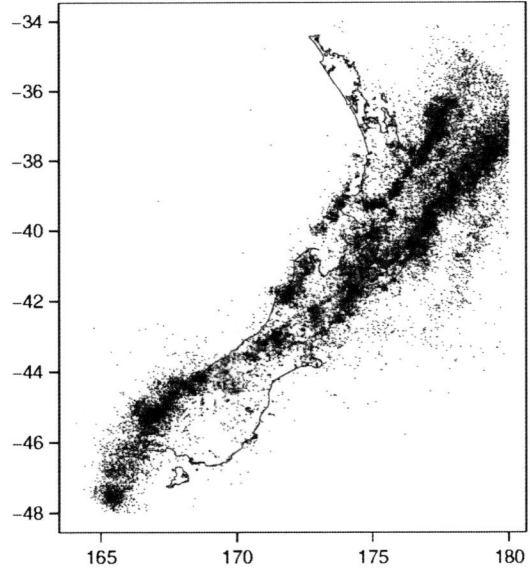

Figure 2
Epicentres of earthquakes in the New Zealand catalogue, 1951–2006, with magnitudes $M > 2.95$ and hypocentral depths $h \leq 45$ km

instantaneously, the years 1964, 1987 and 2000 are the approximate dates of major upgrades of the New Zealand Seismograph Network; the most recent being the transition to the present GeoNet broadband network. We examine the frequency–magnitude distribution of the earthquakes with local magnitude $M_L \geq 3.0$ in each of these eras within the test region and within the buffer region (Fig. 1).

Figure 3 shows frequency-magnitude plots within the test region. For the 1951–1963 period (Fig. 3a), the plot is approximately linear for magnitudes above about 4.2. The deviation from linearity at lower magnitudes is clear evidence of incompleteness up to magnitude 4.0. For the 1964–1976 period (Fig. 3b), the deviation of the plot from linearity suggests a threshold magnitude of completeness slightly below 4.0. On the other hand, the linearity of the plots in Figs. 3c and d suggest that the catalogue is complete, or near complete, for all magnitudes above 3.0 since 1987.

Figure 4 shows corresponding frequency–magnitude plots within the buffer region. For the 1951–1963 period (Fig. 4a), the magnitude threshold of completeness appears to be about 4.8; for the 1964–1976 period (Fig. 4b), it is about 4.2; for the 1987–1989 period (Fig. 4c), it is about 3.9; and for 2000–2006 (Fig. 4d), it is about 3.4. Therefore, in all time periods the catalogue is not as complete in the buffer region as in the test region.

We further examine the change in completeness of the catalogue with time in Figs. 5 and 6.

The numbers of earthquakes in the test region exceeding certain magnitude thresholds, accumulated with time, are plotted in Fig. 5a. This shows that there has been a gradual increase in the rate of accumulation for $M > 2.95$ between 1951 and 2006. The step-like increases are most likely associated with large multiple-earthquake sequences such as the Inangahua aftershock sequence in 1968. Figure 5b can be used to judge the variation of magnitude completeness with time. In this figure, ratios of the number of the earthquakes exceeding certain magnitude thresholds have been plotted. Let $N(M > m)$ be the number of earthquakes exceeding magnitude m in a time interval. Under the assumption of catalogue completeness and a Gutenberg–Richter b value of about 1, the expected value of the ratio $N(M > m + 0.5)/N(M > m)$ is 0.32, shown by the horizontal line

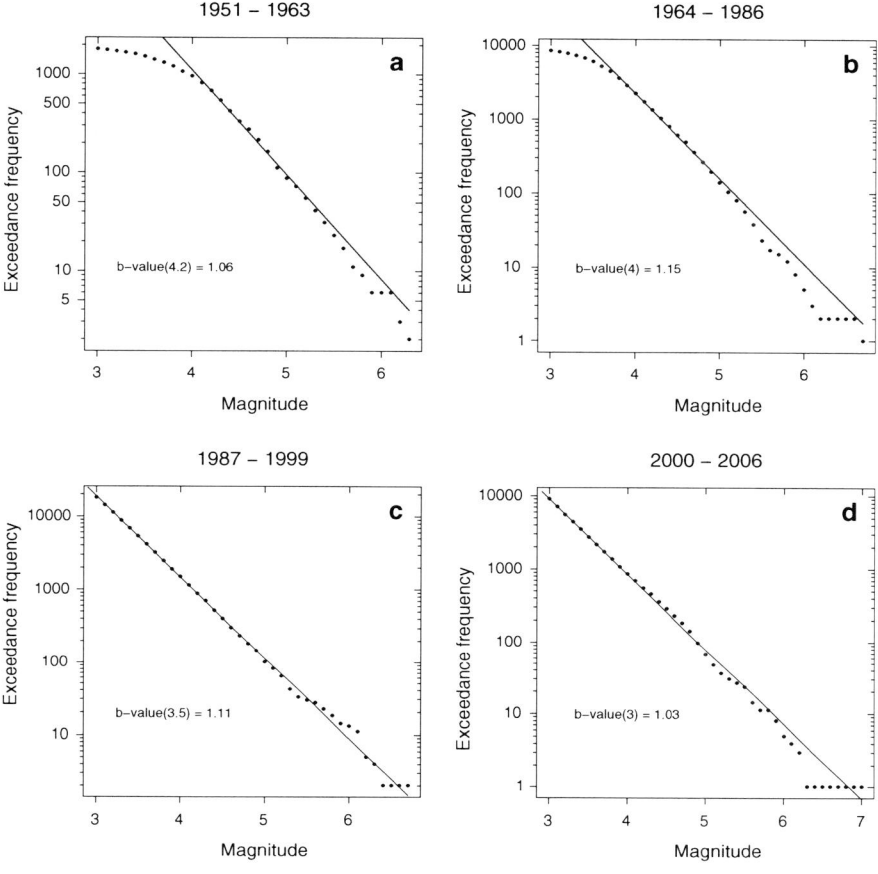

Figure 3
Cumulative frequency versus magnitude for time-period subsets of the New Zealand Earthquake Catalogue (with depths $h \leq 45$ km) within the test region **a** 1951–1963, **b** 1964–1986, **c** 1987–1999, **d** 2000–2006. The Gutenberg–Richter b value listed in each plot corresponds to the line plotted and is calculated using the minimum magnitude threshold given in parentheses

in Fig. 5b. The time when this ratio drops to this level, as shown by the points plotted for 3-year intervals and the associated smooth trend lines, is an indication of the approximate time when the catalogue became complete for magnitude $M > m$.

Based on Fig. 5b, it appears that the catalogue in the test region has been approximately complete at $M > 4.45$ since at least 1951; at $M > 3.95$ since about 1960; at $M > 3.45$ since about 1980; and at $M > 2.95$ since the late 1980s. Being approximately complete implies only that the number of missing earthquakes is statistically small compared with the number actually observed. It means that the catalogue is probably complete over most of the region in question, but not necessarily over each small part of it.

Figure 6 shows the corresponding analysis for the buffer region. Compared to the test region, the increase over time in the rate of accumulation of earthquakes in the lower magnitude bands is much stronger. An extraordinary feature of Fig. 6a is the huge step-up in the cumulative number of earthquakes in all magnitude classes in 1995. This corresponds to the time of the aftershocks of the 1995 Feb. 5 M7.0 East Cape earthquake, many of which occurred in the buffer region.

Based on Fig. 6b, it appears that the catalogue in the buffer region has been nearly complete at $M > 4.45$ since 1951; at $M > 3.95$ since about 1960; at $M > 3.45$ since about 1995; and perhaps at $M > 2.95$ since about 2003. However Fig. 7 displays the frequency magnitude relation in the buffer zone for earthquakes from 2004 to 2006, and it shows that the threshold of completeness is still no lower than 3.4 for this period. There appears to be a deficit of

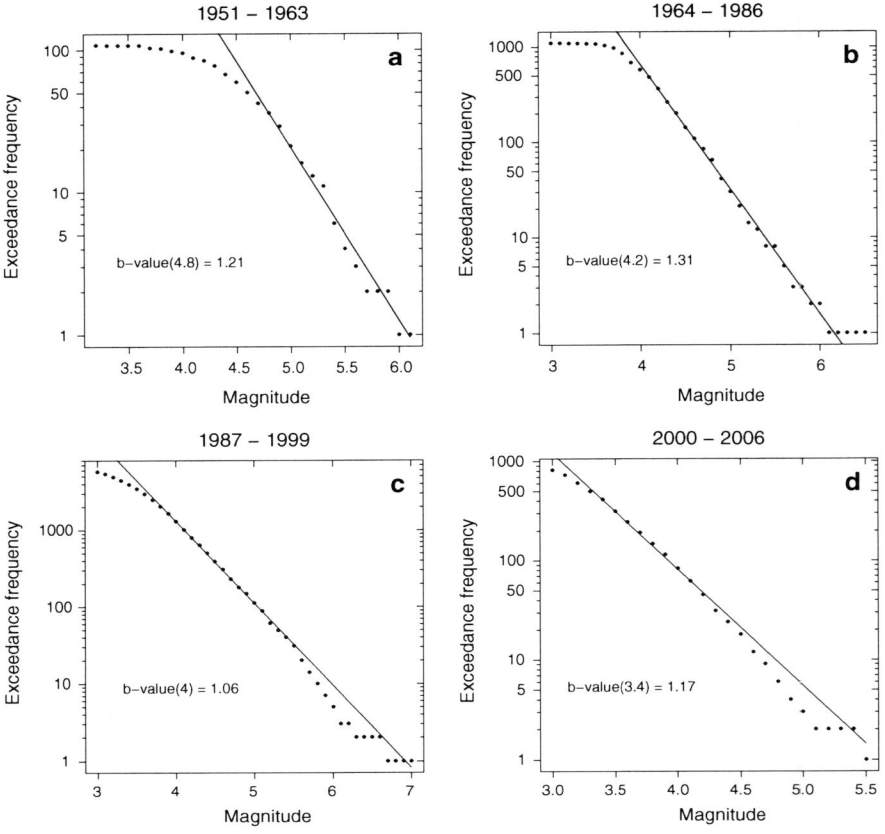

Figure 4
Cumulative frequency versus magnitude for time-period subsets of the New Zealand Earthquake Catalogue (with depths $h \leq 45$ km) within the buffer region **a** 1951–1963, **b** 1964–1976, **c** 1987–1999, **d** 2000–2006. The Gutenberg–Richter b value listed in each plot corresponds to the line plotted and is calculated using the minimum magnitude threshold given in parentheses

at least 250 earthquakes, or about 25% of all earthquakes with $M > 2.95$ in this period. Unless further improvements to the network occur during the testing period, a similar deficit is likely to apply in that period also. This will have an effect on the performance of models which depend on small earthquakes in the buffer region to estimate earthquake occurrence in the test region. This is an effect that needs to be considered in the preparation of models for testing.

5. Other Catalogue Quality Issues

Figure 8 displays scatter plots of hypcocentral depth h against longitude and time. It shows that New Zealand earthquakes occur over a wide range of depths. Hypocentral depths to 300 km are common in the catalogue, and the deepest recorded earthquakes are at about 600 km. The deep earthquakes are mostly associated with the Hikurangi and Fiordland subduction zones. For many shallower earthquakes ($h < \sim 40$ km), the depth is not actually estimated, but rather a depth-restricted solution for the earthquake source is given in the catalogue. Common depth restrictions are $h = 5$, 12 and 33 km. Before the 1980s, depth-restrictions of only 12 or 33 km were used. Figure 8b shows that the number of non depth-restricted solutions increased dramatically in about 1987, but also that the depth-restricted solutions are a significant proportion of the shallow earthquakes right up to the most recent recordings in the catalogue used here, i.e., September 2006. For this reason we must proceed carefully in testing hypocentral depths.

Figure 9 is a histogram of hypocentral depths to 100 km. About 52% of all earthquakes have

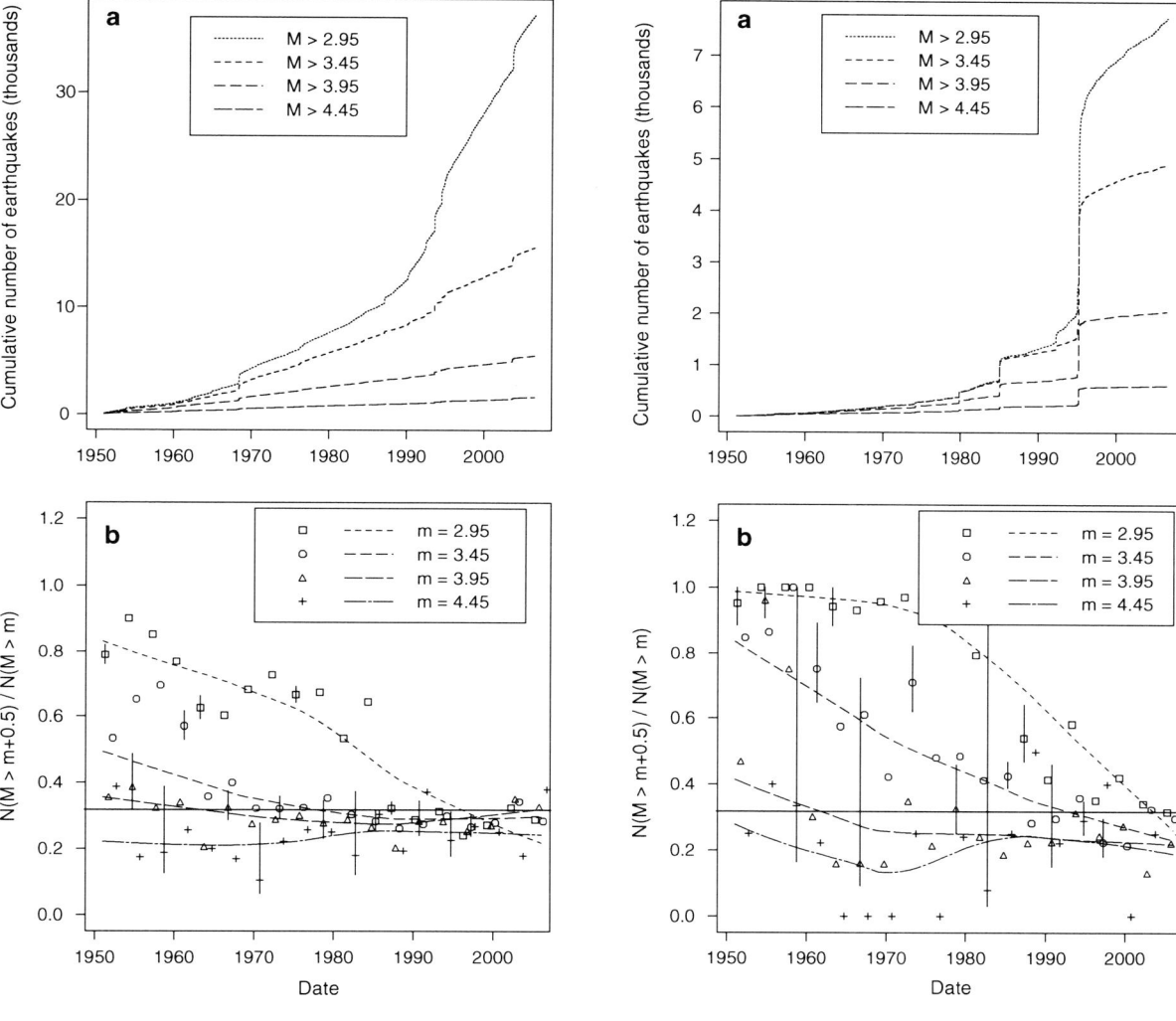

Figure 5
a Cumulative numbers of earthquakes (depths $h \leq 45$ km) within the test region exceeding certain magnitude thresholds as a function of time; **b** Ratio of numbers of earthquakes $N(M > m + 0.5)/N(M > m)$ in 3-year intervals and smooth local fits. The *solid line* shows the expected value of the ratio under catalogue completeness and a Gutenberg–Richter b value of 1. *Error bars* on selected points are approximate 80% confidence intervals

Figure 6
a Cumulative numbers of earthquakes (depths $h \leq 45$ km) within the buffer region exceeding certain magnitude thresholds as a function of time; **b** Ratio of numbers of earthquakes $N(M > m + 0.5)/N(M > m)$ in 3 year intervals and smooth local fits. The *solid line* shows the expected value of the ratio under catalogue completeness and a Gutenberg–Richter b value of 1. *Error bars* on selected points are approximate 80% confidence intervals

$h \leq 40$ km, and only 2% have $40 < h \leq 50$. Figure 9 shows a sharp change in rate of earthquake occurrence versus depth at about $h = 35$ km. Note that the three apparent peaks in frequency versus depth correspond to the three conventional depth restrictions applied to shallow earthquakes, at 5, 12, and 33 km, and should not be taken as evidence that earthquakes occur more frequently in the 0–5, 10–15, and 30–35 km classes than in other shallow depth classes.

Based on extensive discussion with the network operators, we have confidence that any event restricted to a depth of less than 40 km actually occurred in this depth range and it has been agreed by the initial participants of the testing centre that the test region will only include earthquakes within this range; no differentiation of depths will be attempted within the range. This decision has been made in the light of the present quality of the catalogue and the

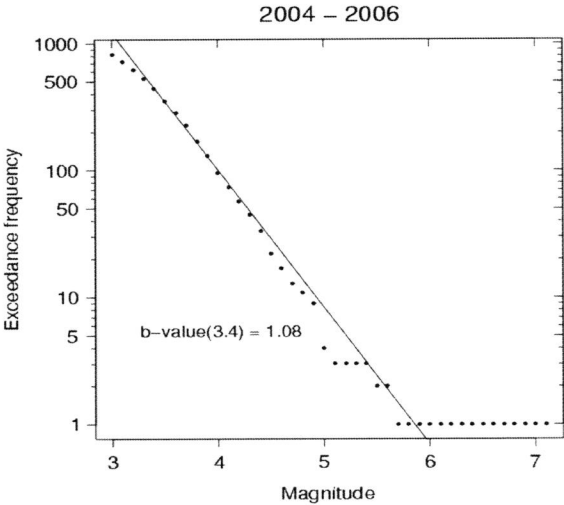

Figure 7
Frequency versus magnitude plot for earthquakes in the buffer region for hypocentral depths $h \leq 45$ km from 2004 to September 2006. The Gutenberg–Richter b value listed corresponds to the line plotted and is calculated using the minimum magnitude threshold given in *parentheses*

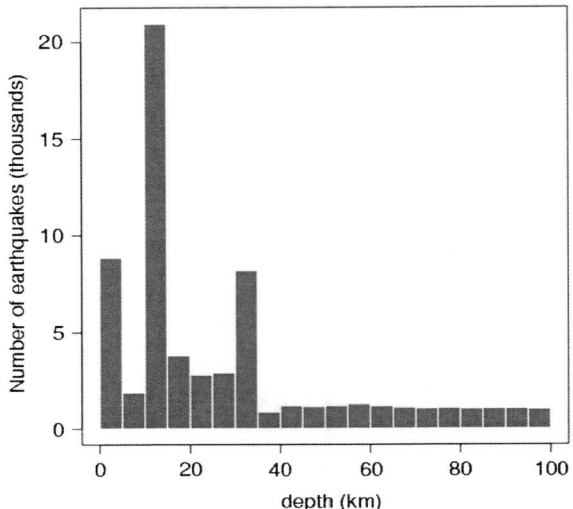

Figure 9
Histograms of hypocental depth h for earthquakes with magnitude $M > 2.95$ in the New Zealand earthquake catalogue, 1951–September 2006, $h \leq 100$ km

present state of modelling of earthquake occurrence. It does not rule out the possibility that at some future time a wider depth range could be included and models which discriminate depth within the 0–40 km range could be tested.

6. Requirements for Model Submission

Testing for 5-year, 6-month, 3-month and 24-hour models will be carried out on the earthquake catalogue beginning from 1 January, 2008, based on the expected number of earthquakes with hypocentral depths $h \leq 40$ km in bins defined by a grid of magnitude values at 0.1 interval spacing, and a grid of latitude and longitude coordinates at 0.1° spacing, with centres inside the testing-region polygon (Appendix 1 and Fig. 1). Models in all classes are invited from the international scientific community. New models may be submitted at any time. Six-month, 3-month and 24-hour models must be installed on Centre computers, as described below;

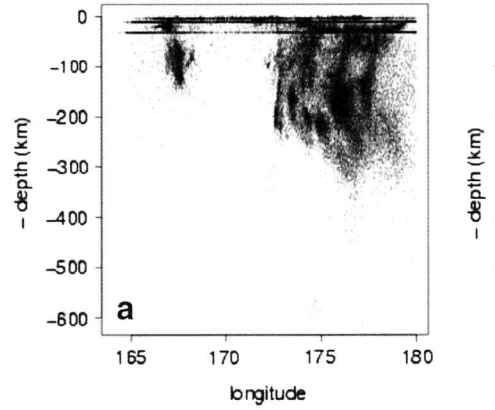
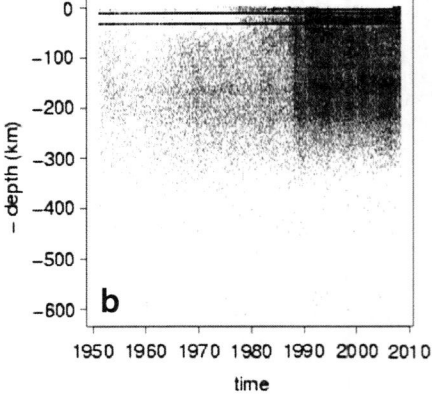

Figure 8
Hypocentral depth h versus **a** longitude; **b** time; for earthquakes with magnitude $M > 2.95$ in the New Zealand earthquake catalogue 1951–September 2006

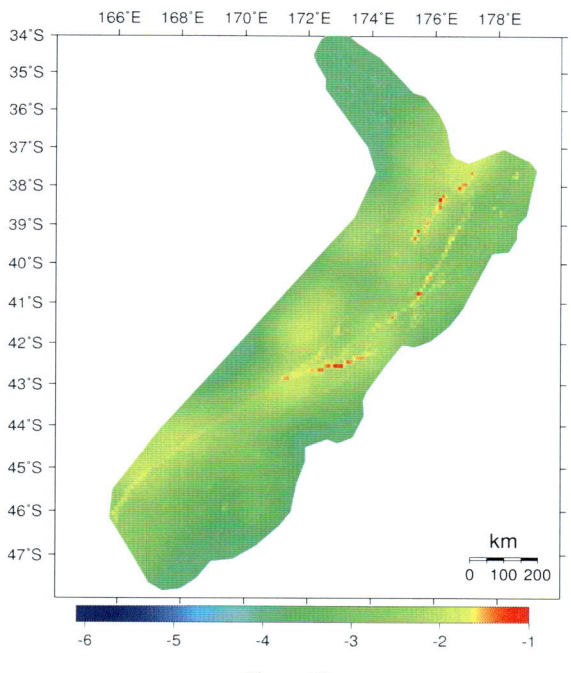

Figure 10
Log$_{10}$ of expected number of earthquakes with magnitude $M \geq 5.0$ in a 5-year period under the NZNSHM model

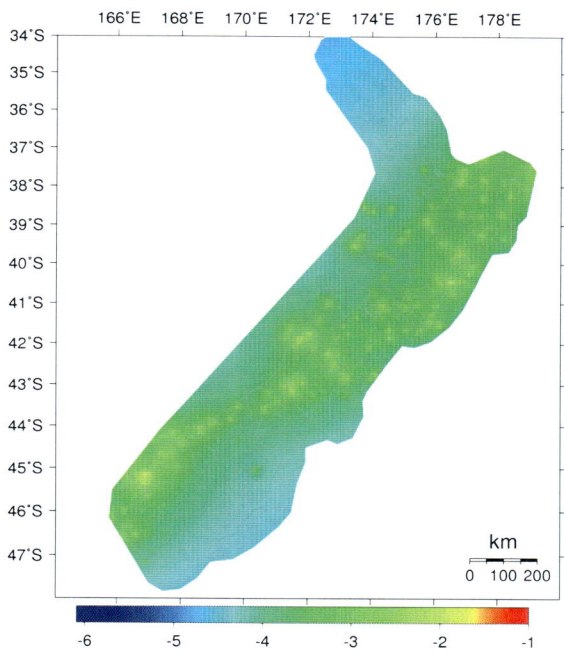

Figure 11
Log$_{10}$ of expected number of earthquakes with magnitude $M \geq 5.0$ in a 3-month period (July–September, 2006) under the PPE model

installations will be done with the assistance of the Testing Centre committee.

6.1. Five-year Models

The 5-year test is designed for time-invariant or quasi time-invariant models. Modellers must supply a file specifying the expected number of earthquakes over a 5-year period in each location and magnitude bin. Magnitude bins are centred on values from 5.0 up to 9.0 in steps of 0.1. Only one depth bin (0–40 km) is being used at present.

6.2. Six-month Models

The 6-month class was added specifically for the available implementation of the M 8 model (HARTE et al., 2007) and models are expected to forecast earthquake occurrence over the next 6-month period. Modellers must supply a computer program which accepts the past earthquake catalogue in the data-collection region (in a format supplied by the testing centre) as input, and outputs a file specifying the expected number of earthquakes over a 6-month period in each location and magnitude bin. The program must be written in such a way that the testing centre can control the input files and specify the period for which the forecast is being made. Magnitude bins are centred on values from 5.0 up to 9.0 in steps of 0.1. Only one depth bin (0–40 km) is being used at present. Six-month models will be supplied with the fully updated and official catalogues from the beginning of 1951 up to 50 days before the start of each 6-month test period.

6.3. Three-month Models

The 3-month test is for intermediate-term forecasting models which use the past earthquake catalogue to forecast the earthquake occurrence over the next 3-month period. Modellers must supply a computer program which accepts the past earthquake catalogue in the data-collection region (in a format supplied by the testing centre) as input, and outputs a file specifying the expected number of earthquakes over a 3-month period in each location and magnitude bin. The program must be written in such a way that the testing centre can control the input files and specify the period for which the forecast is being made.

If required, 3 month models will be supplied with the fully updated and official catalogue from the beginning of 1951 up to 50 days before the start of each 3-month test period. This 50-day information lag is adopted for the 3 month models so as to clearly separate the time-frame being tested from that of the 24-hour models. The models will be tested over successive 3-month intervals up to 5 years or the total period of the tests, for magnitudes greater than 5.0.

6.4. 24-hour Models

The 24-h test is for short-term forecasting models which use the past earthquake catalogue to forecast the earthquake occurrence over the next 24-h period. The minimum magnitude bin for 24-hour models is centred on magnitude 4.0. Otherwise, the test region is the same as for the 5-year and 3-month models. Modellers must supply a computer program which accepts the past earthquake catalogue in the data collection region (in the format supplied by the testing centre) as input, and outputs a file specifying the expected number of earthquakes over a 24-h period in each location and magnitude bin. The program must be written in such a way that the testing centre can control the input files and specify the period for which the forecast is being made.

The 24-hour models may be supplied with the finalised catalogue from the beginning of 1951 up to just before the start of each 24-h test period. They will be tested over a succession of daily intervals up to 5 years or the total period of the tests.

7. Tests of Model Performance

The tests of model performance will initially be the same as those carried out in the CSEP testing centre in California. These tests have been described in detail by SCHORLEMMER et al. (2007), and so are only briefly reviewed here. The tests treat the cell expected values as means of independent Poisson-valued random variables.

7.1. N test

The N test compares the total number of earthquakes expected under the model with the actual number. The N test will reject a model if the total number of earthquakes occurring during the test period is inconsistent with a Poisson random variable with mean N, where N is the total expected number of earthquakes under the model.

7.2. L test

The L test compares the likelihood of the actual earthquake catalogue, i.e., the number of earthquakes occurring in each bin, with the distribution of likelihoods of synthetic catalogues conforming to the model. The model is rejected if the likelihood of the actual catalogue lies outside the distribution of likelihoods of the synthetic catalogues conforming to the model.

7.3. R test

The R test compares the likelihoods of alternative models on the actual data. It tests the statistical significance of any differences by comparing the observed difference with what would be expected if each model, in turn, were the correct one. In order to do this, it generates synthetic earthquake catalogues consistent with each model in turn, and evaluates the likelihood for each model using its own and each other model's set of synthetic catalogues.

In the R test, each model is regarded in turn as the null hypothesis H_0 to be compared against alternatives H_A. Suppose $L(\Omega|H)$ denotes the log likelihood of the actual earthquake catalogue under hypothesis H. The R-statistic is $L(\Omega|H_A) - L(\Omega|H_0)$. A high value of R is favourable to H_A. Model H_0 can be rejected by H_A if R is statistically large compared to the distribution of R-statistics computed using synthetic catalogues consistent with H_0.

7.4. Catalogue Uncertainties

The methodology described in SCHORLEMMER et al. (2007) allows for uncertainties in the published catalogue to be included in the evaluation of model performance. This includes, in principle, both magnitude uncertainties and epicentral location uncertainties. Initially, it is not proposed to specifically allow for such uncertainties in the New Zealand centre testing. However, this is a refinement that could be included later, with the agreement of the participants. Any allowance for uncertainties in the catalogue values has to be made with care, because treatment of uncertainties inevitably involves using a model to generate these uncertainties. Such a model might not be agreed to by all participants. At the very least, any model that is being used to generate catalogue uncertainties must be well understood by all participants.

7.5. Declustering

Some long-term models are designed to forecast only main shocks, and not aftershocks. For such models, tests will be run against a declustered catalogue. The method of REASENBERG (1985), with the default parameter values, will be used to decluster the earthquake catalogue. No declustering will be carried out on the catalogues used for the 3-month and 24-h hour tests; for the 5-year model class, testing will be available with and without declustering.

8. Other Considerations

Inevitably, there is a delay between the occurrence of an earthquake and the inclusion of the final version of its location and magnitude in the earthquake catalogue. This is especially so for the smaller earthquakes, for which the delay may be many months. The test calculations cannot be performed until the final catalogue needed for a given test period is available. Despite this delay, these tests remain truly prospective, because the modellers will have no interaction with the model after the test begins and only data occurring prior to the test initiation will be used to train the models.

The test calculations will be carried out using software developed at the California CSEP testing centre run by the Southern California Earthquake Center. Updates of this software are expected to be released from time to time, and these will normally be incorporated in the New Zealand testing Centre.

If we were to try to define the best performing model, this model would, in the words of SCHORLEMMER et al. (2007), "never be rejected by an R test and would show data consistency in the N and L tests". For technical reasons, if for no other reason, it is unlikely that any model will perform to this ideal standard. Even if a model provides a perfect description of earthquake occurrence, the limited updating of forecasts allowed for under the CSEP testing will prevent it from making perfect forecasts. For example, when a major earthquake occurs, the short-term forecasting models should in theory be instantly updated to estimate the subsequent aftershocks. In practice, no update of expected seismicity can be made until the end of the next 24-h forecasting period. The practical effect of this restriction on updating is that the number of earthquakes in a cell will not conform to either of the assumptions underlying the tests, i.e., deviations from the expected values will be neither Poisson-distributed nor independent between cells.

Therefore, it will be necessary to carry out tests other than those already incorporated in the testing centre software. In particular, we would like to develop tests that emphasize measurement of the information value of the different models, and identify sub-regions where particular models are more or less informative than others. In the longer term, we would like to extend the tests to forecasts of ground-shaking. Such an extension creates a new set of problems, including the modelling of site effects at the locations of strong-motion instruments. The likely process for acceptance of new tests is that they will be promoted through the CSEP collaborative process, and incorporated in software made available to all regional testing centres.

A web page http://www.cseptesting.org/centers/gns has been established for the New Zealand testing centre. This will be used to disseminate and update information about the centre.

Participants submitting models to the New Zealand earthquake forecasting testing centre retain all existing rights to their own models. The centre claims no right to use the models except for testing purposes.

9. Models Installed in the Testing Centre

Three-five-year models, 2–3-month models and two 24-hour models have currently been submitted to the testing centre.

9.1. Five-year Models

The current 5-year models are SUP (Stationary Uniform Poisson), PPE (Proximity to Past earthquakes), and NZNSHM (New Zealand National Seismic Hazard Model).

The SUP model is included as a reference model of least information. This time-invariant model is also available as a reference model for testing over any other time period. The assumption of this model is that seismicity is stationary and spatially homogeneous over the entire test region. It is not a realistic model of seismicity because it does not incorporate either temporal or spatial variation of the rate of earthquake occurrence.

The PPE model is closely based on a model proposed by JACKSON and KAGAN (1999). It is a type of smoothed seismicity model, in which earthquakes are forecast to occur close to the epicentres of past earthquakes above the magnitude threshold m_c for the test region. The PPE model was described in detail by RHOADES and EVISON (2004).

A forecast for the Wellington region based on the quasi-static model of ROBINSON and BENITES (1996) has also been installed in the centre. To create the forecast, a synthetic catalogue of several hundred thousand years was first created using the model and subsequently used as input into the PPE model. The details of 55 known faults, the subduction zone, and 3,000 random small faults are used to create the synthetic catalogue that is only for the Wellington region.

The NZNSHM model is derived from the work of STIRLING et al. (2002). It is a model developed using modern probabilistic seismic hazard analysis techniques and consists of earthquake sources of two types: (1) fault sources; and (2) distributed seismicity sources. The fault source model consists of more than 300 faults where the following parameters are specified for each fault: fault type (e.g., normal or strike-slip), maximum and minimum depth, single event displacement, maximum magnitude, and recurrence interval. These parameters are defined through a combination of field work, modelling and expert judgement. The distributed seismicity sources are based on a smoothed representation of the historical catalogue of earthquakes in New Zealand from 1840 to the present.

The model installed in the testing centre differs from the original NZNSHM model in that the rates of earthquakes applied to a single fault in the original model have been applied to one or more grid cells, in order to meet the grid-based testing requirements of the tests applied within the centre. The faults are transformed to grid cells by projecting the faults to the surface and evenly distributing the fault-based event rate to all grid cells through which the fault passes. Figure 10 shows a 5-year forecast of earthquakes with magnitude $M \geq 5.0$ under the NZNSHM model.

9.2. Six-month Models

The M 8 model (HARTE et al., 2007) implemented in the testing centre is based on the original algorithm of KEILIS-BOROK and KOSSOBOKOV (1990) which uses seismicity patterns to forecast large magnitude events (i.e., magnitude 8). The HARTE et al. (2007) implementation of the algorithm has been adapted to provide synoptic forecasts and to forecast events as small as magnitude 5.0.

9.3. Three-month Models

The current 3-month models are the PPE model and the EEPAS (Every Earthquake a Precursor According to Scale) model.

The PPE model is submitted for testing as a 3-month model as well as a 5-year model. Its role in the 3-month tests is mainly as a reference model which is spatially varying but quasi-time invariant. The only time varying element in the PPE model is due to the augmentation, at 3-month intervals, of the earthquake data-base to include the most recent earthquakes. Because of this updating, and because of the general tendency of earthquakes to cluster in both time and location, it is expected that the PPE model may perform slightly better in the 3-month testing than in the 5-year testing. Figure 11 shows a 3-month forecast of earthquakes with magnitude $M \geq 5.0$ under the PPE model.

The EEPAS model (RHOADES and EVISON, 2004, 2005, 2006; CONSOLE et al., 2006; RHOADES, 2007) is a method of forecasting based on the notion that the precursory scale increase phenomenon (EVISON and RHOADES, 2002, 2004) occurs at all scales in the seismogenic process. Four different versions of EEPAS have been submitted to the New Zealand testing centre. These are EEPAS_0r (a version with equal weighting and restricted parameter optimization), EEPAS_1r (with down-weighting of aftershocks and restricted parameter optimization), EEPAS_0f (with equal weighting and full parameter optimization); and EEPAS_1f (with down-weighting of aftershocks and full parameter optimization). EEPAS_0f is the best performing model in retrospective tests on the past catalogue, however whether the same is true in prospective testing remains to be seen. Figure 12 shows a 3-month forecast of earthquakes with magnitude $M \geq 5.0$ under the EEPAS_0r model.

9.4. 24-hour Models

The current 24-hour models are STEP (Short-Term Earthquake Probability) and a space–time ETAS (Epidemic-Type Aftershock Sequence) model.

The STEP model (GERSTENBERGER, 2003; GERSTENBERGER et al., 2005) is an aftershock model based on the idea of superimposed Omori (OGATA, 1988, 1998) type sequences. The model comprises two components: (1) a background model; and (2) a time-dependent clustering model. The background model can consist of any model that is able to forecast a rate of events for the entire region of interest at all times; for the testing centre the NZNSHM is applied as the background model. The clustering model is based on the work of REASENBERG and JONES (1989) which defines aftershock forecasts based on the a and the b value from the Gutenberg–Richter relationship (GUTENBERG and RICHTER, 1944) and the p value from the modified Omori law (OGATA, 1988, 1998). Figure 13 shows a 24-h forecast of earthquakes with magnitude $M \geq 5.0$ under the STEP model.

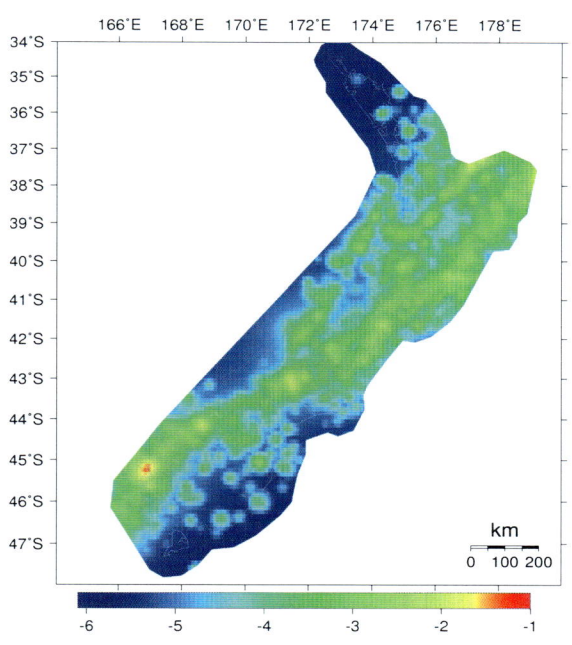

Figure 12
Log_{10} of expected number of earthquakes with magnitude $M \geq 5.0$ in a 3-month period (July–September, 2006) under the EEPAS_0r model

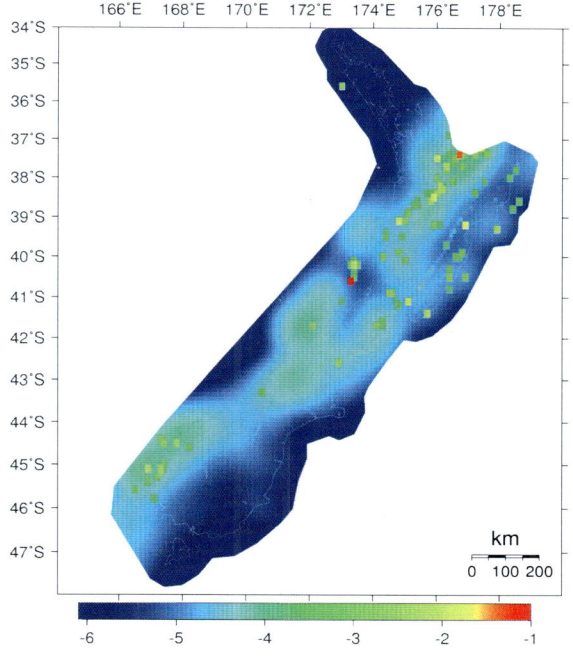

Figure 13
Log_{10} of expected number of earthquakes with magnitude $M \geq 5.0$ in a 24-h period under the STEP model

The ETAS model is a widely used model of earthquake clustering (OGATA, 1988, 1989, 1998; CONSOLE and MURRU, 2001; CONSOLE et al., 2003, 2006). The version of the ETAS model installed in the testing centre is different in several details from most published versions. It is based on the aftershock model used to down-weight aftershocks in the EEPAS model (RHOADES and EVISON, 2004). The spatial distributions of aftershocks in this model follow a bivariate normal distribution with circular symmetry, rather than an inverse power law, which is commonly adopted in other versions of the space–time ETAS model.

10. Conclusion

The establishment of the New Zealand Earthquake Forecast Testing Centre is an important milestone towards the development of usable scientific earthquake forecasts for the New Zealand region. The software and testing methods used in the New Zealand centre are consistent with other testing centres in California and Europe. The New Zealand testing centre, along with the other regional testing centres, provides for more rigorous and transparent testing of a wide range of proposed forecasting models than has occurred in the past. Researchers worldwide are invited to submit their models for testing in the New Zealand centre, as well as the other regional centres.

The models already submitted include representatives of some of the best established and most studied classes of models in existence. However, every model is capable of further development; whether it is a long-term model such as NZNSHM, a medium-term forecasting model such as EEPAS, or a short-term model, such as STEP and ETAS. Moreover, there is scope for improving forecasting capability by attempting to combine the information from essentially different models into hybrid forecasting models (RHOADES and GERSTENBERGER, 2009).

The collaboration that is developing under the umbrella of CSEP is likely to spawn a more rapid development of new testable models of earthquake occurrence, including models based on observations which make use of other data bases than the past earthquake catalogue, and some models that could not be accommodated in the present testing framework. It will therefore be necessary for the testing centre activities to be expanded to respond to the challenge of testing new types of models.

The centre's activities could also be usefully expanded to make the testing software available to researchers when preparing their models for submission. The process of developing a forecasting model is an arduous one which involves extensive computer code development. Errors in large computer codes are easy to create and difficult to find. It is better that coding errors be found before the models are submitted, rather than after years of formal testing. Error detection could be assisted if the researchers were able to retrospectively test their models using the same software that will be used in the prospective tests, before submitting their models for testing.

Acknowledgements

This work was funded by the EQC Research Foundation under project number 06/510, and by the Southern California Earthquake Center. SCEC is funded by NSF Cooperative Agreement EAR-0106924 and USGS Cooperative Agreement 02HQAG0008. The SCEC contribution number for this paper is 1240. Software developed by the California CSEP Earthquake Forecast Testing Center has been extensively used in setting up the New Zealand testing centre. Appreciation is extended to R. Brownrigg for the implementation of the M 8 model. C. Holden, M. Stirling, E. Smith and an anonymous reviewer have provided helpful reviews of the manuscript. Some plots were produced using the Generic Mapping Tools package (WESSEL and SMITH, 1995).

Appendix 1

See Table 1.

Table 1
Polygon defining test region

Latitude	Longitude	Latitude	Longitude	Latitude	Longitude	Latitude	Longitude
−46.1520	165.8468	−35.5322	175.2823	−39.7427	177.8053	−44.8421	171.9297
−45.4971	165.7431	−35.6257	175.6624	−40.1637	177.5288	−45.1696	171.7569
−44.0936	167.3329	−36.0936	176.1118	−41.1462	176.8721	−45.3567	171.6532
−38.8070	173.4505	−36.5146	176.3537	−41.5673	176.4574	−45.7310	171.5495
−37.6374	174.1071	−37.1228	176.4919	−41.9415	175.8698	−46.0117	171.4804
−36.9825	173.8652	−37.2632	176.6302	−42.0819	175.3514	−46.3392	171.1002
−35.4386	172.5173	−37.4035	177.0449	−42.0351	174.9712	−46.8070	170.3399
−35.1579	172.5173	−37.0292	178.1509	−42.5029	174.4873	−47.0877	169.6486
−34.6901	172.2062	−37.3567	178.9804	−43.1579	173.8652	−47.1345	168.9574
−34.5029	172.1371	−37.5906	179.1878	−43.3918	173.7270	−47.5088	168.5081
−34.0819	172.3790	−38.2456	179.0150	−43.7661	173.7615	−47.7427	167.9896
−33.8947	172.3790	−38.7602	178.8767	−44.2807	173.4159	−47.7895	167.4366
−33.8947	173.1048	−38.9942	178.6002	−44.4211	172.9320	−47.6023	166.9873
−34.2222	173.5887	−39.3684	178.5657	−44.3275	172.6210	−46.1520	165.7431
−34.5965	174.2454	−39.6959	178.3237	−44.5146	171.9297		

Appendix 2

See Table 2.

Table 2
Polygon defining data-collection region

Latitude	Longitude	Latitude	Longitude	Latitude	Longitude	Latitude	Longitude
−46.2368	165.1210	−33.4474	173.2776	−39.8158	179.0841	−45.6579	172.2408
−45.9211	165.0173	−33.8158	173.9343	−40.0789	178.6694	−45.9211	172.2062
−45.3947	165.1901	−34.4474	174.9712	−40.2368	178.2546	−46.4474	172.0334
−44.9737	165.5012	−35.1842	175.5933	−40.8684	177.8399	−46.7632	171.5150
−41.7632	169.1993	−35.5000	176.3882	−41.6053	177.3214	−46.8158	171.3767
−38.4474	172.9320	−36.0263	176.7339	−42.0263	176.8721	−47.3947	170.5127
−37.5000	173.3468	−36.2895	176.8721	−42.5000	175.9389	−47.5526	169.8560
−37.1842	173.3122	−36.8684	177.1141	−42.6579	175.1440	−47.6053	169.5449
−35.7105	172.0680	−36.6053	177.9781	−43.2368	174.6601	−47.9737	168.9228
−35.1842	171.8952	−36.7632	179.0150	−43.5526	174.4528	−48.1842	168.0933
−35.1316	171.7224	−37.1842	179.6717	−44.0789	174.4528	−48.2895	167.3675
−34.5000	171.4804	−37.8684	179.7753	−44.7105	173.9689	−48.1316	166.6417
−33.9211	171.6878	−38.9737	179.4988	−44.8684	173.1048	−47.5000	165.9850
−33.5526	172.1717	−39.1316	179.3606	−44.9211	172.6210	−46.2368	165.1210
−33.4474	172.6210	−39.6053	179.1532	−45.3421	172.4827		

REFERENCES

ANDERSON, H., and WEBB, T. (1994), *New Zealand seismicity—Patterns revealed by the upgraded National-Seismograph-Network*, N. Z. J. Geol. Geophys. *37*, 477–493.

CONSOLE, R., and MURRU, M. (2001), *A simple and testable model for earthquake clustering*. J. Geophys. Res. *106*, 8699–8711.

CONSOLE, R., MURRU, M., and LOMBARDI, A. M. (2003), *Refining earthquake clustering models*, J. Geophys. Res. *108*, 2468, doi:10.1029/2002JB002130.

CONSOLE, R., RHOADES, D. A., MURRU, M., EVISON, F. F., PAPADIMITRIOU, E. E., and KARAKOSTAS, V. G. (2006), *Comparative performance of time-invariant, long-range and short-range forecasting models on the earthquake catalogue of Greece*, J. Geophys. Res. *111*, B09304, doi:10.1029/2005JB004113.

EVISON, F. F., and RHOADES, D. A. (2002), *Precursory scale increase and long-term seismogenesis in California and Northern Mexico*, Ann. Geophys. *45*, 479–495.

EVISON, F. F., and RHOADES, D. A. (2004), *Demarcation and scaling of long-term seismogenesis*, Pure Appl. Geophys. *161*, 21–45.

FIELD, E. H. (2007), *Overview of the Working Group for the Development of Regional Earthquake Likelihood Models (RELM)*, Seismol. Res. Lett. *78*(1), 7–16.

GERSTENBERGER, M. (2003), *Earthquake Clustering and Time-dependent Probabilisitic Seismic Hazard Analysis for California*, Dissertation submitted to the Swiss Federal Institute of Technology for the degree of Doctor of Science, Zürich.

GERSTENBERGER, M., WIEMER, S., JONES, L. M., and REASENBERG, P. A. (2005), *Real-time forecasts of tomorrow's earthquakes in California*, Nature *435*, 328–331.

GUTENBERG, B., and RICHTER, C. F. (1944), *Frequency of earthquakes in California*, Bull. Seismol. Soc. Am. *34*, 185–188.

HARTE, D., FENG-DONG, L., VREEDE, M., VERE-JONES, D., WANG, Q. (2007), *Quantifying the M8 algorithm: model, forecast, and evaluation, New Zealand*, J. Geol. Geophys. *50*, 117–130.

JACKSON, D. D., and KAGAN, Y. Y. (1999), *Testable Earthquake Forecasts for 1999*, Seismol. Res. Lett. *70*(4), 393–403.

JORDAN, T. H. (2006), *Earthquake predictability, brick by brick*, Seismol. Res. Lett. *77*(1), 3–6.

KEILIS-BOROK, V.I., and KOSSOBOKOV, V.G. (1990), *Premonitory activation of earthquake flow: algorithm M8*, Phys. Earth. Planet. Inter. *61*, 73–83.

OGATA, Y. (1988), *Statistical models for earthquake occurrences and residual analysis for point processes*, J. Am. Stat. Assoc. *83*, 9–27.

OGATA, Y. (1989), *Statistical models for standard seismicity and detection of anomalies by residual analysis*, Tectonophysics *169*, 159–174.

OGATA, Y. (1998), *Space-time point process models for earthquake occurrence*, Ann. Inst. Stat. Math. *50*, 379–402.

REASENBERG, P. A. (1985), *Second-order moment of central California seismicity, 1969–1982*, J. Geophys. Res. *90*, 5479–5496.

REASENBERG, P. A., and JONES, L. M. (1989), *Earthquake hazard after a mainshock in California*, Science *243*, 1173–1176.

RHOADES, D. A. (2007), *Application of the EEPAS model to forecasting earthquakes of moderate magnitude in Southern California*, Seismol. Res. Lett. *78*(1), 110–115.

RHOADES, D. A., and EVISON, F. F. (2004), *Long-range earthquake forecasting with every earthquake a precursor according to scale*, Pure Appl. Geophys. *161*, 47–71.

RHOADES, D. A., and EVISON, F. F. (2005), *Test of the EEPAS forecasting model on the Japan earthquake catalogue*, Pure Appl Geophys. *162*, 1271–1290.

RHOADES, D. A., and EVISON, F. F. (2006), *The EEPAS forecasting model and the probability of moderate-to-large earthquakes in the Kanto region, Central Japan*, Tectonophysics *417*, 119–130.

RHOADES, D., and GERSTENBERGER, M. (2009), *Mixture models for improved short-term earthquake forecasting*, Bull. Seismol. Soc. Am. *99*, 636–646, doi:10.1785/0120080063.

ROBINSON, R., and Benites, R. (1996), *Synthetic Seismicity Models for the Wellington Region, New Zealand: Implications for the Temporal Distribution of Large Events*, J. Geophys. Res. *101*, 27833–27845.

SCHORLEMMER, D., and GERSTENBERGER, M. C. (2007), *RELM Testing Center*, Seismol. Res. Lett. *78*(1), 30–36.

SCHORLEMMER, D., GERSTENBERGER, M. C, WIEMER, S., JACKSON, D. D., and RHOADES, D. A. (2007), *Earthquake likelihood model testing*, Seismol. Res. Lett. *78*(1), 17–29.

STERN, T. A., STRATFORD, W. R., and SALMON, M. L. (2006), *Subduction evolution and mantle dynamics at a continental margin: Central North Island, New Zealand*, Rev. Geophys. *44*, RG4002, doi:10.1029/2005RG000171.

STIRLING, M. W., MCVERRY, G. H., and BERRYMAN, K. R. (2002), *A new seismic hazard model for New Zealand*, Bull. Seismol. Soc. Am. *92*, 1878–1903.

WESSEL, P., and SMITH, W. H. F. (1995), *New version of the generic mapping tools released*, EOS Trans. Am. Geophys. U. *76*, 329.

(Received August 30, 2008, revised July 6, 2009, accepted July 27, 2009, Published online April 20, 2010)

The Area Skill Score Statistic for Evaluating Earthquake Predictability Experiments

J. Douglas Zechar[1,2] and Thomas H. Jordan[1]

Abstract—Rigorous predictability experimentation requires a statistical characterization of the performance metric used to evaluate the participating models. We explore the properties of the area skill score measure and consider issues related to experimental discretization. For the case of continuous alarm functions and continuous observations, we present exact analytical solutions that describe the distribution of the area skill score for unskilled predictors, and we also describe how a Gaussian distribution with known mean and variance can be used to approximate the area skill score distribution. We quantify the deviation of the exact distribution from the Gaussian approximation by specifying the kurtosis excess as a function of the number of observed target earthquakes. For numerical earthquake predictability experiments that involve discretization of the study region and observations, we explore simulation procedures for estimating the area skill score distribution, and we present efficient algorithms for various experimental scenarios. When more than one target earthquake occurs within a given space/time/magnitude bin, the probabilities of predicting individual events are not independent, and this requires special consideration. Having presented the statistical properties of the area skill score, we describe and illustrate a preliminary procedure for comparing earthquake prediction strategies based on alarm functions.

Key words: Statistical seismology, earthquake predictability, earthquake prediction, Molchan diagram, error diagram.

1. Earthquake Forecasting with an Alarm Function

Earthquake forecasts can be stated in various forms: one may estimate the time of the next major earthquake on a given fault or fault segment; one might predict that a large earthquake will occur within a specified space/time/magnitude range; or one might forecast the future rate of seismicity throughout a geographical region. In practice, predictions of the first type are difficult to evaluate because they may require decades of waiting for large earthquakes and because fault structures often are not precisely defined, making the assignment of an earthquake to a specific fault or fault segment a subjective procedure. If properly specified, the latter two types of experiments can be evaluated formally, and experiments to do so are underway. For example, the Reverse Tracing of Precursors (RTP) algorithm (KEILIS-BOROK *et al.*, 2004; SHEBALIN *et al.*, 2006) has been used to make predictions of moderate to large earthquakes in several regions, and it is currently being tested in various regional settings. As part of the Regional Earthquake Likelihood Model (RELM) working group project (FIELD, 2007, and references therein), a number of research teams have submitted 5-year seismicity rate forecasts in prescribed latitude/longitude/magnitude bins in California. The RELM forecasts are being evaluated within a Collaboratory for the Study of Earthquake Predictability (CSEP) testing center (SCHORLEMMER *et al.*, 2010).

A difficulty arises, however, when comparing forecasts stated in different forms, even when forecasts apply to the same space/time/magnitude domain. For example, RELM likelihood tests require a gridded rate forecast and the tests cannot be used to compare forecasts that are not of this type. One way to address this problem is to consider earthquake forecasts in the most fundamental terms. Most forecast statements can be reduced to an ordering of space/time/magnitude bins by the expected probability of each bin to host a specified future earthquake (or earthquakes). In other words, such forecasts can be translated to a statement similar to the following:

---

[1] Department of Earth Sciences, University of Southern California, Los Angeles, CA 90089, USA. E-mail: zechar@ldeo.columbia.edu; tjordan@usc.edu

[2] Lamont-Doherty Earth Observatory, Columbia University, Palisades, NY 10964, USA.

space/time/magnitude bin r_1 is more likely to host a future earthquake than bin r_2, which in turn is more likely than r_3, and so on. This yields a very general approach for comparing forecasts originally stated in different terms. For example, if we consider an experiment in some study region R comprised of bins $r_1, r_2, \ldots, r_n$, where Forecast A predicts rates of seismicity in each bin and Forecast B predicts the probability of earthquakes in each bin, each forecast provides an ordering of the bins. In this context, a forecast does well when many earthquakes occur in the most highly ranked bins and few earthquakes occur in bins with low ranking. At the conclusion of a predictability experiment, one might compare Forecast A and Forecast B by considering, for example, the ten most highly ranked bins for each forecast and counting the number of earthquakes that occur within these bins, i.e., those that have been successfully predicted. Implicit in this evaluation is the choice of a threshold, below which the rankings are disregarded—this yields a binary prediction. We call any bin above the threshold an alarm, where one or more target earthquakes are expected. Furthermore, we call this form of prediction "alarm-based," and we consider the ranking to be an alarm function. We note that an alarm function need not be stated in terms of rank, but the implicit ordering should be unambiguous. For example, each of the RELM forecasts is an alarm function with values specified by expected rates—the bin with the highest forecasted rate has the top ranking. Likewise, any algorithm that computes a seismicity index—e.g., the Pattern Informatics method (RUNDLE et al., 2002)—provides an alarm function with values specified by the index. Methods such as RTP that forecast individual earthquakes by explicitly declaring alarms are characterized by alarm functions with only two rankings: zero—no earthquake is predicted—and one—one or more earthquakes are predicted.

Alarm functions are multidimensional; they can be defined over geographical space, time, magnitude, focal mechanism, etc. To compare two alarm functions, each must cover the same space/time/magnitude domain, although they need not employ the same discretization—in the notation above, R must be the same but the partitioning of R into bins may be different. The simple threshold testing method described above can be iterated to consider the entire alarm function by varying the threshold from the highest rank to the lowest.

In ZECHAR and JORDAN (2008), we suggested a performance metric—the area skill score—based on alarm functions and a threshold approach to testing. In this paper, we present a statistical characterization of the area skill score and consider the details of its use in discrete earthquake predictability experiments. We begin by considering the Molchan diagram in a discrete, a posteriori context; that is, at the conclusion of a predictability experiment in which the study region has been divided into bins and the number of observed target earthquakes is known. We present the average Molchan diagram behavior for unskilled forecasts and use this to determine analytical expressions for the area skill score distribution. We then consider special cases for which the analytical solutions are not applicable, and we describe efficient algorithms for numerically estimating the area skill score distribution. We also outline a possible testing procedure based on the area skill score and illustrate its use in a hypothetical predictability experiment.

2. Molchan Diagram for Testing Alarm Functions

In ZECHAR and JORDAN (2008), we described the Molchan diagram in terms of a continuum. In this section, we present a discrete analysis, as numerical predictability experiments are treated in discrete terms. An earthquake forecast statement should include advance specification of the class of earthquakes to be predicted—the target earthquakes. At the conclusion of an earthquake predictability experiment, given an alarm function, a threshold, and the observed target earthquake catalog, a number of measures of success based on the contingency table can be computed (MASON, 2003). The Molchan diagram (MOLCHAN, 1991; MOLCHAN and KAGAN, 1992) is a useful diagnostic because it captures two such measures and the tradeoff between them: miss rate, v—the proportion of target earthquakes falling outside all alarms—and the fraction of space/time occupied by alarm, τ. The latter measure requires a reference model $\tilde{q}$ to define the measure of space. The reference model must be a probability density function that estimates the future distribution of target earthquakes; typically, the reference model is inferred from the historical distribution of earthquakes.

Consider an alarm function f defined on a space/time/magnitude region R that can be treated as a set of n discrete, non-overlapping space/time/magnitude bins:

$$R = \{r_1, r_2, \ldots, r_n\}, |R| = n. \quad (1)$$

By applying a threshold λ to the alarm function, we obtain an alarm set:

$$A = \{r_i | f(r_i) > \lambda\}. \quad (2)$$

An example alarm function and a few derived alarm sets are illustrated in Figure 6 of ZECHAR and JORDAN (2008). At the conclusion of a prediction experiment, the number of earthquakes in the observed target earthquake catalog, N, is known. The number of hits, h, is the number of target earthquakes occurring inside A, and the miss rate is:

$$v = \frac{N - h}{N}. \quad (3)$$

The fraction of space–time occupied by alarm is:

$$\tau = \sum_{i=1}^{n} \tilde{q}(r_i) \mathbf{1}_{r_i \in A}. \quad (4)$$

Here, $\mathbf{1}_X$ is a logical function that yields unity if X is true and otherwise yields zero. For any threshold $\lambda \geq \sup\{f(\mathbf{x})\}$, A is the empty set—no alarm is declared—and all events are missed: $(\tau, v) = (0, 1)$. Likewise, for any threshold $\lambda < \inf\{f(\mathbf{x})\}$, all of R is an alarm region—$A = R$—and no events are missed: $(\tau, v) = (1, 0)$. One can choose many different thresholds and obtain what we call a Molchan trajectory, the set of (τ, v) points on $[0, 1] \times [0, 1]$ that completely characterize the performance of the alarm function during the experiment. Without any loss of information, we can reduce this set to only the set of points where one or more new hits occur, points which we call Molchan trajectory jumps. We write this reduced Molchan trajectory as the set of minimum τ values such that some number of hits is obtained. In other words, τ_k is the minimum fraction of space/time that the alarm function must occupy to obtain k hits:

$$\{\tau_k = \inf(T) | v = v_k, k \in [1, N]\}.$$

Here, T is the set of τ values from the complete Molchan trajectory, and we use the following indexed notation to specify the miss rate:

$$v_k = \frac{N - k}{N}. \quad (5)$$

We can also express the Molchan trajectory in terms of miss rate as a stairstep function of τ:

$$v_f(\tau) = \sup\{v_k | H(\tau < \tau_k) = 1\}. \quad (6)$$

Here, H is the Heaviside function. For unskilled alarm functions (i.e., those with random values that do not reflect the distribution of earthquakes and therefore have no predictive skill), we show in Appendix 1 that the expected value for a Molchan trajectory jump is:

$$\langle \tau_k \rangle = \frac{k}{N + 1}. \quad (7)$$

Note that (7) corrects the minor misperception that the descending diagonal between the Molchan trajectory endpoints ($\langle \tau_l \rangle = l/N$) represents the average behavior of unskilled alarm functions for a given experiment; rather, the diagonal should be replaced by a stairstep function starting at $(\tau, v) = (0, 1)$ with stairs of width $1/(N + 1)$ and height $1/N$ (see Fig. 1).

Confidence bounds on the Molchan diagram can be computed using the cumulative binomial distribution (KOSSOBOKOV, 2006; ZECHAR and JORDAN, 2008). Using the Molchan diagram and its confidence bounds to evaluate an entire alarm function, however, can yield ambiguous results. In particular, an alarm function may yield some alarm sets that demonstrate significant skill (i.e., trajectory points outside the confidence bounds) and some alarm sets that demonstrate otherwise (trajectory points well within the confidence bounds). Therefore, ZECHAR and JORDAN (2008) suggested a scalar cumulative measure called the area skill score that depends on multiple Molchan trajectory points. While the statistical power of these metrics is experiment- and alarm function-dependent, we have found that the area skill score tends to be most powerful when considering the entire Molchan trajectory, and in this case the area skill score is at

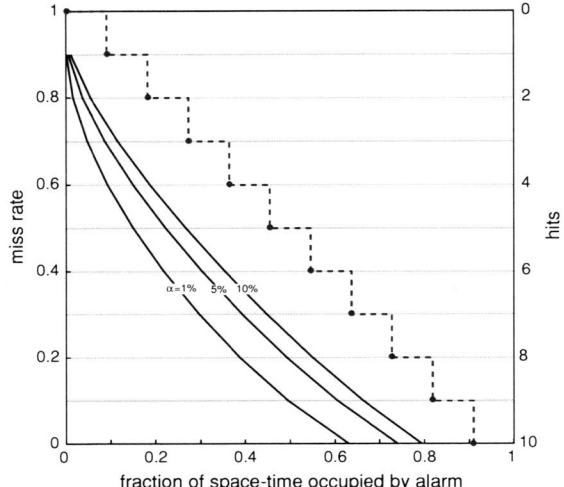

Figure 1
Schematic Molchan diagram for $N = 10$. The *dashed stairstep* represents the long-run average behavior of an unskilled alarm function. Also shown are the 90, 95, and 99% confidence bounds. A color version of this figure is available in the electronic edition

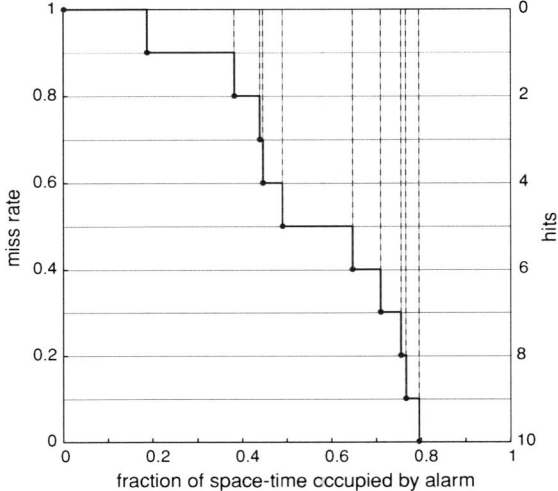

Figure 2
Molchan diagram for $N = 10$, shown here with a sample trajectory based on an unskilled alarm function. The area skill score is the area of the region above the trajectory, shown here as a sum over the vertical strips. The *dots* are the trajectory jumps. A color version of this figure is available in the electronic edition

least as powerful as considering individual points on the Molchan diagram. In the following sections, we describe the area skill score and methods for determining the relevant confidence bounds.

3. Area Skill Score

In ZECHAR and JORDAN (2008), we defined the area skill score for alarm function f:

$$a_f(\tau) = \frac{1}{\tau} \int_0^\tau [1 - v_f(t)] \, dt. \qquad (8)$$

This is the normalized area above the Molchan trajectory v_f up to the given value of τ. For an experiment with N target earthquakes, the area skill score evaluated at $\tau = 1$ measures the predictive skill of f throughout the entire space of the experiment—that is, all N target earthquakes and the entire forecast region R are considered. Evaluating the area skill score of the entire trajectory addresses how well an alarm function estimates the distribution of target earthquakes. In this case, we can write (see also Fig. 2):

$$a_f(1) = \sum_{i=0}^{N-1} (v_i [\tau_{i+1} - \tau_i]). \qquad (9)$$

By substituting (5) into (9) and combining terms, we find

$$a_f(1) = 1 - \frac{1}{N} \sum_{i=1}^{N} \tau_i. \qquad (10)$$

From (10), we note that the area skill score for an alarm function is proportional to the average of its Molchan trajectory jumps $\{\tau_i\}$. Therefore, by substitution of (7), it can be shown that the expected area skill score for unskilled forecasts is $\langle a_f(1) \rangle = \frac{1}{2}$.

4. Area Skill Score Distribution

Hypothesis testing with the area skill score requires knowledge of its distribution for unskilled alarm functions. By unskilled, we mean an alarm function that essentially guesses the future distribution of seismicity, randomly ranking the constituent bins of the study region. In practice, we represent an unskilled alarm function by a multidimensional matrix with values that are uniform random variables on (0, 1]. Owing to experiment discretization and/or the distribution of target earthquakes, it may occur in some experiments that more than one target earthquake occurs in a single forecast bin. The case of discretized experiments and, in particular, the case in which more than one target earthquake may occur in a single bin, are addressed separately in Sect. 6; for the remainder of this section, we consider experiments wherein the reference model is a continuous function and therefore any value of τ can be realized.

When using a continuous reference model, the Molchan trajectory for an unskilled alarm function can be considered as an ordered sequence of independent and identically distributed (i.i.d.) uniform random variables on (0, 1]. Given this, and having shown that the area skill score is proportional to the normalized sum of these variables in (10), we write the additive complement of the area skill score, or the area under the Molchan trajectory:

$$1 - a_f(1) = \hat{a} = \frac{1}{N}[\tau_1 + \tau_2 + \cdots + \tau_N],$$
$$N\hat{a} = u = [\tau_1 + \tau_2 + \cdots + \tau_N]. \quad (11)$$

The area skill score distribution is symmetric about its mean value of ½. Therefore, the distribution of the area skill score is the same as the distribution of the complement $\hat{a}$ and we can obtain the distribution of $\hat{a}$ if we know the distribution of u. The distribution of u—namely, the distribution of the sum of N uniform random variables on (0, 1]—is known (e.g., SADOOGHI-ALVANDI et al., 2007) and, in terms of probability density, is described by the following:

$$p(u) = \frac{1}{(N-1)!} \sum_{k=0}^{\lfloor u \rfloor} (-1)^k \binom{N}{k} (u-k)^{N-1}. \quad (12)$$

Here, $\lfloor u \rfloor$ (the floor function) denotes the largest integer less than or equal to u. The variable u is defined over (0, N] but we seek the distribution of $\hat{a}$, which is defined over [0, 1), so we need to rescale $p(u)$. In general, if we know $p_1(x)$—the probability density of x—and we want to know $p_2(y)$—the probability density of y—where $y = g(x)$, then we can use the following:

$$p_2(y) = \frac{1}{g'(g^{-1}(y))} p_1(g^{-1}(y)). \quad (13)$$

Here g' is the first derivative of g. Such a rescaling yields:

$$p(\hat{a}) = \frac{N}{(N-1)!} \sum_{k=0}^{\lfloor N\hat{a} \rfloor} (-1)^k \binom{N}{k} (N\hat{a} - k)^{N-1}. \quad (14)$$

We can use (14) to compute the cumulative density for any area skill score for arbitrary N, and thereby establish the statistical significance of any area skill score:

$$D(\hat{a}) = \int_0^{\hat{a}} p(\hat{a}) \, d\hat{a}. \quad (15)$$

A straightforward numerical approach to estimate the distribution of the area skill score for a given experiment is brute-force simulation, which we refer to as Simulation Method A: generate a large number of random alarm functions and compute Molchan trajectories and corresponding area skill scores for each random alarm function. This process can become quite computationally cumbersome, particularly as experiment discretization decreases and the number of target earthquakes increases. Fortunately, we can often optimize Simulation Method A. Rather than simulating many random alarm functions and computing a Molchan trajectory for each, we can use Simulation Method B: repeatedly select N uniform random variables on (0, 1], where N is the number of observed target earthquakes. For each simulation, we sort these N values in ascending order and analyze their distribution. To understand the equivalence of Simulation Methods A and B, consider an experiment with N target earthquakes. In Method A, because the alarm functions randomly rank the bins of the study region R, the resultant Molchan trajectory points will be N random samples from (0, 1]; Method B generates these samples directly and yields an equivalent distribution without unnecessary simulations.

We note that, by applying the Central Limit Theorem to the i.i.d. trajectory values, the area skill score distribution asymptotically approaches a normal distribution with mean $\mu = $ ½ and variance σ^2 that depends on N; in Sect. 5, we provide an analytical expression for the variance. MOLCHAN (1990) showed a related tendency to the normal distribution in the case of time prediction with a renewal process model; in that case, the asymptote is temporal rather than directly based on the number of observed target earthquakes. For larger values of N, the normal approximation is computationally simpler than the exact solution provided by (14) or the simulation methods described above. In the following section, we quantify the accuracy of the Gaussian approximation through a discussion of the moments of the area skill score distribution.

5. Moments of the Area Skill Score Distribution

The exact area skill score distribution described in (14) can be better understood by examining its moments. Because the area skill score is the normalized sum of N i.i.d. uniform random variables on the interval (0, 1], we can find all central moments of the area skill score distribution by considering the normalized central moments of the uniform distribution on the same interval. From Sect. 3, we know that the expected value of the area skill score is ½. Therefore, the nth central moment of the area skill score can be written:

$$\hat{\mu}_n = \frac{1}{N}\int_0^1 \left(x - \frac{1}{2}\right)^n dx. \qquad (16)$$

Owing to the symmetry of the area skill score distribution, all odd central moments are zero. It follows from (16) that the second and fourth central moments of the area skill score distribution are:

$$\hat{\mu}_2 = \sigma^2 = \frac{1}{12N}, \qquad (17)$$

and

$$\hat{\mu}_4 = \frac{1}{80N}. \qquad (18)$$

Using (17), we can express the Gaussian approximation to the area skill score distribution, suitable for large N:

$$\tilde{p}(a) = \frac{6N\sqrt{\frac{\pi}{6N}}}{\pi}\exp\left(-6N\left(a - \frac{1}{2}\right)^2\right). \qquad (19)$$

To understand how large N should be to use the Gaussian approximation given by (19) in place of the exact distribution given by (14), we must quantify the difference between the distributions, which have identical central moments for $n \leq 3$. This difference is characterized by the kurtosis excess γ_2, which is defined (ABRAMOWITZ and STEGUN, 1972, p. 928):

$$\gamma_2 = \frac{\hat{\mu}_4}{(\hat{\mu}_2)^2} - 3. \qquad (20)$$

Kurtosis excess characterizes how peaked a distribution is relative to the normal distribution, where a negative value of γ_2 indicates a less peaked distribution. By substitution of (17) and (18) into (20), we find the kurtosis excess of the area skill score distribution to be:

$$\gamma_2 = -\frac{6}{5N}. \qquad (21)$$

By (21), the exact area skill score distribution is shown to be platykurtic—that is, it has a negative kurtosis excess—which indicates "thin tails" relative to the normal distribution. Indeed, this should be the case because the range of the area skill score distribution is [0, 1), whereas the normal distribution has infinite range. From (21), we also note that as the number of observed target earthquakes N increases, the kurtosis excess approaches zero, in agreement with the Central Limit Theorem. Figure 3 shows how the approximation (19) differs from the exact solution (14) for several values of N, and indicates that for N as small as 5, the normal approximation provides a good estimate.

6. Experimental Discretization

In the two previous sections, for the purpose of deriving analytical results, we have considered the distribution of the area skill score only in the case in which the reference model was assumed to be continuous. In practice, however, numerical predictability experiments almost always involve discrete alarm functions, discrete reference models, and discrete observations. Discretization reduces computation time and can be used to address loosely the errors arising from measurement uncertainty (e.g., epicenter uncertainties). Under certain circumstances, despite discretization, the analytical solutions and approximations presented in the previous sections can provide accurate estimates of predictive

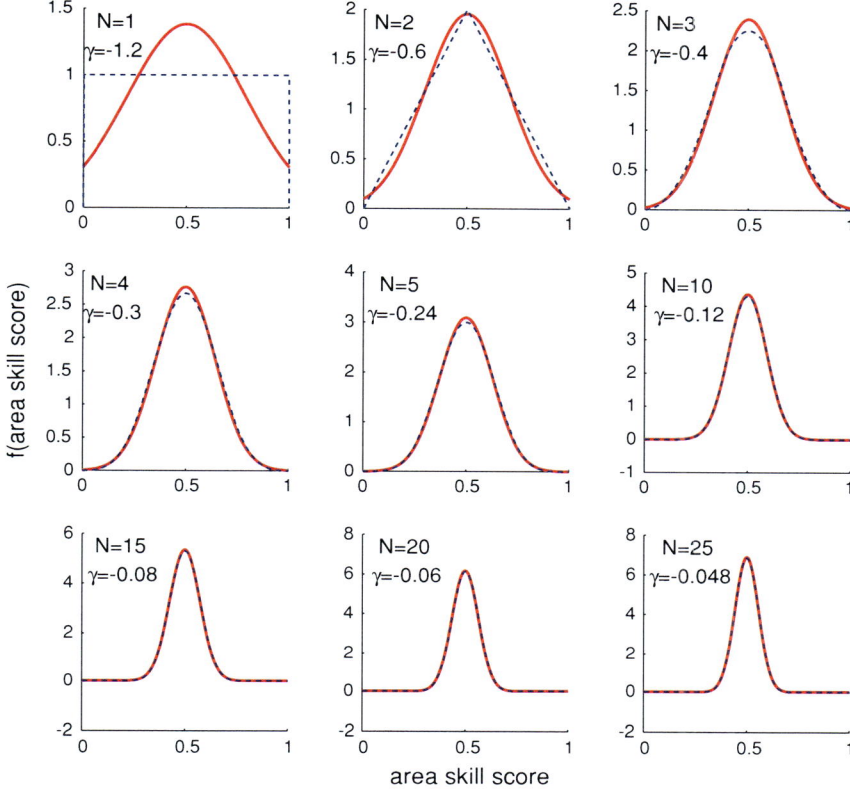

Figure 3
Comparison of exact area skill score probability density given by (14), shown here as *dashed lines*, and the Gaussian approximation given by (19), shown here as *solid lines*. As N increases, the Gaussian approximation quickly approaches the exact density. A color version of this figure is available in the electronic edition

skill. In particular, in the case where no single space/time/magnitude bin contains more than one target earthquake, the analytical solutions presented above become increasingly accurate as experiment cell size decreases toward the continuum limit. In dealing with more coarsely grained experiments, however, we rely on simulation methods; in this section, we discuss some relevant caveats that should be recognized when computing the significance of a given area skill score. To illustrate these caveats, we will refer to the alarm function shown in Fig. 4 and the experiment results shown in Fig. 5.

In the case of an experiment with an alarm function whose values are specified for the discrete study region R, where R is partitioned into n bins (e.g., Fig. 5d), and a uniform reference model is used—that is, the a priori probability of a target earthquake in every bin is assumed to be constant and equal—the only attainable values of τ are members of the following set:

$$\left\{\frac{1}{n}, \frac{2}{n}, \ldots, 1\right\}. \qquad (22)$$

This is illustrated in Fig. 5e. If we use Simulation Method B described in Sect. 4—namely, simulating τ values by selecting random variables from a uniform distribution on (0, 1]—we violate this constraint and therefore may obtain incorrect results, results which become less accurate as n decreases. We can return to Simulation Method A—simulating alarm functions and computing the corresponding Molchan trajectories—or we can employ Simulation Method C, a slight modification of Simulation Method B: rather than drawing N random numbers uniformly distributed on (0, 1], we draw N random integers from a discrete uniform distribution on the set of integers

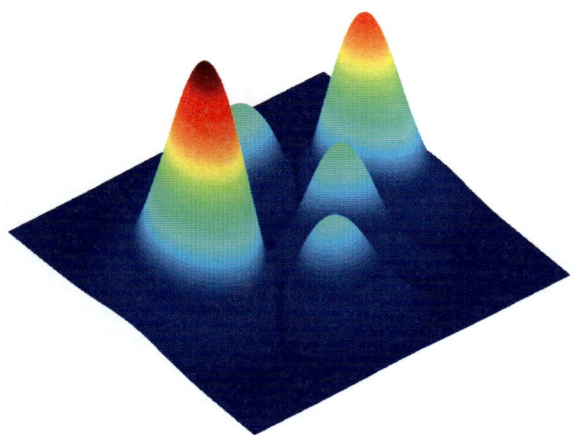

Figure 4
Illustrative alarm function, shown here in continuous form. A color version of this figure is available in the electronic edition

Figure 5
Hypothetical predictability experiments comparing the results between the continuous version (*left column*) and a discretized version (*right column*) of the alarm function shown in Fig. 4. **a** Map view of the continuous alarm function, with four example target earthquakes denoted by *stars*. **b** Molchan trajectory (*filled points*) for the experiment shown in **a**, using a uniform reference model for simplicity and including 90, 95, and 99% confidence bounds for each discrete value of v. **c** Area skill score trajectory corresponding to experiment shown in **a**, again using a uniform reference model for simplicity and including 90, 95, and 99% confidence bounds. **d** Map view of a very coarsely discretized version of the alarm function, shown with the same hypothetical target earthquake distribution. In an effort to show more details of the alarm function, the color scale is based on the natural logarithm of the alarm function values. Note that, due to the discretization, two earthquakes now fall within a single bin (*third column, fourth row*). **e** Molchan trajectory for the experiment shown in **d**, using a uniform reference model for simplicity and including 90, 95, and 99% confidence bounds for each discrete value of v. Note that all values of both τ and v are now discrete, and that the trajectory has infinite slope at $\tau = 5/16$, owing to two target earthquakes being in the same cell. **f** Area skill score trajectory corresponding to experiment shown in **d**, again using a uniform reference model for simplicity and including 90, 95, and 99% confidence bounds. A color version of this figure is available in the electronic edition

$(0, n]$ and divide each by n, letting the resulting quotient represent the τ values.

In principle, we could use Simulation Method C for any arbitrary reference model, where the only attainable values of τ are given by the nonzero sums of the reference model values. If we were to construct the set of all reference model value sums, we would draw N entries from this set to simulate the trajectory from an unskilled alarm function. Constructing this set soon becomes prohibitively expensive, however, particularly when dealing with a reference model specified over thousands of bins. As we prove in Appendix 2, if the reference model has n values, the set of sums has $(2^n - 1)$ elements. As n becomes large, it is more efficient to use Simulation Method A. There is some trade-off here, though. For a fixed reference model alarm function, as n becomes large, the set of attainable τ values approaches the continuum between 0 and 1. In practice, for n on the order of a thousand or more, Simulation Method B offers a good trade-off between approximation accuracy and speed.

There is one important special case remaining: The case in which experiment discretization and the observed target earthquake distribution are such that more than one target earthquake occurs in a single bin (e.g., Fig. 5d). When this happens, the probabilities of correctly predicting these events are not independent—this independence is an implicit assumption in the Simulation Methods A and B. To correct for this, we can examine the target earthquake distribution and construct the simulated unskilled trajectory appropriately using Simulation Method D: for a bin containing more than one earthquake, we draw a random number from $(0, 1]$ and append it to the simulated unskilled trajectory. Rather than appending this random number once and moving on, however, we append the random number $N(r_i)$ times, where $N(r_i)$ is the number of target earthquakes in this bin r_i. Simulation Method D therefore captures the fact that, when a bin is covered by an alarm, all the target earthquakes within the bin are successfully predicted.

In practice, most predictability experiments take place in a discretized, finely gridded framework. In all cases, Simulation Method A will be accurate and appropriate but, as we have pointed out, it can be computationally cumbersome. A careful examination of the experimental discretization, the target earthquake distribution, and the reference model should be conducted prior to evaluation using the area skill score; based on the outcome of this examination, it is likely that one of the alternative simulation methods discussed here is applicable. In the rare case of the predictability experiment in a continuum—for example, the RTP experiment—the analytical solutions are applicable and, as N becomes large, the Gaussian approximation for determining statistical significance is appropriate.

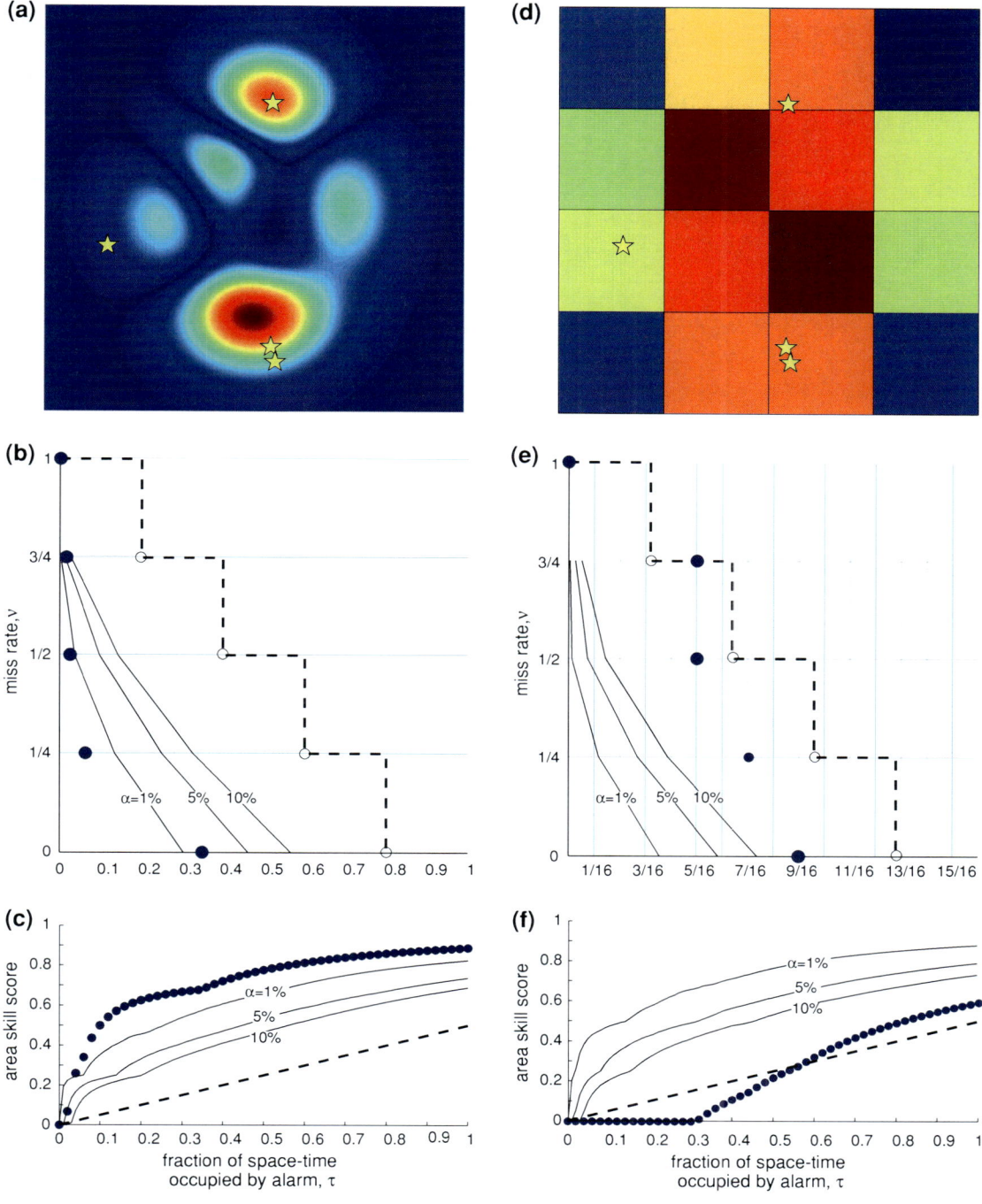

7. Discussion

Imagine we have a set of candidate alarm functions and we want to determine a posteriori whether any had predictive skill in a given experiment. We can begin with a uniform test by choosing a uniform reference model and computing the area skill scores for each alarm function and the corresponding significance using (15). Because earthquakes cluster in space and time and do not occur everywhere, we expect that most candidate alarm functions will incorporate some form of clustering, at least in space,

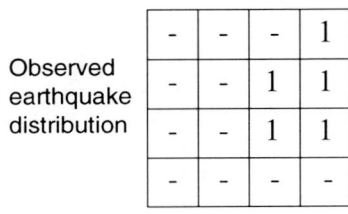

Table 1

Alarm function	Area skill score should exceed	Observed area skill score
Uniform test ($\alpha = 0.05$)		
f_1	0.65	0.8125
f_2	0.65	0.8125
f_3	0.65	0
f_4	0.65	0.8125

Alarm function	Acceptable area skill score range	Observed area skill score
Self test ($\alpha = 0.025$–0.975)		
f_1	0.263–0.675	0.457
f_2	0.263–0.675	0.137
f_4	0.263–0.675	0.534
Round-robin test ($\alpha = 0.025$–0.975), f_1 reference model		
f_4	0.263–0.675	0.457

Figure 6
Hypothetical predictability experiment outcome. The number of observed target earthquakes in each bin is given in the top of the figure: five bins have each hosted one target earthquake, and all other bins have hosted none. Four candidate alarm functions are also depicted, with their alarm function values given in each bin. Summary results for these alarm functions and this observed target earthquake distribution are given in Table 1

In the round-robin test, each surviving alarm function is fixed as the reference model and the area skill scores for all other alarm functions are computed. An alarm function is supported as a good reference model if none of the area skill scores fall in the tails of the distribution for unskilled forecasts. If any one area skill score deviates significantly, the reference alarm function may be considered to be an inappropriate reference model.

To elucidate this testing procedure, we consider the hypothetical predictability experiment outcome in Fig. 6; the corresponding results are shown in Table 1. Note that, for illustrative purposes, alarm function f_1 was constructed such that the five most highly ranked cells are the same cells where target earthquakes occurred and, in each of these cells, the alarm function values are of similar magnitude. All other cells of f_1 have much smaller values. Because this alarm function is very similar to the observed target earthquake distribution, we expect that it will be deemed an appropriate reference model by the tests. Alarm function f_2 is identical to f_1 except in the top-right cell, where the value is 15.1, compared to 1.1 for f_1. The third alarm function, f_3, is a uniform distribution with all values being equal. The final alarm function, f_4, is similar to f_1 with the exception of two cells that have slightly larger values. Recall that alarm function values do not have units, but rather they are used to infer a ranking of the cells and thereby derive alarm sets. When an alarm function is

and will therefore obtain a significantly high area skill score relative to the uniform reference model. Therefore, the goal of further testing is to improve the reference model and thereby distinguish the alarm functions in terms of their predictive skill.

For all those candidate alarm functions that pass the uniform test, we can continue with a test of self-consistency. For alarm function f_1, we take f_1 as the reference model ($\bar{q} \equiv f_1$) and recalculate its Molchan trajectory and area skill score. If f_1 is a reasonable reference model—that is, it approximates the distribution of seismicity well—we expect that its area skill score will not fall in the tails of the corresponding area skill score distribution for unskilled alarm functions. If it does, this indicates that f_1 does not demonstrate predictive skill.

For all those alarm functions that pass this test of self-consistency, we proceed with a round-robin test.

used as the reference model, its values are normalized so that their sum is unity.

As indicated in Table 1, for the observed distribution of target earthquakes and a one-sided hypothesis test using a significance value of $\alpha = 0.05$, three of the four alarm functions pass the uniform test. Because f_3 is itself a uniform distribution, this alarm function obtains hits only when declaring an alarm in every cell (i.e., when $\tau = 1$); therefore, its area skill score is zero and is not considered in subsequent tests. In the self-test, we use a two-sided hypothesis test where we check that the observed area skill score is within the central 95% confidence region of the area skill score distribution described in Sect. 4. Alarm function f_2 obtains an area skill score that is outside this range, while the other remaining alarm functions—f_1 and f_4—pass the test. The reason that f_2 obtains such a low area skill score is because of the exceptionally high alarm function value in the top-right cell—this value corresponds to a prediction that nearly 15 times more events will occur in this cell than in any other. Certainly, this prediction is not supported by the observed earthquake distribution and therefore the self-test indicates that f_2 is not an appropriate reference model.

For the round-robin test, we show results in Table 1 for the case in which f_1 is considered the reference model and we again use a two-sided hypothesis test emphasizing the central 95% confidence region. Given that the area skill score for f_4 relative to the f_1 reference model is well within this confidence region, the results indicate that alarm function f_1 is an appropriate reference model and shows predictive skill in this experiment.

Comparative testing of earthquake prediction strategies is a difficult problem, particularly given the heterogeneity of current forecast models and experiment configurations. The testing procedure outlined in this section is but one way that the area skill score can be employed for assessing predictive skill, and the procedure is now in preliminary use within CSEP experiments. These ongoing experiments ought to provide insight into the utility of the area skill score testing procedure described here, and a detailed examination of the results may suggest further modifications.

8. Conclusion

In this paper we have explored the concept of an alarm function: a general form for specifying earthquake forecasts defined by an ordering of space/time/magnitude bins in terms of their estimated probability to contain future target earthquakes. We described the Molchan diagram and Molchan trajectories, and presented relevant analysis that includes a corrected stairstep-diagonal describing the behavior of unskilled alarm functions. We also emphasized the explicit use of a reference model in computing Molchan trajectories. We presented the exact distribution of the area skill score for predictability experiments in a continuum; we also presented an approximation of the distribution using a Gaussian distribution and described the moments of the exact distribution. We have outlined potential pitfalls regarding experimental discretization and the special case of more than one target earthquake occurring in a single space/time/magnitude bin. We also proposed a testing method to compare and evaluate a set of candidate alarm functions at the end of an earthquake predictability experiment.

Acknowledgments

We thank Yan Kagan and an anonymous reviewer for insightful comments that improved the manuscript. We thank Max Werner for stimulating discussion. This research has been supported by the National Science Foundation via a Graduate Research Fellowship (JDZ) and grant CMG 0621119, and by the Southern California Earthquake Center (SCEC). SCEC is funded by NSF Cooperative Agreement EAR-0106924 and USGS Cooperative Agreement 02HQAG0008. The SCEC contribution number for this paper is 1216.

Appendix 1: Expected Value of Molchan Trajectory Jump $\langle \tau_k \rangle$

We seek $\langle \tau_k \rangle$, the expectation of the kth Molchan trajectory jump of an unskilled alarm function, where expectation is defined as

$$\langle X \rangle = \int_{-\infty}^{\infty} x f(x)\, dx. \qquad (23)$$

Here $f(x)$ is the probability density function; therefore, we need to find the probability density for τ_k.

We can find the probability density by taking the derivative of the cumulative density function. For τ_1, this is the probability that the trajectory has experienced at least one jump prior to reaching τ. In other words, it is the probability of covering τ and obtaining 1, 2, 3,..., or N hits. This probability is given by summing binomial terms:

$$D_{\tau_1}(\tau) = \sum_{j=1}^{N} \binom{N}{j} \tau^j (1-\tau)^{N-j}. \qquad (24)$$

We can express (24) in the following closed form:

$$D_{\tau_1}(\tau) = \begin{cases} 0, & \tau < 0 \\ 1 - (1-\tau)^N, & \tau \in [0,1]. \\ 1, & \tau > 1 \end{cases} \qquad (25)$$

By differentiating (25) with respect to τ, we obtain the probability density:

$$p_{\tau_1}(\tau) = \frac{dD_{\tau_1}(\tau)}{d\tau} = \begin{cases} N(1-\tau)^{N-1}, & \tau \in [0,1] \\ 0 & \text{otherwise} \end{cases}. \qquad (26)$$

Now we can substitute (26) into (23) to obtain the expectation, changing the limits of integration to isolate the region where the probability density is nonzero:

$$\langle \tau_1 \rangle = \int_0^1 \tau p_{\tau_1}(\tau) d\tau = \int_0^1 \tau N(1-\tau)^{N-1} d\tau$$

$$= -\frac{(1-\tau)^N (N\tau+1)}{N+1}\bigg|_0^1$$

$$= \frac{1}{N+1}. \qquad (27)$$

This shows that the first hit is expected to be obtained by unskilled alarm functions when they cover $\frac{1}{N+1}$ of the study region. Similarly, we can express the c.d.f., p.d.f., and expectation for the next jump:

$$D_{\tau_2}(\tau) = 1 - (1-\tau)^N - N\tau(1-\tau)^{N-1}, \qquad (28)$$

$$p_{\tau_2}(\tau) = N(1-\tau)^{N-1}$$
$$- \left[-N\tau(N-1)(1-\tau)^{N-2} + N(1-\tau)^{N-1} \right]$$
$$= N(N-1)\tau(1-\tau)^{N-2}, \qquad (29)$$

$$\langle \tau_2 \rangle = \int_0^1 N(N-1)\tau^2 (1-\tau)^{N-2} d\tau$$
$$= -\frac{(1-\tau)^{N-1}((N-1)\tau(N\tau+2)+2)}{N+1}\bigg|_0^1$$
$$= \frac{2}{N+1}. \qquad (30)$$

These equations show that, on average, unskilled alarm functions obtain two hits once they have covered $\frac{2}{N+1}$ of the study region. Likewise for the following jump, we have

$$D_{\tau_3}(\tau) = 1 - (1-\tau)^N - N\tau(1-\tau)^{N-1}$$
$$- \binom{N}{2}(\tau)^2 (1-\tau)^{N-2}, \qquad (31)$$

$$p_{\tau_3}(\tau) = N(1-\tau)^{N-1} - \left[-N(N-1)\tau(1-\tau)^{N-2} + N(1-\tau)^{N-1} \right] - \left[-\binom{N}{2}(N-2)(\tau)^2 \right.$$
$$\left. \times (1-\tau)^{N-3} + 2\binom{N}{2}\tau(1-\tau)^{N-2} \right]$$
$$= N(N-1)\tau(1-\tau)^{N-2} + \binom{N}{2}$$
$$\times (N-2)(\tau)^2 (1-\tau)^{N-3}$$
$$- N(N-1)\tau(1-\tau)^{N-2}$$
$$= \frac{N(N-1)(N-2)}{2}(\tau)^2 (1-\tau)^{N-3}, \qquad (32)$$

$$\langle \tau_3 \rangle = \frac{N(N-1)(N-2)}{2} \int_0^1 (\tau)^3 (1-\tau)^{N-3} d\tau$$
$$= \frac{N(N-1)(N-2)}{2}(1-\tau)^N$$
$$\times \left(-\frac{2N+3}{N(N+1)} - \frac{\tau}{N+1} \right.$$
$$\left. -\frac{3}{(N-1)(\tau-1)} + \frac{1}{(2-N)(\tau-1)^2} \right)\bigg|_0^1$$
$$= \frac{N(N-1)(N-2)}{2}$$
$$\times \left(\frac{2N+3}{N(N+1)} - \frac{3}{(N-1)} - \frac{1}{(2-N)} \right) \Rightarrow$$

$$\langle \tau_3 \rangle = \frac{3}{N+1}. \tag{33}$$

For an inductive proof, we assume that $\langle \tau_{N-1} \rangle = \frac{N-1}{N+1}$ and compute $\langle \tau_N \rangle$. At this point, we can get a compact expression for the c.d.f. by returning to the original formulation in (24), such that

$$D_{\tau_N}(\tau) = \sum_{j=N}^{N} \binom{N}{j} \tau^j (1-\tau)^{N-j} \tag{34}$$
$$= (\tau)^N.$$

Then,

$$p_{\tau_N}(\tau) = N(\tau)^{N-1}, \tag{35}$$

$$\langle \tau_N \rangle = \int_0^1 N\tau(\tau)^{N-1} d\tau$$

$$= N \int_0^1 (\tau)^N d\tau$$

$$= \left. \frac{N(\tau)^{N+1}}{N+1} \right|_0^1$$

$$\langle \tau_N \rangle = \frac{N}{N+1}. \tag{36}$$

This completes the proof and thus, for all jumps, the expected value of the jump is described:

$$\langle \tau_k \rangle = \frac{k}{N+1}. \tag{37}$$

Appendix 2: Number of Nonzero Sums of a Set's Elements

In the case of a discretized reference model, the set of attainable values of τ is finite and its elements are the nonzero linear combinations of the reference model values with coefficients equal to zero or one. In other words, for a reference model specified in n bins, the set T_n of attainable τ values is comprised of the nonzero sums of the n reference model values. We denote the cardinality of this set $|T_n|$.

Theorem. $|T_n| = 2^n - 1$.

Proof. In the case where $n = 1$, it is clear that there is only one nonzero sum: $T_1 = \{1\}$, $|T_1| = 1$. In the case where $n = 2$, we represent a reference model as the set $\{n_1, n_2\}$. In this case, the set of attainable τ values is $\{n_1, n_2, n_1 + n_2\}$; $|T_2| = 3$. When $n = 3$, we represent a reference model as the set $\{n_1, n_2, n_3\}$ and the set of sums is $\{n_1, n_2, n_3, n_1 + n_2, n_1 + n_3, n_2 + n_3, n_1 + n_2 + n_3\}$; $|T_3| = 7$. We assume that the relation holds for all values of n up to and including $(x - 1)$ and we consider $n = x$. The set T_x will contain all elements of $T_{(x-1)}$ as well as each of these elements added to $n_{(x+1)}$; the only additional sum in T_x is $n_{(x+1)}$. Thus the cardinality $|T_x|$ is twice the cardinality $|T_{(x-1)}|$ plus one additional element:

$$|T_x| = 2|T_{(x-1)}| + 1 = 2\left(2^{(x-1)} - 1\right) + 1 = 2^x - 1.$$

REFERENCES

ABRAMOWITZ, M. and STEGUN, I.A., *Handbook of Mathematical Functions with Formulas, Graphs, and Mathematical Tables* (Dover, New York 1972).

FIELD, E.H. (2007), *Overview of the working group for the development of Regional Earthquake Likelihood Models (RELM)*, Seismol. Res. Lett. 78(1), 7–16.

KEILIS-BOROK, V.I., SHEBALIN, P.N., GABRIELOV, A. and TURCOTTE, D.L. (2004), *Reverse tracing of short-term earthquake precursors*, Phys. Earth Planet. Inter. 145, 75–85.

KOSSOBOKOV, V.G. (2006), *Testing earthquake prediction methods: "The West Pacific short-term forecast of earthquakes with magnitude $M_w HRV \geq 5.8$"*, Tectonophys. 413, 25–31.

MASON, I.B., *Binary events*, In *Forecast Verification* (eds. Jolliffe, I.T. and Stephenson, D.B.) (Wiley, Hoboken 2003) pp. 37–76.

MOLCHAN, G.M. (1990), *Strategies in strong earthquake prediction*, Phys. Earth and Planet. Inter. 61, 84–98.

MOLCHAN, G.M. (1991), *Structure of optimal strategies in earthquake prediction*, Tectonophys. 193, 267–276.

MOLCHAN, G.M. and KAGAN, Y.Y. (1992), *Earthquake prediction and its optimization*, J. Geophys. Res. 97, 4823–4838.

RUNDLE, J.B., TIAMPO, K., KLEIN, W. and SA MARTINS, J.S. (2002), *Self-organization in leaky threshold systems: The influence of near-mean field dynamics and its implications for earthquakes, neurobiology, and forecasting*, Proc. Natl. Acad. Sci. USA. 99, 2514–2521.

SADOOGHI-ALVANDI, S.M., NEMATOLLAHI, A.R. and HABIBI, R. (2007), *On the distribution of the sum of independent uniform random variables*, Stat. Papers doi:10.1007/s00362-007-0049-4.

SCHORLEMMER, D., ZECHAR, J.D., WERNER, M.J., FIELD, E.H., JACKSON, D.D. and JORDAN, T.H. (2010), *First results of the Regional Earthquake Likelihood Models experiment*, Pure Appl. Geophys. doi: 10.1007/s00024-010-0081-5.

SHEBALIN, P.N., KEILIS-BOROK, V.I., GABRIELOV, A., ZALIAPIN, I. and TURCOTTE, D. (2006), *Short-term earthquake prediction by reverse analysis of lithosphere dynamics,* Tectonophys, *413*, 63–75.

ZECHAR, J.D. and JORDAN, T.H. (2008), *Testing alarm-based earthquake predictions*, Geophys. J. Inter. *172*(2), 715—724. doi: 10.1111/j.1365-246X.2007.03676.x

(Received August 7, 2008, revised May 26, 2009, accepted May 26, 2009, Published online March 10, 2010)

Space–Time Earthquake Prediction: The Error Diagrams

G. Molchan[1,2]

Abstract—The quality of earthquake prediction is usually characterized by a two-dimensional diagram n versus τ, where n is the rate of failures-to-predict and τ is a characteristic of space–time alarm. Unlike the time prediction case, the quantity τ is not defined uniquely. We start from the case in which τ is a vector with components related to the local alarm times and find a simple structure of the space–time diagram in terms of local time diagrams. This key result is used to analyze the usual 2-d error sets $\{n, \tau_w\}$ in which τ_w is a weighted mean of the τ components and w is the weight vector. We suggest a simple algorithm to find the (n, τ_w) representation of all random guess strategies, the set D, and prove that there exists the unique case of w when D degenerates to the diagonal $n + \tau_w = 1$. We find also a confidence zone of D on the (n, τ_w) plane when the local target rates are known roughly. These facts are important for correct interpretation of (n, τ_w) diagrams when we discuss the prediction capability of the data or prediction methods.

Key words: Prediction, earthquake dynamics, statistical seismology.

1. Introduction

The sequence of papers (Molchan, 1990, 1991, 1997, 2003) considers earthquake prediction as a decision-making problem. The basic notions in this approach are the strategy, π, and the goal function, φ. Any strategy is a sequence of decisions $\pi(t)$ about an alarm of some type for a next time segment $(t, t + \delta), \delta \ll 1$; $\pi(t)$ is based on the data $I(t)$ available at time t. The goal of prediction is to minimize φ, and the mathematical problem consists in describing the optimal strategy. Molchan (1997) considered the problem under the conditions in which target events form a random point process $dN(t)$ ($N(t)$ is the number of events in the interval $(0, t)$), and the aggregate $\{dN(t), I(t), \pi(t)\}$ is stationary.

Dealing with the prediction of time, Molchan (1997) considered, along with the general case, the situation in which the optimal strategy is locally optimal, i.e., is optimal for any time segment. This case arises when the goal function has the form $\varphi(n, \tau)$, where n, τ are the standard prediction characteristics/errors: n is the rate of failures-to-predict and τ the alarm time rate. The optimal strategy can then be described in much simpler terms, and can be expressed by the conditional rate of target events

$$r(t) = P\{dN(t) > 0 | I(t)\}/dt, \qquad (1)$$

the loss function φ, and the error diagram $n(\tau)$. The last function can be defined as the lower bound of the set $\mathcal{E} = \{n, \tau\}$; this set consists of the (n, τ) characteristics of all the strategies based on $I(t)$. The search for the optimal strategy on a small time segment $(t, t + \delta)$ is reduced to the classical testing of two simple hypotheses such that the errors of the two kinds $(\beta(\alpha), \alpha)$ (Lehmann, 1959) converge to $(n(\tau), \tau)$ as $\delta \downarrow 0$. In statistical applications the curve $1 - \beta(\alpha)$ is known as the ROC diagram or Relative/Receiver Operating Characteristic (Swets, 1973); its limit in the case of the locally optimal strategy gives the curve $1 - n(\tau)$.

The error diagram $n(\tau)$ has proved to be so convenient a tool for the analysis of prediction methods that it began to be also used for the prediction of the space–time of target events. In that case the part of τ is played by a weighted mean of τ over space. To be specific, we divide the space G into nonintersecting parts $\{G_i, i = 1, ..., k\}$ and denote by τ_i the alarm time

---

[1] International Institute of Earthquake Prediction Theory and Mathematical Geophysics, Russian Academy of Sciences, Moscow, Russia. E-mail: molchan@mitp.ru
[2] The Abdus Salam International Centre for Theoretical Physics, SAND Group, Trieste, Italy.

rate in G_i for the strategy π. The space–time alarm is effectively measured by

$$\tau_w = \sum_{i=1}^{k} w_i \tau_i, \quad \sum_{i=1}^{k} w_i = 1, \quad w_i \geq 0, \qquad (2)$$

where the $\{w_i\}$ depend on the prediction goals, e.g., at the research stage of prediction one uses

$$w_i = \text{area of } G_i/\text{area of } G \qquad (3)$$

(Tiampo et al., 2002; Shen et al., 2007; Zechar and Jordan, 2008; Shcherbakov et al., 2007) or

$$w_i = \lambda(G_i)/\lambda(G), \qquad (4)$$

where $\lambda(G)$ is the rate of target events in G (Keilis-Borok and Soloviev, 2003; Kossobokov, 2005). When dealing with the social and economic aspects of prediction, it is advisable to use weights of the form

$$w_i = \int_{G_i} p(g) \mathrm{d}g \bigg/ \int_{G} p(g) \mathrm{d}g, \qquad (5)$$

where $p(g)$ is, e.g., the density of population in G.

The $n(\tau_w)$ diagrams are constructed by analogy with the time error diagram, i.e., as the boundary of the set $\{n, \tau_w\}$ which lies below the diagonal $n + \tau_w = 1$. Quite often properties of $n(\tau)$ are transferred to $n(\tau_w)$ as well. We now mention those properties which, in the case of $n(\tau_w)$, either must be better specified or are wrong.

(a) $n(\tau)$ characterizes the limiting prediction capability of the data $\{I(t)\}$. That means that the minimum of any loss function $\varphi(n, \tau)$ with convex levels $\{\varphi \leq c\}$ is reached at the curve $n(\tau)$;
(b) φ and $n(\tau)$ define the optimal strategy and its characteristics (n, τ);
(c) the diagonal D of the square $[0, 1]^2$, $n + \tau = 1$, is the antipode of $n(\tau)$, because it describes the characteristics of *all* trivial strategies that are equivalent to random guess strategies. Therefore, the maximum distance between $n(\tau)$ and D, i.e., $\max_\tau (1 - n(\tau) - \tau)/\sqrt{2}$, characterizes the prediction potential of $\{I(t)\}$;
(d) $1 - n(\tau)$ is a ROC diagram arising in the testing of simple statistical hypotheses.

Molchan and Keilis-Borok (2008) recently considered the prediction of the space–time of target events under conditions in which where the optimal strategies coincide with the locally optimal ones (the word "locally" now also refers to both space and time). This paper gives a correct extension of the error diagram, which provides the key to the understanding of the information contained in an $n(\tau_w)$ diagram. The present note supplements the above-mentioned study. We refine the structure of the error diagram for space–time prediction and analyze the properties of two-dimensional $n(\tau_w)$ diagrams.

2. The Error Diagram

We quote the main result by Molchan and Keilis-Borok (2008) relevant to the prediction of space–time for target events.

Let $\{G_i\}$ be some partition of G into nonintersecting regions. The prediction of location means the indication of $\{G_i\}$ where a target event will occur. Consequently, the model of target events in G is the stationary random vector point process

$$\mathrm{d}\mathbf{N}(t) = \{\mathrm{d}N_1(t), \ldots, \mathrm{d}N_k(t)\},$$

whose components describe target events in $\{G_i\}$. We shall consider the binary yes/no prediction with the decisions

$$\boldsymbol{\pi}(t) = \{\pi_1(t), \ldots, \pi_k(t)\},$$

of the form

$$\pi_i(t) = \begin{cases} \text{alarm in } G_i \times (t, t+\delta) \\ \text{no alarm in } G_i \times (t, t+\delta) \end{cases}$$

where t takes on values on a lattice at a step δ. The decision $\boldsymbol{\pi}(t)$ is based on the data $I(t)$ that are available at time t.

Under certain conditions, namely, the aggregate $\{\mathrm{d}\mathbf{N}(t), I(t), \boldsymbol{\pi}(t)\}$ is ergodic and stationary, and moreover $P\{\sum_{i=1}^{k} \mathrm{d}N_i(t) > 1\} = o(\mathrm{d}t)$, i.e., the probability of observing more than one target event in any one time instant is vanishingly small, the basic characteristics of the strategy $\pi = \{\boldsymbol{\pi}(t)\}$ are defined as the limit of its empirical means. We have in view the rate of failures-to-predict n and the vector

$$\boldsymbol{\tau} = (\tau_1, \ldots, \tau_k),$$

which determines the alarm time rate in the $\{G_i, i = 1, \ldots, k\}$. The quantities $(n, \boldsymbol{\tau})$ are defined for any small δ. We shall assume that n and $\boldsymbol{\tau}$ have limits as $\delta \downarrow 0$, for which we retain the same notation. The passage to the limit is not a restriction, since the data may reflect a seismic situation with a fixed time delay.

The set of $(n, \boldsymbol{\tau})$ characteristics for different strategies π based on $\{I(t)\} = I$ is a convex subset in the $(k + 1)$-dimensional unit cube, i.e., the error set

$$\mathcal{E}(I) = \{(n, \boldsymbol{\tau})_\pi : \pi \text{ based on } I\} \subseteq [0, 1]^{k+1}, \quad (6)$$

(see Fig. 1). The set $\mathcal{E}$ contains the simplex

$$\mathbf{D} = \{(n, \boldsymbol{\tau}) : n + \sum_{i=1}^{k} \lambda_i \tau_i / \lambda = 1, 0 \leq n, \tau_i \leq 1\}, \quad (7)$$

where $\lambda_i = \lambda(G_i)$. The set (7) describes strategies that are equivalent to the random guess strategies. For indeed, if an alarm is declared in G_i with the rate τ_i, then $\lambda_i \tau_i / \lambda$ will give the rate of random successes in G_i. The equality in (7), i.e., $1 - n = \sum_{i=1}^{k} \lambda_i \tau_i / \lambda$, means that the success rate is identical with the rate of random successes. Such strategies will be called *trivial*.

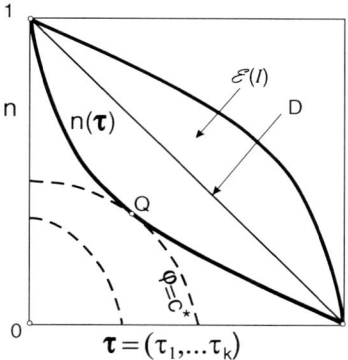

Figure 1
Space–time prediction characteristics: n versus $\boldsymbol{\tau} = (\tau_1, \ldots, \tau_k)$ (the horizontal axis is multidimensional). *Notation*: $\mathcal{E}(I)$ represents all strategies based on the data I, the hyperplane D represents the trivial strategies (random guesses), and the surface $n(\boldsymbol{\tau})$ the optimal strategies (the error diagram). The level sets of the loss function $\varphi(n, \boldsymbol{\tau})$ are shown by *dashed lines*, the characteristic of the optimal prediction is the tangent point Q between $n(\boldsymbol{\tau})$ and the suitable level set of φ.

The boundary of $\mathcal{E}$, viz., $n(\boldsymbol{\tau})$, which lies below the hyperplane (7), will be called the *error diagram*. To describe the properties of $n(\boldsymbol{\tau})$, we define the loss function φ. This will be a function of the form $\varphi(n, \boldsymbol{\tau})$ that is nondecreasing in each argument and for which any level set, $\{\varphi \leq c\}$, is convex.

The following is true (MOLCHAN and KEILIS-BOROK, 2008).

2.1. The minimum of $\varphi(n, \boldsymbol{\tau})$ on $\mathcal{E}$ is reached on the surface $n(\boldsymbol{\tau})$. The point of the minimum, Q, is found as the point where the suitable level $\{\varphi \leq c\}$ is tangent to $n(\boldsymbol{\tau})$ (see Fig. 1). The coordinates of $Q = (n, \boldsymbol{\tau})$ define the characteristics of the optimal strategy with respect to the goal function φ.

2.2. The optimal strategy declares an alarm in $G_i \times (t, t + \delta), \delta \ll 1$ as soon as

$$r_i(t) = P\{N_i(t + \delta) - N_i(t) > 0 | I(t)\} / \delta \geq r_{0i} \quad (8)$$

and declares no alarm otherwise.

2.3. The threshold r_{0i} depends on φ, e.g., if

$$\varphi = a \lambda n + \sum_{i=1}^{k} b_i \tau_i \quad (9)$$

then $r_{0i} = b_i / a$. In the general case one has

$$r_{0i} = -\lambda \frac{\partial \varphi}{\partial \tau_i} \bigg/ \frac{\partial \varphi}{\partial n}(Q).$$

The result described above yields an important corollary:

Corollary 2.4. *The error diagram for space–time prediction in $G = \{G_i\}$ based on $\{I(t)\}$ can be represented as*

$$n(\tau_1, \ldots, \tau_k) = \sum_{i=1}^{k} \lambda_i n_i(\tau_i) / \lambda, \quad (10)$$

where $n_i(\tau)$ is the error diagram for time prediction in G_i based on the same data $\{I(t)\}$.

Proof Consider such a loss function (9) that the hyperplane $\varphi = c$ is tangent to $n(\boldsymbol{\tau})$ at $\boldsymbol{\tau}_0 = (\tau_{01}, \ldots, \tau_{0k})$. The optimal strategy thus has the form (8) with $r_{0i} = b_i / a$ and the errors $(n(\boldsymbol{\tau}_0), \boldsymbol{\tau}_0)$. However, the strategy for time prediction in G_i of the form (8) minimizes the loss function $\varphi_i = a \lambda_i n + b \tau$ (MOLCHAN, 1997). The point of the minimum has the coordinate $\tau = \tau_{0i}$, hence the other coordinate is

$n = n_i(\tau_{0i})$. Consequently, the collective strategy (8) minimizes

$$\sum_{i=1}^{k} \varphi_i = a\lambda \left(\sum_{i=1}^{k} \lambda_i n_i / \lambda \right) + \sum_{i=1}^{k} b_i \tau_i \qquad (11)$$

and has $n = \sum_{i=1}^{k} \lambda_i n_i(\tau_{0i})/\lambda$ as the rate of failures-to-predict. The right-hand side of (11) is identical with $\varphi(n, \tau)$. It follows that (10) is true with $n = n(\tau_0)$, since the strategy (8) also minimizes (9). Since τ_0 is arbitrary, the corollary is proven.

3. The Reduced Error Diagrams

Usually regional error diagrams $n_i(\tau)$ are poorly estimated, so that for practical purposes the result of a space–time prediction is represented by the two-dimensional diagram $n(\tau_w)$, $\tau_w = \sum_{i=1}^{k} w_i \tau_i$ where the weights are $w_i \geq 0$ and $\sum_{i=1}^{k} w_i = 1$. This is obtained from the set of "errors" $\mathcal{E}_w = \{(n, \tau_w)\}$ as its lower boundary.

Relation (10) can be used to analyze the properties of $n(\tau_w)$ diagrams. Later we shall use the following notation: if the set B is the image of $A = \{(n, \tau)\}$ by the mapping

$$\gamma_w : (n, \tau) \to (n, \tau_w), \quad \tau_w = \sum_{i=1}^{k} w_i \tau_i,$$

then $B = A_w$; in particular, the image of τ is τ_w, the image of $\mathcal{E}$ is $\mathcal{E}_w$, while the image of D (see (7)) is D_w.

The following is true (see Appendix for proof).

3.1. $\mathcal{E}_w$ is a convex subset of the square $[0, 1]^2$ that contains the diagonal $\tilde{D} : n + \tau_w = 1$.

3.2. D_w is a convex subset of $\mathcal{E}_w$; D_w degenerates to the diagonal of the unit square, if and only if $w_i = \lambda_i/\lambda$, $i = 1, ..., k$.

3.3. D_w can be obtained as the convex hull of points of the form

$$n = 1 - \sum_{i=1}^{k} \varepsilon_i \lambda_i / \lambda, \quad \tau_w = \sum_{i=1}^{k} w_i \varepsilon_i, \qquad (12)$$

where $\{\varepsilon_i\}$ are all possible sequences of 0 and 1 (see Fig. 2).

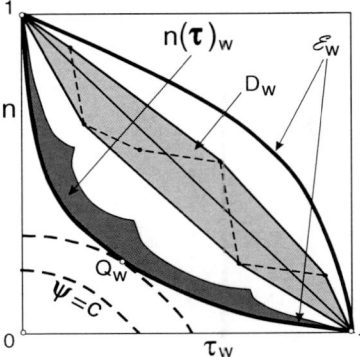

Figure 2
The reduced error diagram: n versus $\tau_w = \sum_{i=1}^{k} \tau_i w_i$. Notation: $\mathcal{E}_w$ contoured by *bold lines* represents all strategies $\mathcal{E}$ in the (n, τ_w) coordinates; the *stippled zone* D_w represents the trivial strategies; the *broken line* within D_w illustrates the method used to construct D_w, see **3.3**; the *filled zone* is the image of the $n(\tau)$ diagram; isolines of the loss function $\varphi = \psi(n, \tau_w)$ are shown by *dashed lines*; φ yields the optimal characteristics $Q_w = (n, \tau_w)$.

In particular, let $w_1 = ... = w_k$ (this will be the case for (3) when G is divided into isometric parts). Then the lower boundary of the convex hull (convex minorant) of the (n, τ_w) points

$$(1, 0), (1 - \lambda_{(k)}/\lambda, 1/k), ...,$$
$$\left(1 - \sum_{i=1}^{p} \lambda_{(k-i+1)}/\lambda, p/k \right), ..., (0, 1)$$

gives the lower boundary of D_w, while the upper boundary of the convex hull (concave majorant) of the points

$$(1, 0), (1 - \lambda_{(1)}/\lambda, 1/k), ...,$$
$$\left(1 - \sum_{i=1}^{p} \lambda_{(i)}/\lambda, p/k \right), ..., (0, 1)$$

gives the upper boundary of D_w. Here, $\lambda_{(1)} \leq ... \leq \lambda_{(k)}$ are the $\{\lambda_i\}$ arranged in increasing order.

3.4. Except for trivial cases, the image of the error diagram $n(\tau)$ is a two-dimensional set (see Fig. 2) with the lower boundary $n(\tau_w)$ and the upper boundary $n^+(\tau_w)$. In the regular case, i.e., $n_i(0) = 1$, $i = 1, ..., k$, one has

$$n^+(x) = \max_{i, \varepsilon} \{\lambda_i/\lambda \cdot n_i(x/w_i - a_i(\varepsilon)) + b_i(\varepsilon)\}, \qquad (13)$$

where

$$\boldsymbol{\varepsilon} = (\varepsilon_1, \ldots, \varepsilon_k), \quad \varepsilon_i = 0, 1,$$
$$a_i(\boldsymbol{\varepsilon}) = \sum_{j \neq i} w_j \varepsilon_j / w_i,$$
$$b_i(\boldsymbol{\varepsilon}) = \sum_{j \neq i} \lambda_j (1 - \varepsilon_j) / \lambda,$$

and the maximum is taken over such i and $\boldsymbol{\varepsilon}$ sequences for which the argument of n_i in (13) makes sense, i.e., is in [0, 1].

If $\{n_i(\tau)\}$ are piecewise smooth and $n_i(0) = 1$, $i = 1, \ldots, k$, then the image of $n(\boldsymbol{\tau})$ degenerates to a one-dimensional curve, if and only if $\{I(t)\}$ is trivial, i.e., $1 - n(\boldsymbol{\tau}) = \sum_{i=1}^{k} \lambda_i \tau_i / \lambda$ and $w_i = \lambda_i / \lambda$, $i = 1, \ldots, k$.

3.5. The curve $n(\tau_w)$ represents those strategies which are optimal for loss functions of the form

$$\varphi(n, \boldsymbol{\tau}) = \psi(n, \tau_w), \quad \tau_w = \sum_{i=1}^{k} w_i \tau_i. \quad (14)$$

To be specific, if $(n, \boldsymbol{\tau}) = Q$ are the optimal prediction characteristics with respect to the goal function of the form (14), then Q_w belongs to the $n(\tau_w)$ diagram. In addition, Q_w is the point at which the curve $n(\tau_w)$ is tangent to the suitable level set of ψ.

3.6. The strategy that optimizes (14) declares an alarm in $G_i \times (t, t + \delta)$ as soon as

$$r_i(t)/w_i \geq c, \quad (15)$$

where the threshold c is independent of G_i and r_i is given by (8). According to **2.3**,

$$c = \lambda \frac{\partial \psi}{\partial \tau_w} \bigg/ \frac{\partial \psi}{\partial n}(Q_w).$$

In particular, if $\varphi = an + b \sum_{i=1}^{k} w_i \tau_i$, then $c = \lambda b/a$. If $w_i = \lambda_i/\lambda$, then (15) will have the form $r_i(t)/\lambda_i \geq c\lambda$, where the left-hand side is known as the probability gain.

3.7. For any point Q in the error diagram we can find such weights $\{w_i\}$ that Q_w will lie in the reduced (n, τ_w) diagram, i.e., any optimal strategy can be represented by a suitable (n, τ_w) diagram. The desired weights are

$$w_i = -\frac{\partial n}{\partial \tau_i}(Q)/c,$$

where c is a normalizing constant. The point Q determines the optimal prediction characteristics with respect to the loss function

$$\varphi = n + c \sum_{i=1}^{k} w_i \tau_i.$$

3.8. The curve $1 - n(\tau_w)$ can be interpreted as a ROC diagram if and only if $w_i = \lambda_i/\lambda$, $i = 1, \ldots, k$.

The ROC property of a (n, τ_w) diagram means that we can treat (n, τ_w) characteristics as errors of the two kinds (β, α) in hypothesis testing: H_1 versus H_0, i.e.,

$$\beta = P(H_0|H_1) = n, \quad \text{and} \quad \alpha = P(H_1|H_0) = \tau_w \quad (16)$$

and $\alpha + \beta = 1$, if the prediction data $\{I(t)\}$ are trivial.

In the case $w_i = \lambda_i/\lambda$ the measures $P(\cdot|H_j)$, $j = 0, 1$ can be specified as follows. Both measures define probabilities for events $\omega = \{I(t), v = i\}$, where v is the random index of a subregion and has the distribution $P(v = i) = \lambda_i/\lambda := p_i$. The measure related to the H_0 hypothesis is

$$P(\mathrm{d}\omega|H_0) = P_0(\mathrm{d}I)p_i, \quad v(\omega) = i, \quad (17)$$

where P_0 is the stationary measure on $I(t)$ induced by the process $\{\mathrm{d}N(t), I(t), \pi(t)\}$. In the H_1 case

$$P(\mathrm{d}\omega|H_1) = r_i(t)/\lambda_i \cdot P(\mathrm{d}\omega|H_0), \quad v(\omega) = i, \quad (18)$$

where $r_i(t)$ is given by (8).

It is better to say that testing H_1 versus H_0 for the case $G = \{G_i\}$ involves two points: A random choice of G_i with probabilities $p_i = \lambda_i/\lambda$, $i = 1, \ldots, k$ and testing H_1 versus H_0 for the relevant subregion. The second point is considered in MOLCHAN and KEILIS-BOROK (2008).

The following is a nontrivial corollary of the previous statement:

3.9. For the regular case, $n_i(0) = 1$, $i = 1, \ldots, k$ and $\{w_i\} = \{\lambda_i/\lambda\}$, one has

$$\int_0^1 f\left(-\frac{\mathrm{d}n_\lambda}{\mathrm{d}\tau}\right) \mathrm{d}\tau = \sum_{i=1}^{k} p_i \int_0^1 f\left(-\frac{\mathrm{d}n_i}{\mathrm{d}\tau}\right) \mathrm{d}\tau, \quad p_i = \lambda_i/\lambda \quad (19)$$

where f is any continuous function and $n_\lambda(\tau)$ is an alternative notation for the $n(\tau_w)$ diagram in the special case $w_i = \lambda_i/\lambda$, $i = 1, \ldots, k$.

If $f = x \log x$, the quantity

$$I_i = \int_0^1 f\left(-\frac{dn_i}{d\tau}\right) d\tau = \int_0^1 \ln\left(-\frac{dn_i}{d\tau}\right) dn_i \quad (20)$$

is known in time prediction as the *information score* (see KAGAN, 2007 and HARTE and VERE-JONES, 2005).

Comments. In the non-regular case, $n_\lambda(0) < 1$, the score (19) is equal to ∞ for unbounded $f(x)$ at $x = \infty$, e.g., $f = x \log x$. Therefore the scores (19), (20) are unstable. (An extensive literature on skill scores can be found in JOLLIFFE and STEPHENSON, 2003; see also MOLCHAN, 1997 and HARTE and VERE-JONES, 2005.) Here we mention only the *area skill score*, AS, which is considered as a stable score (see JOLLIFFE and STEPHENSON, 2003, p. 73 but not ZECHAR and JORDAN (2008) where the same term is used). A linear transformation of AS appears as follows:

$$A = 2\int_0^1 (1 - n_\lambda(\tau) - \tau) d\tau, \quad 0 \le A \le 1. \quad (21)$$

Due to the convexity of $n_\lambda(\tau)$ the area under the integrand is approximated by a triangle from within and by the trapezium from the outside. Therefore

$$H \le A \le H(2 - H),$$

where

$$H = \max_\tau (1 - n_\lambda(\tau) - \tau), \quad 0 \le H \le 1.$$

Thus $\widehat{A} = H(3 - H)/2$ is a good estimate of A, because

$$|A - \widehat{A}| \le H(1 - H)/2 \le 1/8. \quad (22)$$

In the simplest case the distribution of the statistical estimate of H can be found. This circumstance can be used for comparison of a real forecasting method with the simplest one when it is important to take into account the observed number of target events N in the test period $[0, T]$. To be specific, let us consider an alternative forecasting method based on permanent alarm zones $G_\tau \subset G$. We suppose that these zones increase with τ and are normalized by the relation $\lambda(G_\tau)/\lambda(G) = \tau$ for any $0 \le \tau \le 1$. The case of G_τ with a decreasing ratio $\lambda(G_\tau)/$(area of G_τ) is most important for practice (see TIAMPO et al., 2002).

Suppose that N_τ is the number of target events in the space–time volume $G_\tau \times [0, T]$. Then the estimate of H for the alternative forecasting method is

$$\widehat{H} = \max_\tau (N_\tau/N - \tau) = \max_\tau (1 - \widehat{n}(\tau) - \tau),$$

where $\widehat{n}(\tau)$ is the convex minorant of the points $((1 - N_\tau/N), \tau)$. The quantity $\widehat{H}$ has the same distribution as the one-sided Kolmogorov–Smirnov statistic D^+_N (BOLSHEV and SMIRNOV, 1983), provided the target events in space–time form a Poissonian process with the measure $\lambda(dg)\, d\tau$.

3.10. In practical prediction the normalized rates of target events $p_i = \lambda_i/\lambda$ are unknown. At best we know their estimates $\widehat{p}_i$ and a confidence zone Z_q. For this reason the trivial strategies will be represented on the (n, τ_w) plane by a region D_w that depends on the true values $\{p_i\}$, see **3.3**. Combination of all D_w with $\{p_i\}$ from Z_q forms a confidence zone U_q for the trivial strategies on the (n, τ_w) plane. Moreover, U_q has the same confidence level as Z_q.

Let Z_q be given by

$$\widetilde{\chi}^2 := \sum_{i=1}^k \frac{(p_i - \widehat{p}_i)^2}{\widehat{p}_i} = \sum_{i=1}^k \frac{p_i^2}{\widehat{p}_i} - 1 < q. \quad (23)$$

Then the lower bound of U_q is the convex minorant of (n, τ_w) points:

$$n(\boldsymbol{\varepsilon}) = f\left(\sum_{i=1}^k \widehat{p}_i \varepsilon_i\right), \quad \tau_w(\boldsymbol{\varepsilon}) = \sum_{i=1}^k w_i \varepsilon_i,$$

where $\boldsymbol{\varepsilon} = (\varepsilon_1, \ldots, \varepsilon_k)$ are arbitrary sequences of 0 and 1, f is monotone nonincreasing function in the interval $(0, 1)$:

$$f(x) = \begin{cases} 1 - x - \sqrt{qx(1-x)}, & (1+q)x < 1 \\ 0, & (1+q)x > 1. \end{cases} \quad (24)$$

In the particular case $w_1 = \ldots = w_k = 1/k$, the set $\{(n(\boldsymbol{\varepsilon}), \tau_w(\boldsymbol{\varepsilon}))\}$ can be reduced to

$$\left\{ f\left(\sum_{i=1}^j \widehat{p}_{(k-i+1)}\right), j/k \right\}, \quad j = 1, \ldots, k,$$

where $\widehat{p}_{(1)} \le \ldots \le \widehat{p}_{(k)}$ are the $\{\widehat{p}_i\}$ arranged in increasing order.

The estimates of $\{\lambda_i/\lambda\}$ are usually derived using smaller events than the target ones (KOSSOBOKOV,

2005). In that case it can be assumed that the number of events N_- used for estimating $\{\lambda_i\}$ is large. If the estimates $\hat{p}_i$ are unbiased and $\hat{p}_i N_- \gg 1$, then the distribution of $\tilde{\chi}^2 \cdot N_-$ is approximately χ^2 with $f = k - 1$ degrees of freedom. Hence

$$q \approx (k - 1 + \rho\sqrt{2(k-1)})/N_-,$$

where $\rho \leq 2.5$, if $k \geq 5$ and the confidence level of Z_q is greater then 97.5%.

The quantity $h = \max_\varepsilon(1 - n(\varepsilon) - \tau_w(\varepsilon))$ characterizes the distance between the lower bound of U_q and the diagonal $n + \tau_w = 1$. Let us consider the most interesting case $w_i = \hat{p}_i$. Due to convexity (24) and the inequality

$$1 - f(x) - x \leq 0.5\sqrt{q},$$

the points $(n(\varepsilon), \tau_w(\varepsilon))$ are vertices of the lower bound of U_q and $h \leq 0.5\sqrt{q}$. Using the numbers $k = 11$ and $N_- = 2000$, typical for the prediction of $M = 6$ in Italy, we have $h \leq 0.05$. This value we can consider as a correction to the H score due to inaccuracy of the local rates of target events.

4. Conclusion and Discussion

1. *Results.* In the case of time prediction, the error set $\mathcal{E}$ is organized as follows: All trivial strategies concentrate at the diagonal $n + \tau = 1$ of the square $[0, 1]^2$, while the optimal strategies are on the lower boundary of $\mathcal{E}$, viz. $n(\tau)$. In the case of time-space prediction, the two-dimensional images of $\mathcal{E}$, i.e., $\mathcal{E}_w$, are organized differently: the diagonal $n + \tau_w = 1$ does not include all trivial strategies, and the (n, τ_w) diagram does not include all optimal strategies (see Fig. 2).

Nevertheless, $n(\tau_w)$ is a convenient tool to visualize such optimal strategies as are suitable for a trade-off between n and τ_w. However, if $\{w_i\} \neq \{\lambda_i/\lambda\}$, then the distance of $n(\tau_w)$ from the diagonal $n + \tau_w = 1$ does not tell us anything about the prediction potential of the relevant strategies. To learn something about this potential, we need the image of trivial strategies D on the (n, τ_w) plane. The lower boundary of D_w may be very close to the ideal strategy with the errors $(0, 0)$.

Let us consider an example. The relative intensity (RI) method (TIAMPO *et al.*, 2002) predicts the target event in that location where the historical seismicity rate, $f(g)$, is the highest, $f > c$. The RI is a typical example of a trivial strategy occasionally employed as an alternative to meaningful prediction techniques (see, e.g., MARZOCCHI *et al.*, 2003). By the RI method, $\tau_i = 1$ if $f > c$ in the i-th bin and $\tau_i = 0$ otherwise. If $\{w_i = \lambda_i/\lambda\}$, then

$$1 - n = \int_{f > c} f(g) dg = \tau_w,$$

i.e., $n + \tau_w = 1$ for any level c. If $w_i = |G_i|/|G|$, where $|G|$ is the area of G, then the curve $n(\tau_w)$ can be obtained by using (12) (see also ZECHAR and JORDAN, 2008). The curve passes close to (0,0), if most of the target events occur in a relatively small area, say, λ_1/λ is close to 1 and w_1 is close to 0.

One gets a unique set of weights by choosing $w_i = \lambda_i/\lambda$ (see **3.2**, **3.8**). It is only in this particular case that all trivial strategies are projected onto the diagonal $\tilde{D} : n + \tau_w = 1$, and $1 - n(\tau_w)$ is a ROC curve. Besides, the projection on the (n, τ_w) plane preserves the relative distance between any strategy and the set of trivial strategies. To be more specific, the following relations are true:

$$1 - n - \sum_{i=1}^{k} \tau_i \lambda_i/\lambda = \frac{\rho(Q, D)}{\rho(O, D)} = \frac{\rho(Q_w, \tilde{D})}{\rho(O_w, \tilde{D})}$$
$$= 1 - n - \tau_w \qquad (25)$$

(MOLCHAN and KEILIS-BOROK, 2008). Here, ρ is the Euclidean distance, e.g., $\rho(O, D)$ is the distance from $Q = (n, \tau)$ to the hyperplane $D = \{n + \sum \tau_i \lambda_i/\lambda = 1\}$, and $O = (0, 0...0)$ corresponds to the ideal strategy. The right-hand side of (25) is known in the contingency table analysis as the HK skill score (HANSSEN-KUIPER, 1965). Consequently, when $\{w_i = \lambda_i/\lambda\}$, the quantity $H = \max_{\tau_w}(1 - n(\tau_w) - \tau_w)$ gives the greatest relative distance between the optimal and the trivial strategies.

The choice of $\{w_i\}$ at the research stage instead of $\{\lambda_i/\lambda\}$ is justified by difficulties in the manner of estimating the $\{\lambda_i\}$. This justification is illusory, however. One must know the lower boundary of D_w in order to answer the question of how nontrivial the $n(\tau_w)$ diagram is. But this again requires knowledge of the $\{\lambda_i\}$ (see (12) and Fig. 2). In this connection the result of section 3.10 is of interest because it describes the lower confidence boundary for D_w when statistical estimates of $\{\lambda_i/\lambda\}$ are known only.

2. *The relation to the SDT.* In recent years the studies in earthquake prediction actively used the Signal Detection Theory (SDT) developed in the late 1980s in the atmospheric sciences (see, e.g., JOLLIFFE and STEPHENSON, 2003 and the references therein). The main object of study in this theory is a warning system, which characterizes the state of hazard by a scalar quantity ξ. The system is tested by making $K \gg 1$ trials in which the i-th event $\{\xi > u\}$ is interpreted as an alarm, $\widehat{x}_i = 1$, otherwise $\widehat{x}_i = 0$. The results are compared with observations x = Yes or No with respect to a target event. Any dependence between the members of the sequence $\{(\widehat{x}_i, x_i)\}$ is ignored a priori. It is required only that the rate of target events (x = Yes) should be $0 < s < 1$. This condition is essential for getting an acceptable estimate for the simultaneous distribution of $(\widehat{x}_i, x_i)$. Note that $s = 0$ in our approach.

Two problems are formulated: Assessing the prediction performance and choosing the threshold u in a rational manner. The first problem is attacked using the 2×2 contingency table of forecasts and the second by using the ROC diagram related to the hypothesis testing about the conditional distribution of ξ given x = Yes and given x = No.

In our terminology this situation is one with discrete "time" where the data I in a trial are given by ξ. Therefore, the SDT is equivalent to the analysis of the time prediction of earthquakes using a specified precursor/algorithm, even though the prediction of large earthquakes involves $s \ll 1$. The ROC/$n(\tau)$ diagram then quantifies the predictive potential of a precursor, ξ in this case. All meaningful strategies are functions of ξ, hence reduce to choosing the level u.

In the case of any data, $I(t)$, $n(\tau)$ characterizes the prediction performance of $\{I(t)\}$ and gives the lower bound to ROC curves for any algorithm based on $\{I(t)\}$. The studies of MOLCHAN (1990, 1997) answer the question of how ξ should be constructed for the original prediction data and why the relation to hypothesis testing arises at all.

The gist of the matter lies in the fact that the 2×2 contingency table is defined by three parameters (n, τ, s), and the program of prediction optimization is formulated, explicitly or implicitly, in terms of that table. As a result, we have to deal with local optimal strategies only. When real time is incorporated in the SDT framework, there arise additional parameters that are important for seismological practice, e.g., the rate of connected alarms (alarm clusters) v. The optimization of the loss function $\varphi = an + b\tau + cv$ at once gets us beyond the SDT framework and its tools. The strategies that optimize φ are not locally optimal, and can be found from Bellman-type equations (MOLCHAN and KAGAN, 1992; MOLCHAN, 1997).

The use of the SDT approach in space–time prediction imposes a rather unrealistic limitation: The spatial rate of target events must be homogeneous. Otherwise, the ROC diagram loses its meaning and becomes a (n, τ_w) diagram (see Fig. 2).

Acknowledgements

This work was supported by the Russian Foundation for Basic Research through grant 08-05-00215. I thank D.L. Turcotte for useful discussions, which have stimulated the writing of the present paper.

Appendix

We are going to prove the statements **3.1–3.10**.

Proof for 3.1, 3.2 Obviously, the projection γ_w preserves the property of convexity. Therefore, $\mathcal{E}_w$ and D_w are convex at the same time as are $\mathcal{E}$ and D. If D_w degenerates to the diagonal $\widetilde{D}: n + \tau_w = 1$, then the simplex D is given by any of the two equations: $n + \sum_{i=1}^{k} w_i \tau_i = 1$ and $n + \sum_{i=1}^{k} \lambda_i \tau_i/\lambda = 1$. Hence $w_i = \lambda_i/\lambda$.

Proof of 3.3. The simplex D is the convex hull of (n, τ) points of the form $Q(\varepsilon) = (1 - \sum \lambda_i \varepsilon_i / \lambda, \varepsilon_1, \ldots, \varepsilon_k)$, where $\varepsilon_i = 0, 1$. Accordingly, D_w is the convex hull of the $Q_w(\varepsilon)$, see (12).

Proof of 3.4. This statement follows intuitively from dimensionality considerations: The k-dimensional surface $n(\tau)$ with $k > 1$ is projected onto the (n, τ_w) plane, hence its image cannot be single-dimensional in the generic case.

In order to prove (13), we note that a convex function on the simplex $S_n = \{\sum_{i=1}^k \tau_i w_i = u, 0 \leq \tau_i \leq 1\}$ reaches its maximum at one of the edges, specifically, at a point of the form

$$\tau = (\varepsilon_1, \ldots, \varepsilon_{i-1}, x, \varepsilon_{i+1}, \ldots, \varepsilon_k), \quad \varepsilon_j = 0; 1.$$

The use of (10) gives (13).

Suppose the upper and lower boundaries of the image of $n(\tau)$ are identical and the $\{n_i(\tau)\}$ are piecewise smooth functions. Consider all $\tau = (\tau_1, \ldots, \tau_k)$ for which

$$\sum_{i=1}^k \lambda_i n(\tau_i)/\lambda = n_0, \quad \sum_{i=1}^k w_i \tau_i = \tau_w, \quad n_0 = n(\tau_w),$$

where τ_w is fixed.

Varying, e.g., τ_1 and τ_2, we have after differentiation:

$$\lambda_1 n_1'(\tau_1)\tau_1' + \lambda_2 n_2'(\tau_2) = 0, \quad \tau_1' = -w_2/w_1. \quad (A1)$$

If τ_1, τ_2 are points of smoothness of $n_i(\tau)$, $i = 1, 2$, then repeated differentiation of (A1) will give

$$\lambda_1 n_1''(\tau_1)(w_2/w_1)^2 + \lambda_2 n_2''(\tau_2) = 0.$$

However, $n_i''(\tau_i) \geq 0$, $i = 1, 2$. Hence $n_i''(\tau_i) = 0$, i.e., $n_i(\tau)$ are locally linear at all points of smoothness. Since $n_i(\tau)$ are piecewise smooth, it follows that for any discontinuous point τ_1 of $n_1(\cdot)$ one can find a point τ_2 where $n_2(\cdot)$ will be smooth. Consequently, when n_1 is discontinuous at τ, one should replace $n_1'(\tau_1)$ with $n_1'(\tau_1 + 0)$ and $n_1'(\tau_1 - 0)$ in equation (A1). But then we have from (A1) that $n_1'(\tau)$ is continuous at τ_1; hence all $n_i(\tau)$ are linear. Taking the boundary conditions $n_i(0) = 1$ and $n_i(1) = 0$ into account, we have $n_i(\tau) = 1 - \tau$. However, in that case one has $\mathcal{E} = D$, and, in virtue of **3.2**, $w_i = \lambda_i/\lambda$.

Proof of 3.5. Let Q_w be the point where the convex set $\{\psi \leq c\}$ is tangent to the convex curve $n(\tau_w)$. The function ψ reaches its minimum at the point Q_w on $\mathcal{E}_w$, because the sets $\{\psi \leq c\}$ increase with increasing c. Since $Q_w \in \mathcal{E}_w$, the pre-image $Q = (n, \tau) \in \mathcal{E}$. At this point $\varphi(Q) = \psi(Q_w)$ reaches its minimum on $\mathcal{E}$, hence Q belongs to the surface $n(\tau)$.

Proof of 3.6 follows from **2.3**.

Proof of 3.7. Let $Q = (n_0, \tau_{01}, \ldots, \tau_{0k})$ belong to $n(\tau)$. If $w_i = -\frac{\partial n}{\partial \tau_i}(Q)/c$, then the equation

$$n + c \sum_{i=1}^k w_i \tau_i = n_0 + c \sum_{i=1}^k w_i \tau_{i0} \quad (A2)$$

defines the tangent plane to $n(\tau)$. Since $n(\tau)$ is convex and decreasing, it follows that $w_i \geq 0$ and $\mathcal{E}$ lie on the same side of the plane (A2). Consequently, a strategy having the characteristics $Q = (n_0, \tau_{01}, \ldots, \tau_{0k})$ optimizes the losses $\varphi = n + c \sum_{i=1}^k w_i \tau_i$. Using **3.5**, we complete the proof.

Proof of 3.8. By (10) and (16) one has

$$\beta = n = \sum_{i=1}^k \lambda_i/\lambda \cdot n_i(\tau_i), \quad \alpha = \tau_w = \sum_{i=1}^k w_i \tau_i.$$

In the trivial case of $I(t)$, one has $n_i(\tau) = 1 - \tau$ and $\alpha + \beta = 1$. Hence

$$\beta = 1 - \sum_{i=1}^k \lambda_i/\lambda \cdot \tau_i, \quad \alpha = \sum_{i=1}^k w_i \tau_i = 1 - \beta,$$

i.e., $w_i = \lambda_i/\lambda$, $i = 1, \ldots, k$.

Suppose that $\{w_i\} = \{\lambda_i/\lambda\}$. The likelihood ratio of measures (17) and (18) at the point $\omega = (J(t), j)$ is

$$L(\omega) = P(d\omega|H_1)/P(d\omega|H_0) = r_j(t)/\lambda_j.$$

Accepting the hypothesis H_1 as soon as $L(\omega) > c$ and H_0 otherwise, one has

$$\alpha = \int_{L > c} P(d\omega|H_0) = \sum_{j=1}^k E\mathbf{1}_{(r_j/\lambda_j > c)} \cdot \lambda_j/\lambda$$

$$= \sum_{j=1}^k \tau_j \lambda_j/\lambda = \tau_w,$$

$$\beta = \int_{L > c} L(w) P(d\omega|H_0)$$

$$= \sum_{j=1}^k E r_j/\lambda_j \cdot \mathbf{1}_{(r_j/\lambda_j < c)} \cdot \lambda_j/\lambda$$

$$= \sum_{j=1}^k n_j(\tau_j) \lambda_j/\lambda = n.$$

Here we have used **2.1** and **2.2**.

Proof of 3.9. Let us consider the following testing problem: H_1 versus H_0 with the errors $\beta = P_1(L < c)$ and $\alpha = P_0(L \geq c)$ where $L(\omega) = dP_1/dP_0$ is the likelihood ratio. Obviously

$$E_0 f(L) := \int f(L(\omega))\mathrm{d}P_0(\omega) = \int f(c)\mathrm{d}F_L(c),$$

where F_L is the distribution of L with respect to the measure P_0. But $\mathrm{d}\beta = c\,\mathrm{d}F(c)$ and $\mathrm{d}\alpha = -\mathrm{d}F(c)$. Therefore

$$E_0 f(L) = \int_0^1 f\left(-\frac{\mathrm{d}\beta}{\mathrm{d}\alpha}\right)\mathrm{d}\alpha.$$

Applying this relation to the case (16), (17), (18), one has

$$\int_0^1 f\left(-\frac{\mathrm{d}n_\lambda}{\mathrm{d}\tau}\right)\mathrm{d}\tau = E_0 f(L) = \sum_{i=1}^k Ef\left(\frac{r_i(t)}{\hat{\lambda}_i}\right)p_i$$

$$= \sum_{i=1}^k Ef(L_i)p_i = \sum_{i=1}^k p_i \int_0^1 f\left(-\frac{\mathrm{d}n_i}{\mathrm{d}\tau}\right)\mathrm{d}\tau$$

Here L_i is the likelihood ratio $\mathrm{d}P_1/\mathrm{d}P_0$ for G_i.

Proof of 3.10. In virtue of **3.3** the set D_w is a convex hull of the points (12). Let $\tau_w(\varepsilon) = \sum_{i=1}^r w_i$. We need to find the maximum of $y = \sum_{i=1}^r p_i$ under the conditions

$$\sum_{i=1}^k p_i^2/\hat{p}_i \leq 1 + q, \qquad (A3)$$

$$\sum_{i=1}^k p_i = 1, \quad p_i \geq 0. \qquad (A4)$$

Because y is a linear function, it reaches its maximum at the boundary of the region (A3). Consequently, we will consider y as a function of the variables $(p_2 ..., p_{k-1})$ given by the equation

$$\sum_{i=1}^k p_i^2/\hat{p}_i = 1 + q \qquad (A5)$$

where $p_1 = y - p_2 - ... - p_r$, $p_k = 1 - y - p_{r+1} - ... - p_{k-1}$. The point of maximum of y is formally defined by the following equations

$$\partial y/\partial p_i = 0, \quad 1 < i < k.$$

This gives $p_i = c_1 \hat{p}_i, i = 1, ..., r$; $p_j = c_2 \hat{p}_j, j = r+1, ..., k$. Taking (A5, A4) into account, we get two equations for c_1 and c_2. Finally we have

$$y = \hat{y} + \sqrt{q\hat{y}(1-\hat{y})}, \quad \hat{y} = \sum_{i=1}^r \hat{p}_i,$$

$$p_i = y\hat{p}_i/\hat{y}, \quad i \leq r,$$

$$p_j = (1-y)\hat{p}_j/(1-\hat{y}), \quad j > r.$$

The conditions (A5, A4) hold, if $0 \leq y \leq 1$, that is, when $\hat{y}(1+q) \leq 1$. Otherwise we have to consider the vector

$$(\hat{p}_1/\hat{y}, ..., \hat{p}_r/\hat{y}, 0, ..., 0).$$

This vector satisfies (A3, A4) and gives the maximum possible value for y, $y = 1$.

The proof is complete.

References

BOLSHEV, L. N. and SMIRNOV, V. N. *Tables of Mathematical Statistics* (Nauka, Moscow 1983).

HARTE, D., and VERE-JONES, D. (2005), *The entropy score and its uses in earthquake forecasting*, Pure Appl. Geophys. *162*, 1229–1253.

HANSSEN, A. W., and KUIPER, W. J. A. (1965), *On the relationship between the frequency of rain and various meteorological parameters*, Modedeelingen en Verhandelingen, Royal Notherlands Meteorological Institute, 81.

JOLLIFFE, I. T. and STEPHENSON, D. B. (eds.), *Forecast Verification: A Practitioner's Guide in Atmospheric Science* (John Wiley & Sons, Hoboken 2003).

KAGAN, Y. Y. (2007), *On earthquake predictability measurement: information score and error diagram*, Pure Appl. Geophys. *164*, 1947–1962.

KEILIS-BOROK, V. I. and SOLOVIEV, A. A. (eds.), *Nonlinear Dynamics of the Lithosphere and Earthquake Prediction* (Springer-Verlag, Berlin-Heidelberg 2003).

KOSSOBOKOV, V. G. (2005), *Earthquake prediction: principles, implementation*, Perspect. Comput. Seismol. *36-1*, 3–175, (GEOS, Moscow).

LEHMANN, E. L., *Testing Statistical Hypotheses* (J. Wiley & Sons, New York 1959).

MARZOCCHI, W., SANDRI, L., and BOSCHI, E. (2003), *On the validation of earthquake-forecasting models: The case of pattern recognition algorithms*, Bull. Seismol. Soc. Am. *93*, 5, 1994–2004.

MOLCHAN, G. M. (1990), *Strategies in strong earthquake prediction*, Phys. Earth Planet. Inter. *61*(1–2), 84–98.

MOLCHAN, G. M. (1991), *Structure of optimal strategies of earthquake prediction*, Tectonophysics *193*, 267–276.

MOLCHAN, G. M. (1997), *Earthquake prediction as a decision making problem*, Pure Appl. Geophys. *149*, 233–247.

MOLCHAN, G. M., *Earthquake prediction strategies: A theoretical analysis*. In *Nonlinear Dynamics of the Lithosphere and Earthquake Prediction* (eds. Keilis-Borok, V.I. and Soloviev, A.A.) (Springer-Verlag, Berlin-Heidelberg 2003), pp. 209–237.

MOLCHAN, G. M., and KAGAN, Y. Y. (1992), *Earthquake prediction and its optimization*, J. Geophys. Res. *97*, 4823–4838.

MOLCHAN, G. M. and KEILIS-BOROK, V. I. (2008), *Earthquake prediction: Probabilistic aspect*, Geophys. J. Int. *173*, 1012–1017.

SHCHERBAKOV, R., TURCOTTE, D. L., HOLLIDAY, J. R., TIAMPO, K. F., and RUNDLE, J. B. (2007), *A Method for forecasting the locations of future large earthquakes: An analysis and verification*, AGU, Fall meeting 2007, abstract #S31D-03.

SHEN, Z.-K., JACKSON, D. D., and KAGAN, Y. Y. (2007), *Implications of geodetic strain rate for future earthquakes, with a five-year forecast of M 5 Earthquakes in Southern California*, Seismol. Res. Lett. *78*(1), 116–120.

SWETS, J. A. (1973), *The relative operating characteristic in psychology*, Science *182*, 4116, 990–1000.

TIAMPO, K. F., RUNDLE, J. B., MCGINNIS, S., GROSS, S., and KLEIN, W. (2002), *Mean field threshold systems and phase dynamics: An application to earthquake fault systems*, Europhys. Lett. *60*(3), 481–487.

ZECHAR, J.D. and JORDAN, Th., H. (2008), *Testing alarm-based earthquake predictions*, Geophys. J. Int. *172*, 715–724.

(Received July 15, 2008, revised December 18, 2008, accepted March 5, 2009, Published online March 10, 2010)

Identifying Seismicity Levels via Poisson Hidden Markov Models

K. Orfanogiannaki,[1] D. Karlis,[2] and G. A. Papadopoulos[1]

Abstract—Poisson Hidden Markov models (PHMMs) are introduced to model temporal seismicity changes. In a PHMM the unobserved sequence of states is a finite-state Markov chain and the distribution of the observation at any time is Poisson with rate depending only on the current state of the chain. Thus, PHMMs allow a region to have varying seismicity rate. We applied the PHMM to model earthquake frequencies in the seismogenic area of Killini, Ionian Sea, Greece, between period 1990 and 2006. Simulations of data from the assumed model showed that it describes quite well the true data. The earthquake catalogue is dominated by main shocks occurring in 1993, 1997 and 2002. The time plot of PHMM seismicity states not only reproduces the three seismicity clusters but also quantifies the seismicity level and underlies the degree of strength of the serial dependence of the events at any point of time. Foreshock activity becomes quite evident before the three sequences with the gradual transition to states of cascade seismicity. Traditional analysis, based on the determination of highly significant changes of seismicity rates, failed to recognize foreshocks before the 1997 main shock due to the low number of events preceding that main shock. Then, PHMM has better performance than traditional analysis since the transition from one state to another does not only depend on the total number of events involved but also on the current state of the system. Therefore, PHMM recognizes significant changes of seismicity soon after they start, which is of particular importance for real-time recognition of foreshock activities and other seismicity changes.

Key words: Poisson Hidden Markov Models, seismicity levels, transition probabilities.

1. Introduction

Traditional approaches for the stochastic modelling of the earthquake occurrence in time include memoryless models such as the random or Poisson model (e.g., Utsu, 1969; Gardner and Knopoff, 1974; Lomnitz, 1974; Kagan and Jackson, 1991) as well as the negative binomial one, which provides a better description of the clustering properties of seismicity than the random model does (e.g., Rao and Kaila, 1986; Dionysiou and Papadopoulos, 1992). For the time distribution of seismic sequences following or preceding strong main shocks, that is for aftershocks and foreshocks, power-law decay of the number of events, n, with the time from the main shock origin time was proposed by Omori (1894) for aftershocks and by Mogi (1963) for foreshocks. Especially for aftershocks, more elaborate models were developed by Utsu (1961) and Ogata (1988). The complexity of earthquake processes, however, requires more complex approaches which accept some manner of memory of the past state(s) of the system. The aim is to describe adequately future states of the system bearing more predictive capacity with respect to the memoryless model.

One of the well-known models with memory is the Markov chain which describes the dependency between observations collected in successive time intervals. The Markovian property predicts that in a process with a series of known past states, the next state depends only on the current state of the process and not on the previous ones. There are many applications of Markov chains on seismicity problems with pure Markov models being used directly on the data (Vere-Jones, 1966; Knopoff, 1971; Hagiwara, 1975; Patwardhan et al., 1980; Fujinawa, 1991; Suzuki and Kiremidjian, 1991; Tsapanos and Papadopoulou, 1999; Nava et al., 2005; Herrera et al., 2006; Rotondi and Varini, 2006). Very few applications of Hidden Markov Models (HMM) on earthquake problems have been published and only

---

[1] Institute of Geodynamics, National Observatory of Athens, Lofos Nymfon, 11810 Thissio, Athens, Greece. E-mail: korfanogiannaki@gmail.com

[2] Department of Statistics, Athens University of Economics and Business, Athens, Greece.

for modeling continuous data for intervals between earthquakes (GRANAT and DONNELLAN, 2002; CHAMBERS et al., 2003; EBEL et al., 2007). Recently, ORFANOGIANNAKI et al. (2006, 2007) extended the approach for seismicity patterns by applying HMM to discrete data for the number of earthquakes in each interval.

Formally, a Markov chain consists of a set of states, a set of initial probabilities to determine which one will be the starting state and a transition probability matrix defining the transitions between states. Similarly to Markov chains, HMM consist of a set of states, a set of initial probabilities and a transition probability matrix although in this case the state is unobserved. In addition, each state in HMM is associated with a probability distribution to which the unobserved state refers. Many probability distributions may be used depending on the nature of the data. Poisson, binomial, negative binomial, Gaussian and exponential are some of the distributions that have already been used in HMM. HMM may be applied to both continuous and discrete data. In the present paper we focus on discrete valued HMMs and particularly on Poisson Hidden Markov Models (PHMMs).

Among a variety of different models for discrete valued time series PHMMs offer certain advantages in the seismic context. Mainly, the PHMMs provide the possibility to estimate the sequence of unobserved states of the seismogenic system that underlies the data and, hence, to reveal unknown properties of the mechanism that generated the data. PHMMs were originally developed and applied in the biometric field (e.g., ALBERT, 1991; LEROUX and PUTERMAN, 1992). To our knowledge such a model has not been applied before in discrete valued time series of seismicity, that is in data sets consisting of earthquake event counts. Because of the discrete nature of the seismicity data, the Poisson distribution is selected as the observation distribution and, therefore, PHMM is derived.

The aim of the present paper is to introduce PHMM as candidate model for the description of seismicity patterns in the time domain. The possible application to the space-size domains goes beyond the scope of this paper. We are mainly interested in testing the model with real seismicity data sets and understanding its performance in recognizing short-term precursory seismicity patterns, such as foreshock activity preceding main shocks. The highly seismogenic area of Killini, Ionian Sea, Greece, was selected to perform the test.

2. Method

2.1. Definitions and Notation

HMMs are discrete time stochastic processes that consist of two parts. The first part is an unobserved finite state Markov chain $\{C_t : t \in N\}$ on m states. The second part is an observed sequence of nonnegative integer valued stochastic processes $\{S_t : t \in N\}$ such that, for all positive integers T, conditionally on $C^{(T)} = \{C_t : t = 1, 2, \ldots, T\}$ the random variables $S_1, \ldots, S_T$ are independent, where T is the length of the observational sequence. If we assume that for every point of time t, the conditional distribution of C_t given the state of S_t at time t is Poisson, then we derive the PHMM. In this case, the marginal distribution of S_i since the state is not observed is a finite mixture from the parametric Poisson family, given by:

$$p(s_i) = \sum_{j=1}^{m} a_j f(s_i | \lambda_j)$$

where $a_j > 0$, $j = 1, \ldots, m$, $\sum_{j=1}^{m} a_j = 1$ are the mixture proportions and $f(s|\lambda) = \exp(-\lambda)\lambda^s/s!$, $s = 0, 1, \ldots, \lambda \geq 0$, i.e., the probability function of the Poisson distribution with parameter λ. Each component of the mixture distribution corresponds to one state. Note that the interpretation of the states is not an easy task. It is done through a one-to-one correspondence between the data and the estimated sequence of states.

We denote the transition probabilities by γ_{ij}, i.e., γ_{ij} is the probability to move from state i, at time $t-1$, to state j, at time t, for any state i, j and for any time t, i.e., $\gamma_{ij} = P(C_t = j \mid C_{t-1} = i)$. Thus, in order to estimate the parameters of the PHHM model we need to estimate the transition probability matrix and the λ's. The procedure for parameters estimation using maximum likelihood is described in MACDONALD and ZUCCHINI (1997) using an EM type algorithm and is omitted here to save space.

2.2. Model Selection

The EM algorithm considers m (the number of states) as known and fixed. In applications, one of the challenging points is to estimate m. A way of selecting m is to find an order that balances the improvement of the log-likelihood with the number of components being fitted. Two such measures are the Akaike's Information Criterion (AIC) (AKAIKE, 1974) and the Bayesian Information Criterion (BIC) (SCHWARZ, 1978). We use the following versions of AIC and BIC as model selection criteria for PHHM, namely $AIC(m) = -2l(m) + 2m^2$ and $BIC(m) = -2l(m) + m^2 \log T$, $l(m)$ denotes the maximized log-likelihood for a model with m components; T was defined above. Therefore, we choose the number of components to be the number that minimizes $AIC(m)$ and $BIC(m)$.

2.3. Time Unit and Time Interval Selection

Preliminary runs of the model with actual data showed that the selection of the parameters of the model, that is the number of states as well as the respective activity rates are, as one may expect, sensitive to the selection of the time unit (or counting interval) τ.

We applied PHMMs in the selected catalogue of the test area for two distinct time units. Our main interest is to investigate the capability of PHMMs to describe short-term seismicity patterns in the time domain, such as initiation of foreshock or aftershock activity before or after the main shock, respectively. Therefore, we selected time unit equal to 1 day (=24 h) which makes the model sensitive enough in catching short-term seismicity variations. In fact, the routine practice of seismicity analysis followed in most seismograph centers updates earthquake catalogues by adding new events on a daily basis. On the other hand, there is need to check whether the selection of a longer time interval affects the results, since it would be more appropriate for the investigation of long-term seismicity changes. To this aim we selected an alternative τ equal to 1 month (= 30 days) and we repeated the PHMM application to the aggregated data set of the test area.

Important effects in the results are also expected by changing the total length of the time interval examined. From a statistical point of view it is common that a decrease in the time length produces smaller counts and, thus, usually, decreases the overdispersion. In the literature of mixture models it has been shown that small overdispersion usually needs few components to be captured, which in our field implies fewer states of a PHMM.

3. The Test Area

3.1. Geotectonics

We consider data from the seismically active area of Killini, Ionian Sea, Greece (Fig. 1). The test area of approximately 80 km × 50 km dimensionally, is situated on the narrow inner shelf of the northwestern side of the Hellenic Arc occupying the strait between the Island of Zakynthos in the Ionian Sea and the Killini Peninsula of NW Peloponnisos on the Greek mainland. This is part of the continental side of the plate boundary segment where the Mediterranean lithosphere moves from about SW to NE towards the Eurasian plate and subducts beneath it. From a geotectonic point of view the area between the Island of Zakynthos and the Killini Peninsula belongs to the Ionian Zone. For reasons of brevity this area is called in this paper the "Killini area".

3.2. Seismicity

In the instrumental era of seismicity, that is in about the last 100 years, several strong main shocks occurred in the Killini area. In the last 20 years or so three strong earthquakes occurred on 16.10.1988 ($M_w = 5.9$), 26.03.1993 ($M_w = 5.6$), and 02.12.2002 ($M_w = 5.6$), (PAPADOPOULOS et al., 1993, 2003; ROUMELIOTI et al., 2004); moment magnitudes, M_w, are according to the Harvard CMT solutions (http://www.seismology.harvard.edu). In adjacent earthquake sources earthquake magnitudes reach about 7.0. The most recent examples are the 17.01.1983, ($M_w = 6.9$) and 18.11.1997 ($M_w = 6.6$) large shocks, which occurred to the west of Cephalonia Island and to the south of Zakynthos Island, respectively (see upper

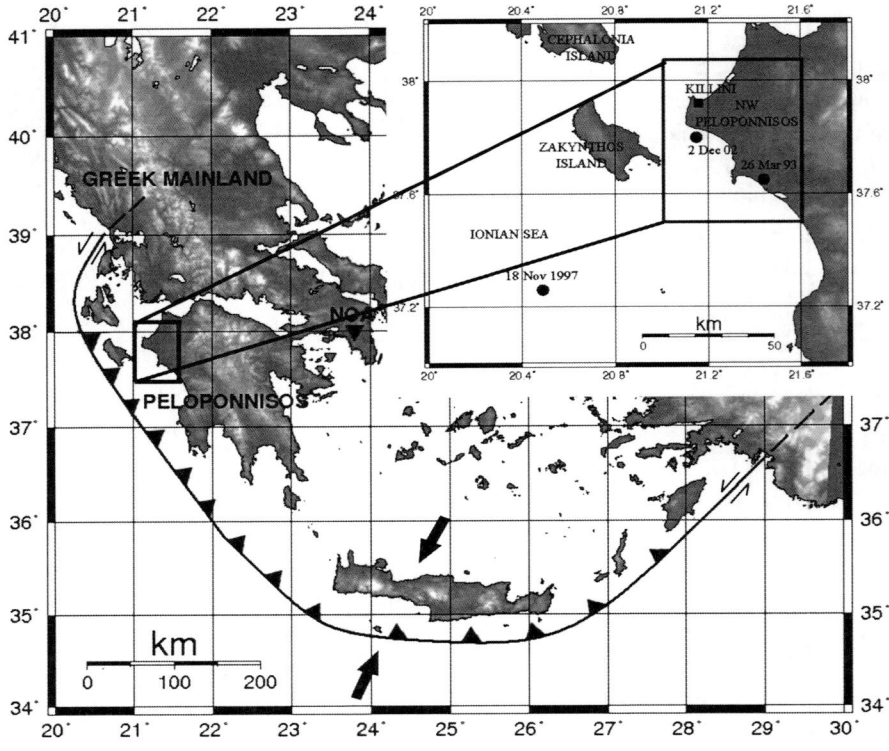

Figure 1
Seismotectonic setting of the test area. The main feature is the active underthrusting of the African lithosphere beneath the Aegean Sea region along the Hellenic Arc. *Arrows* show plate motion. *Rectangle* represents the test area of Killini which can be seen enlarged in the upper right part of the figure

right part of Fig. 1). The Killini area appears to be prone to the systematic incidence of foreshock sequences preceding main shocks (PAPADOPOULOS et al., 2003) and, therefore, it makes an interesting case to check the capability of the PHMM approach to identify precursory seismicity patterns, such as foreshock activity.

3.3. *Data Selection*

For the selection of the time interval to apply PHMMs in the test area, two conditions must be fulfilled. The first is that at least two strong earthquakes, separated in time by several years, should be included in the earthquake catalogue. In this way we may ensure that some seismicity clusters associated with the strong earthquakes will be present in the catalogue. The second condition is that the magnitude cut-off which ensures data completeness over a certain magnitude should not be very high,

with the aim that abundant earthquake events exist in the catalogue. From this point of view, it is of importance that the year 1964 signifies the initiation of the systematic seismicity monitoring in Greece by the modern national seismograph system of NOAGI (Institute of Geodynamics, National Observatory of Athens) established in accordance with the WWSSS. Completeness tests based on the magnitude-frequency or GUTENBERG and RICHTER (1944) (G-R) diagram, performed over the catalogues of NOAGI (2007) and GLAUTH (Geophysical Laboratory of the Aristotelian University of Thessaloniki, 2007) have shown that before 1964 the data are vastly incomplete. G–R indicates that spanning 1990 to 2006 the data set in the NOAGI (2007) catalogue for the Killini area is complete for $M_L \geq 3.2$ (Fig. 2).

The time interval from 1990 to 2006 inclusive was finally selected for examination, given that the catalogue contains two strong earthquakes, those of 26.03.1993 ($M_w = 5.6$) and 02.12.2002 ($M_w = 5.6$),

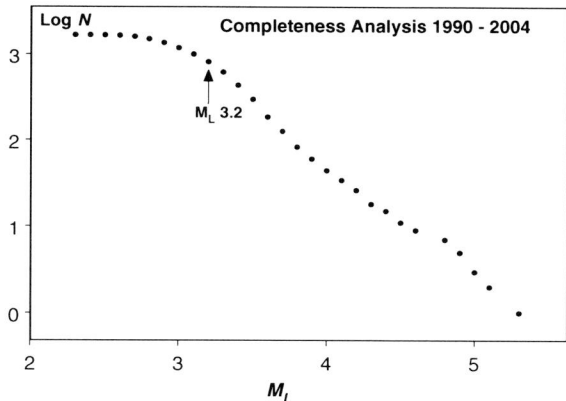

Figure 2
Magnitude-frequency relation for the Killini area catalogue of the period 1990–2006 inclusive. N = cumulative number of events M_L = local magnitude. Completeness is taken for the part of the curve which exhibits best fitting, that is for $M_L \geq 3.2$

associated with foreshock and aftershock sequences. In addition, the aftershock activity of the large ($M_w = 6.6$), relatively distant earthquake of 18.11.1997 was partially extended within the Killini area and, therefore, an additional cluster of seismicity is expected to be contained in the catalogue of the area. Focal parameters of the three strong main shocks involved in this examination are listed in Table 1.

Finally, note that for the current analysis we assume stationarity of the series considered.

3.4. Seismicity Analysis

The investigation of foreshock activity occurring before main shocks in the Killini area has been based on a procedure of seismicity analysis which aims to discriminate between the mean background seismicity, r, and the foreshock activity, r_f. The background seismicity is determined by the mean seismicity rate prevailing in time intervals free of aftershocks and foreshocks. Therefore, aiming to determine r reliably, a residual earthquake catalogue for the Killini area was produced from the NOAGI earthquake catalogue, for the time interval between 01.01.1990 and 31.12.2006, by removing dependent events on the basis of a Greek version of the GARDNER and KNOPOFF (1974) algorithm (LATOUSSAKIS and STAVRAKAKIS, 1992). In the residual earthquake catalogue produced, the mean seismicity rate (in events/day) was calculated for events of magnitude $M \geq m_c$, where m_c is the magnitude cut-off (Fig. 3). The significant increase of the seismicity rate before a Killini main shock, at least at the 95% level, is defined as foreshock activity. This process consists of a base-tool to compare the PHMM results. The rates estimated from PHMMs and the ability of PHMMs to recognize abnormal seismic activity are tested against the statistical results obtained from traditional analysis.

In traditional analysis the mean background seismicity rate in the declustered catalogue spanning 1990 to 2006 inclusive, is only $r = 0.06$ events/day (Table 2). At the beginning of February 1993 the process entered a period of gradually increasing seismicity up to the main shock of 26.03.1993 (Fig. 4, upper panel). The seismicity rate was equal to $r_f = 0.43$ events/day during the entire foreshock period from 01.02.1993 to 26.03.1993 (Table 2).

Table 1

Focal parameters of the earthquakes examined in this paper

Year	Month	Day	Time	ϕ_N	λ_E	M_w
2002	12	02	04:58:56	37.80	21.15	5.6
1997	11	18	13:07:41	37.58	20.57	6.6
1993	03	26	11:58:18	37.65	21.44	5.6

Parameters are according to the NOAGI determinations (http://www.gein.noa.gr), magnitudes are taken by the CMT solutions data base of Harvard (http://www.seismology.harvard.edu)

ϕ_N = geographic latitude, λ_E = geographic longitude, M_w = moment magnitude

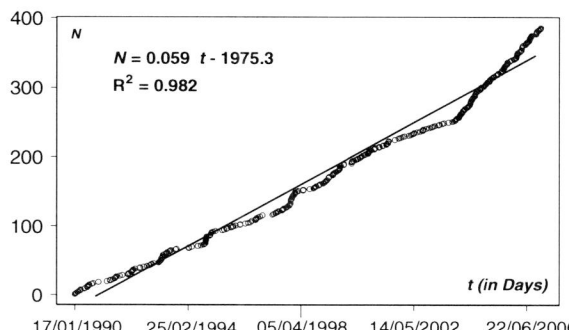

Figure 3
Background seismicity rate (= 0.06 events/day) in the declustered catalogue of the period from 1990 to 2006 inclusive. N = cumulative number of events, t = time, R = correlation coefficient

Table 2

Seismicity rates for the time intervals examined

No.	n	Time			Interval			Rate	a
		Year	Month	Day	Year	Month	Day	(events/day)	
1	386	1990	01	01	2006	12	31	0.06	–
2	26	1993	02	01	1993	03	26	0.43	0.01
3	11	1997	05	01	1997	10	03	0.06	–
4	56	2002	09	14	2002	11	26	0.70	0.01

No = code number, n = number of events, a = significance level of the difference between the rate of the corresponding time interval and the background seismicity rate according to the t test

Assuming that r and r_f are independent, the significance of difference between them was tested using the t test and it was found significant at 99% probability level. As regards the 1997 main shock, no gradual acceleration of the activity was observed in the period preceding the main shock; the last 6 months prior to the main shock occurrence the mean seismicity rate remained equal to 0.06 events/day, which does not deviate from the mean background seismicity rate (Table 2).

Features of the foreshock activity that preceded the 02.12.2002 main shock were similar to the features of the 1993 sequence. More precisely, about 2.5 months before the main shock occurrence the seismicity process started to accelerate (Fig. 4, lower panel). The mean seismicity rate, r_f, for the foreshock period that lasted from 14.09.2002 to 26.11.2002 was 0.7 events/day. The t test showed that the difference between r and r_f is significant at 99% probability level.

4. Results of PHMMs

4.1. Counting Interval $\tau = 1$ day

In order to select the best model, the PHMM was fitted with $m = 2, \ldots, 5$ with fixed initial state (the first) each time. For counting interval $\tau = 1$ day both AIC and BIC select the model with four states. The different values of the log-likelihood for different number of states are also reported (Table 3). For the selected model the transition probability matrix can be seen in Table 4. The sequence of the unobserved states is estimated using the Viterbi algorithm once the observational sequence is given.

To illustrate the results we plot seismicity states determined by the PHMM approach as a function of time. To track the natural sequence of counts, the daily number of earthquake events against time is also plotted. This plot indicates clearly the presence of three clusters of earthquakes (Fig. 5a). The first and third are clusters around the times of the two strong main shocks which occurred on 26.03.1993 ($M_w = 5.6$) and 02.12.2002 ($M_w = 5.6$). The second is a cluster of seismicity which is due to aftershocks of the distant large main shock ($M_w = 6.6$) of 18.11.1997. Although this earthquake took place well outside the Killini area, the seismicity that followed it was extended into the Killini area.

The time plot of seismicity states determined by the PHMM (Fig. 5b) not only repeats the same seismicity features but, in addition to this, quantifies the seismicity level and, at the same time, underlies the degree of strength of the serial dependence of the events at any point of time. To show more clearly the importance of this we focused on time windows of only a few months in either sides of the main shock occurrence. As is expected, the time plots of event counts (Figs. 6a, 7a, 8a) show that the three main shocks of 26.03.1993, 18.11.1997 and 02.12.2002 share the common feature of being followed by aftershock activity which gradually decreases with time. The two main shocks of 26.03.1993 and 02.12.2002 are preceded by precursory activity that is incidence of short-term foreshocks, by about 1.5 and 0.5 months, respectively. Such a feature is not evident before the large, relative distant 1997 main shock, even though its aftershock sequence was partially extended within the Killini area. In fact, a very weak, temporary increase of the event counts by the beginning of November 1997 does not imply initiation of precursory activity before the 18.11.1997 main shock. The time plots of seismicity states derived from the PHMM approach (Figs. 6b, 7b, 8b) signify very clearly the onset of a precursory phase of activity, with the transition from state 1 to state 2 or even to state 3, in all three earthquake sequences. Before the 26.03.1993 and 02.12.2002 main shocks the precursory activity started on 10.02.1993 and 14.09.2002, respectively. In the case of the

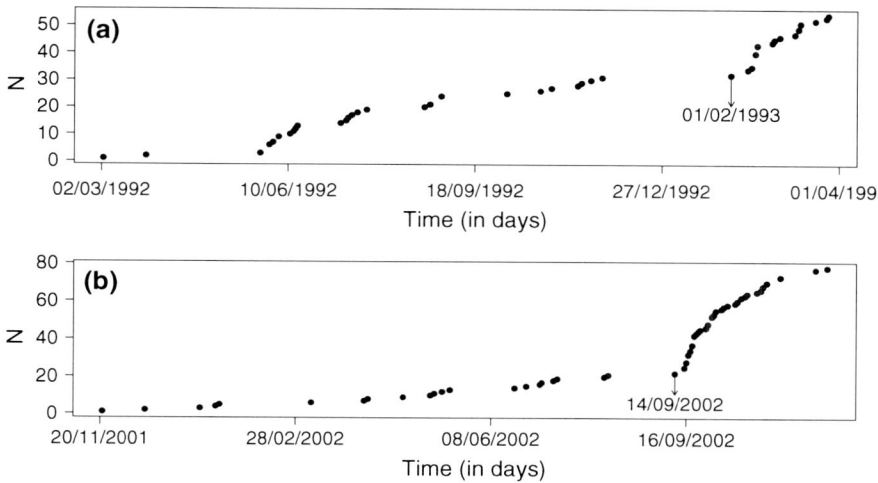

Figure 4
Cumulative number of events, N, versus time, t, for the time interval between 01.02.1993 and 26.03.1993 (*upper panel*) and between 14.09.2002 and 02.12.2002 (*lower panel*). The *arrow* indicates the onset of the foreshock activity before the 26.03.1993 main shock and the 02.12.2002 main shock, respectively

Table 3
Comparison of models on the basis of AIC and BIC for counting interval $\tau = 1$ day and $\tau = 1$ month

τ	m	d	Log-likelihood	LRT	p value	AIC	BIC
1 day	2	4	−2154.638	264.880	<0.01	4317.276	4343.711
	3	9	−2022.198	73.880	<0.01	4062.397	4121.875
	4	16	−1985.258	7.006	0.09	4002.516	4108.255
	5	25	−1981.755			4013.510	4178.727
1 month	2	4	−529.459	179.408	<0.01	1066.918	1079.690
	3	9	−439.755	49.168	<0.01	897.509	926.246
	4	16	−415.171	25.138	<0.01	862.341	913.429
	5	25	−402.602	0.510	0.26	855.203	935.027
	6	36	−402.347			876.694	991.640

(m is the number of components, while d denotes the number of estimated parameters). We also report values of the LRT statistic and the associated p values

18.11.1997 main shock, the weak activity noted by the beginning of November 1997 was recognized as significant enough by the PHHM approach and, therefore, the system jumped from state 1 to state 3 on 01.11.1997. Thereafter, the system remained either at state 2 or 3 until the occurrence of the main shock.

A common feature that the three sequences share is that after the transition to higher state the seismogenic system never returns to state 1 before the occurrence of the main shock. The persistence of the system to remain at higher states at any point of time before the main shock is interpreted not only by the increased seismicity rate but also by the strong serial dependence of the events during the precursory stage. Moreover, it is noteworthy that the transition probabilities from state 1 to state 2 and from state 1 to state 3 are of the order of only 0.0005 and 0.0065, respectively. On the contrary, the transition probability for the system to remain at state 1 is 0.993. This means that if the system is in state 1, once the foreshock activity starts it does not follow the most probable state.

Not all periods of increased seismicity in the Killini data are associated with main shocks. Small clusters of swarm type are observed as well. PHMMs identify these clusters and associates them with states 2 and 3. In the swarm type of activity the system

Table 4

Transition probability matrix determined for the best model for counting interval $\tau = 1 = 1$ day

	1	2	3	4
1	0.993	0.0005	0.0065	0
2	0.029	0.92	0.05	0.001
3	0	0.48	0.5	0.02
4	0	0	0.36	0.64

remains in states 2 and 3 only for a very short period of time while in the foreshock sequences the system persists to remain at higher states until the occurrence of the main shock.

4.2. Counting Interval $\tau = 1$ month

The procedure followed for the selection of the best model for counting interval $\tau = 1$ day was repeated over the same data set for counting interval $\tau = 1$ month. AIC and BIC (Table 3) select the model with five and four components, respectively. For the model with 5-components, the Poisson seismicity rates as well as the transition probability matrix are reported in Tables 5 and 6, respectively. We also report standard errors for the λ's based on the Hessian matrix derived numerically from the maximized log-likelihood. However, note that since some of the estimated probabilities lie on the boundary of the parametric space, the derived standard errors must be handled with caution.

The time plot of event counts for $\tau = 1$ month (Fig. 9a) shows three peaks of seismicity associated with the strong main shocks of 1993, 1997 and 2002. However, the time plot of states (Fig. 9b) is incapable of revealing the short-term foreshock activity that preceded the three main shocks as is shown in the time plot of states for $\tau = 1$ day (Figs. 6b, 7b, 8b).

If we divide the one month intensities by 30 we get a parameter estimate in counts per day. The results are shown in the third row of Table 5 and they are comparable to the one-day intensities. We observe that the values we obtain for states 1 and 2 are very close to the rates determined for counting interval $\tau = 1$ day. The rate we obtain for state 3 is about 3 times smaller from the corresponding daily rate. However, the value we get for state 4 is extremely higher from the corresponding daily one. This fact strengthens what is also shown in Fig. 9; that the question of whether to use 4 or 5 states for $\tau = 1$ month is somewhat of an artifact of the numerical values of the counts for the three main shocks (2 counts near 80/month and 2 counts near 40/month). This example implies that the selection of the counting interval is directly dependent on the nature of the seismicity patterns, e.g., short-term, long-term

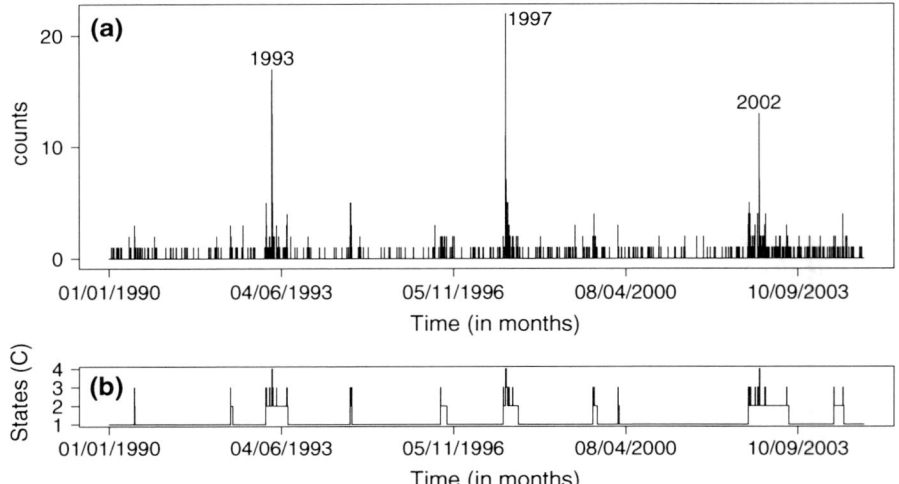

Figure 5
Time distribution of: **a** the event counts and **b** seismicity states determined by the PHMM approach for counting interval $\tau = 1$ day. The positions in time of the three mainshocks of 1993, 1997 and 2002 are shown

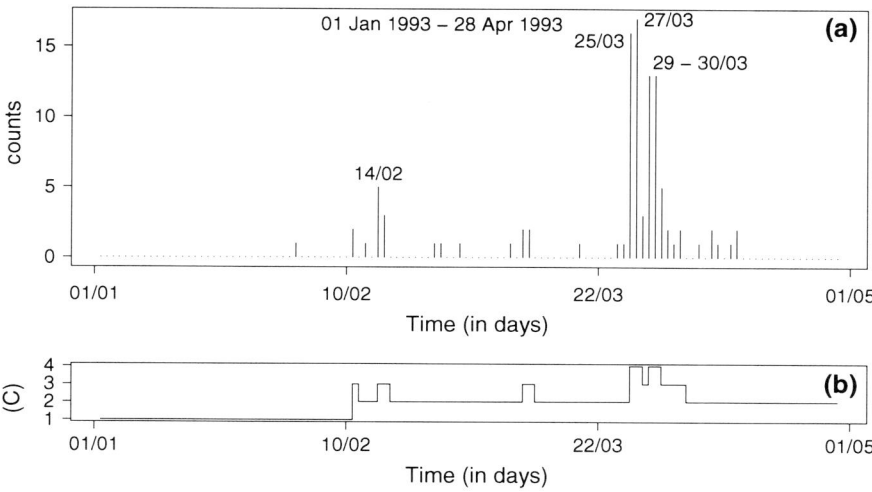

Figure 6
Part of the time distributions shown in Fig. 5 around the time of occurrence of the main shock of 27.03.1993: **a** event counts, **b** PHMM seismicity states

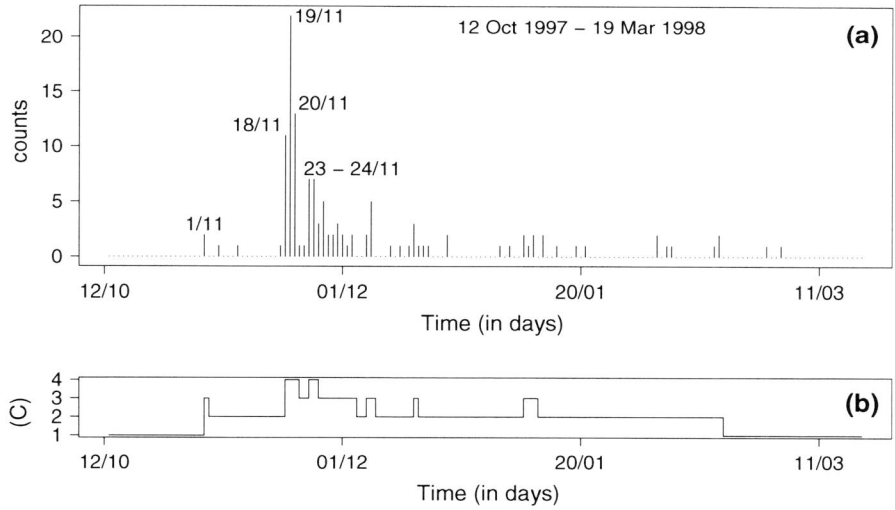

Figure 7
Part of the time distributions shown in Fig. 5 around the time of occurrence of the main shock of 18.11.1997: **a** event counts, **b** PHMM seismicity states

or medium-term seismicity variations, while the selection of the number of states is affected by some extreme values included in the data.

4.3. Model Selection

The Likelihood Ratio Test (LRT) has also been applied to test hypotheses concerning the number of components. Namely, to check whether the improvement on the log-likelihood is statistically significant, we have tested the null hypothesis H_0: the number of components is k against the alternative H_1: the number of components is $k + 1$. Such a test can be seen as complementary to the AIC and/or BIC and does not necessarily provide the same number of components nonetheless since it takes into account the variability in the log-likelihood it can be useful. It is well known in the mixture likelihood (see e.g.,

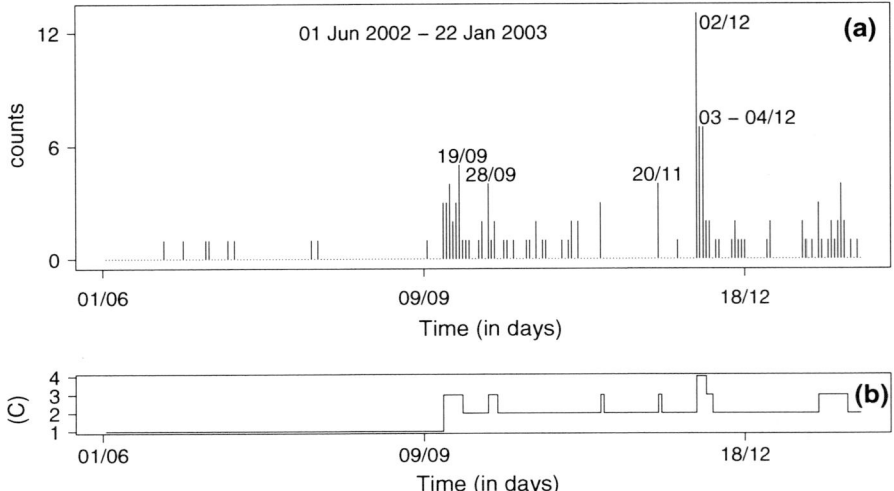

Figure 8
Part of the time distributions shown in Fig. 5 around the time of occurrence of the main shock of 02.12.2002: a event counts, (PHMM) seismicity states

McLachlan and Peel, 2000) that standard asymptotic results for LRT are not valid and thus bootstrap versions of the test must be used. We run bootstrap LRT by simulating 1,000 samples from the null hypothesis and comparing the values of the LRT from the simulated samples with the observed one. We found that for $\tau = 1$ day four components are sufficient, while for $\tau = 1$ month the LRT resulted in five components. Both results are in agreement with the results based on AIC. The values of LRT are shown in Table 3, along with the p values. The significance level used was 5%.

5. Conclusions

The performance of the PHMMs in describing short-term precursory seismicity changes preceding strong main shocks has been tested for the first time. Application in the complete part ($M_L \geq 3.2$) of the seismicity catalogue of the highly seismogenic area of Killini, Ionian Sea, Greece, for the period from 1990 to 2006 successfully reached the following results:

1. PHMMs is an adequate approach of the time distribution of seismicity. In fact, simulations for counting interval $\tau = 1$ day indicate excellent fit between the true and the simulated data. For counting interval $\tau = 1$ month the resemblance between the true and the simulated data is also quite good.
2. For both counting intervals $\tau = 1$ day and $\tau = 1$ month the model with four components and five components was selected, respectively.
3. The plot of counts of the real catalogue against time indicates clearly the existence of three seismicity clusters associated with three strong main shocks. The first and third main shock occurred within the Killini area, while the second one occurred in an adjacent area.
4. The time plot of seismicity states determined by the PHMM approach not only reveals the above seismicity features but, in addition to this, quantifies the seismicity level and, at the same time, underlies the degree of strength of the serial dependence of the events at any point of time.
5. Time plots of seismicity states signify very clearly the onset of precursory foreshock activity before the three earthquake sequences of 1993, 1997 and 2002, with the transition from state 1 to state 2 or even to state 3. When the system is in state 1, the transition probability to remain at state 1 is 0.995. The transition probabilities from state 1 to state 2 and from state 1 to state 3 are of the order of only 0.0005 and 0.0065, respectively. This means that after the initiation of the precursory foreshock activity the

Table 5

Poisson seismicity rates determined for counting intervals $\tau = 1$ day and $\tau = 1$ month

τ	1	2	3	4	5
1 day	0.0595	0.2319	1.7838	11.5862	–
St.err	0.0043	0.0332	0.1908	1.1356	–
1 month	1.7745	5.8882	17.5183	38.0001	79.5001
St.err	0.1343	0.5619	1.5771	4.3594	6.3053
1 month/30	0.0592	0.1963	0.584	1.27	2.65

Standard errors were derived using the Hessian of the maximized log-likelihood

Table 6

Transition probability matrix determined for the best model for counting interval $\tau = 1$ month

	1	2	3	4	5
1	0.92	0.036	0.03	0.0066	0.0074
2	0.26	0.71	0	0.03	0
3	0.18	0.69	0	0	0.13
4	0	0	1	0	0
5	0	0	1	0	0

system does not follow the most probable state, which geophysically reflects changes in the seismogenic system during the foreshock activity.

6. The persistence of the system to remain at higher states at any point of time from the initiation of the foreshock activity until the main-shock occurrence bears a clear geophysical message. It is interpreted not only by the increased seismicity rate but also by the strong serial dependence of the events during the precursory stage.

7. The selection of the counting interval is directly dependent on the nature of the seismicity pattern we need to study, e.g., long-term, medium-term, short-term seismicity changes, that is it depends on the "resolution" that we require from the PHMM approach to provide. Further tests with different time units are required for the selection of an "optimum" counting interval.

8. Traditional analysis, based on the determination of highly significant changes of seismicity rates, recognized foreshock activities before the 1993 and 2002 main shocks but failed to recognize precursory activity before the 1997 main shock. This is due to the relatively low number of events that preceded the 1997 main shock. From this point of view the PHMM performs better than the traditional analysis since the transition from one state to another does not only depend on the total number of events involved but also on the current state of the system. Therefore, the PHMM recognizes significant changes of seismicity as soon as they start, which is of particular importance for real-time recognition of foreshock activities and other seismicity changes.

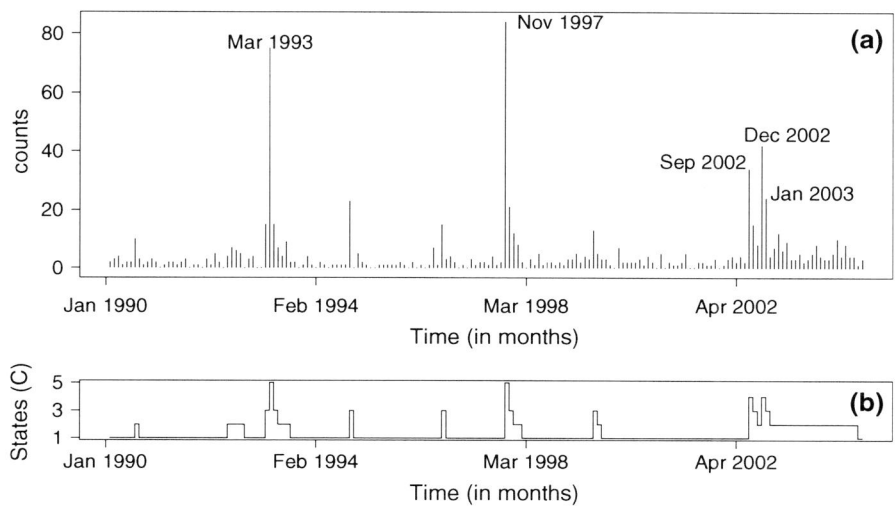

Figure 9
Time distribution of the event counts (**a**) and of the seismicity states (**b**) determined by the PHMM approach for counting interval $\tau = 1$ month. The positions in time of the three main shocks of 1993, 1997 and 2002 are shown

Finally, we would like to mention a few topics of current research. The present paper only slightly tackles the issue of the good predictability offered by the model in the sense that we used the same data to fit the model and evaluate the results. Thus some sort of overfitting is expected. It is ongoing research to see the predicting potential of the model in future observations not yet seen at the time of prediction. Moreover the value of τ is also important for obtaining better information. While we believe that $\tau = 1$ day makes sense it would be interesting to examine the dynamics in other time intervals.

Acknowledgments

This paper is part of the research project Seismic eArly warning For EuRope (SAFER) supported by the CEC, contract n. 036935, 6th Framework Programme, FP6-2005-Global-4, Reduction of seismic risks. We express sincere thanks to two anonymous referees whose comments helped us considerably to improve the initial manuscript.

REFERENCES

AKAIKE, H. (1974), *A new look at the statistical model identification*, IEEE Trans. Automat. Control *AC-19*, 716–723.

ALBERT, P. S. (1991), *A two-state Markov model for a time series of epileptic seizure counts*, Biometrics *47*, 1371–1381.

CHAMBERS, D., EBEL, J. E., KAFKA, A. L., and BAGLIVO, J. A. (2003), *Hidden Markov approach to modeling interevent earthquake times*, Eos Trans. AGU, *84*(46), Fall Meet. Suppl., Abstract S52F–0179.

DIONYSIOU, D. D. and PAPADOPOULOS, G. A. (1992), *Poissonian and negative binomial modelling of earthquake time series in the Aegean area*, Phys. Earth Planet. Int. *71*, 154–165.

EBEL, J. E., CHAMBERS, D. W., KAFKA, A. L., and BAGLIVO, J. A. (2007), *Non-Poissonian earthquake clustering and the hidden markov model as bases for earthquake forecasting in California*, Geophys. Res. Lett. *78*, 57–65.

FUJINAWA, Y. (1991), *A method for estimating earthquake occurrence probability using first- and multiple-order Markov chain models*, Nat. Hazards *4*, 7–22.

GARDNER, J. K. and KNOPOFF, l. (1974), *Is the sequence of earthquakes in Southern California, with aftershocks removed, Poissonean?*, Bull. Seism. Soc. Am. *64*, 1363–1367.

GRANAT, R. and DONNELLAN, A. (2002), *A hidden Markov model-based tool for geophysical data exploration*, Pure Appl. Geophys. *159*, 2271–2283.

GUTENBERG, B. and RICHTER, C. F. (1944), *Frequency of earthquakes in California*, Bull. Seismol. Soc. Am. *34*, 149–168.

HAGIWARA, Y. (1975), *A Stochastic Model of earthquake occurrence and the accompanying horizontal land deformation*, Tectonophysics *26*, 91–101.

HERRERA, C., NAVA, F. A., and LOMNITZ, C. (2006), *Time-dependent earthquake hazard evaluation in seismogenic systems using mixed Markov Chains: An application to the Japan area*, Earth Plan. Space *58*, 973–979.

KAGAN, Y. Y. and JACKSON, D. D. (1991), *Long-term earthquake clustering*, Geophys. J. Int. *104*, 117–133.

KNOPOFF, L. (1971), *A stochastic model for the occurrence of main-sequence earthquakes*, Revs. Geophys. *9*, 175–188.

LATOUSSAKIS, J. and STAVRAKAKIS, G. N. (1992), *Times of increased probability of earthquakes of ML $\geq$ 5.5 in Greece diagnosed by algorithm M 8*, Tectonophysics *210*, 315–326.

LEROUX, B. G. and PUTERMAN, M. L. (1992), *Maximum-penalized-likelihood estimation for independent and Markov-dependent mixture models*, Biometrics *48*, 545–558.

LOMNITZ, C., Global Tectonics and Earthquake Risk (Elsevier, Amsterdam 1974).

MACDONALD, I. L. and ZUCCHINI, W., Hidden Markov models and other models for discrete-valued time series (Chapman and Hall, London 1997).

MCLACHLAN, G. and PEEL, D., Finite Mixture Models (Wiley, NY 2000).

MOGI, K. (1963), *The fracture of a semi-infinite body caused by an inner stress origin and its relation to the earthquake phenomena (second paper)*, Bull. Earthq. Res. Inst. *41*, 595–614.

NAVA, F. A., HERRERA, C., FREZ, J., and GLOWACKA, E. (2005), *Seismic hazard evaluation using Markov Chains: Application to the Japan area*, Pure Appl. Geophys. *162*, 1347–1366.

OGATA, Y. (1988), *Space-time point-process models for earthquake occurrences*, Ann. Inst. Statist. Math. *50*, 379–402.

OMORI, F. (1894), *On the aftershocks of earthquakes*, J. Coll. Sci. Imp. Univ. Tokyo, 7–111.

ORFANOGIANNAKI, K., KARLIS, D., and PAPADOPOULOS, G. A. (2006), *Identification of temporal patterns in the Sumatra rupture zone using Poisson Hidden Markov models*, Geophys. Res. Abstr. *8*, 01141.

ORFANOGIANNAKI, K., KARLIS, D., and PAPADOPOULOS, G. A. (2007), *Application of Poisson Hidden Markov models on foreshock sequences: The case of Samos, October 2005 and Zakynthos, April 2006, Earthquakes*, XIIth Int. Conf. Appl. Stochastic Models and Data Analysis, Chania, Crete, Greece, May 29–June 1, 2007, 142.

PAPADOPOULOS, G. A., FOUNDOULIS, D., and GRIVAS, K. (1993), *The 26 March 1993 earthquake in the area of Pyrgos, NW Peloponnesus, Greece*, 2nd Workshop on Statistical Models and Methods in Seismology—Applications on Prevention and Forecasting of Earthquakes, Eur. Seismol. Comm., Cephalonia, 2–5 June, 1993, 208–217.

PAPADOPOULOS, G. A, KARASTATHIS, V., KONSTANTINOU, K., GANAS, A., and ORFANOGIANNAKI, K. (2003), *Foreshock activity preceding the 2 December 2002 Killini shock and its relation to the crustal heterogeneity*, The 1st International Workshop on Earthquake Prediction, Eur. Seismol. Com., Athens, 6–7 November, 2003.

PATWARDHAN, A. S., KULKARNI, R. B., TOCHER, D. (1980), *A semi-Markov model for characterizing recurrence of great earthquakes*, Bull. Seism. Soc. Am. *70*, 323–347.

RAO, N. M. and KAILA, K. L. (1986), *Application of the negative binomial to earthquake occurrences in the Alpide-Himalayan belt*, Geophys. J. R. Astr. Soc. *85*, 283–290.

ROTONDI, R. and VARINI, E. (2006), *Bayesian analysis of marked stress release models for time-dependent hazard assessment in the western Gulf of Corinth*, Tectonophysics *423*, 107–113.

ROUMELIOTI, Z., BENETATOS, Ch., KIRATZI, A., STAVRAKAKIS, G., and MELIS, N. (2004), *Evidence for sinistral strike-slip faulting in western Peloponnese (Greece): A study of the 2 December 2002 (M 5.5) Vartholomio earthquake sequence*, Tectonophysics *387*, 65–79.

SCHWARZ, G. (1978), *Estimating the dimension of a model*, Ann. Stat. *6*, 461–464.

SUZUKI, S. and KIREMIDJIAN, A. S. (1991), *A random slip rate model for earthquake occurrences with Bayesian parameters*, Bull. Seism. Soc. Am. *81*, 781–795.

TSAPANOS, T. M. and PAPADOPOULOU, A. A. (1999), *A discrete Markov model for earthquake occurrences in Southern Alaska and Aleutian Islands*, J. Balk. Geophys. Soc. *2*, 75–83.

UTSU, T. (1961), *Statistical study on the occurrence of aftershocks*, Geophys. Mag. *30*, 521–605.

UTSU, T. (1969), *Aftershocks and earthquake statistics (I) —Some parameters which characterize an aftershock sequence and their interrelations*, J. Fac. Sci. Hokkaido Univ., Ser. VII *3*, 121–195.

VERE-JONES, D. (1966), *A Markov Model for aftershock occurrence*, Pure Appl. Geophys. *64*, 31–42.

(Received September 18, 2008, revised June 11, 2009, accepted June 17, 2009, Published online April 20, 2010)

Pure Appl. Geophys. 167 (2010), 933–958
© 2010 Birkhäuser / Springer Basel AG
DOI 10.1007/s00024-010-0089-x

Pure and Applied Geophysics

Distribution of Seismicity Before the Larger Earthquakes in Italy in the Time Interval 1994–2004

S. GENTILI[1]

Abstract—The Region–Time–Length (RTL) algorithm has been applied to different instrumental catalogues to detect seismic quiescence before medium-to-large earthquakes in Italy in the last two decades. RTL performances are sensitive to the choice of spatial and temporal parameters. The method for automatic parameters selection developed by Chen and Wu has been applied to twelve Italian earthquakes with magnitude greater than 5. The limits of the method in constructing maps of seismic quiescence before the earthquake are demonstrated, and a simple improvement is proposed. Then a new technique, namely RTL$_{surv}$, is proposed for routine surveys of the Italian seismicity. RTL$_{surv}$ has been applied to all the earthquakes with magnitude greater than 4 in the Italian area in the time interval 1994–2004; four different sub-areas have been identified, with different characteristics in the level of recorded seismicity. One subarea—Tyrrhenian Sea—was characterized by a too low level of recorded seismicity for the application of the method. In the other three subareas a seismic quiescence was detected before at least the 66% of the earthquakes with magnitude greater or equal to 4 and all the earthquakes with magnitude greater than 5.

Key words: Region–Time–Length, seismic quiescence, improved RTL, precursory seismic activity, Italy, earthquake forecasting.

1. Introduction

Temporal seismic observations have shown trends of seismic quiescences and foreshock activation preceding large events. Many studies focussed on the quiescence occurring during the phase of seismic energy accumulation before moderate and large earthquakes (SCHOLZ, 1988; WYSS and HABERMANN,

---
[1] Dipartimento Centro Ricerche Sismologiche, Istituto Nazionale di Oceanografia e di Geofisica Sperimentale, Via Treviso 55, 33100 Cussignacco, UD, Italy. E-mail: sgentili@inogs.it

1988, WIEMER and WYSS, 1994), or on the increase of seismicity (BUFE and VARNES, 1993; BREHM and BRAILE, 1998). The RTL analysis is a statistical method developed by SOBOLEV and TYUPKIN (1996, 1997) to detect seismic anomalies preceding isolated large earthquakes.

The RTL has been previously applied to large earthquakes in Kamchatka and Caucasus (SOBOLEV and TYUPKIN, 1997, 1999; HUANG, 2004), Greece (SOBOLEV *et al.*, 1997), Japan (HUANG and SOBOLEV, 2002; HUANG *et al.*, 2001; HUANG and NAGAO 2002, HUANG, 2004, 2006), Turkey (HUANG *et al.*, 2002), Taiwan (CHEN and WU, 2006) and China (JIANG *et al.*, 2004; RONG and LI, 2007). Using this technique, earthquakes in Italy have been studied by DI GIOVAMBATTISTA and TYUPKIN (1999, 2000, 2004) and by GENTILI and BRESSAN (2007).

In this paper, I present new results for the Italian area, analyzing the medium-to-large earthquakes of two catalogues of instrumental seismicity; the first one compiled at the national scale, and the second one referred to NE Italy (see Sect. 3). A preliminary analysis has been done on $M > 5$ earthquakes using the method for the automatic parameters calibration developed by CHEN and WU (2006). In this paper, I analyze the space of the parameters to show how their choice influences the spatial mapping of RTL before the earthquake. A simple improvement of the Chen and Wu method is then proposed.

The choice of the time window to perform RTL mapping is one of the most critical points of the method. It is generally done *a posteriori* (e.g., DI GIOVAMBATTISTA and TYUPKIN, 2000; HUANG and SOBOLEV, 2002), selecting a time window where RTL function for the examined earthquake assumes low values. In this paper, I propose a new method for routine surveys of the Italian seismicity, named RTL$_{surv}$; it takes into account the seismicity of the last years before a predefined time, without an *ad hoc* choice of the time window.

This survey has been applied to all the $M > 4$ earthquakes of the national and regional catalogues in the time period 1994–2004. The performance of the method in terms of quiescence detection and spatial location of the anomalies has been evaluated with respect to earthquake magnitudes.

2. The RTL Algorithm and its Improved Versions

The RTL algorithm was originally proposed by SOBOLEV *et al.*, (1997, 1996) and it has been described in detail elsewhere (see e.g., DI GIOVAMBATTISTA and TYUPKIN, 2004). Briefly, it represents the deviations from the background seismicity. Quiescence is outlined by a decrease of an RTL function, and activation of seismicity by an increase of RTL.

The RTL value at a given test site (x, y) at time t, is defined as the product of the epicentral function R, the temporal function T and the source-site function L, divided by their standard deviations (Eq. 1).

$$\text{RTL}(x,y,t) = \frac{R(x,y,t)}{\sigma_R} \cdot \frac{T(x,y,t)}{\sigma_T} \cdot \frac{L(x,y,t)}{\sigma_L}. \quad (1)$$

These functions are defined as:

$$R(x,y,t) = \sum_{i=1}^{n} \exp\left(-\frac{r_i}{r_0}\right) - R_s(x,y,t), \quad (2)$$

where r_i is the distance between the test site (x, y) and the ith earthquake, $2r_0$ is the search distance and the summation is performed on the n events considered in the time window $(t - 2t_0, t)$, having magnitude in the interval (M_{min}, M_{max}), with $M_{min} = M_c$, the level of the catalogue completeness.

In the temporal term

$$T(x,y,t) = \sum_{i=1}^{n} \exp\left(-\frac{t-t_i}{t_0}\right) - T_s(x,y,t), \quad (3)$$

t_i is the time of occurrence of the events preceding the time of the forecast. Finally, the source-site function is

$$L(x,y,t) = \begin{cases} \sum_{i=1}^{n}\left(\frac{l_i}{r_i}\right) - L_s(x,y,t) & \text{if } r_i > \varepsilon \\ \sum_{i=1}^{n}\left(\frac{l_i}{\varepsilon}\right) - L_s(x,y,t) & \text{if } r_i \leq \varepsilon \end{cases} \quad (4)$$

where ε is the accuracy of the epicenter location and l_i is the size of the source of the selected earthquakes; l_i is usually calculated from empirical relationships between earthquake magnitude and source size. R_s, T_s, and L_s are linear trend corrections.

The RTL function is calculated after declustering the earthquake data set.

Various improvements of the method or different interpretations of the results have been proposed. In JIANG *et al.*, (2004) and CHEN and WU (2006), both decreases and increases of RTL are considered earthquake precursors, while for most other authors (see e.g., HUANG and SOBOLEV, 2002) only a decrease (seismic quiescence) is considered a reliable precursor.

One of the open problems in the application of the method is the choice of the free parameters r_0, t_0 and M_{max}. HUANG (2006) on the large Tottori earthquake ($M = 7.3$) and GENTILI and BRESSAN (2007) on a set of moderate earthquakes in NE Italy and Western Slovenia, show that the results are stable for a large range of chosen parameters. CHEN and WU (2006) propose a method for the choice of the most stable value of r_0 and t_0, described in detail in the following. In some recent papers no M_{max} is imposed, to avoid the introduction of potential artificial changes by the two cutoff magnitudes M_{min} and M_{max} (HUANG and SOBOLEV, 2002; HUANG, 2004, 2006; CHEN and WU, 2006), or because the M_{max} is demonstrated to not influence the results (GENTILI and BRESSAN, 2007). In previous papers $M_{max} \approx M_{seq} - 2$, where M_{seq} is the magnitude of the tested main shock (SOBOLEV and TYUPKIN, 1997), or it is set to a given value, like DI GIOVAMBATTISTA and TYUPKIN (1999, 2000, 2004), who set $M_{max} = 3.8$ independently of the magnitude of the tested main shock. In this paper, I exclude M_{max} from the RTL calculation.

Another open problem is how to map RTL in space, in order to detect the seismic quiescence zone. The region in the neighborhood of the epicenter—its dimension changes from author to author, ranging from 150 × 250 km in DI GIOVAMBATTISTA and

TYUPKIN (2004) to 1,900 × 1,800 km in HUANG et al., (2001)—is sampled by a grid, and, for every node of the grid, the RTL is computed for the same time window containing the minimum value of the RTL at the epicenter of the analyzed earthquake. The value mapped is in some cases the minimum of the RTL in the time window (RTL$_{min}$—see e.g., SOBOLEV 2000, DI GIOVAMBATTISTA and TYUPKIN, 2000, 2004). In most recent papers, the Q parameter is mapped, which is defined as the mean of the RTL in the considered time window (HUANG et al., 2002; HUANG, 2004, 2006). In both cases, this analysis must be done *a posteriori*, when the position of the epicenter of the earthquake is known. This is a drawback, since it lowers the interest of RTL as an earthquake predictor. Section 4.3 describes the method proposed in this paper to solve this problem.

3. Data-Set pre-Processing

Two earthquake catalogues have been used in this paper, pertaining to the regional and national scale. The first one is the regional catalogue released by the Seismic Network of Northeastern Italy, managed by the Seismological Research Center (CRS) of OGS—National Institute of Oceanography and Experimental Geophysics. The catalogue (hereinafter defined as NEI catalogue) reports 16,200 events with $M > 0$ from May 1977 (Fig. 1a). A full description of the characteristics of the monitoring is reported in PRIOLO et al. (2005), and the list of events located using HYPO71 (LEE and LAHR, 1975) is public and available on the web site http://www.crs.inogs.it/bollettino/RSFVG/RSFVG.en.html. The magnitude adopted in this catalogue is the duration magnitude obtained using the REBEZ and RENNER (1991) formula. The mean location uncertainty is 1.5 km.

The second catalogue covers the whole country, and it is the one proposed by LOLLI and GASPERINI (2006) (L&G catalogue). It is a compilation obtained by integrating with uniform criteria different catalogues, namely the PFG catalogue (POSTPISCHL, 1985) that reports macroseismic and instrumentally derived records from 1000 to 1980, the CSTI catalogue (CSTI Working Group, 2001, 2004; instrumental seismicity from 1981 to 1996), the CSI catalogue (CASTELLO

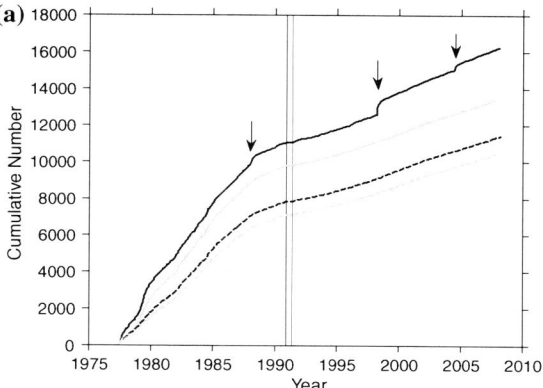

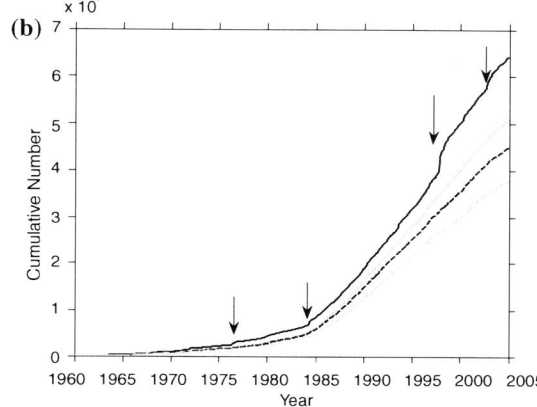

Figure 1
Cumulative number of earthquakes with $M_D > 0$ in **a** the NEI catalogue and **b** the L&G catalogue. *Vertical lines* the time window of data recording interruption for the NEI catalogue; *black continuous line* catalogue before the declustering; *black arrows* increases of seismicity; *grey continuous line* catalogue after the modified Knopoff algorithm declustering; *black dashed line* catalogue after the Reasenberg declustering using default parameters; *grey dashed line* catalogue after the Reasenberg declustering using Lolli and Gasperini parameters

et al., 2005; earthquakes in 1981–2002) and two years (2003–2004) of instrumental bulletins of the Istituto Nazionale di Geofisica e Vulcanologia (INGV) available at site http://legacy.ingv.it/~roma/reti/rms/bollettino. The resulting catalogue (Fig. 1b) reports 64,200 events with $M > 0$ from January 1960 to December 2004 and can be downloaded by anonymous ftp at the address: ibogfs.df.unibo.it and it is in the directory LOLLI/AFT2005 (LOLLI and GASPERINI, 2006). The magnitude listed is local magnitude. The mean location uncertainty is 7.5 km (LOLLI and GASPERINI 2003).

3.1. Declustering

RTL applies to declustered catalogues, where both foreshocks and aftershocks are removed. In order to decluster the L&G and NEI catalogues, I have applied and compared the performances of two alternative algorithms: that by KNOPOFF (2000) and that by REASENBERG (1985), the latter implemented in Zmap software (WIEMER, 2001). The whole Zmap software package can be downloaded from the Swiss Federal Institute of Technology ZÜrich (ETH Zurich) web pages (http://www.earthquake.ethz.ch/software/).

Knopoff algorithm, for $M_L < 6.4$, is a windowing algorithm for cluster identification. It means that for every event in the catalogue a space time-window is defined and any earthquake within the window is deemed as cluster event. The extension in time and in space of the window depends on the magnitude of the earthquake. In applying the algorithm to the L&G catalogue, I used the original table proposed by KNOPOFF et al., (2000), extended to smaller magnitudes as in PACE et al., (2006) in Italian catalogues declustering:

$$\begin{aligned}\text{Log}(T) &= 0.725 M_L - 2.007, \\ \text{Log}(D) &= 0.347 M_L - 0.567,\end{aligned} \quad (5)$$

where T and D are the time and space windows expressed in days and kilometers, respectively, and M_L is the local magnitude. For NEI catalogue declustering, since the listed magnitude is the duration magnitude, I adopted the relationships obtained by GENTILI and BRESSAN (2008) for the area covered by the catalogue:

$$\begin{aligned}\text{Log}(T) &= 0.33 M_D + 0.42, \\ \text{Log}(D) &= 0.41 M_D - 1,\end{aligned} \quad (6)$$

where M_D is the duration magnitude. For simplicity, since the algorithm is the same for NEI and L&G catalogues, even if the relations (5) and (6) are different, this algorithm will be referred in the following simply as modified Knopoff algorithm.

The Reasenberg algorithm defines a seismic sequence as a chain of events linked to each other by spatial and temporal windows. The window's extension in time and space depends on a set of parameters and on the seismic moment of the largest and the most recent event. In particular, the spatial extent of the main shock is taken coincident with its source dimension; that is (KANAMORI and ANDERSON 1975):

$$r = \left(\frac{7 M_0}{16 \Delta \sigma}\right)^{1/3}, \quad (7)$$

where $\Delta \sigma$ is the stress drop, assumed 3 MPa for all the earthquakes, and M_0 is the seismic moment. The spatial extent of the most recent event is its source dimension scaled by a parameter Q. A location uncertainty is considered in the calculation. The temporal window is defined as a function of the time t after the beginning of the sequence:

$$\tau = \frac{-\ln(1-P)t}{10^K}, \quad (8)$$

where P is related to the probability of observing one or more events and K is related to the main shock magnitude. An upper and lower limit for τ ($\tau_{\min}$ and $\tau_{\max}$, respectively) are input parameters. Since the characteristics of the catalogues and of the seismicity in California, for which the algorithm was developed, are different, I tested the declustering using both the default parameters (Q = 10, $\tau_{\min} = 1$, $\tau_{\max} = 10$, P = 0.95) and the ones proposed for Italy by LOLLI and GASPERINI (2003) (Q = 20, $\tau_{\min} = 2$, $\tau_{\max} = 10$, P = 0.99). LOLLI and GASPERINI (2003) double the spatial extent of the window.

I used the following conversion rule between local magnitude and seismic moment (LOLLI and GASPERINI 2003; GASPERINI and FERRARI, 2000):

$$\text{Log}_{10}(M_0) = 1.22 M_L + 17.7. \quad (9)$$

In order to find a relation between duration magnitude and seismic moment valid for Northeastern Italy, I used the data of moment magnitude listed in FRANCESCHINA et al., (2006) and BRESSAN et al., (2007) obtaining the following regression (see Fig. 2):

$$\begin{aligned}\text{Log}_{10}(M_0) &= 1.3 M_D + 9.3, \\ \sigma(\text{Log}_{10}(M_0)) &= 0.3,\end{aligned} \quad (10)$$

where $\sigma(.)$ is the standard deviation.

In Fig. 1a, b the cumulative number of earthquakes with magnitude >0 for the two catalogues are compared with their declustered version after applying (1) the declustering obtained by applying KNOPOFF

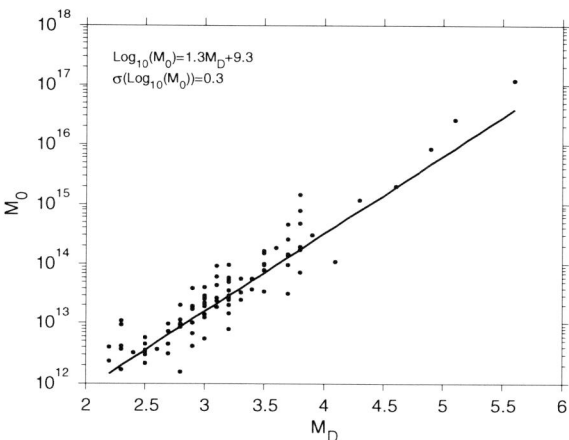

Figure 2
Seismic moment M_0 as a function of duration magnitude M_D in NE Italy area. Data from FRANCESCHINA *et al.*, (2006) and BRESSAN *et al.*, (2007)

(2000) algorithm, (2) the Reasenberg algorithm with default parameters, and (3) the Reasenberg algorithm with Lolli and Gasperini parameters.

The NEI catalogue (Fig. 1a) presents a clear increase in the number of earthquakes corresponding to the seismic sequences of the April 12th, 1998 $M_D = 5.6$ and the July 12th, 2004 $M_D = 5.1$ Kobarid earthquakes. A smaller increase of the cumulative number of earthquakes can be seen also due to the seismic sequence of the February 1st, 1988 $M_D = 4.1$ Mena earthquake (for more details on the sequences in this area, see GENTILI and BRESSAN, 2008). All the declustering procedures were able to eliminate these increases. In addition, in both the original and declustered catalogue it is possible to see a decrease in slope of the curve in 1988. This is due to the change of the data acquisition method. The seismic data acquisition before January 1, 1988 was analog, all the data were collected and there was no triggering algorithm. After this date, the seismicity was acquired with digital signal processing; the signal was acquired when the ratio (short term average)/(long term average) passed a given threshold. Therefore, smaller magnitude earthquakes were neglected (MARCELLINI and MILANI, 2003). The seismicity recording was interrupted on December 3, 1990 and started again on May 21, 1991 (GENTILI and BRESSAN, 2007). The interruption is marked by two vertical lines. The main difference in the results of the declustering procedures is in the number of events in the declustered catalogues. While the modified Knopoff algorithm removes about 2,800 events out of the 16,200 with $M_D > 0$ from the catalogue, Reasenberg algorithm removes 4,800 and 5,800 events using default and Lolli and Gasperini parameters, respectively. In order to understand which is the best declustering procedure, I verified their performances on two seismic sequences and a swarm, whose events are listed in BRESSAN *et al.*, (2007). I found that the best procedure is the KNOPOFF (2000) algorithm. The mainshocks of the sequences are the April 12th, 1998 $M_D = 5.6$, Kobarid and the February 14th, 2002 $M_D = 4.9$ Sernio Mountain earthquakes. The swarm was composed of three sequences whose main shocks were the January 27th, 1996 $M_D = 3.5$, the February 27th, 1996 $M_D = 3.8$ and the April 13th, 1996 $M_D = 4.3$ Claut earthquakes. The results are the following:

- KNOPOFF (2000) algorithm correctly removes from the catalogue all the dependent earthquakes of the seismic sequences; all the main shocks are correctly recognized.
- Reasenberg algorithm using default parameters fails in removing two aftershocks in Sernio sequence and one in the last sequence of the Claut swarm. In addition, it wrongly removes the April 13th, 1996 $M_D = 4.3$ Claut main shock from the catalogue.
- Reasenberg algorithm using Lolli and Gasperini parameters fails in removing two aftershocks in the Sernio sequence. In addition, it wrongly removes the April 13th, 1996 $M_D = 4.3$ Claut main shock from the catalogue.

The same analysis was performed on the L&G catalogue (see Fig. 1b). The L&G catalogue presents a remarkable increase of the number of earthquakes due to the cluster of earthquakes in Umbria-Marche which began on September 26th, 1997. Minor increases can be seen corresponding to the seismic sequence of the May 6th, 1976 $M_L = 6.1$ Friuli earthquake, of the April 29th, 1984 $M_L = 5.2$ Gubbio/Valfabbrica earthquake and of the September 6th, and October 31st, 2002 $M_L = 5.6$ Palermo and $M_L = 5.6$ Molise earthquakes. All the declustering procedures were able to eliminate these increases. A change of slope of the cumulative number of earthquakes can be seen around the year 1985 in both the original and the declustered catalogues. This is due to a change in the network acquisition system in 1984 (LOLLI and GASPERINI, 2003) and an increase of the number of seismic stations in 1986 that improved the detection capability of some Italian regions (e.g., in Sicily: see DI GIOVAMBATTISTA and TYUPKIN, 2004). In this case also, the main difference among the declustered catalogues is the number of earthquakes. While the modified Knopoff algorithm removes about 13,400 events out of the 64,200 with $M_L > 0$ from the catalogue, Reasenberg algorithm removes 19,200 and 26,300 events using default and Lolli and Gasperini parameters, respectively. In order to make a comparison between declustering the methods, I verified their performances on Umbria-Marche seismicity in 1997 using the data from SELVAGGI et al., (2002) in order to have a more accurate location of the earthquakes ipocenters; the data set was composed of 646 earthquakes. The modified Knopoff algorithm detected fifteen independent events, recognizing all the distinct ruptures of fault segments, while Reasenberg algorithm recognized 8 and 4 independent events using default and Lolli and Gasperini parameters, respectively. These results and the final choice of the modified Knopoff algorithm are in agreement with the ones obtained by PACE et al., (2006) on the same data.

3.2. Completeness

The magnitude of completeness has been evaluated for the two catalogues by using the Entire Magnitude Range method (EMR) (WOESSNER and WIEMER, 2005) and a bootstrapping method for uncertainties evaluation. The software adopted was Zmap. The analysis was performed on declustered catalogues, since during large clusters or aftershock sequences a different policy of recording with respect to standard background seismicity is generally applied (see also SCHORLEMMER and WOESSNER, 2008). The magnitude of completeness for the NEI catalogue is $M_D = 1.9 \pm 0.4$ using modified KNOPOFF (2000) declustering method, 2.3 ± 0.07 and 2.3 ± 0.05 using Reasenberg method with standard and Lolli and Gasperini parameters, respectively. The magnitude of completeness for L&G catalogue is $M_L = 2.1 \pm 0.09$ using modified KNOPOFF 2000 declustering method, 2.1 ± 0.1 and 2.1 ± 0.09 using Reasenberg method with standard and Lolli and Gasperini parameters, respectively.

However, the study of the whole catalogue hides the changes in time and space of M_c, depending on the seismic network available. Figure 3 shows the changes in time of M_c for both catalogues, calculated using Zmap software. It is possible to notice an increase of M_c for NEI catalogue in the time interval 1985–1990 and a successive stabilization on larger

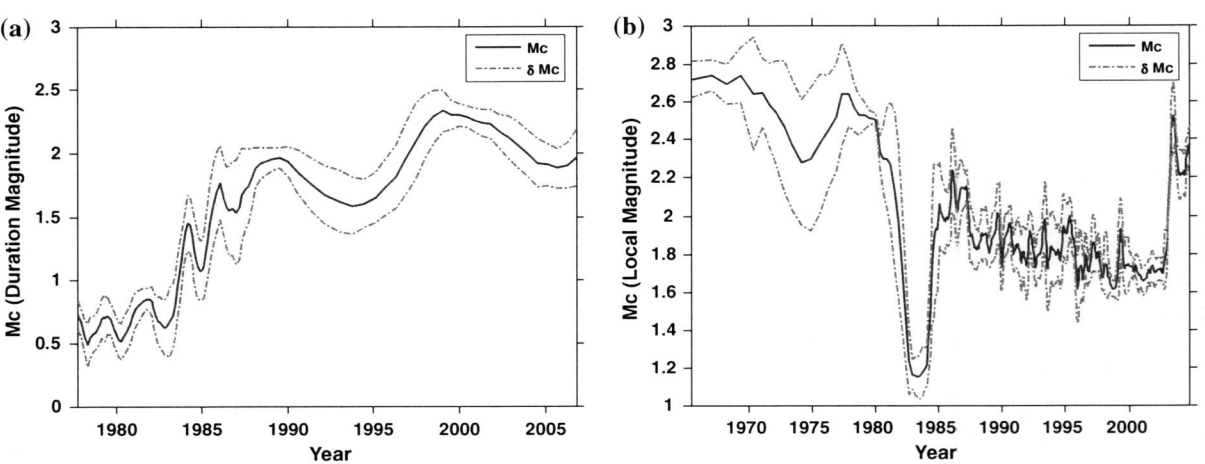

Figure 3
Changes with time of the magnitude of completeness for **a** NEI catalogue (duration magnitude) **b** Lolli and Gasperini catalogue (local magnitude)

values, due to the change in network acquisition system (see also Sect. 3.1). The L&G catalogue is characterized by a general decrease of the completeness magnitude from 1986 to the end of 2002. An increase of M_c is clearly detectable in 2003, when the instrumental bulletin data were merged with pre-existing catalogues.

In order to manage homogeneous catalogues, both in terms of network structure and of acquisition system, without interruptions in data recording, in a mostly overlapping time period, the analysis is performed for NEI catalogue from May 21st 1991 until the end of 2004 and for L&G catalogue from January 1st 1986 until the end of 2004.

3.3. Selection of Target Events

The RTL analysis needs a seismicity recording time interval, before the target event, in order to become stable. The duration of this time interval depends on the earthquake rate in the region under study and on the completeness magnitude of the catalogue. For studies in Italy, this time interval ranges between 6 years for the 1996 $M_D = 4.3$ Claut earthquake (GENTILI and BRESSAN, 2007) to 12 years for the 2002 $M_L = 5.6$ earthquake (DI GIOVAMBATTISTA and TYUPKIN, 2004). Table 1 shows all the earthquakes in Italy and in the surrounding regions with $M > 5$, from January 1st, 1994 to December 31st, 2004, listed in NEI and L&G catalogue; the February 14th, 2002 $M_D = 4.9$ earthquake is analyzed as a $M_L > 5$ earthquake, even if L&G catalogue lists it as $M_L = 4.9$, because its local magnitude is considered in this paper to be $M_L = 5.1$ in accord with FRANCESCHINA et al., (2006). All the earthquakes analyzed are preceded by a time interval of at least seven years of available data in the catalogue adopted for RTL analysis (see later).

In order to apply the RTL method for retrospective earthquake forecasting with reliable results, the following selection rules have been applied:

Table 1

All the earthquakes with $M_L > 5$ listed in L&G and NEI catalogues

Date (UTC)	Time (UTC)	Zone	Lat. (deg)	Lon. (deg)	Depth (km)	M_L (L&G)	M_D (NEI)
1994/01/05	13:24:08	South Tyrrhenian Sea	38.97	15.41	307	5.7	–
1995/09/30	**10:14:34**	**Gargano (Apulia)**	**41.79**	**15.97**	**28**	**5.4**	–
1996/10/15	**09:56:00**	**Reggio Emilia**	**44.80**	**10.68**	**1**	**5.5**	–
1997/09/26	**00:33:13**	**Umbria Marche**	**43.02**	**12.89**	**4**	**5.6**	–
1997/09/26	09:40:27	Umbria Marche	43.01	12.85	10	5.8	–
1997/10/06	23:24:53	Umbria Marche	43.03	12.85	4	5.4	–
1997/10/12	11:08:37	Umbria Marche	42.91	12.92	0	5.1	–
1997/10/14	15:23:11	Umbria Marche	42.90	12.90	7	5.5	–
1998/03/26	16:26:17	Umbria Marche	43.15	12.81	45	5.4	–
1998/04/03	07:26:37	Umbria Marche	43.19	12.76	2	5.3	–
1998/04/12	**10:55:33**	**Kobarid**	**46.32**	**13.68**	**15**	**5.6**	**5.6**
1998/05/18	17:19:11	South Tyrrhenian Sea	39.06	15.02	219	5.4	–
1998/09/09	**11:28:00**	**Calabrian-Lucanian Apennines**	**40.06**	**15.95**	**29**	**5.6**	–
2001/07/17	**15:06:15**	**Merano**	**46.70**	**11.07**	**0**	**5.3**	**5.2**
2002/02/14	**03:18:03**	**Sernio Mountain**	**46.43**	**13.10**	**11**	**4.9**	**4.9**
2002/09/06	**01:21:29**	**Palermo**	**38.38**	**13.65**	**27**	**5.6**	–
2002/10/31	**10:32:59**	**Molise**	**41.72**	**14.89**	**25**	**5.4**	–
2002/11/01	15:09:02	Molise	41.74	14.84	21	5.3	–
2003/03/29	**17:42:16**	**Adriatic Sea**	**43.11**	**15.46**	**10**	**5.4**	–
2004/05/05	13:39:43	Aeolian Islands	38.51	14.88	240	5.3	–
2004/07/12	**13:04:07**	**Kobarid**	**46.30**	**13.63**	**5**	**5.2**	**5.1**
2004/11/24	**22:59:39**	**Garda Lake**	**45.56**	**10.57**	**5**	**5.2**	**4.9**

The earthquakes outlined in boldface are the ones analyzed in this paper

Lat. and *Lon.* north latitude and east longitude of the main shock (degrees)

M_L (L&G) local magnitude as listed in L&G catalogue

M_D (NEI) duration magnitude as listed in NEI catalogue

1. In cases of swarms or seismic sequences only the first shock with magnitude greater than 5 is considered.
2. Deeper earthquakes (deeper than 200 km earthquakes) are not analyzed.

The first choice is made because RTL has been developed for detecting seismic anomalies preceding isolated large earthquakes. A quiescence preceding a large earthquake precedes also its seismic sequence or the other earthquakes composing a seismic swarm. RTL does not allow a discrimination between one or more shocks following a seismic quiescence. According to point 1, six Umbria-Marche earthquakes and the November 1st, 2002 Molise earthquake (in italic in Table 1) are rejected from the target events list. The second choice is done to avoid confusion between shallower seismicity quiescence and deeper events; this choice causes the rejection of three other earthquakes from the target events list: the $M_L = 5.7$ January 5th, 1994 and the $M_L = 5.4$ May 18th, 1998 South Tyrrhenian Sea earthquakes and the $M_L = 5.3$ May 5th, 2004 Aeolian Islands earthquake. The twelve earthquakes analyzed are listed in boldface in Table 1. In order to avoid artifacts due to the mixing of the two different kinds of magnitude listed in NEI and L&G catalogues, which can cause fake quiescence patterns (see also HABERMANN, 1987 and WYSS, 1991), the analysis for each earthquake is done using only the catalogue that better covers the epicentral area. In particular, Sernio Mountain and the two Kobarid earthquakes are analyzed by using NEI catalogue, while for the other nine earthquakes L&G catalogue is adopted.

4. RTL Analyses

In order to analyze the parameter dependence of RTL, the algorithm has been retrospectively tested on $M > 5$ earthquakes, using both the CHEN and WU (2006) method and standard values. The comparison has been also performed with results in the literature, when available. The parameters' choice affects both the duration of the detected seismic quiescence and the spatial extension of the quiescence area. A simple improvement of the CHEN and WU (2006) method, that reduces the spatial extension of quiescence area, is proposed. In addition, a survey method for the entire Italian area, based on RTL, is presented.

4.1. Retrospective Forecasting of $M > 5$ Earthquakes

The RTL has been calculated at the epicenter of the twelve earthquakes listed in bold in Table 1. When L&G was the reference catalogue, the relation adopted by DI GIOVAMBATTISTA and TYUPKIN (1999) (PAPADOPOULUS and VOIDOMATIS, 1987) for l_i evaluation has been taken:

$$\text{Log}_{10}(l) = 0.44 M_L - 1.289, \quad (11)$$

where l is the size in km of the seismic source and M_L is the local magnitude. The accuracy of the epicenter location was given as $\varepsilon = 7.5$ km (LOLLI and GASPERINI, 2003). The analysis started on January 1st, 1986 and lasted until midnight of the day before the considered main shock.

If the NEI catalogue was used, an additional conversion rule was invoked before applying Eq. (11):

$$M_L = 1.2 M_D - 0.73, \quad (12)$$

where M_L is the local magnitude and M_D is the duration magnitude (GENTILI and BRESSAN 2008). The accuracy of the epicenter location adopted was the mean of value of horizontal errors given by the location procedure (HYPO71) given as $\varepsilon = 1.5$ km. The time interval considered was from May 21st, 1991 to midnight of the day before the considered main shock.

In order to avoid artifacts due to heterogeneities in the completeness threshold in the catalogues in different regions of Italy, M_c has been evaluated for each earthquake, by the EMR method implemented in Zmap software, using a search radius of 100 km from the epicenter, by analyzing also its variation in time, from the "starting date of the catalogue" (January 1st, 1986 for the L&G catalogue, May 21st, 1991 for NEI) until the date of the studied events. The highest value reached by M_c has been set as minimum magnitude for computation. The M_c adopted for each earthquake is listed in Table 2.

Table 2

Analyzed earthquakes and adopted completeness magnitude

Date	Zone	Catalogue	Magnitude	M_c
1995/09/30	Gargano (Apulia)	L&G	M_L	2.5
1996/10/15	Reggio Emilia	L&G	M_L	2.0
1997/09/26	Umbria Marche	L&G	M_L	2.0
1998/04/12	Kobarid	NEI	M_D	2.2
1998/09/09	Calabrian-Lucanian Apennines	L&G	M_L	2.2
2001/07/17	Merano	L&G	M_L	2.1
2002/02/14	Sernio Mountain	NEI	M_D	2.4
2002/09/06	Palermo	L&G	M_L	2.3
2002/10/31	Molise	L&G	M_L	2.3
2003/03/29	Adriatic Sea	L&G	M_L	2.7
2004/07/12	Kobarid	NEI	M_D	2.3
2004/11/24	Garda Lake	L&G	M_L	2.2

Catalogue catalogue adopted for the analysis

Magnitude type of magnitude listed

M_c completeness magnitude inside a circle of radius 100 km centered at the epicenter in the magnitude units of the adopted catalogue

Table 3

The t_0 and r_0 values selected by the CHEN and WU (2006) method and the corresponding peak values $p(t_0)$ and $p(r_0)$

Zone	t_0 [years]	$p(t_0)$	r_0 [km]	$p(r_0)$
Gargano (Apulia)	1.1	47	15	76
Reggio Emilia	0.4	92	55	43
Umbria Marche	0.6	93	35	100
Kobarid 1998	0.6	100	60	98
Calabrian-Lucanian Apennines	0.9	55	45	93
Merano	1.1	53	65	92
Sernio Mountain	0.6	100	25	100
Palermo	1.4	45	30	99
Molise	1.4	77	15	86
Adriatic Sea	1.0	72	25	52
Kobarid 2004	0.4	100	40	92
Garda Lake	1.7	22	55	67

The peak values $p(t_0)$ and $p(r_0)$ are expressed as %

The t_0 and r_0 parameters have been estimated using the search method developed by CHEN and WU (2006):

- The t_0 parameter was varied in the range (t_{0min}, t_{0max}) with step Δt_0.
- For each value $\hat{t}_0$ of t_0, a set of RTL($\hat{t}_0$, r_0) was calculated, varying the value of r_0 in the range (r_{0min}, r_{0max}) with step Δr_0.
- The RTLs were considered two by two and their correlation coefficients at a significance level of 0.05 (BENDAT and PIERSOL, 2000) were evaluated.
- The percentage $p(\hat{t}_0)$ of correlation coefficients over a given threshold θ was evaluated.
- The value of t_0 corresponding to the peak value of the distribution of $p(t_0)$ was selected as the best one.
- An analogous approach has been applied to r_0 parameter.

The method finds the values of the parameters for which the RTL is more stable, eliminating or reducing anomalies correlated with the choice of parameters (CHEN and WU, 2006).

The tested time t_0 has been varied in this paper from 0.2 to 2 years with steps of 0.1 year; the tested radii from 15 to 80 km with steps of 5 km. Accordingly with CHEN and WU (2006), the threshold θ has been set equal to 0.8 for t_0 estimation and equal to 0.5 for r_0 estimation. Different thresholds have been used because changes in t_0 influence the RTL more than changes in r_0 (see also HUANG, 2004, 2006 and GENTILI and BRESSAN, 2007). The results are summarized in Table 3. The values of t_0 range from 0.4 to 1.4 years, while r_0 ranges from 15 to 65 km. For most sequences the estimated coefficient t_0 has more than the 50% of correlation coefficients greater than 0.8 and the estimated r_0 has more than 65% of the correlation coefficients greater than 0.5.

In Fig. 4 the obtained RTL is shown by a solid thick line. The RTL obtained with fixed parameters ($r_0 = 30$ km, $t_0 = 0.5$ years, used by DI GIOVAMBATTISTA and TYUPKIN (2004) for Palermo earthquake) given in the literature is plotted too (thin grey line in Fig. 4). It is possible to see that the results in Fig. 4 for different choices of the parameters are generally qualitatively similar, even if the duration of quiescence and the minimum value reached by RTL may change. Three other tests have been done in order to verify the stability of the method.

The first test is done by changing the magnitude threshold used for calculation; the test is repeated for both magnitudes greater than the completeness magnitude and for smaller magnitudes, under the hypothesis that the completeness magnitude is overestimated. In Table 4 the maximum and the minimum magnitude threshold for which the quiescence is detectable are listed, together with the completeness

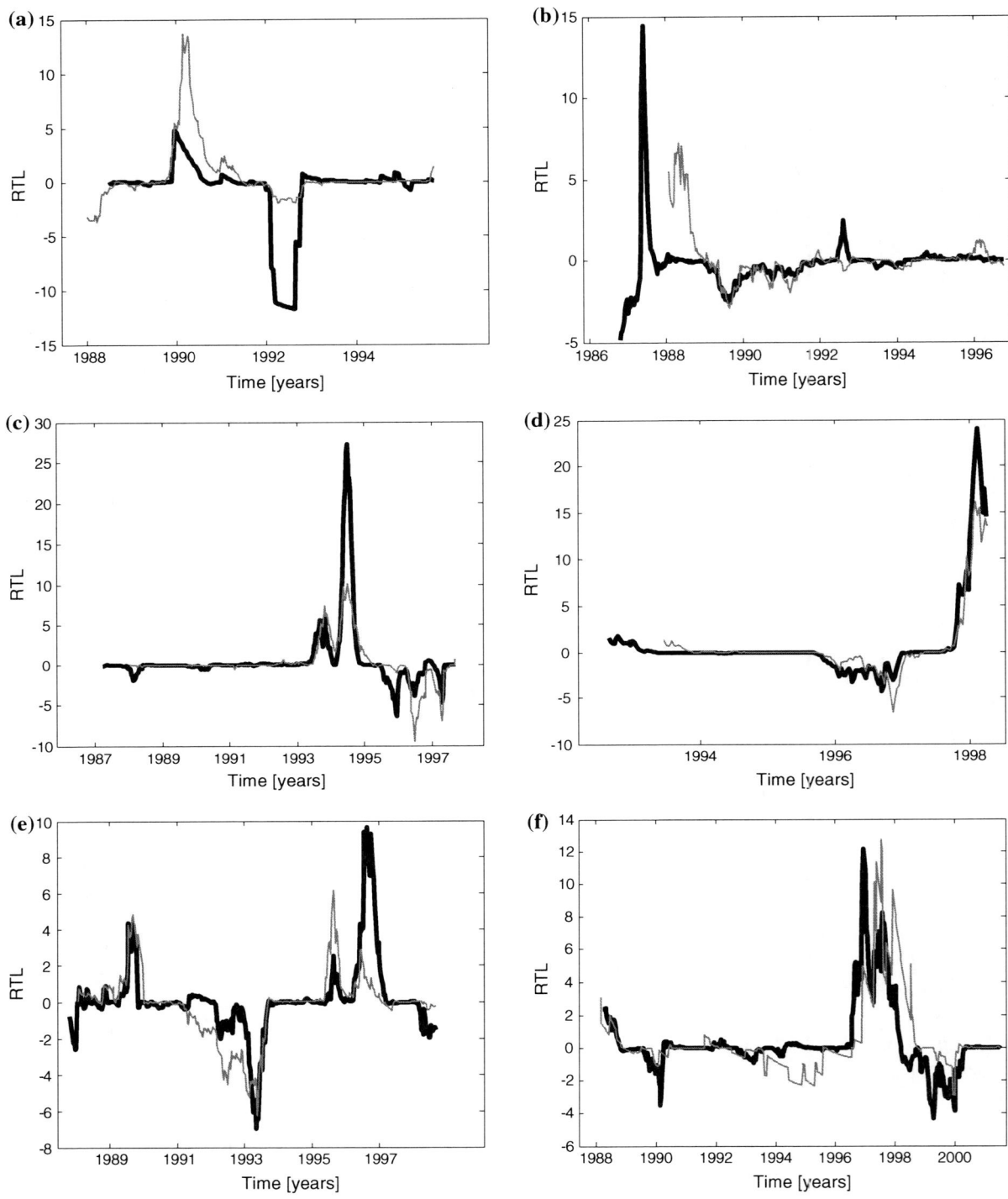

Figure 4
RTL calculated at the epicenters of analyzed earthquakes using the parameters listed in Table 3 (*thick line*) and standard parameters $r_0 = 30$ km $t_0 = 1$ year (*thin grey line*) until midnight before the main shock. Adopted declustering method: modified Knopoff algorithm. Epicenter location: **a** Gargano (Apulia), **b** Reggio Emilia, **c** Umbria Marche, **d** Kobarid (1998), **e** Calabrian-Lucanian Apennines, **f** Merano, **g** Sernio Mountain, **h** Palermo, **i** Molise, **l** Adriatic Sea, **m** Kobarid (2004), **n** Garda Lake

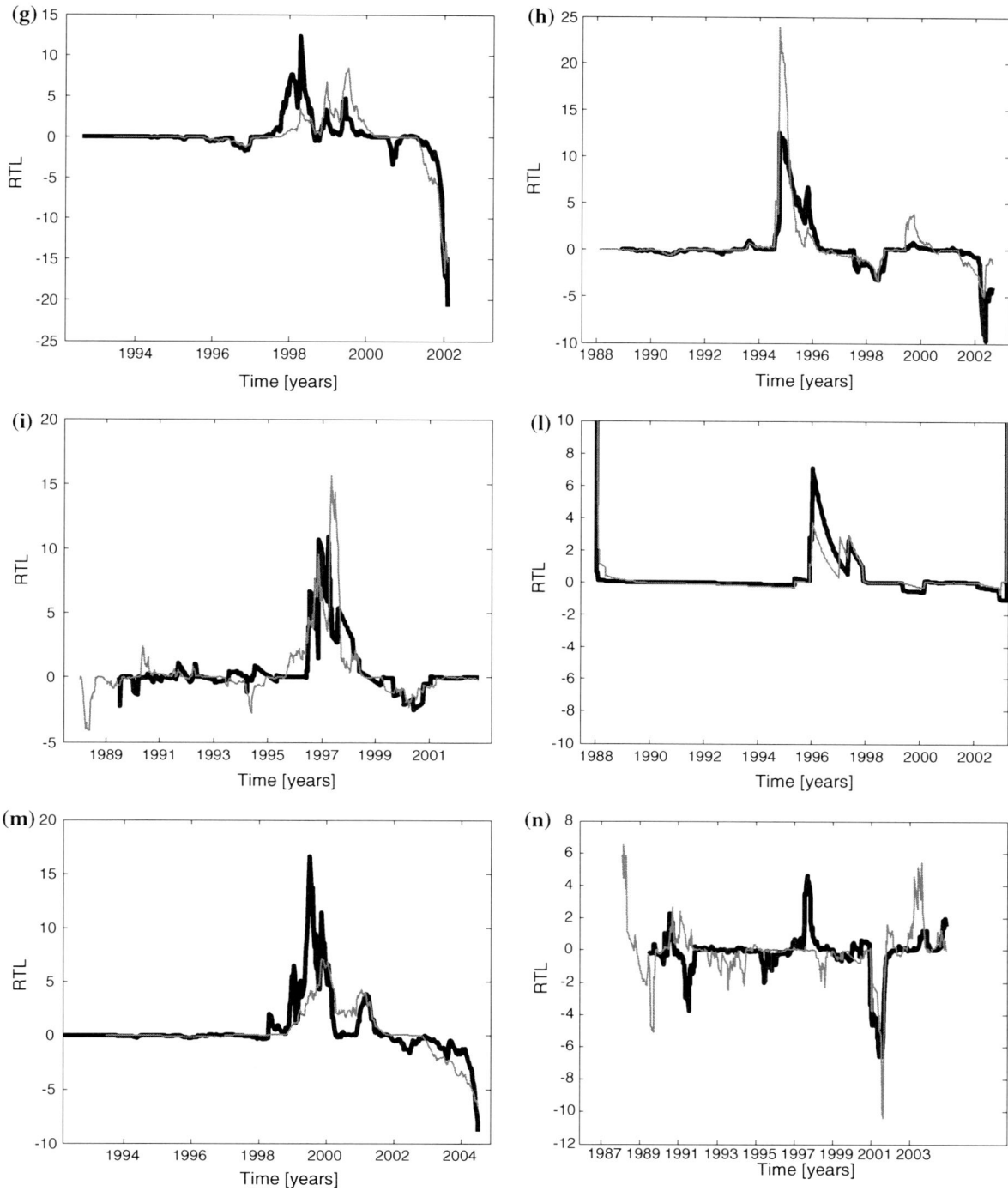

Figure 4
continued

Table 4

Minimum (M_{min}) and maximum (M_{max}) magnitude threshold used for calculation for which the quiescence before the analyzed earthquakes is detected

Zone	M_c	M_{min}	M_{max}
Gargano (Apulia)	2.5	0	2.8
Reggio Emilia	2.0	–	–
Umbria Marche	2.0	0	3.5
Kobarid	2.2	0	3.3
Calabrian-Lucanian Apennines	2.2	0	3.5
Merano	2.1	1.5	3.1
Sernio Mountain	2.4	0	3.1
Palermo	2.3	0	3.1
Molise	2.3	0	2.8
Adriatic Sea	2.7	0	3.1
Kobarid	2.3	0	3.2
Garda Lake	2.2	0	3.2

Adopted magnitude threshold (completeness magnitude M_c) is shown for comparison

magnitude. Tests on the Reggio Emilia earthquake are not done, because no seismic quiescence is detected. For most earthquakes, the minimum magnitude is 0 (with no magnitude threshold). The only case in which it is larger than 0 is Merano earthquake; however, also in that case, the minimum magnitude is well below the completeness magnitude. The maximum magnitude coincides in all cases with the maximum magnitude for which the number of earthquakes used to calculate RTL is not too small to supply reliable results. I have considered the RTL to be reliable if at least 50 earthquakes are used for calculation. The previous test outlines a good stability of the RTL also for different choices of completeness magnitude.

The second test is done to verify the performances of RTL using the L&G catalogue in the NE area. The three $M > 5$ earthquakes in the area, analyzed by NEI catalogue, present a well-defined seismic quiescence some years before (KOBARID, 1998, Fig. 4d) or immediately before the main shock (Sernio Mountain and Kobarid 2004: Fig. 4g and m) that is stable under a wide interval of magnitudes (see Table 4) but also t_0 and r_0 values (see Table 3 and Sect. 4.2). The same earthquakes are analyzed by applying Chen and Wu procedure for the choice of r_0 and t_0 parameters but using the L&G catalogue. The results are shown in Fig. 5. Comparing Fig. 4d, g and m with the corresponding Fig. 5a, b, c it is possible to see that the RTL applied to the L&G catalogue is not able to detect seismic quiescence in the Sernio Mountain case (Fig. 5b); it detects a seismic sequence in the Kobarid 1998 case but distant from the time of the main shock (Fig. 5a); only in the Kobarid 2004 case (Fig. 5c) are the results qualitatively similar. This is probably due to inaccuracy of earthquake location and magnitude determination of national catalogues in NE area, which is not well covered by the national network (a map of the national seismic network can be found in the Centro Nazionale Terremoti of the Istituto Nazionale di Geofisica e Vulcanologia—INGV web pages http://www.cnt.ingv.it).

The last test on RTL stability is done changing the declustering method. In particular, the declustering is performed by using the Reasenberg declustering method with the parameters of Lolli and Gasperini and the same r_0 and t_0 parameters of Fig. 4. The corresponding RTL are shown in Fig. 6. Using Reasenberg declustering method, the RTL are qualitatively similar to the ones obtained by modified Knopoff method for most of the analyzed earthquakes, with the only exception the 1998 Calabrian-Lucanian earthquake (Figs. 4e and 6e), for which the quiescence preceding the earthquake is less relevant than using Knopoff method—RTL values in the range $(-0.7,0)$—and 2003 Adriatic sea earthquake (Figs. 4l and 6l), where the RTL could be evaluated only for $r_0 \geq 30$ due to the small number of earthquakes after the declustering.

Table 5 summarizes the results, in terms of quiescence and successive activation stage duration (when an activation occurred), and of time shift before the analyzed shock, for the RTL calculated using the Knopoff declustering method and the parameters listed in Table 3. If more than one quiescence or activation phase occurred, the last one is considered. Only three of the earthquakes considered present an activation phase after the quiescence. However, 11 of the 12 analyzed earthquakes present a seismic quiescence. The duration of the quiescence ranges from 0.6 years to 3.0 years. The interval between the start of the quiescence to the earthquake is of the order of years, ranging from 0.6 to 4.0 years. This confirms the results already discussed in SOBOLEV et al., (1997, 1996) and HUANG and SOBOLEV (2002) for which RTL should be

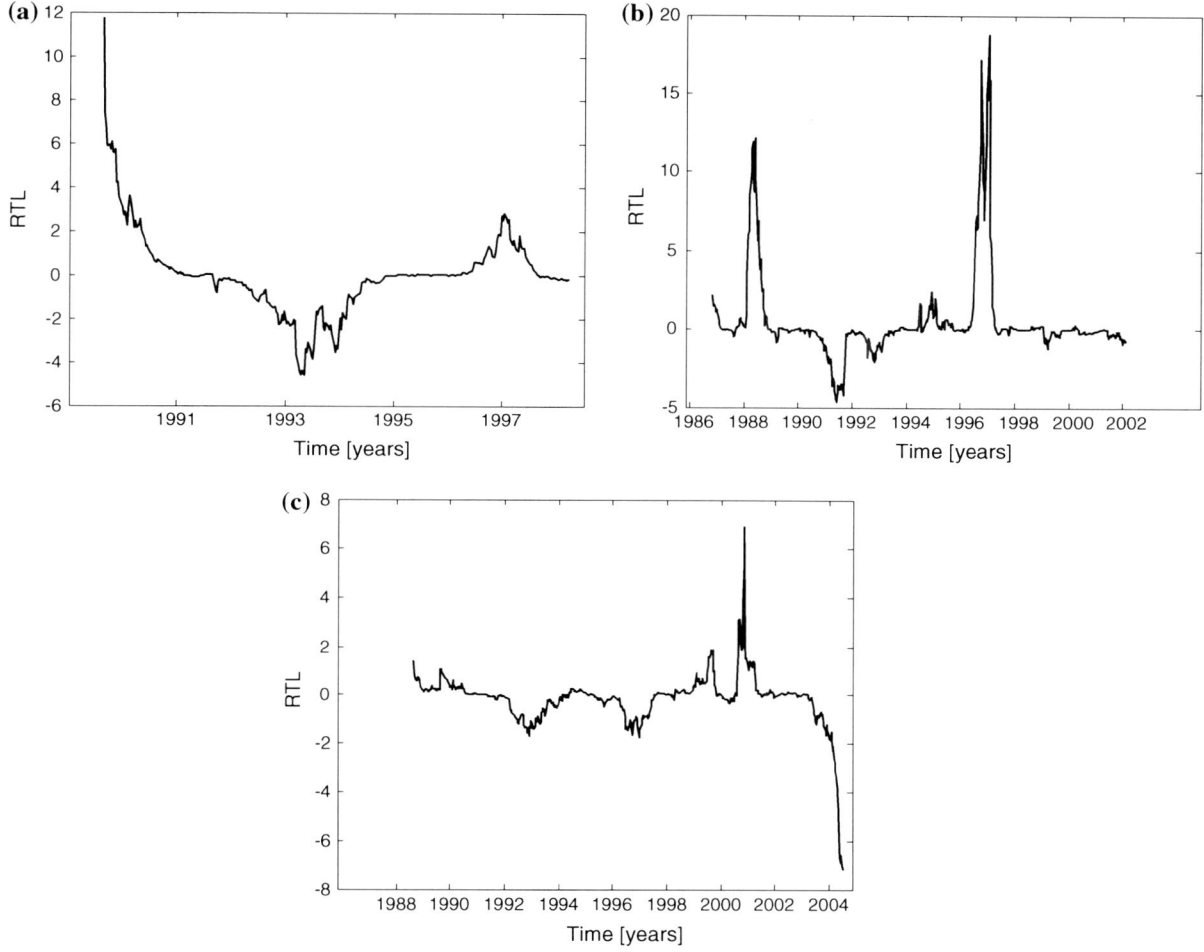

Figure 5
RTL calculated at the epicenters of NE Italy earthquakes using L&G catalogue and the parameters r_0 and t_0 calculated by the Chen and Wu method on the same catalogue. Epicenter location: **a** Kobarid (1998) $r_0 = 35$ km $t_0 = 1.8$ years, **b** Sernio Mountain $r_0 = 55$ km $t_0 = 0.4$ years, **c** Kobarid (2004) $r_0 = 30$ km $t_0 = 1.3$ years

considered an intermediate term precursor (range of years) and is not useful as a short-term precursor (months to weeks range).

Six of the analyzed earthquakes (Reggio Emilia, Umbria Marche, Kobarid 1998, Sernio Mountain, Palermo, Kobarid 2004) already have been studied using RTL with empirical choices of the parameters (DI GIOVAMBATTISTA and TYUPKIN, 1999, 2000, 2004; GENTILI and BRESSAN, 2007). Since the parameters here are generally different from the ones adopted in the literature, the catalogues used for all of Italy are different (in particular in Lolli and Gasperini catalogue the magnitude is revised) and the declustering method is different, the comparison only can be qualitative.

RTL does not present any quiescence before the Reggio Emilia earthquake, neither in this paper (see Fig. 4b) nor in DI GIOVAMBATTISTA and TYUPKIN (1999). An activation stage from the last months of 1995 until the earthquake day (October 15th, 1986) can be seen in DI GIOVAMBATTISTA and TYUPKIN (1999), but not in this paper. The RTL curves before Umbria-Marche (Fig. 4c) and the Palermo (Fig. 4h) earthquakes are qualitatively similar to the ones in the literature (DI GIOVAMBATTISTA and TYUPKIN 2000 and 2004, respectively), with an RTL decrease followed

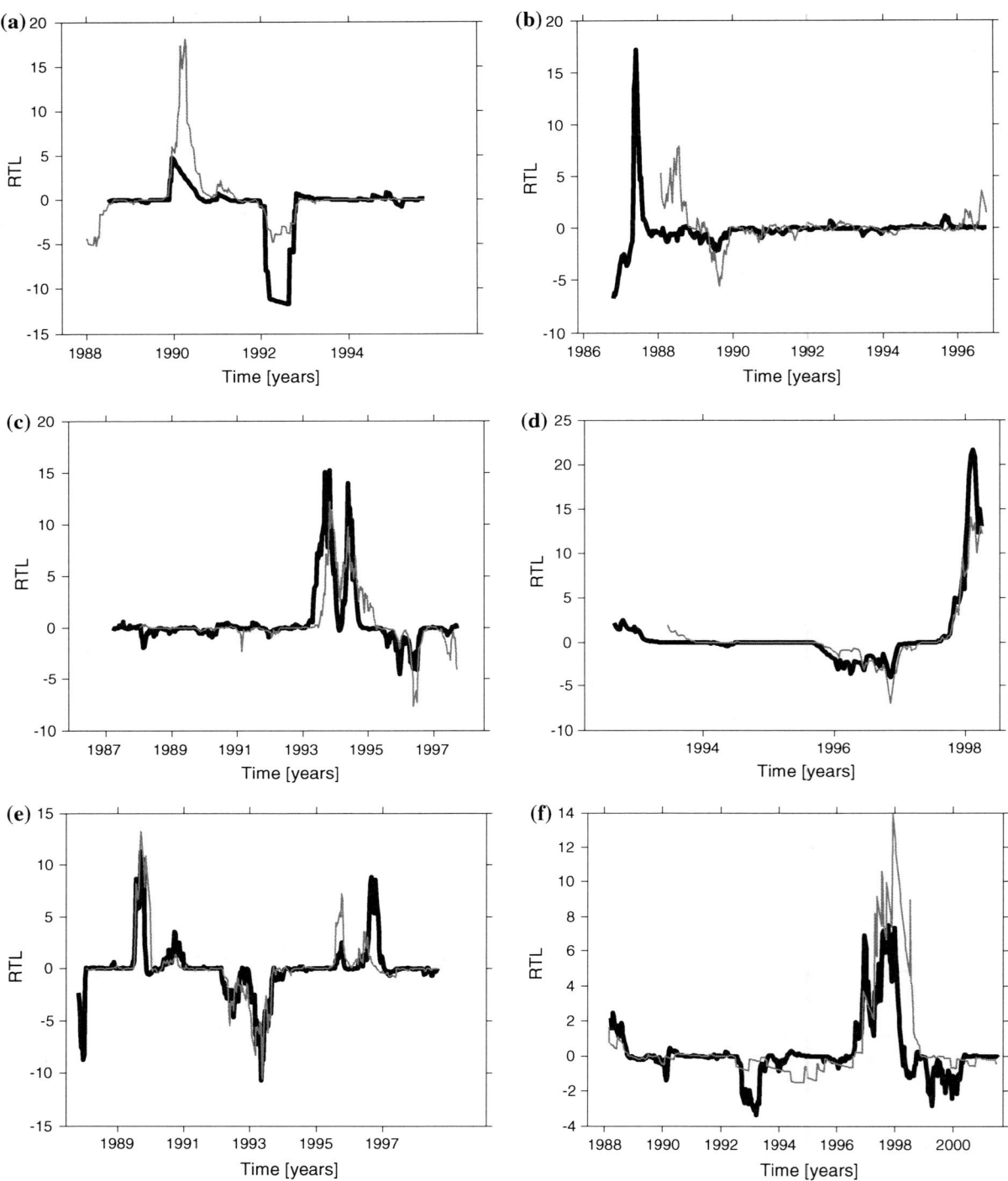

Figure 6
RTL calculated at the epicenters of analyzed earthquakes using the parameters listed in Table 3 (*thick line*) and standard parameters $r_0 = 30$ km $t_0 = 1$ year (*thin grey line*) until midnight before the main shock. Adopted declustering method: Reasenberg declustering algorithm with Lolli and Gasperini parameters. Epicenter location: **a** Gargano (Apulia), **b** Reggio Emilia, **c** Umbria Marche, **d** Kobarid (1998), **e** Calabrian-Lucanian Apennines, **f** Merano, **g** Sernio Mountain, **h** Palermo, **i** Molise, **l** Adriatic Sea, **m** Kobarid (2004), **n** Garda Lake

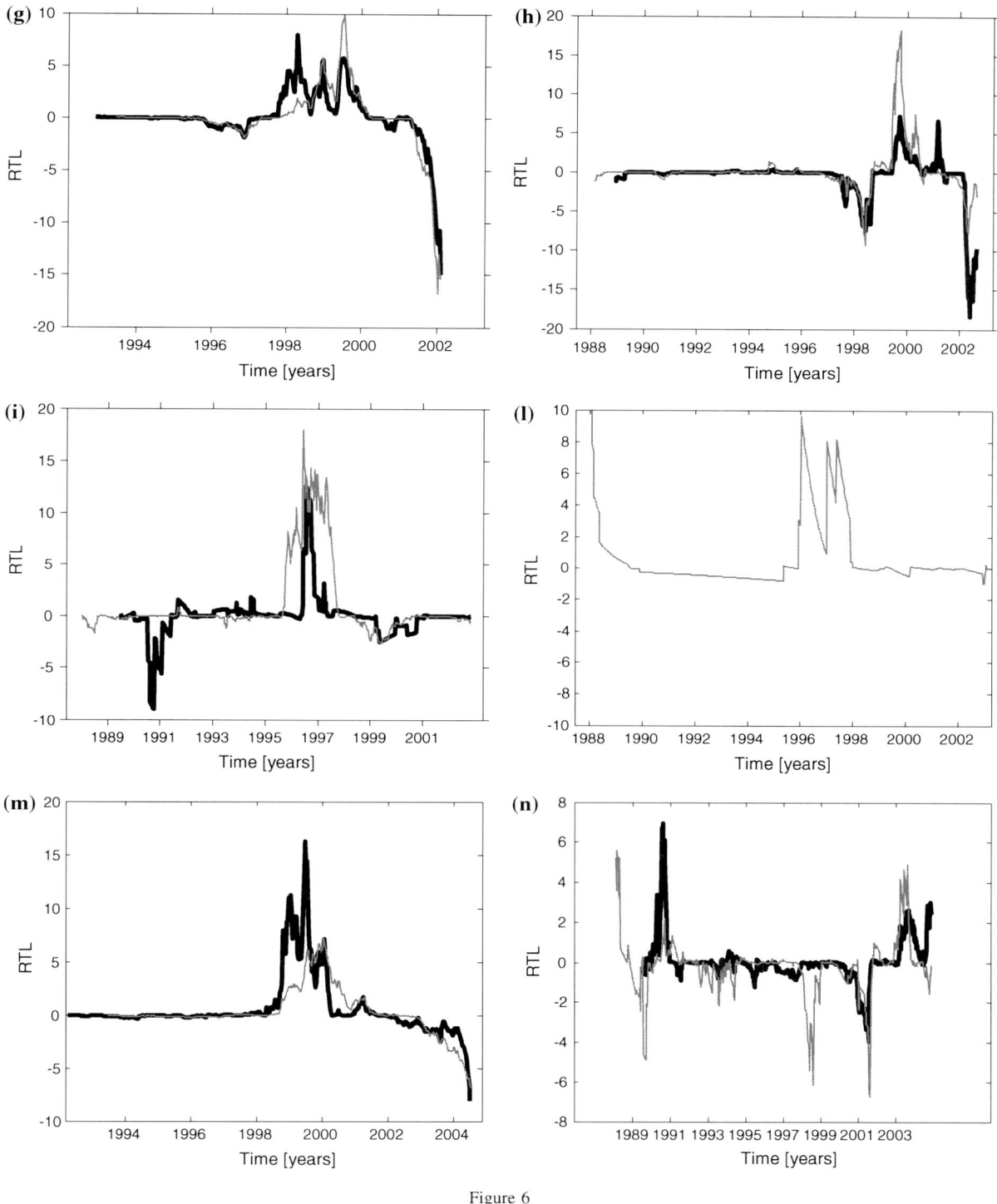

Figure 6 continued

Table 5

The characteristics of quiescences and activation phases detected by the RTL obtained at the epicenter of the analyzed earthquakes, using the parameters listed in Table 3

Zone	Quiescence duration (years)	Quiescence shift (years)	Activation duration (years)	Activation shift (years)
Gargano (Apulia)	1.1	2.9	–	–
Reggio Emilia	–	–	–	–
Umbria Marche	2.0	0.3	–	–
Kobarid 1998	1.6	1.0	0.7	0
Calabrian-Lucanian Apennines	0.6	0	–	–
Merano	2.2	1.2	–	–
Sernio Mountain	0.6	0	–	–
Palermo	1.4	0	–	–
Molise	1.4	1.8	–	–
Adriatic Sea	1.5	0.03	0.03	0
Kobarid 2004	3.0	0	–	–
Garda Lake	1.2	2.8	0.5	0

by a recovery stage before the main shock. Regarding earthquakes analyzed by the NEI catalogue, (Kobarid 1998, Sernio Mountain and Kobarid 2004: Fig. 4d, g, m, respectively) the resulting RTL curves are qualitatively similar to the results presented in the literature (GENTILI and BRESSAN 2007), with a quiescence followed by an activation of seismicity preceding the main shock in the Kobarid 1998 case (Fig. 4d), and a decrease of RTL until the midnight before the mainshock for the Sernio and Kobarid 2004 earthquakes (Fig. 4g and m).

4.2. Spatial Distribution of RTL

The quiescence region has been mapped in this paper by the Q parameter (HUANG et al., 2002) i.e., the mean of the value of RTL in a selected time window. In order to avoid noisy results, only values of $Q \leq -0.5$ are considered. The time window chosen for each analysis is listed in Table 6. The completeness magnitude is evaluated for the entire area shown in the map by the EMR method by also analyzing the variation in time. The highest value reached has been used for computation and is listed

Table 6

Adopted parameters in RTL map construction by applying (i) the Cheng and Wu method (ii) Threshold on r_0, (iii) minimum radius. For each choice, the corresponding quiescence anomaly linear dimension is listed

Zone	M_c	Analysis time interval (year/month/day)	Cheng and Wu		Threshold on r_0		Minimum radius	
			r_0 (km)	Quiescence anomaly dimension (km)	r_0 (km)	Quiescence anomaly dimension (km)	r_0 (km)	Quiescence anomaly dimension (km)
Gargano (Apulia)	2.5	1991/10/25–1992/10/26	15	~100	15	~100	10	~50
Reggio Emilia	2.1	1995/10/16–1996/10/15	55	–	30	–	–	–
Umbria Marche	2.2	1995/07/02–1996/07/01	35	~300	30	~300	10	~150
Kobarid 1998	2.2	1995/12/01–1996/11/30	60	~350	30	~250	10	~150
Calabrian-Lucanian Apennines	2.4	1997/09/10–1998/09/09	45	~500	30	~450	30	~450
Merano	2.1	1999/01/28–2000/01/27	65	~800	30	~200	30	~200
Sernio Mountain	2.4	2001/02/14–2002/02/13	25	~200	25	~200	10	~150
Palermo	2.3	2001/09/06–2002/09/05	30	~500	30	~500	10	~100
Molise	2.3	1999/10/29–2000/10/28	15	~100	15	~100	15	~100
Adriatic Sea	2.7	2002/03/17–2003/03/16	25	~150	25	~150	20	~50
Kobarid 2004	2.4	2003/07/12–2004/07/11	40	~350	30	~300	10	~100
Garda Lake	2.2	2000/12/28–2001/12/27	55	~600	30	~300	30	~300

in the Table (column 2). Note that M_c presented in Table 6 is generally larger than the one listed in Table 2, due to the larger area considered in the computation. A first test has been done by using the values selected in the previous section (see Table 3) for the evaluation of RTL at the epicenter of the main shocks; the radius r_0 adopted (column 4) is compared with the linear dimension of the quiescence zone (column 5). It is easy to see how the area covered by the quiescence is generally larger for larger values of the radius r_0, independently of the earthquake analyzed, while the time t_0 is not influential. This is not surprising, since enlarging the radius of the RTL causes each point to be affected by the influence of a larger area causing a blurring of the map image. In some cases (see e.g., the Merano earthquake) the area characterized by seismic sequence is so large that RTL loses utility for seismic risk assessment. For this reason, the obvious question is whether it is necessary to use such large radii like e.g., for the Merano earthquake, or if with smaller ones it is possible to obtain reliable results.

In this paper I propose two possible approaches to the problem. The first approach consists simply in setting a threshold T on r_0, and considering the r_0 corresponding to the peak value of $p(r_0)$, for $r_0 \leq T$, as the best parameter for RTL mapping. I set the threshold to $r_0 = 30$ km, that is the value adopted by DI GIOVAMBATTISTA and TYUPKIN (2004) for Palermo earthquake; 30 km is also the minimum radius for which RTL supplies stable results for low seismicity areas like e.g., the Merano earthquake area. The results are listed in columns 6 and 7 of Table 6. It is possible to see that the quiescence is still detectable for the same 11 earthquakes for which it was found using the Chen and Wu method, but in five cases the quiescence area extent is smaller. It is particularly interesting in the Merano earthquake case, where the linear dimensions of the quiescence area passes from 800 km to 200 km.

The second approach merges the stability requirements addressed by the Chen and Wu method and the additional requirement that r_0 is as small as possible, in order to reduce the dimension of the quiescence area. The algorithm adopted in this case is the following:

1. The RTL is calculated at the epicenter of the future earthquake, using r_0 and t_0 parameters obtained by the Chen and Wu method.
2. If a quiescence is detected the analysis time interval is selected (that is one listed in column 3 of Table 6).
3. The r_0 parameter is recursively reduced, maintaining t_0 unchanged, evaluating again the RTL.
4. The minimum value of r_0 for which the quiescence is still detectable is selected as the correct parameter for mapping.

This simple method allows study of a quiescence that is stable on a wide range of parameters and on the other side it allows smaller quiescence areas. The results are listed in Table 6 (columns 8 and 9) and the corresponding maps are presented in Fig. 7. The linear dimensions of the quiescence regions are smaller than the ones obtained by the CHEN and WU method in 91% of the cases and range from 50 to 450 km.

All the earthquakes except Reggio Emilia are inside the quiescence zone, even if only a few of them coincide with the minimum of Q, while in three cases (Umbria Marche: Fig. 7c, Merano: Fig. 7f, Palermo: Fig. 7h) the earthquake epicenters are close to the borders of the quiescence region. This result is coherent with other results in literature (HUANG et al., 2001, 2002; SOBOLEV, 2000).

Of the twelve earthquakes with magnitude >5 in Italy only for the Umbria Marche and Palermo earthquakes is the RTL map supplied in literature (DI GIOVAMBATTISTA and TYUPKIN, 2000 and 2004, respectively). For Umbria Marche the time interval for the analysis adopted by Di Giovambattista and Tyupkin is different; they chose the time interval September 5th, 1996–September 5th, 1997, that has no overlap with the time period chosen in this paper (July 2nd, 1995–July 1st, 1996). The reason for a different choice can be ascribed to the fact that the detected minimum in RTL (seismic quiescence) changes its duration and shape depending on the parameters (see Fig. 4c). The area characterized by the quiescence ranges in the Di Giovambattista and Tyupkin paper from approximately 42° to 44° in latitude and in longitude from less than 11° to 14.5° and covers Italy from one coast to the other. The earthquake epicenter is approximately at the center of the area. The

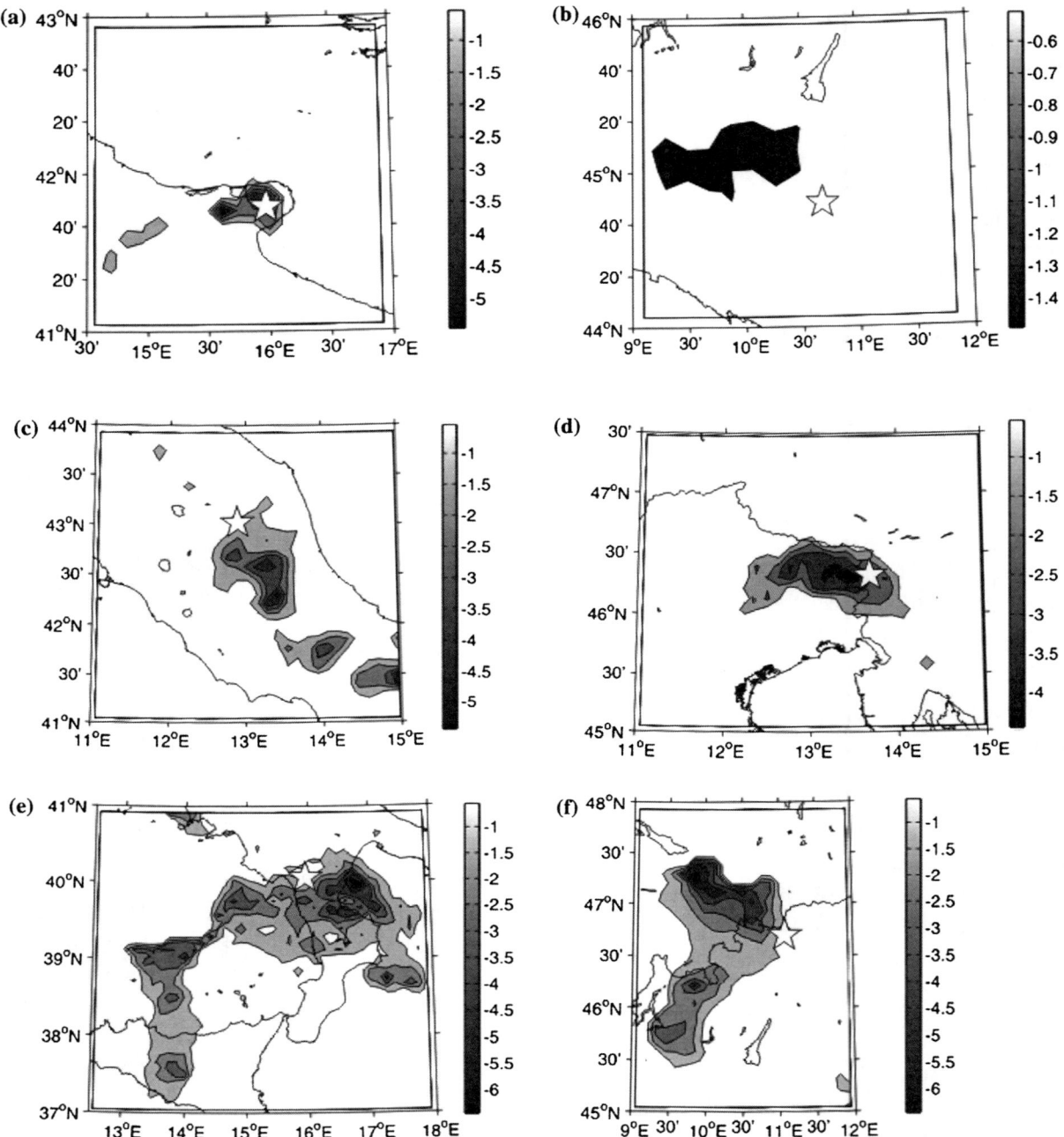

Figure 7
Map of the RTL using the Q method (*grayscale*), t_0 obtained by the Cheng and Wu method and r_0 using minimum radius method (see Table 6). The white star represents the epicenter location for: **a** Gargano (Apulia), **b** Reggio Emilia, **c** Umbria Marche, **d** Kobarid (1998), **e** Calabrian-Lucanian Apennines, **f** Merano, **g** Sernio Mountain, **h** Palermo, **i** Molise, **l** Adriatic Sea, **m** Kobarid (2004), **n** Garda Lake

mapping method they adopted is RTL_{min}. In this work the Q method is adopted and a large quiescence area is evident at lower latitude, with the earthquake epicenter on the northern border of the quiescence area. The latitude range is approximately [41°–43°] after choosing an r_0 equal to 35 or 30 km (see Fig. 8a

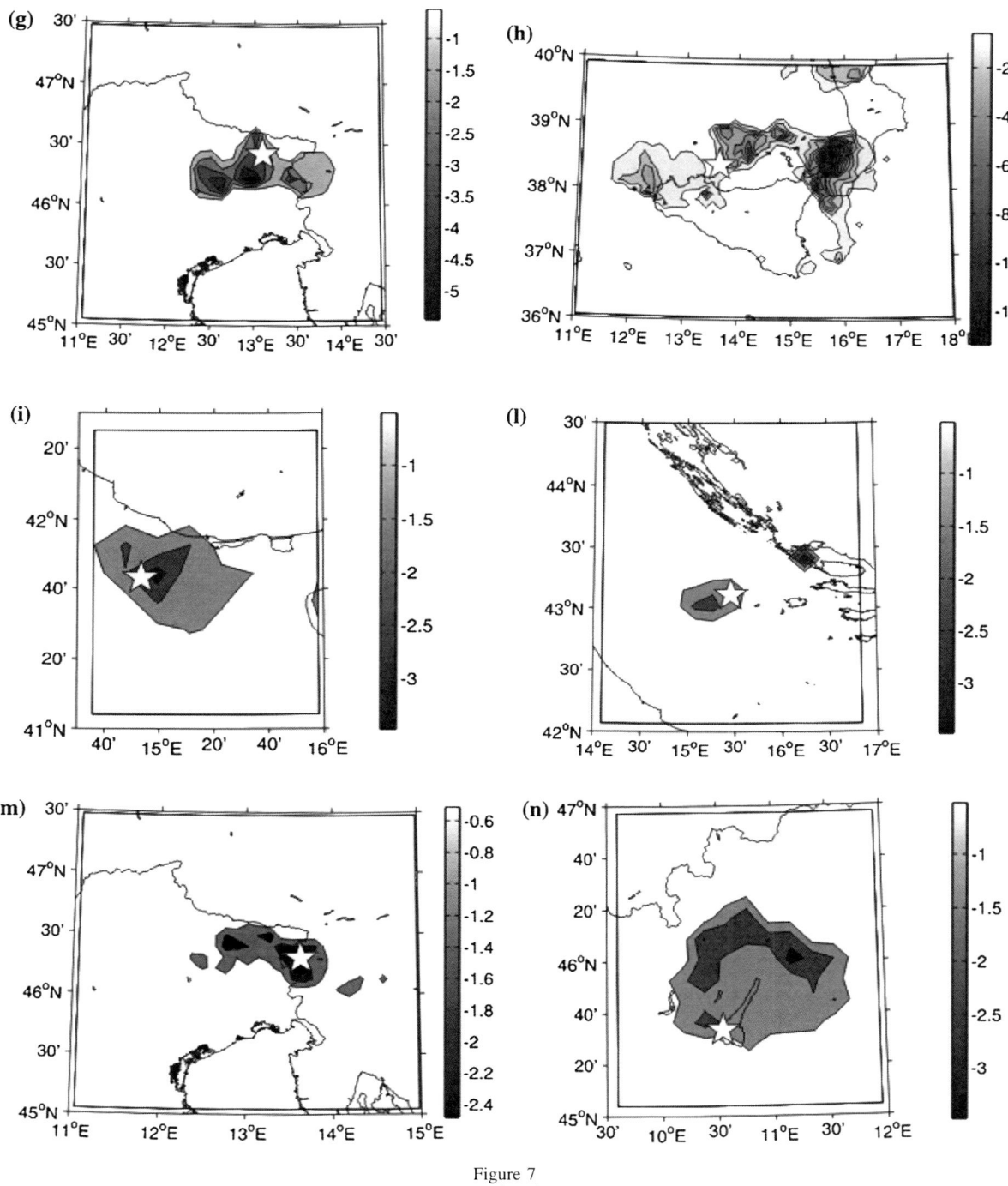

Figure 7
continued

and b, respectively) and [42°–43°] choosing an r_0 equal to 10 km (see Fig. 7c). In addition, the western coast is not included in the quiescence. With r_0 equal to 30 or 35 km the extension ranges from 12° to 15° in longitude and with r_0 equal to 10 km 12.5°–13.5°, not including the eastern coast.

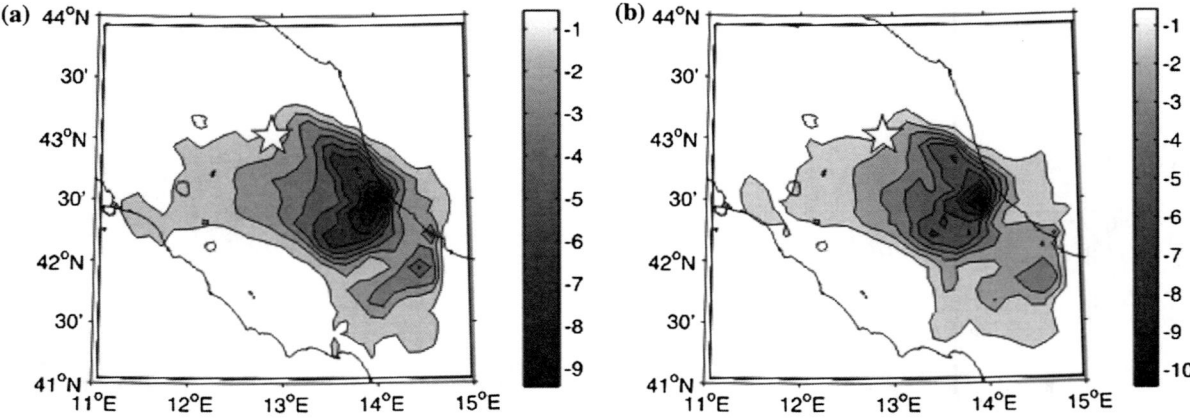

Figure 8
Map of the RTL using the Q method, at the epicenter of Umbria Marche earthquake using $t_0 = 0.6$ years and **a** $r_0 = 35$ km, **b** $r_0 = 30$ km

The map presented by Di Giovambattista and Tyupkin (2004) for Palermo earthquake was obtained in the same time period used in this paper. However, only a small study area is analyzed, and the spatial extension of the quiescence areas cannot be compared.

4.3. Development of RTL_{surv}

Previously, quiescence zones had been found for most of the earthquakes analyzed, but the time interval adopted was chosen as the year when the RTL at the epicenter of the earthquake had a minimum. This is an *a posteriori* choice that can be done only after the epicenter is known. In addition, also the r_0 and t_0 parameters depend on RTL calculated at the epicenter and, therefore, on epicenter position. In this paper, and based on the promising results shown above, I propose a new approach for a continuous survey of the catalogue that can be done before the occurrence of the earthquake.

In developing the proposed method I considered the following facts:

1. Most considered earthquakes are preceded by one or more quiescences.
2. From the start of the quiescence to the earthquake there is a time interval ranging from 0.6 to 4 years (see Table 4). These results are similar to many others in the literature (see Sobolev and Tyupkin 1997, 1999; Sobolev 2000; Di Giovambattista and Tyupkin, 2000, 2004; Huang et al., 2001, 2002; Huang, 2006; Chen and Wu, 2006, a partial review of previous results can be found in Huang, 2004), and therefore can be considered general.
3. Only 1/4 of the analyzed earthquakes present an activation of seismicity, while the 92% present a quiescence. Therefore, only the quiescence and not the activation can be regarded as a useful precursor in this study (see also Huang, 2004 and references within).
4. Even if the quiescence can be detected for different values of r_0 and t_0, its beginning and its end time are parameter dependent.
5. RTL mapping methods existing in literature need an a-priori choice of the time period to be analyzed. Wrong choices of the period, if RTL_{min} method is adopted, may move the analysis into a time period where there is no quiescence, and supply non-negative values in the area where the epicenter will be. The Q method is even more sensitive to the choice of the time window, because if it contains both the quiescence and a successive activation, negative and positive values of the RTL can cancel each other, supplying positive or null values to the RTL map in the area of the future earthquakes.

The last point seems to indicate a slightly better performance of RTL_{min} instead of Q. However, the advantage of the Q method is that it takes into account the whole analyzed time period, while

$\mathrm{RTL}_{\min}$ is sensitive to short duration spikes of the signal that can be due to casual fluctuation or inaccuracy in earthquakes locations.

Summarizing all the previous results and considerations I have developed a method that:

1. Considers all the potential times of quiescence before the earthquake.
2. Neglects activation phases.

The method, named $\mathrm{RTL}_{\mathrm{surv}}$, is calculated at every node of a grid of the investigated area and defined in the following way:

$$\mathrm{RTL}_{\mathrm{surv}}(j) = \frac{1}{N}\sum_{i=1}^{N} \mathrm{RTL}(i,j)\vartheta(\mathrm{RTL}(i,j)), \quad (13)$$

where

$$\vartheta(\mathrm{RTL}(i,j)) = \begin{cases} 1 & \text{if } \mathrm{RTL}(i,j) < 0, \\ 0 & \text{otherwise}. \end{cases}$$

$\mathrm{RTL}(i,j)$ is the RTL calculated at node j at time i, $\mathrm{RTL}_{\mathrm{surv}}(j)$ is calculated at the jth node of the grid and N in the number of data points of RTL in the time interval chosen for the analysis. In this work, a time step of 10 days is chosen. The grid cells are of 10×10 km. In accord with the previous results, the time interval chosen is four years long and ends with the time where the forecast starts. The forecast is valid for six months, and then it is repeated. An earthquake is considered corresponding to a quiescence zone if its distance from the nearest quiescence cell is smaller or equal to $2\varepsilon + L/2$, where L is the cell dimension (10 km in this case). In order to automatically avoid inclusion of unstable results in the maps, for each cell j, RTL, and therefore $\mathrm{RTL}_{\mathrm{surv}}(j)$, is calculated only if at least 50 earthquakes are localized inside a radius of $2r_0$ from the node in the time period from the start of the catalogue until the time of calculation (see also Sect. 4.1). In addition, cells with $\mathrm{RTL}_{\mathrm{surv}} > -0.5$ are not considered as quiescence cells (see Sect. 4.2).

5. Alarms and False Alarms

Due to the bad performances of RTL using L&G catalogue in NE Italy (see Sect. 4.1), the analysis in this area has been done by using NEI catalogue. The method has been tested separately on declustered L&G and NEI catalogues on two contiguous regions shown in Fig. 9a. The analysis of the L&G catalogue is aimed at detecting seismic quiescence before the earthquakes with $M_L \geq 4$ and depth < 100 km in the time interval January 1st, 1994–December 31st, 2004; the analysis of the NEI catalogue is aimed at detecting seismic quiescence before $M_D \geq 3.9$—corresponding approximately to $M_L \geq 4$: see Eq. (12)—earthquakes with depth <100 km in the time interval January 1st, 1998–December 31st,

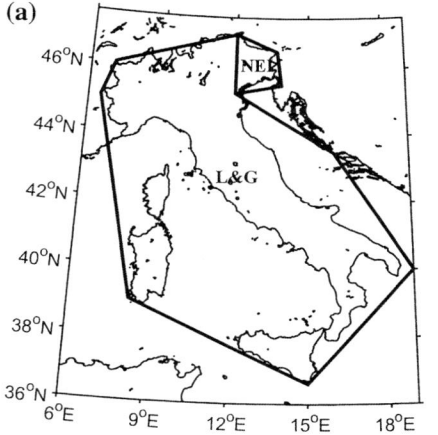 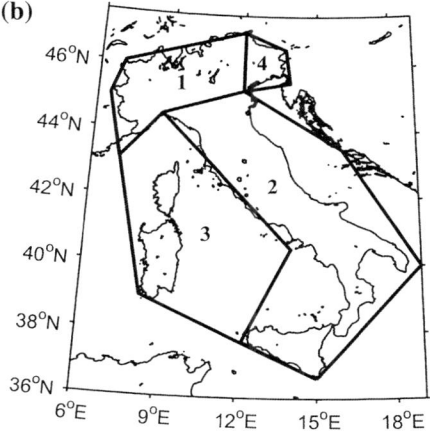

Figure 9
Map of Italy with **a** regions analyzed by L&G and NEI catalogues, **b** draft subdivision into macro-areas as a function of the number of recorded earthquakes

2004. The forecast has been done every 6 months and is valid for 6 months or more. The value of the completeness magnitude for L&G catalogue has been set to $M_L = 2.4$ until the end of 2002 and $M_L = 2.6$ for the last two years because the instrumental bulletin has a higher completeness magnitude (see Fig. 3b); for the NEI catalogue a completeness value of $M_D = 2.4$ has been adopted (see Fig. 3a).

One open problem is the choice of parameters r_0 and t_0. For t_0 I choose simply the mean of the values listed in Table 3, that is $t_0 = 0.9$ years. Ideally, for the choice of r_0, a method like that of CHEN and WU, perhaps with a threshold on the maximum value of r_0, should be applied. However, besides being computationally intensive, this choice would cause different vales of r_0 in different regions of the same map, and not all the points would be weighted evenly. On the other side, selecting the same value of r_0 for large regions, like, e.g., all Italy, engenders some problems. Small values of r_0 can lead to unstable RTL in regions where few earthquakes are recorded, due to the small number of data available. Conversely, large values of r_0 cause a large quiescence area, reducing the value of the method. The performances have been tested for both the L&G and NEI catalogues in the respective analysis areas by varying the value of r_0 from 10 to 30 km with steps of 5 km. The percentage of earthquakes preceded by a quiescence (found earthquakes) with $M_L \geq 4$ is plotted in Figs. 10a and 11a respectively, and compared with the percentage of quiescence area with respect to the entire area analyzed by the catalogue (Figs. 10b and 11b). An approximately linear increase of the quiescence area with the radius is detected for both catalogues. While using the NEI catalogue all the earthquakes with magnitude greater than 4 are found also with the minimum tested radius ($r_0 = 10$ km, see Fig. 11a), the percentage of found earthquakes is smaller using the L&G catalogue and increases approximately linearly with radius.

Figure 12 shows the distribution of found earthquakes (white symbols) and of earthquakes for which no seismic quiescence was detected (lost earthquakes—black symbols) setting $r_0 = 10$ km (Fig. 12a) and $r_0 = 30$ km (Fig. 12b) in the whole time period 1994–2004. Stars represent earthquakes with magnitude $M_L > 5$, dots earthquakes with magnitude $4 < M_L \leq 5$. Figure 12a shows that, using $r_0 = 10$ km, a quiescence is detected for a large percentage of the earthquakes (66% of the earthquakes with magnitude greater than 4, all the earthquakes with magnitude greater than 5) in the central and the southern part of the Italian peninsula and Sicily (zone 2 in Fig. 9b); in addition, a quiescence is detected before all earthquakes in the Northeastern area of Italy (zone 4 in Fig. 9b; the one

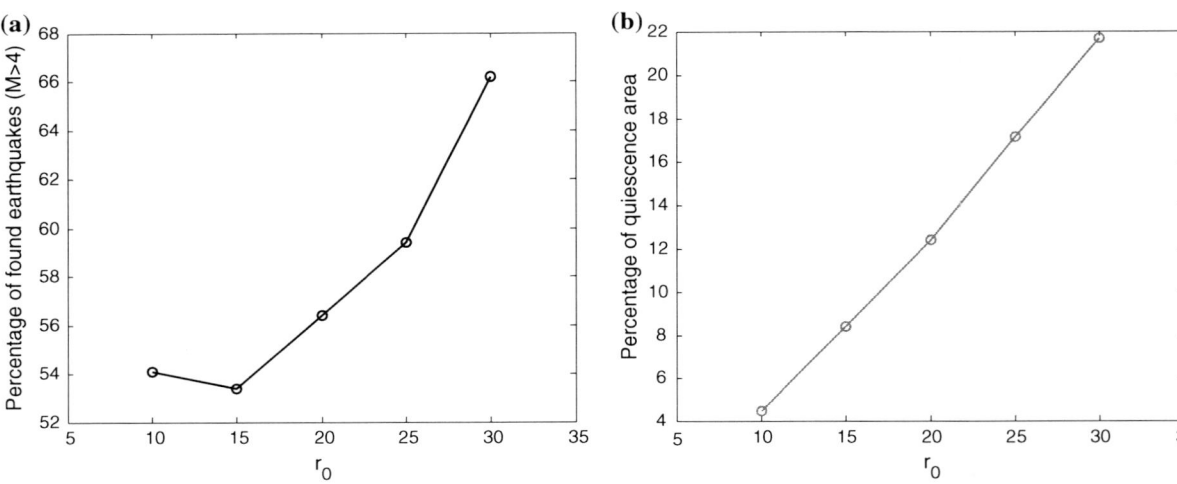

Figure 10
Behaviour of r_0 for the L&G catalogue (see Fig. 9a): **a** Percentage of earthquakes with $M_L \geq 4$ for which a precursory quiescence is detected as function of parameter r_0 adopted **b** percentage of the area characterized by a detected seismic quiescence as function of parameter r_0 adopted

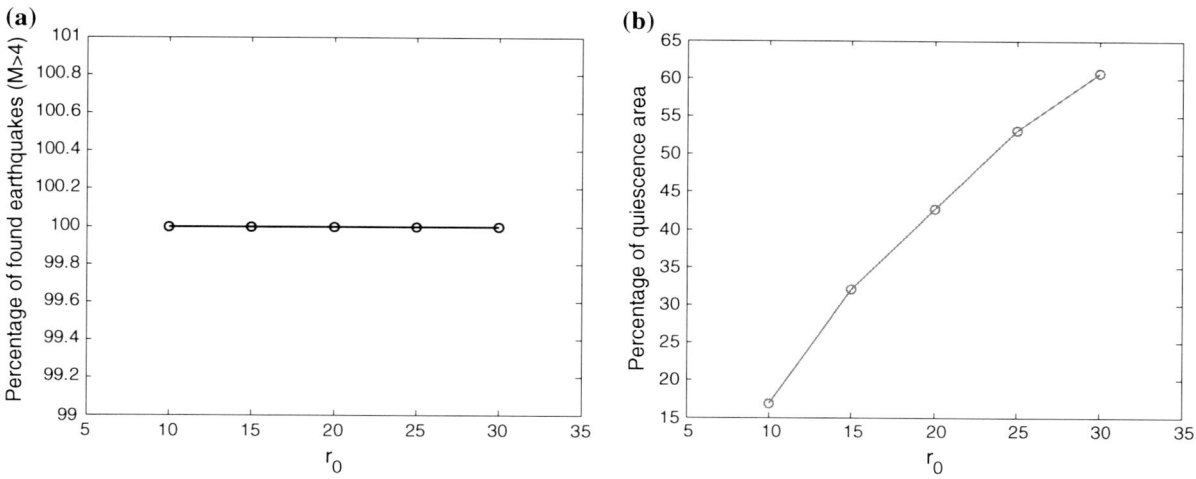

Figure 11
Behaviour of r_0 for the NEI catalogue (see Fig. 9a): **a** Percentage of earthquakes with $M_D \geq 3.9$ (corresponding approximately with $M_L \geq 4$) for which a precursory quiescence is detected as function of parameter r_0 adopted, **b** percentage of the area characterized by a detected seismic quiescence as function of parameter r_0 adopted

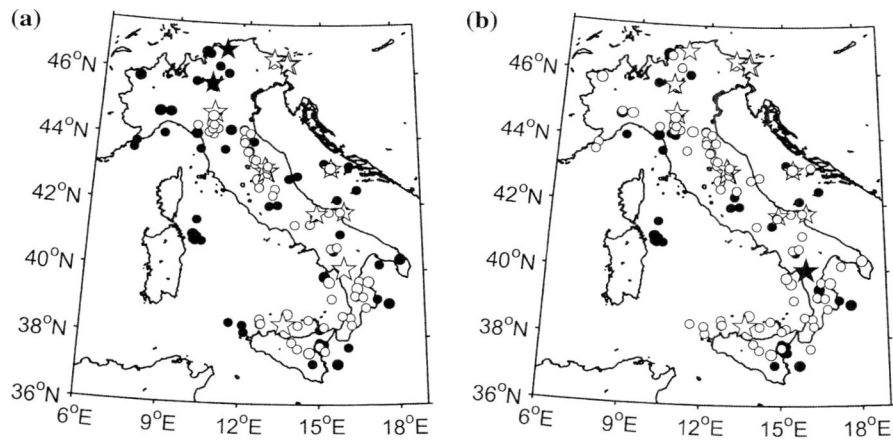

Figure 12
Map of the earthquakes preceded by a detected quiescence (*white symbols*) and not (*black symbols*). Stars correspond to earthquakes with $M_L > 5$, circles to earthquakes with $4 \leq M_L \leq 5$. **a** $r_0 = 10$ km, **b** $r_0 = 30$ km

analyzed by the NEI catalogue). For the rest of Italy no quiescence is detected before the earthquakes. This is due to the instability of RTL for $r_0 = 10$ km in regions where the recorded seismicity is low; the value of RTL$_{surv}$ in these areas could not be evaluated. For zone 3 of Fig. 9b, containing Sardinia and Corsica Islands and the most of Tyrrhenian Sea, no reliable value of r_0 can be applied to obtain a stable RTL, due to the very small number of recorded earthquakes—in a rectangle with vertices (N latitude, E longitude) = [(42, 9),(42, 11),(40, 11), (40, 9)]—around the larger magnitude seismicity in the area, the number of recorded earthquakes from 1994 to 2004 is 14. For most of the northern part of Italy (zone 1 in the Fig. 9b) it is necessary to use $r_0 = 30$ km to obtain the same performance that is obtained in the central-south Italy (zone 2) with $r_0 = 10$ km (see Fig. 12b). An analysis by macro-areas is therefore adopted with the RTL$_{surv}$ method, neglecting the area 3, where no precursor detection is possible using RTL, until a larger number of earthquakes are recorded in that area.

Table 7 shows the obtained performances of the method in terms of percentage of area of the seismic quiescence (alarm area) with respect to the whole area of the region analyzed, and in terms of percentage of found earthquakes, depending on the magnitude. The analysis is done by evaluating the RTL_{surv} every six months and considering the earthquakes which occurred during the six months following the forecast. In order to have a more complete analysis of the method performance, it would be necessary to understand what percentage of the alarm area should be considered a true alarm (the earthquake happens in the alarm area) and what a false alarm. In order to do this, the concept of "false alarm" should be studied in detail. A draft definition of false alarm could be "a false alarm is a detected quiescence region where no earthquake with $M \geq 4$ occurs during the six months following the forecast". However, since the extension of the quiescence areas grows with the increase of the radius r_0, different quiescence areas merge together, and the number of false alarms defined in the previous way decreases, even if the overall alarm area increases. For this reason, I propose an alternative definition of true and false alarm, analyzing separately all the cells of a grid: "A quiescence cell is considered a true alarm if an earthquake with $M_L \geq 4$ occurs within a distance $2\varepsilon + L/2 + 2r_0$ during the six months following the forecast; otherwise, it is a false alarm". This definition takes into account the influence of r_0 on RTL and the inaccuracy of location due to the catalogue and the grid discretization. Table 7 lists the percentage of true and false alarms for zones 1, 2 and 4. From the table, it can be inferred that the percentage of false alarms is high with respect to the whole alarm area. However, if the whole monitored region is considered, the false alarm area is smaller or equal to the 15.5% (see Table 7); considering that and at least 66% of $M \geq 4$ and 100% of $M > 5$ earthquakes are found, a good correlation between seismic quiescence and large to medium earthquakes can be deduced.

6. Conclusions

In this paper, the seismicity quiescence before the larger Italian earthquakes from 1994 to the end of 2004 has been analyzed. In particular, the RTL algorithm has been applied. The CHEN and WU (2006) method allows me to obtain r_0 and t_0 parameters for RTL calculated at the epicenter of large earthquakes. Using these parameters on 12 $M > 5$ earthquakes which occurred in Italy in the time period from January 1994 to December 2004, a quiescence phase is detected in 92% of the cases, while only 25% of the earthquakes are preceded by an activation phase. The duration of the quiescence and the time shift from the end of the quiescence to the earthquake time have been evaluated and range from 0.6 to 3 years and 0 to 2.9 years, respectively, with a time interval from the start of the quiescence to the main shock ranging from 0.6 to 4 years. The obtained parameters also have been applied to the mapping of the quiescence region before the earthquake. The quiescence regions appear in some cases very large, due to the high value of r_0. For this reason, a simple method for the best choice of the parameters is proposed and tested on the $M > 5$ earthquakes. The results are improved for 10 of the 11 cases in which the earthquake was preceded by a quiescence, and remain unchanged in the other. The quiescence regions detected by this method are characterized by linear dimensions ranging from 50 to 450 km with a mean value of 164 km.

A survey method based on RTL for the entire Italy analysis is proposed. The aim of the method, named RTL_{surv}, is to make a survey of the Italian area, in

Table 7

Performances of the RTL_{surv} method on three zones shown in Fig. 9b

Zone number	r_0 (km)	Alarm area (%)	True alarm (%)	False alarm (%)	Found earth. $M_L > 4$ (%)	Found earth. $M_L > 5$ (%)
1	30	18.4	2.9	15.5	66.7	100
2	10	8.7	1.2	7.5	66.0	100
4	10	16.9	1.9	14.9	100.0	100

order to detect seismic quiescence zones that may give rise to large earthquakes. The advantage of the method is that it is not necessary to define *a priori* the analysis time window. RTL_{surv} has been tested on declustered catalogues for the earthquakes with $M_L \geq 4$ in the time interval January 1st, 1994–December 31st, 2004. The Italian area was divided into four zones with different characteristics of the recorded seismicity. One of the zones (North and Central Tyrrhenian Sea) is characterized by too low a number of recorded earthquakes for the analysis. For each of the other three zones a parameter r_0 was found for which at least the 66% of the earthquakes with $M_L > 4$ and all the earthquakes with $M_L > 5$ were preceded by a detected seismic quiescence. The overall alarm area ranges from 8.7 to 18.4% of the whole analyzed region area, depending on the zone analyzed and on parameter r_0.

Even if the results found in this paper are encouraging, I wish to emphasize which are the limits of the method:

1. The time shift between the detection of a quiescence and the earthquake can be long and therefore it has been necessary to select a long time analysis for the RTL_{surv} (4 years). This fact, together with the impossibility of discriminating between an isolated $M \geq 4$ main shock and a swarm of $M \geq 4$ earthquakes during even tens of months, means that after a $M \geq 4$ earthquake the surrounding region may remain an alarm area for a long time, even if no earthquake follows, generating false alarms.

2. The area covered by the quiescence zone can be large, especially in regions characterized by a low recorded seismicity, not allowing us to indications for a reasonable economic planning of construction, timely preparation for potential damage and correct land use.

Acknowledgments

I would like to thank Gianni Bressan, Pier Luigi Bragato and Laura Peruzza for their valuable discussions and suggestions. This research has benefited from funding provided by the Italian Presidenza del Consiglio dei Ministri, Dipartimento della Protezione Civile (DPC). Scientific papers funded by DPC do not represent its official opinion and policies. The management of the Seismometric Network of Friuli-Venezia Giulia is financially supported by the Civil Protection of the Regione Autonoma Friuli-Venezia Giulia. I am grateful to the entire OGS staff of the Dipartimento Centro Ricerche Sismologiche at Udine for managing the network.

REFERENCES

BENDAT, J. S. and PIERSOL, A. G., *Random Data: Analysis and Measurements Procedures* (John Wiley and Sons, New York 2000).

BREHM, D. J. and BRAILE L. W. (1998), *Intermediate-term prediction using precursory events in the New Madrid seismic zone*, Bull. Seismol. Soc. Am. *88*, 564–580.

BRESSAN, G., KRAVANJA, S. and FRANCESCHINA, G., (2007), *Source parameters and stress release of seismic sequences occurred in the Friuli-Venezia Giulia region (Northeastern Italy) and in Western Slovenia*, Phys. Earth Planet. Inter. *160*, 192–214.

BUFE, C.G. and VARNES D.J. (1993), Predictive modeling of the seismic cycle of the greater San Francisco Bay region, J. Geophys. Res. *98*, 9871–9883.

CASTELLO, B., SELVAGGI, G., CHIARABBA, C. and AMATO, A. (2005), *Catalogo della sismicità italiana, -CSI 1.0 (1981–2002)*.

CHEN, C. C, and WU, Y. X. (2006), *An improved region–time–length algorithm applied to the 1999 Chi-Chi, Taiwan earthquake*, Geophys. J. Int. *166*, 1144–1147.

CSTI WORKING GROUP (2001), *Catalogo strumentale dei terremoti Italiani dal 1981 al 1996, Version 1.0*. CD-ROM, Clueb, Bologna, also available at: http://ibogfs.df.unibo.it/user2/paolo/www/gndt/Versione1_0/Leggimi.htm.

CSTI WORKING GROUP (2004), *Catalogo strumentale dei terremoti Italiani dal 1981 al 1996, Version 1.1*, available at: http://ibogfs.df.unibo.it/user2/paolo/www/gndt/Versione1_1/Leggimi.htm.

DI GIOVAMBATTISTA, R. and TYUPKIN, Y. S. (1999), *The fine structure of the dynamics of seismicity before $M \geq 4.5$ earthquakes in the area of Reggio Emilia (Northern Italy)*, Annali di Geofisica *42*, 897–909.

DI GIOVAMBATTISTA, R. and TYUPKIN, Y. S. (2000), *Spatial and temporal distribution of the seismicity before the Umbria-Marche September 26, 1997 earthquakes*, J. Seismol. *4*, 589–598.

DI GIOVAMBATTISTA, R. and TYUPKIN, Y. S. (2004), *Seismicity patterns before the $M = 5.8$ 2002, Palermo (Italy) earthquake: seismic quiescence and accelerating seismicity*, Tectonophysics *384*, 243–255.

FRANCESCHINA, G., KRAVANJA, S. and BRESSAN, G. (2006), *Source parameters and scaling relationships in the Friuli-Venezia Giulia (Northeastern Italy) region*, Phy. Earth Planet. Inter. *154*, 148–167.

GASPERINI, P. and FERRARI, G. (2000), *Deriving numerical estimates from descriptive information: The computation of earthquake parameters*, Ann. Geofis. *43*, 729–746.

GENTILI, S. and BRESSAN, G. (2007), *Seismicity patterns before $M_D \geq 4.1$ earthquakes in the Friuli-Venezia Giulia (northeastern Italy) and western Slovenia areas*, Bollettino di Geofisica Teorica ed Applicata *48*, 33–51.

Gentili, S. and Bressan, G. (2008), *The partitioning of radiated energy and the largest aftershock of seismic sequences occurred in the northeastern Italy and western Slovenia*, J. Seismol. *12*, 343–354.

Habermann, R. E. (1987), *Man-made changes of seismicity rates*, Bull. Seism. Soc. Am. *77*, 141–159.

Huang, Q. (2004), *Seismicity pattern changes prior to large earthquakes-An approach of the RTL algorithm*, Terres. Atmos. Oceanic Sci. *15*, 469–491.

Huang, Q. (2006), *Search for reliable precursors: A case study of the seismic quiescence of the 2000 western Tottori prefecture earthquake*, J. Geophys. Res. *111*, B04301, doi:10.1029/2005JB003982.

Huang, Q. and Nagao, T. (2002), *Seismic quiescence before the 2000 M = 7.3 Tottori earthquake*, Geophys. Res. Lett. *29*, doi: 10.1029/2001GL013835.

Huang, Q., Öncel, A. O. and Sobolev, G. A. (2002), *Precursory seismicity changes associated with the Mw = 7.4 1999 August 17 Izmit (Turkey) earthquake*, Geophys. J. Int. *151*, 235–242.

Huang, Q. and Sobolev, G. A. (2002), *Precursory seismicity changes associated with the Nemuro Peninsula earthquake, January 28, 2000*, J. Asian Earth Sci. *21*, 135–146.

Huang, Q. Sobolev, G. A. and Nagao, T. (2001), *Characteristics of the seismic quiescence and activation patterns before the M = 7.2 Kobe earthquake, January 17, 1995*, Tectonophysics *337*, 99–116.

Jiang, H., Hou, H., Zhou, H. and Zhou, C. (2004), *Region-time-lenght algorithm and its application to the study of intermediate-short term earthquake precursor in North China*, Acta Seismologica Sinica *17*, 164–176.

Kanamori, H. and Anderson, D. L. (1975), *Theoretical basis of some empirical relations in seismology*, Bull. Seism. Soc. Am. *65*, 1073–1095.

Knopoff, L. (2000), *The magnitude distribution of declustered earthquakes in Southern California*, Proc. Natl. Acad. Sci. USA *97*, 11880–11884.

Lee, W. H. K. and Lahr, J. C. (1975), *HYPO71 (revised): A computer program for determining hypocenter, magnitude and first motion pattern of local earthquakes*, U.S. Geol. Surv. Open File Rep. 75–311.

Lolli B., and Gasperini, P. (2003), *Aftershocks hazard in Italy Part I: Estimation of time-magnitude distribution model parameters and computation of probabilities of occurrence*, J. Seismol. *7*, 235–257.

Lolli, B. and Gasperini, P. (2006), *Comparing different models of aftershock rate decay: The role of catalog incompleteness in the first times after main shock*, Tectonophysics *423*, 43–59.

Marcellini, M. and Milani, D. (2003), *Valutazione della sismicità temporale del Friuli-Venezia*, Giulia Internal Report CNR Istituto per la dinamica dei processi ambientali, Milan, Italy.

Pace, B., Peruzza, L., Lavecchia, G. and Boncio, P. (2006), *Layered seismogenic source model and probabilistic seismic-hazard analyses in Central Italy*, Bull. Seism. Soc. Am. *96*, 107–132.

Papadopoulos, G. A. and Voidomatis, P. (1987), *Evidence of periodic seismicity in the inner Aegean seismic zone*, Pure Appl. Geophys. *125*, 613–628.

Postpischl, D. (ed.) (1985), *Catalogo dei terremoti italiani dal l'anno 1000 al 1980*, Quaderni della Ricerca Scientifica, 114 2B, CNR, Rome.

Priolo, E., Barnaba, C., Bernardi, P., Bernardis, G., Bragato, P. L., Bressan, G., Candido, M., Cazzador, E., Di Bartolomeo, P., Duri, G., Gentili, S., Govoni, A., Klinc, P., Kravanja, S., Laurenzano, G., Lovisa, L., Marotta, P., Michelini, A., Ponton, F., Restivo, A., Romanelli, M., Snidarcig, A., Urban, S., Vuan, A. and Zuliani, D. (2005), *Seismic monitoring in Northeastern Italy: A ten-year experience*, Seismolo. Res. Lett. *76*, 451–460.

Reasenberg, P. A. (1985), *Second-order moment of Central California seismicity, 1969-1982*, J Geophys Res. *90*, 5479–5495.

Rebez, A. and Renner, G. (1991), *Duration magnitude for the northeastern Italy seismometric network*. Boll. Geof. Teor. Appl. *33*, 177–186.

Rong, D., Li, Y. (2007), *Estimation of characteristic parameters in region-time-length Algorithm and Its Application*, Acta Seismologica Sinica *20*, 265–272.

Scholz, C. H. (1988), *Mechanisms of seismic quiescences*, Pure Appl. Geophys. *126*, 701–718.

Schorlemmer D. and Woessner J. (2008), *Probability of detecting an earthquake*, Bull. Seism. Soc. Am. *98*, 2103–2117.

Selvaggi, G., Deschamps, A., Ripepe, M. and Castello, B. (2002), *Waveforms, arrival times and locations of the 1997 Colfiorito (Umbria-Marche, Central Italy) aftershocks sequence*, Quad. Geofis. *21*, 6 CDROMs.

Sobolev, G. A. (2000), *Precursory phases of large Kamchatkan earthquakes*, Volc. Seis. *21*, 497–509.

Sobolev, G. A. and Tyupkin Y. S., *New method of intermediate-term earthquake prediction*. In European Seismological Commission XXV General Assembly *Seismology in Europe*, Reykjavik, Iceland, 9–13 September 1996, pp. 229–234.

Sobolev, G. A. and Tyupkin Y. S. (1997), *Low seismicity precursors of large earthquakes in Kamchatka*,. Volc. Seis. *18*, 433–446.

Sobolev, G. A., Tyupkin, Y. S. and Zavialov, A. (1997), *Map of Expectation Earthquakes Algorithm and RTL Prognostic Parameter: Joint Application*. The 29th General Assembly of IASPEI, Thessaloniki, Greece, Abstracts, 77.

Sobolev, G. A. and Tyupkin, Y. S. (1999), *Precursory Phases, seismicity precursors and earthquake prediction in Kamchatka*, Volc. Seis. *20*, 615–627.

Wiemer, S. (2001), *A software package to analyze seismicity: ZMAP*, Seism. Res. Lett. *72*(2), 373–382.

Wiemer, S. and Wyss, M. (1994), *Seismic quiescence before the Landers (M = 7.5) and Big Bear (M = 6.5) 1992 earthquakes*, Bull. Seism. Soc. Am. *84*, 900–916.

Woessner, J. and Wiemer, S. (2005), *Assessing the quality of earthquake catalogues: Estimating the magnitude of completeness and its uncertainty*, Bull. Seism. Soc. Am. *95*, 684–698.

Wyss, M. (1991), *Reporting history of the central Aleutians seismograph network and the quiescence preceding the 1986 Andreanof Island earthquake*, Bull. Seism. Soc. Am. *81*, 1231–1254.

Wyss, M. and Habermann, R. E. (1988), *Precursory seismic quiescence*, Pure Appl. Geophy. *126*, 701–718.

(Received August 11, 2008, revised June 24, 2009, accepted July 21, 2009, Published online March 10, 2010)

Predicting the Human Losses Implied by Predictions of Earthquakes: Southern Sumatra and Central Chile

MAX WYSS[1]

Abstract—Predictions of earthquakes worldwide by the M8-MSc algorithm, which defines locations of Times of Increased Probability (TIPs), have been tested for nearly two decades, and the authors claim a high rate of success. Thus, it might be appropriate to ask what the consequences in terms of human losses may be if the expected earthquakes should occur. The loss estimating tool QUAKELOSS also has been tested in real-time mode during the last five years with success. Therefore, it is reasonable to estimate the order of magnitude of human losses if great earthquakes should occur in TIPs. Here I compare the consequences if M 8.5 earthquakes should happen in the current TIPs of southern Sumatra and central Chile (KOSSOBOKOV and SOLOVIEV, 2008, centers at 4.75S/102.625E and 31.25S/71.77 W, respectively). The selection of the attenuation function is calibrated by matching theoretically calculated intensities and fatalities to the observed values in historic earthquakes. In both areas, the standard attenuation function I use is applicable. The results show that in southern Sumatra fatalities are expected to number fewer than 1,000 (possibly as much as a factor of 5 fewer), whereas they are likely to be larger than 1,000 (possibly as much as a factor six) in central Chile. These figures, however, do not account for possible tsunami effects. The difference is due to two factors. The earthquake sources are farther offshore, and there are only small settlements along the coast in southern Sumatra, whereas along the Chilean coast, large harbor cities are located in the northern part of the TIP area. Regardless of TIP predictions, large earthquakes are to be expected along the Chilean coast. Therefore, it seems advisable to implement mitigating measures in La Serena and Coquimbo, where most of the victims are expected.

Key words: Predicting losses in earthquakes, earthquake risk in Sumatra, earthquake risk in Chile.

1. Introduction

Every earthquake prediction carries with it implied consequences. I propose here that one should evaluate these consequences quantitatively. It is unpleasant to calculate expected fatalities in case a prediction comes true. However, failing to estimate the consequences does not make them go away. One might argue that predictions of earthquakes, as well as loss estimates, are so uncertain that it is not worthwhile to attempt either of them. Here, I advocate the position that in both fields large uncertainties exist, but that in both fields some tools have been tested long enough to warrant attempts at order of magnitude estimates of expected human losses.

The M8-MSc algorithm has now been tested for about twenty years (KEILIS-BOROK *et al.*, 1988; KOSSOBOKOV *et al.*, 1997), and the authors claim a high rate of success (KOSSOBOKOV and SOLOVIEV, 2008). Their predictions are regularly posted and updated on their website (http://www.mitp.ru/restricted_global/predlist1.html) and transmitted by e-mail to interested seismologists. Thus, it seems reasonable to accept the notion that M8+ earthquakes are more likely in the areas defined in the current TIPs of this algorithm than elsewhere. The areas covered by the M8 TIPs are too large for the exercise proposed here because the locations of the expected earthquake are not restricted well enough. However, the areas defined as TIPs by the M8-MSc algorithm can be used to define an earthquake source specific enough for loss estimates. From more than a dozen current M8-MSc TIPs worldwide, I select two that are located along populated coasts: Southern Sumatra and Central Chile.

The loss estimating tool QUAKELOSS has been tested for six years in real-time now (WYSS, 2004; WYSS and ZIBZIBADZE, 2009). After any 'significant' earthquake worldwide, we send an e-mail alert containing an estimate of the expected losses to the Swiss

---

[1] World Agency of Planetary Monitoring and Earthquake Risk Reduction, 2 Rue de Jargonnant, 1207 Geneva, Switzerland. E-mail: wapmerr@maxwyss.com

rescue team, OCHA (UN Office for the Coordination of Humanitarian Affairs) and the interested community. The predefined minimum magnitude of a 'significant' earthquake is M 6 in most areas (lower in Europe and higher in sparsely populated areas). These alerts constitute forward predictions because they are issued about 30 min after the respective earthquakes, at a time when the losses are unknown. These loss calculations can only be order of magnitude estimates because of the many uncertainties that enter the calculations. However, this inaccuracy is acceptable to users like disaster managers and rescue teams because, as a first step, they need to know only the answer to the question: Has this earthquake caused a disaster or not? In 380 real-time loss estimates in five years we have made three mistakes, when judged by the criterion of having correctly answered the aforementioned question.

I have also attempted loss estimates for hypothetical scenario earthquakes that were not predicted but where a large potential is recognized in general. In March 2005, I published loss estimates in seven scenarios for M 8.1 earthquakes in the Himalayas (WYSS, 2005). For one of these scenarios, the assumed epicenter was located in the Indian part of Kashmir. It predicted within a factor of 2 the losses sustained in the M 7.5 earthquake of October, 2005 that occurred in the Pakistani part of Kashmir (WYSS, 2006). Based on this success and the success of the real-time alerts I feel it is reasonable to attempt order of magnitude estimates of losses due to earthquakes expected in TIPs.

A further motivation for a study such as this is testing the validity of our loss estimates in forward mode. Estimating losses before they have occurred will afford an opportunity to compare them to what will happen eventually along these plate boundaries.

2. Method

The method used to calculate human losses consists of the following steps. Given the location and magnitude of the earthquake, the QUAKELOSS program calculates the intensity of shaking at the appropriate distance for every settlement in the database. Then the probability of all damage grades is calculated for each of the building classes according to the respective fragility curves. In a third step, the number of fatalities and injured in three severity classes is calculated using a casualty matrix. This method follows closely the approach of SHACKRAMANIAN et al., (2000) with a few modifications. Currently, we are constructing a second generation tool for estimating losses, QLARM2, which is open source and may be tested upon request by interested seismologists and engineers. Details of the method can be found in TRENDAFILOSKI et al., (2009).

3. Calibration

To validate our estimates, I compared reports of shaking and losses for historic earthquakes in the two regions studied with the values calculated using our loss estimating tool. The information on the intensities of shaking and the losses for historic earthquakes is seldom complete. Sometimes a macroseismic map is available. In other cases, intensities are known for only a few locations. The number of injured are seldom given, but the total number of fatalities is usually available, although it may only be a minimum estimate. I wish to match with our calculations any of these parameters that are available, considering the possibility to adjust the attenuation function to reach agreement.

In southern Sumatra, magnitude 8.5 and 7.9 earthquakes occurred on 12 September, 2007. For both, our real-time loss estimates were correct within the rather wide margins I allowed due to the uncertainty of the epicentral distance from shore. A recalculation of the losses with the final parameters of these quakes showed that our standard attenuation function is appropriate for southern Sumatra.

In Central Chile, the large earthquakes available for calibration are (1) 1946, M 8.2 with 25 fatalities reported, (2) 1971, M 7.5 with 90 fatalities reported, and (3) 1985, M 7.8 with 177 fatalities, 2,575 injured and maximum intensity reported. I did not use the M 7.6 earthquake of 1997 because it was located within the down-dipping slab (PARDO et al., 2002), not along the thrust interface, which means that a different attenuation function is appropriate for that event. I found that when using the standard attenuation function, the intensities and number of injured observed were well matched. In cases (1) and (2) the number of fatalities were also matched (within fewer than 70), but in case (3) the theoretically calculated fatalities were overestimated by a factor of 3, approximately. Changing the attenuation function would not bring a better match overall. Thus, I accepted the standard attenuation function as valid.

The QUAKELOSS program uses the SHEBALIN (1968) relationship as the standard attenuation function to predict shaking intensity as a function of magnitude and hypocentral distance. This relationship (with the constants $b = 1.5$, $c = 4.5$, $e = 3.5$) is quite similar to the ECOS intensity relationship derived by FÄH et al., 2003 from a central European macroseismic intensity database.

4. Results

4.1. Southern Sumatra

Figure 1 shows the location of the TIP by a dashed line. The rectangles show schematically the rupture areas of the two earthquakes in September 2007 (LORITO et al., 2008). The logic by which I selected the hypothetical epicenter was the following. (1) Given the fact that the M 8.4 earthquake ruptured part of the TIP area, a position in the part of the TIP area that has not yet ruptured seems most probable for the next earthquake. (2) For the distance from shore, I selected the possibility closest to shore with the intent to calculate a worst case scenario. The distance selected is the average distance from shore of the two events in September 2007. For the magnitude, I selected 8.5 as the worst plausible one. For depth, 20 km was assumed.

The mean damage state in all settlements experiencing shaking of intensity V and larger is shown by a color code (Fig. 1). The estimated numbers of fatalities and injured are given in Table 1.

4.2. Central Chile

In central Chile, no parts of the TIP have ruptured as recently as in southern Sumatra. Approximate

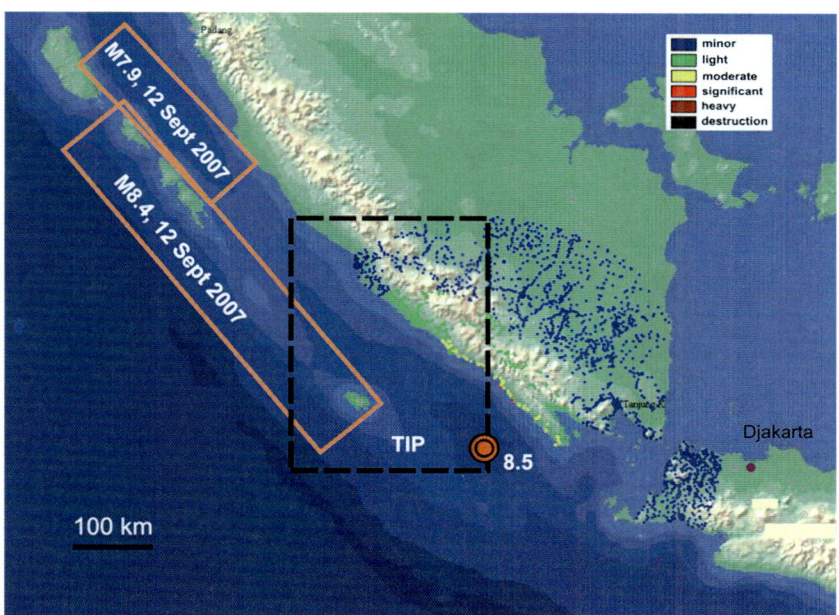

Figure 1
Map of mean damage state in settlements estimated for an earthquake with M 8.5 (epicenter marked by *ring*) located at the edge of the TIP in Southern Sumatra (outlined by *dashed line*). The size of the *dots* is proportional to population. Approximate extent of the two earthquake ruptures of 2007 are shown by *solid orange rectangles*. Injuries and fatalities are only expected in settlements with *red* and *yellow* colors (intensity expected VII and above, and mean damage moderate to significant). The *blue* settlements outline the area in which intensity V is expected

Table 1

Estimated human losses that may be expected in the worst cases if the TIPs defined by KOSSOBOKOV et al., (http://www.mitp.ru/restricted_global/predlist1.html) in Southern Sumatra and Central Chile should produce M 8.5 earthquakes

Scenario	Name	Lat. (deg).	Lon. (deg).	M	F_{min}	F_{max}	I_{min}	I_{max}
1	Sumatra	−5.93	103.74	8.5	200	700	600	2,000
2	Northern Chile	−29.70	−71.50	8.5	3,000	6,500	6,500	13,000
3	Central Chile	−31.25	−71.77	8.5	900	1,900	2,000	4,000

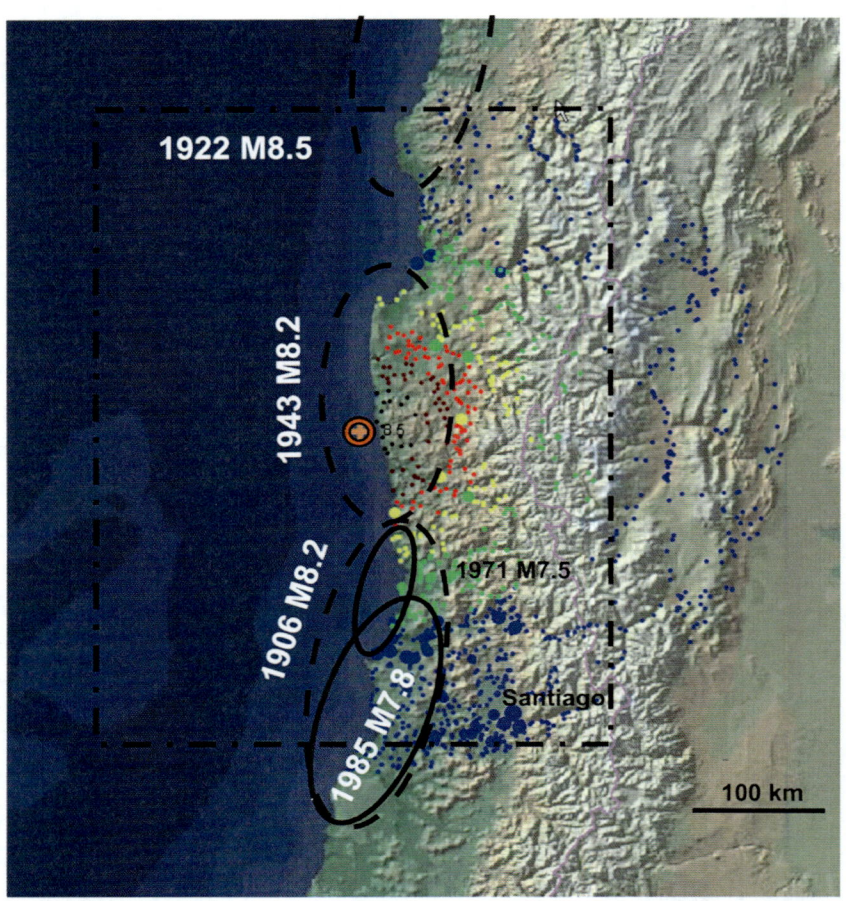

Figure 2
Map of mean damage state in settlements estimated for an earthquake with M 8.5 (epicenter marked by *ring*) located at the center of the TIP in Central Chile (scenario 2, Table 1). Most fatalities would be expected in settlements with *black* and *brown* colors (color legend same as in Fig. 1). The TIP area is outlined by a *dash-dotted rectangle*, approximate rupture extents of historic earthquakes are shown by *dashed lines* for events before 1970, by *solid lines* for more recent events

outlines of historic ruptures since 1906 are shown in Fig. 2 (BARRIENTOS, 1995; MCCANN *et al.*, 1979; NISHENKO, 1991). Given the distribution of these ruptures, the southern and northern parts of the TIP area are least and most likely to produce a large earthquake, respectively, if one assumes that plate motions steadily load elastic energy along the plate boundary. Thus, I propose the scenario shown in Fig. 3 as the most probable and the one with an epicenter at the center of the TIP (Fig. 2) as a second choice.

The hypothetical epicenters are placed offshore at a distance in keeping with recent large earthquakes

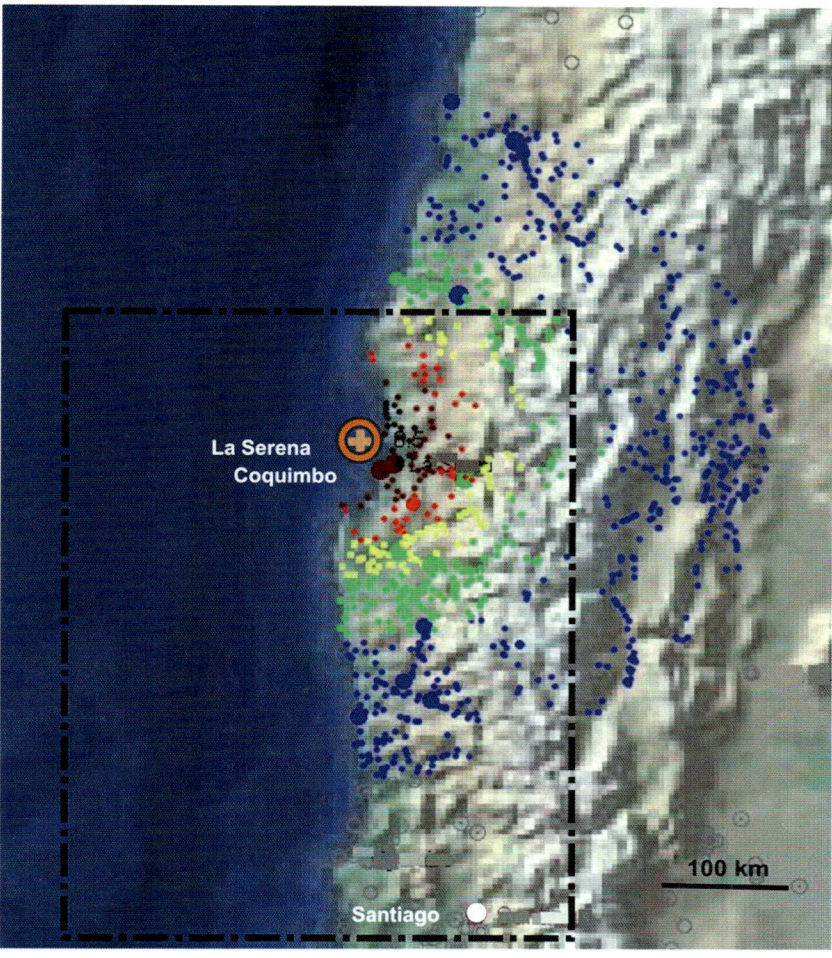

Figure 3
Map of mean damage state for an M 8.5 earthquake at 30 km depth (*ring* marks the epicenter) estimated (scenario 2, Table 1). The legend for damage state is the same as in Fig. 1. Most fatalities are expected in settlements shown in *black* and *brown* colors, in the settlements marked by *green* and *blue* colors no casualties, only moderate damage is expected. The *dash-dotted rectangle* outlines the TIP area

along the South American subduction zone. The magnitude of 8.5 is selected as the largest plausible, given that the 1922 event was of this magnitude, leading to worst case estimates. The expected numbers of fatalities and injured for the selected two hypothetical earthquakes within the TIP off Central Chile are listed in Table 1. The hypocentral depths for scenarios two and three were 25 and 30 km, respectively.

5. Discussion

The result is firm that a further rupture of the TIP in southern Sumatra may result in moderate human losses only. Firstly, it is backed by the recent experience in the M 8.5 and M 7.9 earthquakes of September 2007, where only 25 and zero fatalities were reported, respectively. Secondly, changes of position up or down along the coast do not influence the losses considerably because the coast is populated relative uniformly by small settlements (Fig. 1). The epicenter might well be further offshore than selected in Fig. 1, in which case zero fatalities may result.

The conclusion that more than 1,000 fatalities are likely in an M 8.5 earthquake in the subduction zone of Central Chile cannot be avoided. Here the rupture areas are closer to land, and there is a larger population at risk. In Central Chile, the assumed epicentral position up and down along the coast strongly influences the loss estimates. Selecting an epicenter at the center of the TIP (scenario 3, Table 1, Fig. 2) leads to the most benign case in Central Chile because the settlements in this section of the coast are small. The most likely epicenter (scenario 2, Table1, Fig. 3) leads to the worst case because two large cities, La Serena and Coquimbo, are located in the northern part of the TIP.

One of the factors that may reduce losses below the numbers estimated is the hour of day in which the earthquake will occur. Here I assume the worst case: 1 AM at night, when most people are indoors. The numbers of casualties could be substantially reduced, if the earthquake happened during morning hours, when many people are out of doors on their way to work in the cities or at work in the fields in the countryside.

The hypocentral depth assumed influences the results only slightly in Sumatra because the epicenter is at a considerable distance from shore. This means that moderate differences in depth cause only a minor change in distance the waves travel to the settlement. However, in Chile the assumed depth influences the loss estimates because changes in depth map with little reduction into the distance traveled by the waves. If the main energy release were at 25 km (instead of 30 km) in scenario 2, then approximately 20% more casualties would have to be expected.

Victims due to a possible tsunami are not included in the estimates presented here. The only parameters causing casualties considered are the intensity of the strong ground motions and the resistance of buildings to shaking.

6. Conclusions

I advocate that it is useful, even necessary, to attempt to predict human losses in cases where an increased probability of large and great earthquakes has been defined. However, one must recognize that these are order of magnitude estimates, subject to many uncertainties. I propose that the comparison of the loss potential in the two TIPs of Southern Sumatra and Central Chile demonstrates the usefulness of loss estimates. In the case of Sumatra, the probability of a major disaster is low, whereas in the other case, Chile, it is substantial.

Considering the fact that it is only a matter of time until the subduction zone off La Serena and Coquimbo will rupture in a large to great earthquake, it would seem worthwhile to take mitigating measures in these two cities.

Acknowledgments

This work was carried out with the support of the Swiss Agency for Development and Cooperation but does not necessarily reflect the opinion of this agency.

REFERENCES

BARRIENTOS, S. E. (1995), *Dual seismogenic behavior: The 1985 Central Chile earthquake*, Geophys. Res. Letts. 22, 3541–3544.

FÄH, D. et al. (2003), *Earthquake catalogue of Switzerland (ECOS) and the related macroseismic database*, Eclog. geolog. Helvet. 96, 219–236.

KEILIS-BOROK, V. I. et al. (1988), *Intermediate term prediction of occurrence times of strong earthquakes*, Nature 335, 690–694.

KOSSOBOKOV, V. and SOLOVIEV A. (2008), *Forecast/prediction of extreme events: Fundamentals and prerequisites of verification*, Paper presented at EUG General Assembly, Vienna.

KOSSOBOKOV, V. G. et al. (1997), *Testing an earthquake prediction algorithm*, Pure Appl. Geophys. 149, 219–248.

LORITO, S. et al. (2008), *Source process of the September 12, 2007, M_w 8.4 southern Sumatra earthquake from tsunami tide gauge record inversion*, Geophys. Res. Lett. 35, doi:10.1029/2007GL032661.

MCCANN, W. E. et al. (1979), *Seismic gaps and plate tectonics: Seismic potential for major plate boundaries*, Pure Appl. Geophys 117, 1082–1147.

NISHENKO, S. P. (1991), *Circum-Pacific seismic potential: 1989-1999*, Pure Appl. Geophys. 135, 169–259.

PARDO, M. et al. (2002), *The October 15, 1997 Punitaqui earthquake ($M_w = 7.1$): A destructive event within the subducting Nazca plate in Central Chile*, Tectonophys. 345, 199–210.

SHAKHRAMANIAN, M. A. et al. (2000), *Assessment of the seismic risk and forecasting consequences of earthquakes while solving problems on population rescue*, Moscow.

SHEBALIN, N. V. *Methods of engineering seismic data application for seismic zoning*. In *Seismic Zoning of the USSR*, ed. S. V. MEDVEDEV, pp. 95–111, (Science, Moscow 1968).

TRENDAFILOSKI, G., et al. (2009), *Constructing city models to estimate losses due to earthquakes worldwide: Application to Bucharest, Romania*, Earthquake Spectra 25(3), 665–685.

WYSS, M. (2005), *Human losses expected in Himalayan earthquakes*, Nat. Hazards 34, 305–314.

WYSS, M. and M. ZIBZIBADZE (2009), *Delay times of worldwide global earthquake alerts*, Nat. Hazards, doi:10.1007/s11069-009-9344-9.

Wyss, M. *Real-time prediction of earthquake casualties*, Proc. Internat. Conf. *Disasters and Society–From Hazard Assessment to Risk Reduction*, eds. D. Malzahn and T. Plapp. Logos Publishers, Univ. Karlsruhe (2004).

Wyss, M. *The Kashmir M 7.6 shock of 8 October 2005 calibrates estimates of losses in future Himalayan earthquakes*, Proc. Conf. Internat. Community on *Information Systems for Crisis Response and Management*, (Newark 2006).

(Received June 24, 2008, revised April 2, 2009, accepted June 17, 2009, Published online April 22, 2010)

Space- and Time-Dependent Probabilities for Earthquake Fault Systems from Numerical Simulations: Feasibility Study and First Results

Jordan Van Aalsburg,[1,2] John B. Rundle,[1,2] Lisa B. Grant,[3] Paul B. Rundle,[1,2] Gleb Yakovlev,[2] Donald L. Turcotte,[4] Andrea Donnellan,[5] Kristy F. Tiampo,[6] and Jose Fernandez[7]

Abstract—In weather forecasting, current and past observational data are routinely assimilated into numerical simulations to produce ensemble forecasts of future events in a process termed "model steering". Here we describe a similar approach that is motivated by analyses of previous forecasts of the Working Group on California Earthquake Probabilities (WGCEP). Our approach is adapted to the problem of earthquake forecasting using topologically realistic numerical simulations for the strike-slip fault system in California. By systematically comparing simulation data to observed paleoseismic data, a series of spatial probability density functions (PDFs) can be computed that describe the probable locations of future large earthquakes. We develop this approach and show examples of PDFs associated with magnitude $M > 6.5$ and $M > 7.0$ earthquakes in California.

Key words: Earthquakes, forecasting, California seismicity, earthquake hazard.

1. Introduction

In a series of reports, the Working Group on California Earthquake Probabilities (WGCEP) have computed probabilities of major earthquakes on California faults over a 30-year period.[1,2] These forecasts are used to set insurance rates and by emergency response planners and policymakers. A review of the reports (Field, 2007) describes common features, differences and assumptions of these studies. Field (2007) concludes by advocating the use of numerical simulation-based approaches to the problem of multi-decadal earthquake forecasting. An analogy may be drawn to weather and climate forecasting. Weather and seismicity are both complex, chaotic phenomena. Current weather patterns are routinely extrapolated to forecast several days into the future. These forecasts utilize numerical simulations of atmospheric behavior. Here we develop a similar approach by using *Virtual California*, a topologically realistic numerical simulation of strike-slip faults in California, to develop a series of spatial probability density functions (PDFs) that describe the probable locations of future large earthquakes.

2. The WGCEP Approach

As summarized by Field (2007), the WGCEP approach has been to (1) define a series of geological fault segments; (2) use paleoseismic and other data to determine the mean earthquake recurrence interval on each segment; (3) assume a set of statistical distributions to describe the recurrence statistics; (4) compute the probability of multi-segment ruptures,

---

[1] Department of Physics, University of California, Davis, USA. E-mail: jvan@cse.ucdavis.edu
[2] Computational Science and Engineering Center, University of California, Davis, USA.
[3] Program in Public Health, University of California, Irvine, CA 92697-3957, USA.
[4] Geology Department, University of California, Davis, USA.
[5] Jet Propulsion Laboratory, Pasadena, CA, USA.
[6] Department of Earth Science, University of Western Ontario, London, ON, Canada.
[7] Instituto de Astronomía y Geodesia (CSIC-UCM), Facultad CC Matemáticas, Ciudad Universitaria, Plaza de Ciencias, 3, 28040 Madrid, Spain.

[1] http://www.pubs.usgs.gov/of/2003/of03-214/WG02_OFR-03-214_ExecSummary.pdf.
[2] http://www.wgcep.org/.

assuming statistical independence of fault segments; and (5) adjust the results to reflect the time-dependence of the earthquake cycle (FIELD, 2007). The result is a set of probabilities for the occurrence of earthquakes $M > 6.5$ over the next thirty years.

The WGCEP approach assumes that earthquakes occur on geologically-defined fault segments, that earthquake ruptures rarely jump between fault segments, and that earthquake clustering can be discounted (FIELD, 2007). However, earthquake clustering is an established consequence of earthquake dynamics (MARCO et al. 1996; ZHUANG et al., 2004) and there are recent examples of earthquake ruptures jumping between fault segments, for example the 1992 M 7.3 Landers earthquake (WALD and HEATON 1994) and the 2002 M 7.9 Denali earthquake (EBERHART-PHILLIPS et al., 2003). The methods for including uncertainty in the modeled probabilities are problematic (PAGE and CARLSON, 2006).

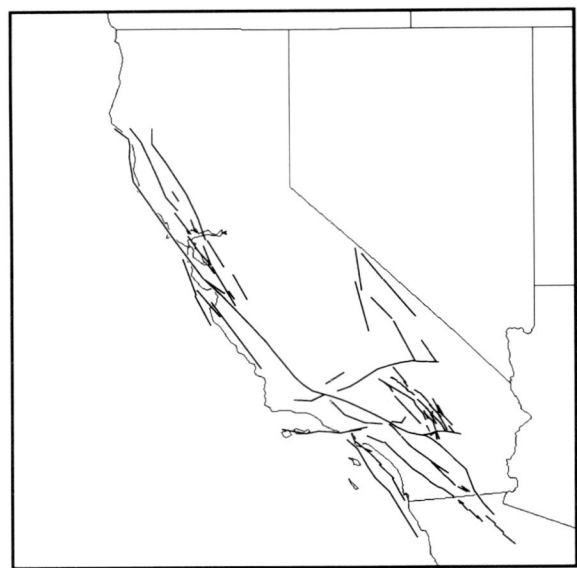

Figure 1
Map of California with the faults used in the Virtual California simulations as shown

3. The Virtual California Simulation Approach

Here we propose a method for computing probabilities using the type of simulation-based approach (RUNDLE, 1988; RUNDLE et al., 2001, 2002, 2004, 2005, 2006; VAN AALSBURG et al., 2007) advocated by FIELD (2007). Virtual California (VC) is a topologically realistic numerical simulation of earthquakes occurring on the San Andreas fault system. It includes the major strike-slip faults in California (Fig. 1). The approach using simulations such as VC is similar to the WGCEP approach. It begins with a series of faults divided into interacting fault elements, and uses paleoseismic and other data to set the frictional properties on each element. We then conduct a series of numerical simulations that attempt to reproduce the statistics and variability of the actual fault system. We search through the simulations to identify sequences of earthquakes that optimally represent the known earthquake history; and use the simulation data to measure the statistics and probabilities for future earthquake occurrence in space and time. The result is a set of probabilities for the occurrence of earthquakes of any size larger than the cutoff over user-selected future time-intervals. The probabilities determined by the simulations are time-dependent, implicitly include the effects of fault interactions, and are based on the same published data available to the WGCEP.

For this study, the VC fault model is composed of 650 fault boundary elements, each of 10 km width and 15 km depth. Elastic dislocation theory is applied to model fault element interactions. VC is a "back-slip" model. The accumulation of a slip deficit on each element is prescribed using available paleoseismic and instrumental data so that the long-term rate of slip is matched, on average, by the observed rate of stress accumulation on the faults (SAVAGE and PRESCOTT, 1978; RUNDLE and KANAMORI, 1987; RUNDLE, 1988). The mean recurrence time of earthquakes is determined using available data, to define friction law parameters. The friction law has several parts, including Mohr–Coulomb stick–slip properties; small amplitude, stable aseismic slip that increases as stress increases; and a stress-rate dependent failure criterion based upon laboratory studies of the functional form of the dynamic stress intensity factor. Fault interactions lead to complexity and statistical variability. Earthquake triggering, or initiation, is controlled by friction coefficients along with the space- and time-dependent stresses on fault elements which are computed by boundary element methods. Historical earthquakes that have moment magnitudes $m \geq 5.0$ during the last 200 years are used to prescribe the friction coefficients. A consequence of the minimum size of the fault elements is that the simulations do not generate earthquakes having magnitudes less than about $m \geq 5.8$. Therefore, additional parameters must be selected by systematic tuning of the model, followed by a search for sequences of events that optimally reproduce the known history of large earthquakes. Similar to the WGCEP approach, accuracy of results in the simulation approach are explicitly constrained by the limited availability of historic and instrumental data on large earthquakes occurring on faults in the model.

4. VC and Assimilation of Paleoseismic Data

Virtual California is an example of a fault simulator, other examples can be found as published by WARD (1992, 1996, 2000), RICHARDS-DINGER and DIETERICH (2008), and ROBINSON (2004). The topology of *VC* faults is shown in Fig. 1. The San Andreas fault (SAF) is the longest continuous fault and the greatest source of seismic hazard in California. Paleoseismic data from the SAF system provide an unparalleled opportunity for documenting and understanding the multi-cycle rupture history of a major active fault. Paleoseismic data consist of geologic observations of faulting from paleo-earthquakes. Data most commonly reported include characteristics of surface ruptures, number of rupture "events" during a Holocene or Quaternary time interval (resulting in "average recurrence interval" for paleo-earthquakes), date of the most recent earthquake and/or sequence of paleo-earthquakes (with uncertainty), and measurements of surface displacement from paleo-earthquakes (GRANT, 2007).

The relatively rich paleoseismic data set from the SAF system provides an unparalleled opportunity for comparison with results of simulations. The best paleo-earthquake record in North America is from the Wrightwood site on the SAF in southern California (FUMAL *et al.*, 2002; BIASI *et al.*, 2002; WELDON *et al.*, 2004, 2005). There are records of multiple ruptures at several other sites, including ten events at Pallett Creek (BIASI *et al.*, 2002; SIEH *et al.*, 1989), and Bidart Fan (GRANT *et al.*, 2005), also on the southern SAF. The record of paleo-earthquakes at these sites, which ruptured most recently in A.D. 1857, has formed the primary data set for probabilistic assessments of future southern San Andreas fault earthquakes, and for testing models of fault behavior and earthquake recurrence (WELDON *et al.* 2005; BIASI *et al.*, 2002). Paleo-earthquake data are also available from the northern SAF and other faults in the SAF system, such as the San Jacinto and Garlock.

Paleoseismic data were compiled and formatted for assimilation into VC simulations in an initial feasibility study (VAN AALSBURG *et al.*, 2007; GRANT, 2007). For this study, we used the same data set as VAN AALSBURG *et al.* (2007). Our goal is to obtain the statistical distribution of waiting times for simulated large earthquakes on specified faults and fault elements of the SAF system. We advance the *VC* model in 1 year increments, and simulate 40,000 years of earthquakes on the SAF system. Average slip on the fault elements and average recurrence intervals are tuned to match observed average rupture intervals at paleoseismic study sites. Due to fault interactions, slip events in the simulations display highly complex behavior, with no obvious regularities or predictability. For distinct groups of fault elements, the Weibull distribution represents the statistics of the largest earthquakes in a number of cases reasonably well, with fits to the empirical distribution functions having regression coefficients in excess of 0.99 (YAKOVLEV *et al.*, 2006).

5. "Data Scoring" Methods

VAN AALSBURG *et al.* (2007) describe a "data scoring" method for identifying time windows in a simulation record that are most similar to the actual paleoseismic record. In any simulation, there are intervals of simulated data that resemble the recent past few hundred years of earthquakes, and periods that are different. If we identify the intervals of simulation data that optimally resemble the recent past, we might hypothesize that the time intervals following these optimal intervals might then possibly characterize future activity on the actual San Andreas fault system.

Two data sets are used for scoring: VC simulation data, and paleoseismic data from the natural SAF system. For simulation data, we are interested in the event—the location corresponding to the latitude and longitude of the fault element, time of slip measured in simulation years, and the amount of slip. The analysis which follows used a catalog containing 200,000 events spanning 40,000 simulation years. The second data of paleoseismic sites and as discussed above, consists of observations dating back 1000 years on the SAF system in California. Unlike simulation data, paleo-earthquake times are known only within a time window ranging from a few years to several hundred years. This data is stored in XML format similar to VC simulation data, with each paleoseismic site containing one or more events defined by a minimum and maximum time value. There are 21 paleoseismic sites used to score the Virtual California catalog and a total of 119 observed events (see VAN AALSBURG et al., 2007; Table 1).

The first step in scoring is to associate paleoseismic sites with fault elements in the VC model. The association can be as single site-element pair (nearest-neighbor) or can include all VC elements within a specified radius (long-range neighborhood). The variable-range neighborhood is implemented because VC is only a simple representation of actual faults.

To score a particular simulation year, we consider the "current time" t_{sim} in the simulation record to represent the "present day", $t = 2009$. We then compare the time history prior to t_{score}, i.e., $t < t_{sim}$, to the known history from paleoseismic data. The scoring algorithm proceeds as follows: (1) For each paleoseismic site, we examine each Virtual California element and compare its slip times to the slip times recorded at that paleoseismic site. (2) A score is assigned based on the method described above, using the scoring function defined below. (3) If a *Virtual California* element occurred within a time window, the total score is incremented using one of the methods described above. The score for a particular simulation year is combined contributions from each paleoseismic site.

In this study, the VC simulation data is scored using a unit-height Gaussian function. The time $t_{P,j}(x)$ are the time of the jth paleoseismic event in years before "actual present" $t = 2009$ at the site x. Time $t_{S,i}(x)$ is the time of the ith "simulation paleoseismic event" in years before "simulation present" t_{sim} at the "simulation paleosite" x. $\sigma_{P,j}^2(x)$ is the quoted squared error of the actual paleoseismic event at the paleosite x. At each value of "simulation present time" t, we compute a score for that year for fault element i by summing over all paleoseismic events by using the event scoring function:

At location x, the contribution to the score $S_{i,j}(t, x)$ from the ith simulation event at time $t_{S,i}$, with respect to the jth paleoseismic event at time $t_{P,j}$, is given by

$$S_{i,j}(t,x) = \exp\left[-\left[t_{S,i}(x) - t_{P,j}(x)\right]^2 / \sigma_{P,j}^2(x)\right] \quad (1)$$

This scoring function assigns a higher score to events which occur closer to the mean paleoseismic value, and a smaller score for simulation events further removed in time from the actual paleoseismic event. A Gaussian is constructed for each paleoseismic event, centered about the mean event date so that about 90% of its area lies within the error bounds. This diminishes the importance of simulation events that occur far from the mean time of the actual paleo-earthquake. VAN AALSBURG et al. (2007) describe this procedure in more detail, and show examples of a "high scoring" time and a "low scoring" time.

The scoring system does not invoke a penalty if there are more VC earthquakes near a paleoseismic site than there are observed paleo-earthquakes. The rationale for this choice is that not all earthquakes can

Table 1

Fault probabilities and fault lengths for the next $M > 6.5$ earthquake corresponding to the spatial probabilities shown in Fig. 2a

Fault	Eq. probability (%)	Fault length (km)
Bartlett Springs	12.2	85.0
Calaveras	74.3	154.0
Concord-Green Valley	1.4	55.0
Death Valley	5.4	248.0
Greenville	0.7	73.0
Maacama	2.0	179.0
Rodgers Creek	2.0	62.0
San Gregorio	0.7	89.0
Sargent	0.7	53.0
San Andreas South	0.7	580.0

Faults not listed in the table had less than 0.1% probability of occurrence for the next $M > 6.5$ earthquake be observed using paleoseismic techniques, and thus the paleoseismic data represent a minimum number of paleo-earthquakes (GRANT, 2002).

6. Forecasting Feasibility Methods

Once a simulation time history has been scored year by year to obtain the time series Score(t_{sim}), we use the scored simulation in a forecasting experiment to forecast large earthquakes. The basic principle is that the higher the score at time t_{sim}, the more closely the seismic history leading up to t_{sim} resembles the actual seismic history of California. The assumption is that the seismic activity at "future" time intervals for times $t > t_{sim}$ will more closely resemble the seismic future in California if the score value is high. In addition, if we stack the time series data from "future" intervals, we can further surmise that the statistics of these stacked intervals may represent the statistics of future events on the real fault system. By using only the set of high-scoring times, together with their immediate future time intervals, we optimize VC to forecast seismic activity on the SAF system.

We select the "high scoring" years by applying a *decision threshold*. What constitutes a high score varies by method, radius of neighborhood, etc. Typically, we select a score so that approximately the top 0.37% of the simulated events are "high scoring". For each of these events we then compute the time until the next large events, either $m > 6.5$ or $m > 7.0$ (VAN AALSBURG et al., 2007). Although the paleoseismic data have been used as part of the procedure to set the model friction parameters and long-term offset rates (e.g., RUNDLE et al., 2001), using them to score the data is not redundant. The friction parameters use only the long-term, average recurrence intervals. The data scoring use the details of inter-event times, meaning that the scoring algorithm uses the variability of the data, rather than just long-term average rates.

7. Results

VAN AALSBURG et al. (2007) present several scoring algorithms, including the unit-height Gaussian scoring function described in Eq. 1, and showed examples of temporal cumulative probability functions (CDFs) obtained by stacking data from high scoring years. These temporal CDFs represented the probability of the next event larger than $M > 7.0$ as a function of time until the next event. The CDFs can generally be characterized as having a Poisson appearance because statistics from many fault elements were stacked. The median time to the next event was found to be generally around eight years. Applied to the present time (2009), this would indicate a 50% probability of an $M > 7.0$ event in California by 2016. VAN AALSBURG et al. (2007) also gave examples of the fit to the paleoseismic data set for a representative low-scoring simulation year, and a representative high-scoring simulation year.

Here we focus on identifying the probable locations of the next $M > 6.5$ and $M > 7.0$ earthquakes in California that may occur on the fault system shown in Fig. 1. To compute these locations, we use the top 0.37% of the highest scoring years as determined from the scoring function Eq. 1. Using these 148 highest-scoring years as "the present", we then determine the boundary element(s) that participate in the "next" $M > 6.5$ or $M > 7.0$ events. A boundary element is considered to have participated if it is within 40 km of the latitude–longitude coordinates of an actual, observed paleoseismic event. Compiling these statistics, we obtain results shown in Figs. 2a, b and 3a, b. These results demonstrate feasibility of the method only, and should not be taken as a statistically validated forecast.

Examination of Fig. 2a indicates that most of the probability for the next $M > 6.5$ event is associated with the Calaveras fault in northern California, with lower probability scattered among other faults in northern California, including the Rodgers Creek and Green Valley—Bartlett Springs fault system. Figure 2b indicates that most of the probability for the next $M > 7.0$ earthquake is associated with the Carrizo section of the San Andreas fault, the Garlock fault, the northern San Andreas fault, the Hunting Creek-Berryessa fault, and to a lesser extent the Hayward and Rodgers Creek faults. The probability for each of these faults is given in Tables 1 and 2.

In Fig. 3a and b, we address the question: "During the fixed time interval consisting of the next thirty years from now, on which fault locations are at least 1

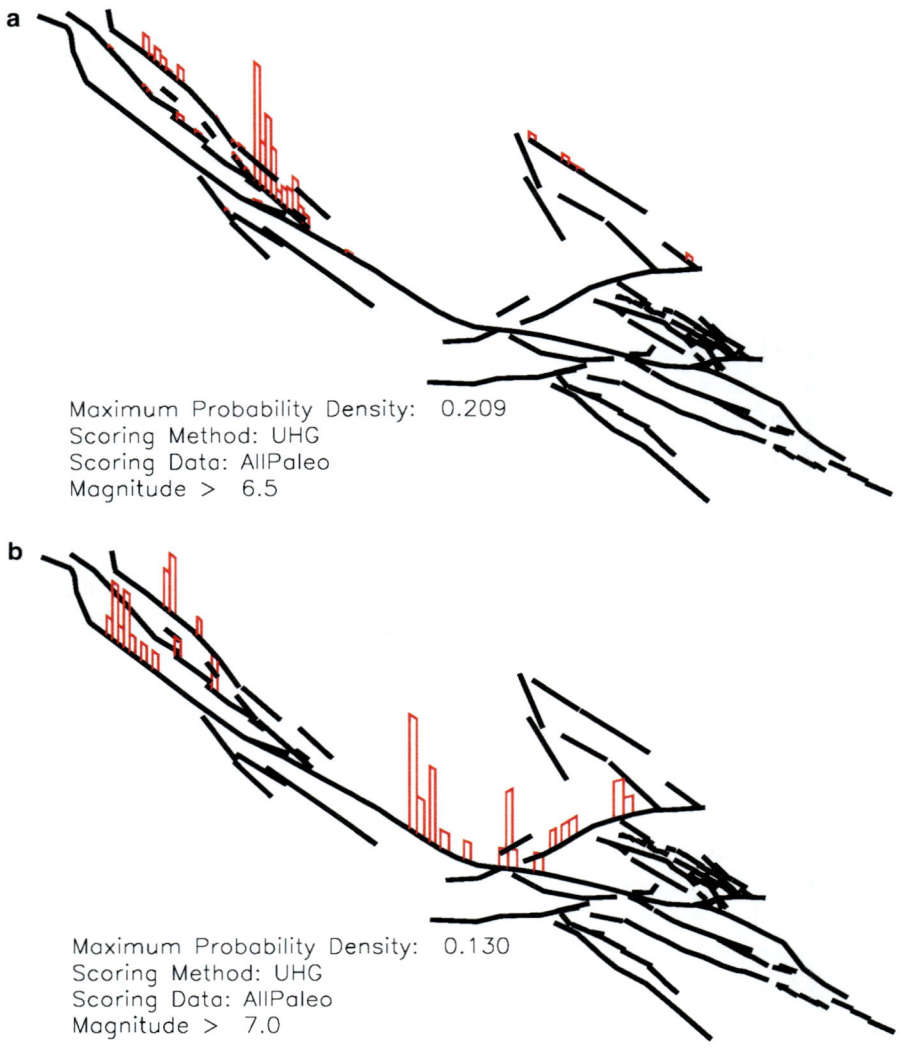

Figure 2
a Map showing the fault boundary element relative probabilities for participation in the next $M > 6.5$ earthquake. The corresponding fault probabilities are tabulated in Table 1. **b** Map showing the fault boundary element relative probabilities for participation in the next $M > 7.0$ earthquake. The corresponding fault probabilities are tabulated in Table 2

$M > 6.5$ (Fig. 3a) or at least 1 $M > 7.0$ events most likely to occur"? Figure 3a shows that the relative probability of $M > 6.5$ earthquakes is widely distributed spatially among many faults (30-year probability per $M > 6.5$ event). Figure 3b shows that for $M > 7.0$ earthquakes, probability is concentrated on the northern San Andreas fault between San Francisco and Mendocino, on the Carrizo section of the southern San Andreas fault, the Garlock and White Wolf faults, the northern San Andreas fault, the Rodgers Creek-Maacama faults, and the Hunting Creek-Berryessa faults. The probabilities for each fault are given in Tables 3 and 4.

In Fig. 4a and b, we address the question: "On the northern and southern San Andreas fault, *when* during the next thirty years are $M > 7$ earthquakes most likely to occur?" On the northern San Andreas fault, we focus on the spatial locations identified in Fig. 3b as being most likely to participate in a $M > 7$ earthquake. These locations can be recognized as having the red vertical bars, along the fault from Mendocino down to San Francisco. Figure 4a indicates that the

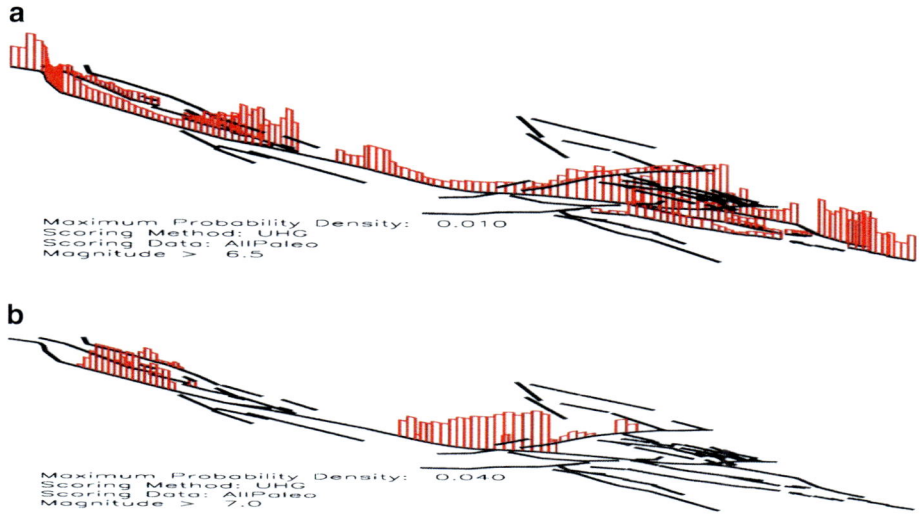

Figure 3
a Map showing the relative probabilities that at least 1 $M > 6.5$ earthquake will occur on the boundary element during the next 30 years. The corresponding fault probabilities are tabulated in Table 3. *Probability bars* are only plotted if the corresponding integrated fault probabilities are larger than 2%. **b.** Map showing the relative probabilities that at least 1 $M > 7.0$ earthquake will occur on the boundary element during the next 30 years. The corresponding fault probabilities are tabulated in Table 4

Table 2

Relative spatial probabilities that the next $M > 7.0$ earthquake will occur on a fault, corresponding to the spatial probabilities shown in Fig. 2b

Fault	Eq. probability (%)	Fault length (km)
Bartlett Springs	10.9	85.0
Hayward	4.3	111.0
Hunting Creek—Berryessa	2.2	59.0
Rodgers Creek	2.2	62.0
San Andreas North	23.9	467.0
San Andreas South	37.0	580.0
Garlock	13.0	234.0
White Wolf	6.5	47.0

Fault lengths are also listed. Faults not listed in the table had less than 0.1% probability of occurrence for the next $M > 7.0$ earthquake

Table 3

Relative spatial probabilities that at least 1 $M > 6.5$ earthquake will occur on a fault during the next 30 years (30-year probability per $M > 6.5$ event)

Fault	Eq. probability (%)	Fault length (km)
Calaveras	7.0	154.0
Hayward	2.3	111.0
Maacama	2.5	179.0
San Andreas North	15.9	467.0
San Andreas South	25.9	580.0
San Jacinto	7.1	291.0
Elsinore	3.2	236.0
Imperial Valley	11.1	162.0
Garlock	1.9	234.0
Brawley	1.9	52.0

Probabilities correspond to those shown in Fig. 3a, and fault lengths are also indicated

Faults not listed in the table had less than a 2% relative probability of occurrence for a $M > 6.5$ earthquake during the next 30 years

highest probability years are years 9 and 17 counting forward from the present, corresponding to 2018 and 2026. On the southern San Andreas fault, the most likely locations for an $M > 7$ earthquake during the next thirty years can be recognized by the vertical red bars located from the the Carrizon south to Fort Tejon. Figure 4b indicates that the most probable year for such an earthquake is year 26 counting forward from present, or 2035. However, on both the northern and southern San Andreas fault, there remains significant, although lesser, probabilities for such an event in other years.

Demonstrating the accuracy of a forecast is a very difficult problem.[3] Figure 4a and b also partially

---
[3] http://www.bom.gov.au/bmrc/wefor/staff/eee/verif/verif_web_page.html.

Table 4

Relative spatial probabilities that at least 1 M > 7.0 earthquake will occur on a fault during the next 30 years (30-year probability per M > 7.0 event)

Fault	Eq. probability (%)	Fault length (km)
Bartlett Springs	3.1	85.0
Hayward	0.6	111.0
Hunting Creek—Berryessa	0.6	59.0
Maacama	2.5	179.0
Rodgers Creek	0.6	62.0
San Andreas North	32.6	467.0
San Andreas South	54.0	580.0
Garlock	5.3	234.0
White Wolf	0.8	47.0

Probabilities correspond to those shown in Fig. 3a, and fault lengths are also indicated

Faults not listed in the table had less than a 0.1% relative probability of occurrence for a $M > 7.0$ earthquake during the next 30 years

answer the question: "For times identified as "optimal" during a *VC* simulation, do similar pasts imply similar futures?" This question bears on the accuracy of forecasts. The basic assumption in this paper is that similar pasts do imply similar futures. If this is not the case, use of simulations for earthquake forecasting will probably not be possible. Here, we have used a long history of simulations to identify optimal times whose preceding activity is similar to the actual paleoseismic events preceding the present, 2009. If the events following these optimal simulation times appear to be only a random sequence of earthquakes, uniformly distributed over the thirty year interval, this would suggest that past activity is not correlated with future activity. In that case, our proposed technique would probably not be useful.

Figure 4a and b appear to indicate that while there is a lower level background of random times, due to statistical variations, there are nonetheless a few times that stand out as preferred occurrence times for future large earthquakes. For Fig. 4a (northern San Andreas fault), these are years 9 and 17. For Fig. 4b (southern San Andreas fault), year 26 stands out. As the simulation model, including faults, average recurrence times, average long-term slip rates, and other model data are more closely matched to the actual San Andreas fault system data, it is possible that statistical variation will be reduced. Because there are nonetheless a few preferred times for future earthquakes that stand out above the relatively uniform background probability, Fig. 4a and b suggest the conclusion that similar earthquake pasts seem to be at least somewhat correlated with similar earthquake futures.

8. Discussion and Conclusions

We have presented a general method for using numerical earthquake fault system simulations to compute spatial forecast probabilities for earthquakes having magnitudes above a given, threshold. Our method utilizes catalogs of simulated earthquakes from the model *Virtual California*, together with a data scoring algorithm that identifies parts of simulation catalogs most similar to recent earthquake history in California as determined by paleoseismology. Optimal parts of the simulation catalogs are then used to compute statistical forecasts for future large events. While our results are preliminary, the probabilities we compute show the power of the method.

Our method can be compared to the recent methods developed by the Working Group on California Earthquake Probabilities (2002, 2008) (see notes [1, 2]), The WGCEP assume that coherent geological fault segments exist and rupture repeatedly as a unit (characteristic earthquake assumption), that earthquake ruptures do not generally jump from one fault to another, that earthquake ruptures typically obey either Brownian Passage Time or log-normal statistics, and that earthquakes on different fault segments are independent and uncorrelated.

In contrast, *VC* is a physically, rather than statistically, motivated model that assumes earthquake faults interact elastically, that friction retards slip on fault surfaces, and that faults typically slip at their observed, long-term rates. *VC* uses topologically realistic models of fault systems to generate catalogs of simulated major earthquakes that can then be analyzed statistically for patterns and other information. Here we show how these simulated catalogs can be used in earthquake forecasting. While the average intervals between paleoearthquakes are used to assign the frictional parameters on the model faults, the

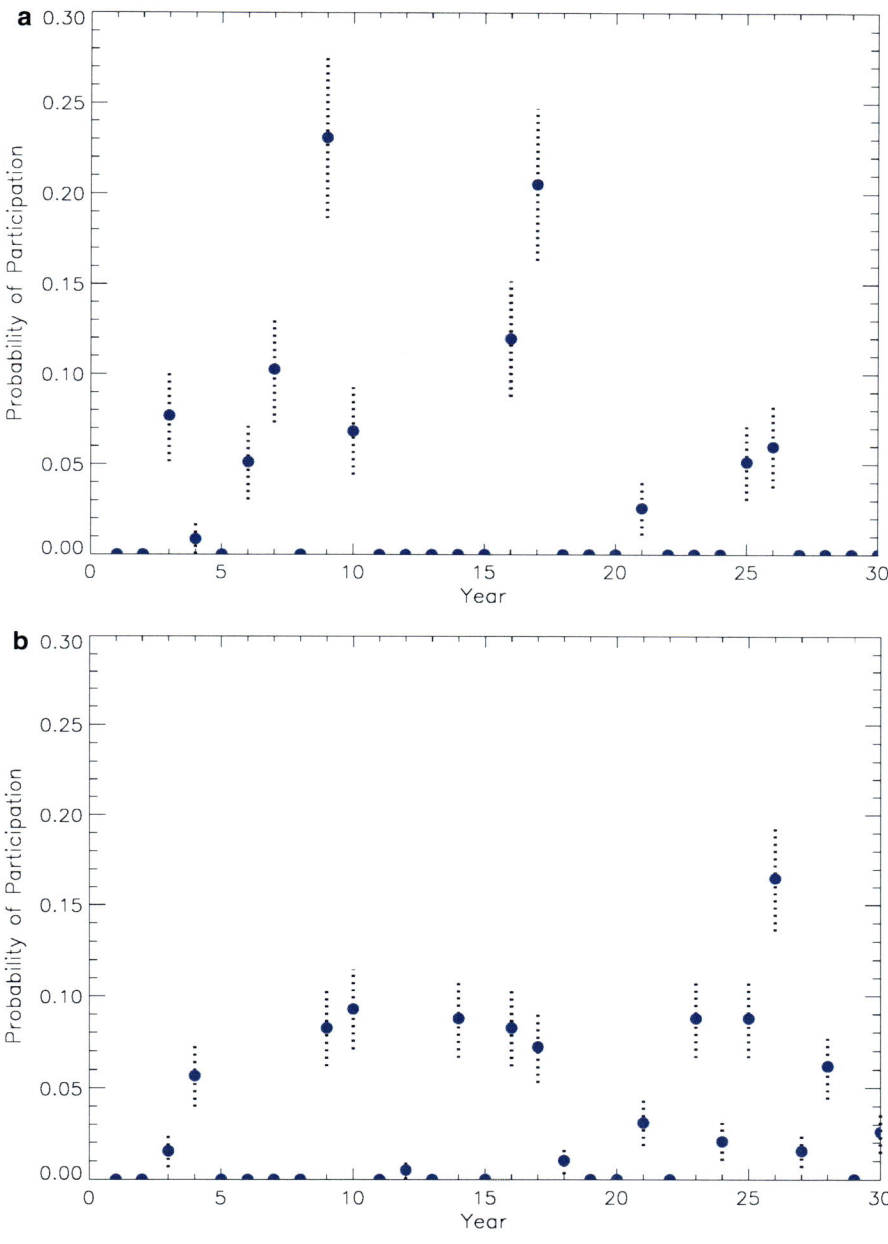

Figure 4
a Probability density function for the times during the next thirty years beginning from present (January 1, 2009) when a $M > 7.0$ earthquake is most likely to occur on the Northern San Andreas fault. Location on the fault corresponds to the high probability region on the NSAF shown in Fig. 3b. *Vertical bars* on data points indicate the 1σ Poisson counting uncertainty. **b** Probability density function for the times during the next thirty years beginning from present (January 1, 2009) when a $M > 7.0$ earthquake is most likely to occur on the Southern San Andreas fault. Location on the fault corresponds to the high probability region on the NSAF shown in Fig. 3b. *Vertical bars* on data points indicate the 1σ Poisson counting uncertainty

variability of the paleoearthquake occurrence times are used to determine which parts of the simulated catalogs are optimal for use in forecasting.

Finally, it is of interest to compare our forecasts to paleoseismic observations published by GRANT and SIEH (1994), GRANT (1996), and AKCIZ et al. (2009).

Their work suggested that recent ruptures of the SAF in the Carrizo were clustered in time ("uncharacteristic earthquakes") rather than more regularly spaced in time ("characteristic earthquakes"). So while the long-term average recurrence time might be several centuries (Sieh and Jahns, 1984; WGCEP 1988 and 1995), their data showed evidence for as many as four major earthquake ruptures between 1218 A.D. and 1510 A.D. The recent work by Akciz et al. (2009) reveals shorter average intervals.

The results shown in Fig. 2b suggest that the next major $M > 7.0$ earthquake could occur on the Carrizo reach of the SAF, possibly within thirty years from 2009. Under the "characteristic earthquake" scenario, with the most recent major rupture having occurred in 1857, it would be unlikely for another major rupture to occur in the near future. However, under a temporally clustered, "uncharacteristic earthquake" scenario, a major rupture in the Carrizo Plane might be expected in the near future.

With respect to $M > 6.5$ earthquakes, the most likely fault to rupture appears to be the Calaveras fault. Evidence from Coulomb stress transfer calculations (Reasenberg and Simpson, 1992) indicates that although the 1989 M 7.1 Loma Prieta earthquake might have raised the stress on the Calaveras fault by less than 1 bar, the seismicity rate nevertheless declined in the years following 1989. For that reason, the high probability on the Calaveras fault as shown in Fig. 2a is somewhat unexpected, if direct stress transfer from the Loma Prieta earthquake is assumed to be a triggering event.

Acknowledgments

This work has been supported by grants from the Computational Technologies Program of NASA's Earth-Sun System Technology Office to the Jet Propulsion Laboratory, the University of California, Davis, and the University of California, Irvine (JBR, PBR, JVA, GY, AD, LG), and by grant ATM 0327558 from the National Science Foundation (DLT).

Open Access This article is distributed under the terms of the Creative Commons Attribution Noncommercial License which permits any noncommercial use, distribution, and reproduction in any medium, provided the original author(s) and source are credited.

References

Akciz, S. O., Grant Ludwig, L., and Arrowsmith, J. R. (2009), *Revised dates of large earthquakes along the Carrizo section of the San Andreas Fault, California, since A.D. 1310 ± 30*, J. Geophys. Res. *114*, B01313, doi:10.1029/2007JB005285.

Biasi, G. P., Weldon, R. J. II, Fumal, T. E., and Seitz, G. G., (2002), *Paleoseismic event dating and the conditional probability of earthquakes on the southern San Andreas fault, California*, Bull. Seism. Soc. Am. *92*, 7, 2,761–2,781.

Eberhart-Phillips, D. et al. (2003), *The 2002 Denali Fault Earthquake, Alaska: A Large Magnitude, Slip-Partitioned Event*, Science *300*, 1113–1118, doi:10.1126/science.1082703.

Field, E. H. (2007), *A summary of previous working groups on California earthquake probabilities*, Bull. Seismol. Soc. Am. *97*, 1033–1053.

Fumal, T. E., Weldon, R. J. II, Biasi, G. P., Dawson, T. E., Seitz, G. G., Frost, W. T., and Schwartz, D. P. (2002), *Evidence for large earthquakes on the San Andreas fault at the Wrightwood, Californa, paleoseismic site: A.D. 500 to present*, Bull. Seismol. Soc. Am. *92*, 7, 2,726–2,760.

Grant, L. B., and Sieh K. E. (1994), *Paleoseismic evidence of clustered earthquakes on the San Andreas fault in the Carrizo Plain, California*, J. Geophys. Res., *99*, 6819–6841.

Grant, L. B. (1996), *Uncharacteristic earthquakes on the San Andreas fault*, Science, *272*, 826–827.

Grant, L. B., *Paleoseismology*, Chapter 30. In *IASPEI International Handbook of Earthquake and Engineering Seismology* (W. H. Lee, H. Kanamori, and P.C. Jennings, eds.) (Internatl. Assoc. Seismol. Phys. Earth's Inter. 2002) v. 81A, 475–489.

Grant, L., *Paleoseismology*. In *Treatise on Geophysics* (G. Schubert, ed.), Volume 4, *Seismology* (H. Kanamori, ed.), (Elsevier, 2007), 567–589.

Grant, L. B., Gould, M. M., Donnellan, A., McLeod, D., Yun-An Chen, A., Sung, S.-S., Pierce, M., Fox, G. C., and Rundle P. (2005), *A Web services-based universal approach to heterogeneous fault databases*, Comput. Sci. Eng. *7*, 4, 51–57.

Marco, S., Stein, M., Agnon A., and Ron, H. (1996), *Long-term earthquake clustering: A 50,000-year paleoseismic record in the Dead Sea Graben*, J. Geophys. Res. *101*, 6179–6192.

Page, M. T., and Carlson, J. M. (2006), *Methodologies for earthquake hazard assessment: Model uncertainty and the WGCEP-2002 forecast*, Bull. Seismol. Soc. Am. *96*, 1624–1633.

Reasenberg, P. A., and Simpson, R. W. (1992), *Response of regional seismicity to the static stress change produced by the Loma Prieta earthquake*, Science *255*, 5052, 1687–1690.

Richards-Dinger, K., and Dieterich, J. H., (2008), *A regional scale earthquake simulator for faults with rate- and state-dependent frictional properties*, Eos Trans. AGU *87*(52), Fall Meet. Suppl., Abstract S34A-08.

Robinson, R. (2004), *Potential earthquake triggering in a complex fault network: the northern South Island, New Zealand*, Geophys. J. Int. *159*, 734–738.

Rundle, J. B., and Kanamori, H. (1987), *Applications of an inhomogeneous stress (patch) model to complex subduction zone* earthquakes: A discrete interaction matrix approach, J. Geophys. Res. *92*, 2606–2616.

RUNDLE, J. B. (1988), *A physical model for earthquakes, 2, Application to southern California*, J. Geophys. Res. *93*, 6255–6274.

RUNDLE, J. B., RUNDLE, P. B., DONNELLAN, A., and FOX, G. (2004), *Gutenberg-Richter statistics in topologically realistic system-level earthquake stress-evolution simulations*, Earth Planets Space *56*, 761–771.

RUNDLE, J. B., RUNDLE, P. B., DONNELLAN, A., TURCOTTE, D. L., SCHERBAKOV, R., LI, P., MALAMUD, B. D., GRANT, L. B., FOX, G. C., MCLEOD, D., YAKOLEV, G., PARKER, J., KLEIN, W., and TIAMPO K. F. (2005), *A simulation-based approach to forecasting the next great San Francisco earthquake*, Proc. Natl. Acad. Sci. *102*, 15363–15367. http://www.pnas.org/cgi/doi/10.1073/pnas.0507528102.

RUNDLE, J. B., TIAMPO K. F., KLEIN W., and MARTINS J. S. S. (2002), *Self-organization in leaky threshold systems: The influence of near-mean field dynamics and its implications for earthquakes, neurobiology, and forecasting*, Proc. Natl. Acad. Sci. USA *99*, 2514–2521., Suppl. 1.

RUNDLE, P. B., RUNDLE, J. B., TIAMPO, K. F., MARTINS, J. S. S., MCGINNIS, S., and KLEIN, W. (2001), *Nonlinear network dynamics on earthquake fault systems*, Phys. Rev. Lett. *87*, 148501.

RUNDLE, P. B., RUNDLE, J. B., TIAMPO, K. F., DONNELLAN, A., and TURCOTTE, D. L. (2006), *Virtual California: Fault model, frictional parameters, applications*, Pure Appl. Geophys. *163*, 1819–1846, doi:10.1007/s00024-006-0099-x.

SAVAGE, J. C., and PRESCOTT W. H. (1978), *Asthenosphere readjustment and the earthquake cycle*, J. Geophys. Res. *83*, 3369–3376.

SIEH, K. E., and JAHNS, R. H. (1984), *Holocene activity of the San-Andreas Fault at Wallace-Creek, California*, Geolog. Soc. Am. Bull. *95*, 883–896.

SIEH, K., STUIVER, M., and BRILLINGER, D. (1989), *A more precise chronology of earthquakes produced by the San Andreas fault in southern California*, J. Geophys. Res. *94*, B1, 603–623.

VAN AALSBURG, J., GRANT, L. B., YAKOVLEV, G., RUNDLE, P. B., RUNDLE, J. B., YAKOVLEV, G., TURCOTTE, D. L, and DONNELLAN, A. (2007), *A feasibility study of data assimilation in numerical simulations of earthquake fault systems*, Phys. Earth Planet Int. *163*, 149–162.

WALD, D. J., and HEATON, T. H. (1994), *Spatial and temporal distribution of slip for the 1992 Landers, California, earthquake*, Bull. Seismol. Soc. Am. *84*, 668–691.

WARD, S. N. (1992), *An application of synthetic seismicity in earthquake statistics: The Middle America Trench*, J. Geophys. Res. *97*, 6675–6682.

WARD, S. N. (1996), *A synthetic seismicity model for southern California: cycles, probabilities, hazards*, J. Geophys. Res. *101*, 22393–22418.

WARD, S. N. (2000), *San Francisco Bay Area earthquake simulations: A step toward a standard physical earthquake model*, Bull. Seismol. Soc. Am. *90*, 370–386.

WELDON, R. J. II, SCHARER, K. M., FUMAL, T. E., and BIASI, G. P. (2004), *Wrightwood and the earthquake cycle: what a long recurrence record tells us about how faults work*, GSA Today *14*, 9, 4–10.

WELDON, R. J. II, FUMAL, T. E., BIASI, G. P., and SCHARER, K. M. (2005), *Past and future earthquakes on the San Andreas Fault*. Science *308*, 5724, 966–967.

WORKING GROUP ON CALIFORNIA EARTHQUAKE PROBABILITIES (WGCEP). (1988), *Probabilities of large earthquakes occurring in California on the San Andreas fault*, U.S. Geol. Surv. Open-File Rept., 62 pp.

WORKING GROUP ON CALIFORNIA EARTHQUAKE PROBABILITIES (WGCEP). (1995), *Seismic hazards in southern California: probable earthquakes, 1994–2024*, Bull. Seismol. Soc. Am. *85*, 379–439.

YAKOVLEV, G., TURCOTTE, D. L., RUNDLE, R. B., and RUNDLE P. B. (2006), *Simulation based distributions of earthquake recurrence times on the San Andreas fault system*, Bull. Seismol. Soc. Am. *96*, 1995–2007.

ZHUANG, J., OGATA, Y., and VERE-JONES, D. (2004), *Analyzing earthquake clustering features by using stochastic reconstruction*, J. Geophys. Res., *109*, B05301, doi:10.1029/2003JB002879.

(Received August 2, 2008, revised February 25, 2009, accepted March 11, 2009, Published online April 23, 2010)

Spatial Separation of Large Earthquakes, Aftershocks, and Background Seismicity: Analysis of Interseismic and Coseismic Seismicity Patterns in Southern California

EGILL HAUKSSON[1]

Abstract—We associate waveform-relocated background seismicity and aftershocks with the 3-D shapes of late Quaternary fault zones in southern California. Major earthquakes that can slip more than several meters, aftershocks, and near-fault background seismicity mostly rupture different surfaces within these fault zones. Major earthquakes rupture along the mapped traces of the late Quaternary faults, called the principal slip zones (PSZs). Aftershocks occur either on or in the immediate vicinity of the PSZs, typically within zones that are ±2-km wide. In contrast, the near-fault background seismicity is mostly accommodated on a secondary heterogeneous network of small slip surfaces, and forms spatially decaying distributions extending out to distances of ±10 km from the PSZs. We call the regions where the enhanced rate of background seismicity occurs, the seismic damage zones. One possible explanation for the presence of the seismic damage zones and associated seismicity is that the damage develops as faults accommodate bends and geometrical irregularities in the PSZs. The seismic damage zones mature and reach their finite width early in the history of a fault, during the first few kilometers of cumulative offset. Alternatively, the similarity in width of seismic damage zones suggests that most fault zones are of almost equal strength, although the amount of cumulative offset varies widely. It may also depend on the strength of the fault zone, the time since the last major earthquake as well as other parameters. In addition, the seismic productivity appears to be influenced by the crustal structure and heat flow, with more extensive fault networks in regions of thin crust and high heat flow.

Key words: Seismicity, California, faults, aftershocks, interseismic seismicity, fault damage zones, San Andreas fault system, evolution of fault zones, earthquake interaction.

1. Introduction

We analyze the Southern California seismicity located in the vicinity of late Quaternary faults to answer the question whether large earthquakes, aftershocks, and background seismicity occur within the same parts of fault zones that are often several kilometers wide. Establishing this spatial relationship is the first step towards understanding the difference in source physics between large and small earthquakes.

In the 1920s Dr. Harry Wood of the Carnegie Institute in Pasadena (today known as the Caltech Seismological Laboratory) proposed that a seismic network should be installed in southern California (WOOD, 1916). He argued that recording the more frequent small earthquakes would help us understand future large damaging earthquakes. The Southern California Seismic Network (SCSN), now a joint project of Caltech and USGS, has been in operation since then and recorded more than 400,000 both small and large earthquakes. During the same time period, geologists have collected data on late Quaternary faults in southern California (e.g., FRANKEL et al., 2002). We synthesize both data sets in an attempt to answer some of the questions regarding how seismicity and faults are related as proposed by Dr. Wood in 1916.

We analyze the southern California earthquake catalog from 1981 to 2005, and the data set of fault segments or principal slip zones (PSZs) of the Southern California Earthquake Center Community Fault Model (SCEC/CFM) (PLESCH et al., 2007). The PSZs accommodate the major earthquakes and thus could have different material properties as well as strength than the adjacent crust. Commonly, the thickness of faults is considered to be smaller than 10 m, with maximum shear on the PSZ occurring on the outer surfaces of fault cores (CHAMBON et al.,

---

[1] Seismological Laboratory, Division of Geological and Planetary Sciences, California Institute of Technology, Pasadena, CA 91125, USA. E-mail: Hauksson@gps.caltech.edu

2006). The thickness of the principal slip zone is thought to be small or only 1- to 10-mm wide (SIBSON, 2003). Their lengths may extend from 10 to 100 s of kilometers while their depths usually extend to 15 km or as much as 25 km. The relationship between large earthquakes and late Quaternary faults is obvious when a large earthquake ruptures the surface (e.g., SIEH et al., 1993). Because the background seismicity has no surface rupture, we use the location of the hypocenter relative to the fault surface to infer the spatial relationship between the seismicity and the PSZs.

Several recent studies have attempted to associate earthquakes and faults in southern California. Analyzing seismicity around a few strike-slip faults, WESNOUSKY (1990) inferred that the background seismicity rate adjacent to late Quaternary faults in southern California was controlled by cumulative fault offset. Focusing on the 1992 Landers aftershocks, LIU et al. (2003), studied the relationship of the aftershocks to the main shock PSZs to estimate the size of the seismic damage zone. They showed that the Landers aftershocks formed a narrow spatial distribution around the PSZ, although only a small fraction of aftershocks seemed to be caused by slip on the PSZ. More recently, WESSON et al. (2003) developed a Bayesian technique for associating historical and instrumental seismicity with faults in the San Francisco Bay area, California. WOESSNER and HAUKSSON (2006) used the Bayesian statistics technique to associate the southern California background seismicity to the CFM, and synthesized the overall statistical patterns. They showed that ~40% of earthquakes occur within 2 km and 60% within 4 km, and found evidence for larger earthquakes being preferentially located closer to the major fault zones.

To identify spatial alignments, we plot the background seismicity in colors as a function of distance from the nearest fault segment (Fig. 1). The bright red to yellow alignments of seismicity such as parts of the San Andreas fault, the San Jacinto fault, the southern Sierra Nevada, and the aftershock zones of

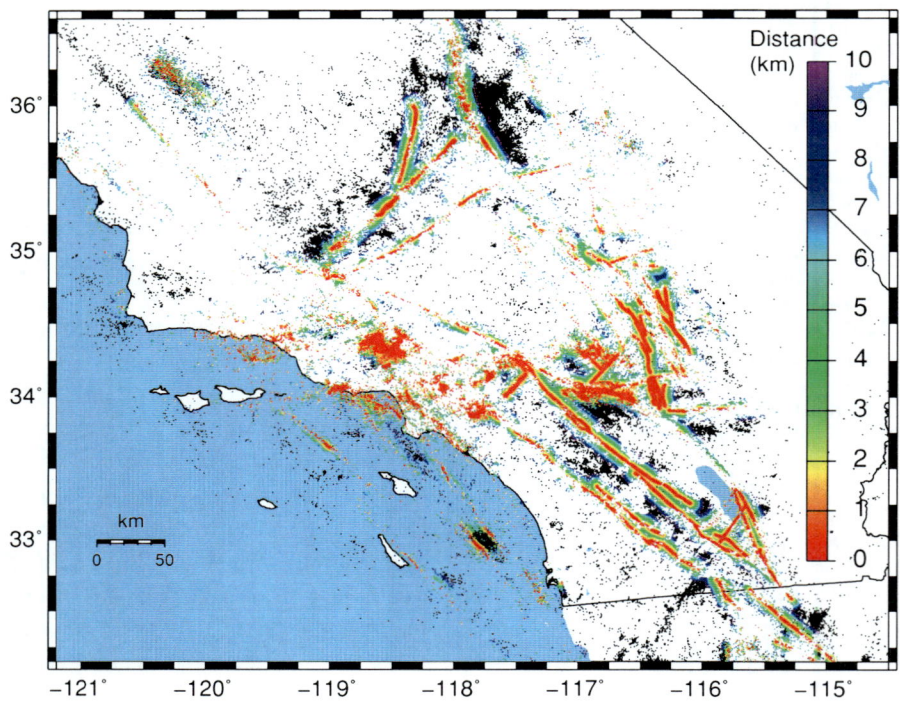

Figure 1
Late Quaternary fault traces are not plotted on this map. Map showing the southern California relocated seismicity from 1981 through 2005. Each epicenter is colored to show the distance from the nearest mapped late Quaternary fault segment (SCEC/CFM), with color bar from 0 to 10 km. Earthquakes in the distance range of 10–20 km are plotted in black. Note how the major faults are illuminated by the adjacent seismicity

the 1992 Landers and 1999 Hector Mine earthquakes, identify where the seismicity is concentrated near PSZs. In contrast, several both high and low slip-rate faults located in the Western Transverse Ranges as well as the Mojave Desert are not surrounded by significant seismicity, illustrating that the relationship between background seismicity and faults is complex.

There are many factors that may influence the seismicity distribution around the PSZ of a fault. The strength of the fault and varying loading of fault zones may affect the width of the seismicity distribution near the PSZs. The geometrical shape and productivity of a seismicity distribution may depend on where within the seismic cycle the fault segment happens to be. For instance, if the fault just had a main shock the seismic damage zone may be dominated by aftershocks. Alternatively, if the fault is late in the seismic cycle it could have returned to normal background seismicity. The age of the fault, initial growth of complexity, and subsequent smoothing of the fault (e.g., SAGY et al., 2007) may influence the seismicity distribution adjacent to the PSZs. Similarly, external effects such as triggering or regional stress release caused by other earthquakes may influence the seismicity. Thus, synthesizing the seismicity with the fault zone properties may provide new understanding of which of these processes are more important than others.

2. Earthquake Data

We analyze the earthquake catalog from the Southern California Seismic Network, a joint project of the USGS and Caltech. LIN et al. (2007) relocated this catalog to improve the earthquake hypocenters by using absolute travel times and cross-correlation differential travel times as well as double-difference type location techniques. We determined the statistical properties (mean and standard deviation) of the seismicity distributions next to each fault segment using a routine from PRESS et al. (1997).

We also analyzed the Southern California Earthquake Center community fault model (SCEC/CFM 3.0), which is a model of fault segments rather than many segments daisy-chained into whole faults (PLESCH et al., 2007). The SCEC/CFM consists of 162 principal slip zones (PSZ) of late Quaternary faults that are mapped in three dimensions (3-D) (Fig. 2). The SCEC/CFM fault model is based mostly on geological data, although in some instances seismicity data are included. The segmentation is somewhat subjective, for instance, the southern San Andreas fault consists of more than four segments, and the Garlock fault consists of only one segment. Short fault segments are more common than long ones. The three-dimensional (3-D) shapes of principal slip surfaces are defined in the SCEC/CFM representation.

A subset of 75 SCEC/CFM fault segments has measured or assigned slip-rates, which have varying error bars (FRANKEL et al., 2002; S. Perry, written communication, 2007). Because both the locations of the seismic stations and the locations of the fault segments are based on global positioning measurements (GPS), it becomes possible to evaluate their relative positions. A. Plesch (written communication, 2007) provided the Euclidian measurements of distances from each hypocenter to the nearest principal slip surface in the SCEC/CFM.

The uncertainties and biases in the data used in this study could affect the results variously. For instance, the PSZs may be incorrectly mapped at the surface or field data incorrectly digitized. Alternatively, a PSZ could be assigned the wrong dip or could have a more complex shape than can be inferred at the surface. In these cases, the mean of the cluster would be offset and in some cases the distribution could be artificially skewed. The earthquake location mean absolute horizontal and depth errors are ~ 0.2 and ~ 0.4 km, respectively, while the relative errors are a factor of ten smaller (LIN et al., 2007). Thus, in extreme cases, the hypocenters may be mislocated by up to at most 1 km horizontally and 2 km in depth, although this is unlikely. Because the velocity contrasts across faults in southern California are usually small, we do not expect systematic location biases but rather random errors, which will have insignificant effects. If faults are closely spaced, some seismicity may be assigned to one fault rather than the other, which causes minor artificial biases in evaluating the seismicity for individual segments.

To average out some of the uncertainties in the data sets and to analyze how the different faulting and

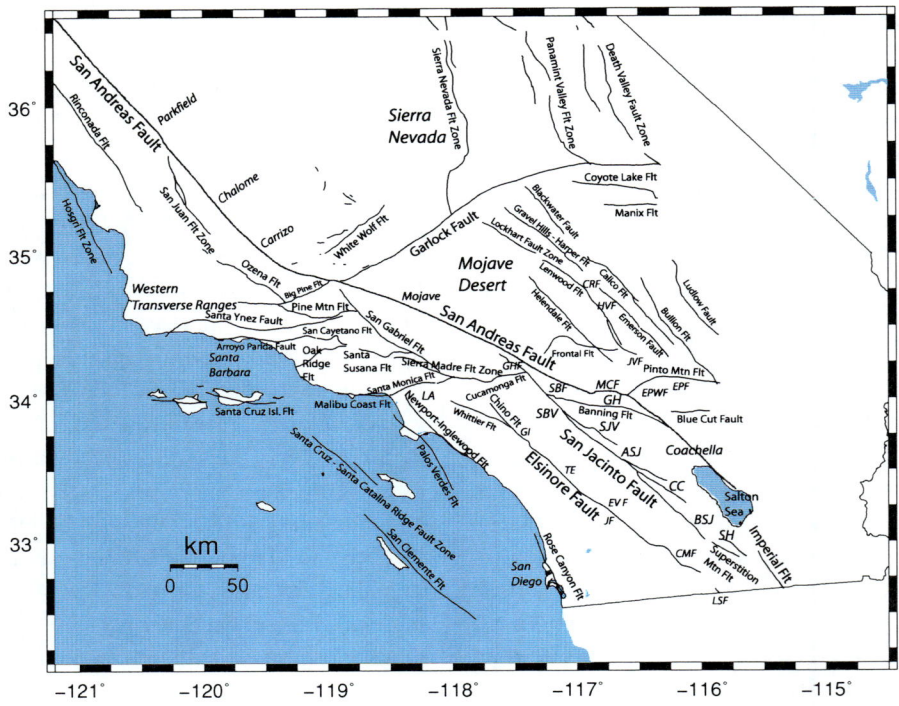

Figure 2
Simplified fault map of southern California based on the JENNINGS (1994) map. *ASJ*—Anza San Jacinto, *BSJ*—Borrego San Jacinto, *CC*—Coyote Creek, *CMF*—Coyote Mountain fault, *CRF*—Camp-Rock fault, *EPF*—Eureka Peak fault, *EPWF*—Eureka Peak West fault, *EVF*—Earthquake Valley fault, *GH*—Garnet Hill, *GHF*—Glen Helen Fault, *GI*—Glenn Ivy, *HVF*—Homestead Valley fault, *JF*—Julian Fault, *JVF*—Johnson Valley fault, *LA*—Los Angeles, *LSF*—Laguna Salada fault, *SBV*—San Bernardino Valley, *SH*—Superstition Hill, *SJV*—San Jacinto Valley, and *TE*—Temecula Elsinore

seismicity parameters vary, we have divided the CFM faults into five groups (Table 1). The first and second groups are defined based on the slip rate. The first group of high slip-rate faults has segments with fast slip rates (≥ 6 mm/year) such as the San Andreas and San Jacinto faults. The second group of low slip-rate faults has slip rates of <6 mm/year. The separation slip-rate of 6 mm/year is chosen somewhat arbitrarily. The third group consists of fault segments with aftershocks and includes aftershock sequences during the time period covered by the catalog (1981–2005). There are 15 aftershock defined fault segments which accommodated some of the large aftershock sequences such as the 1986 Palm Springs, 1987 Superstition Hill, 1990 Upland, 1992 Landers, 1994 Northridge, and 1999 Hector Mine.

The fourth group, which we call "defined by seismicity group," consists of 5% of the CFM fault segments, which are defined mostly by using seismicity. These seismicity distributions often exhibit swarm-like behavior. The seismicity clusters in the regions where these faults are defined form linear trends, which suggest the presence of fault segments. The separation of the seismicity defined segments from the other groups of CFM segments avoids the possible circular reasoning of inferring relationships between seismicity and faults that are defined based on seismicity. Only three of these segments have assigned slip-rates. The fifth group, which is called "unconstrained seismicity group," consists of segments located near the edges or outside the SCSN monitoring region, which is about 20% of the CFM fault segments. The hypocenters near these segments are not well constrained and thus may exhibit unexpected biases.

Several segments that are difficult to categorize and could be in either the aftershock or seismicity defined groups were assigned to the aftershock group.

Table 1

Subdividing the SCEC/CFM Fault segments into Groups

Fault type	Count	Count in %	Count with slip rate
Fast fault slip rate (≥6 mm/year)	13	8	13
Slow fault slip rate (<6 mm/year)	95	58	39
Had main-shock rupture and aftershocks in the last 25 years	15	9	10
Defined mostly by seismicity	8	5	3
Seismicity distribution unconstrained	32	20	10
Total	162	100	75

Also, two high slip rate segments, the Parkfield segment of the San Andreas fault and the Brawley seismic zone are unusual because although they both have high slip rates of 34 and 20 mm/year, respectively, neither was assigned to the high slip-rate group. The Parkfield segment is outside the monitoring region of the SCSN and also had a M 6 main shock-aftershock sequence in 2004. It was assigned to the unconstrained seismicity group. The Brawley seismic zone has a high slip rate however its geometrical shape is mostly based on seismicity. It was assigned to the defined-by-seismicity group. We also calculated the a-value and b-value Gutenberg–Richter parameters for each of the fault groups using the *zmap* software (WIEMER, 2001).

3. Results

Spatially clustered distributions of seismicity exist near the PSZs of all late Quaternary faults in southern California. We call these regions of small slip surfaces where these distributions are located seismic damage zones. These zones extend from the PSZs out to horizontal distances of ±10 km. Most of the seismic damage zones are complex and the seismicity does not cluster at the PSZs, except for aftershocks. Instead, the PSZs are often illuminated by changes in the depth distributions of seismicity, with different depth distributions of seismicity on each side of the fault. In addition, often the rate of seismicity can be higher on one side than the other of the PSZs. These patterns of seismicity suggest that the PSZs are acting more as material discontinuities than zones of weakness where background seismicity is preferentially accommodated.

3.1. Seismicity Patterns Near Selected PSZs

We have analyzed the interseismic seismicity of the three high slip-rate strike-slip faults (San Andreas, San Jacinto and Elsinore faults). We have also analyzed the pre-seismicity as well as the aftershock patterns of the 1992 Landers earthquake to illustrate the complex relationships between the seismicity and the PSZs.

3.1.1 Southern San Andreas Fault

The cross-section histograms and the fault normal depth sections for the San Andreas segments illustrate the complex seismicity distributions in the immediate vicinity of the PSZs (Fig. 3). The central segments exhibit lower rates of background seismicity than the south segments. In general, the Chalome, Carrizo, Mojave, and San Bernardino segments exhibit a somewhat peaked level of activity within ±4 km distance of the PSZs, and more distributed activity on the west side. They also exhibit different depth distributions on either side of the PSZs.

In the south, the damage zone seismicity is distributed over ±10-km-wide zones and the shapes of the histograms are very complex. The Banning, Mill Creek, Garnet Hill, and Coachella segments show distributions with the predominant activity on the east side of the PSZs. The complexity in the seismicity distribution is in part related to the multiple strands of the San Andreas fault through Banning and San Gregonio Pass.

Several segments, such as the Carrizo, Mojave, San Bernardino, and Mill Creek segments show evidence of seismic quiescence with a small decrease in the histogram values near the PSZs,

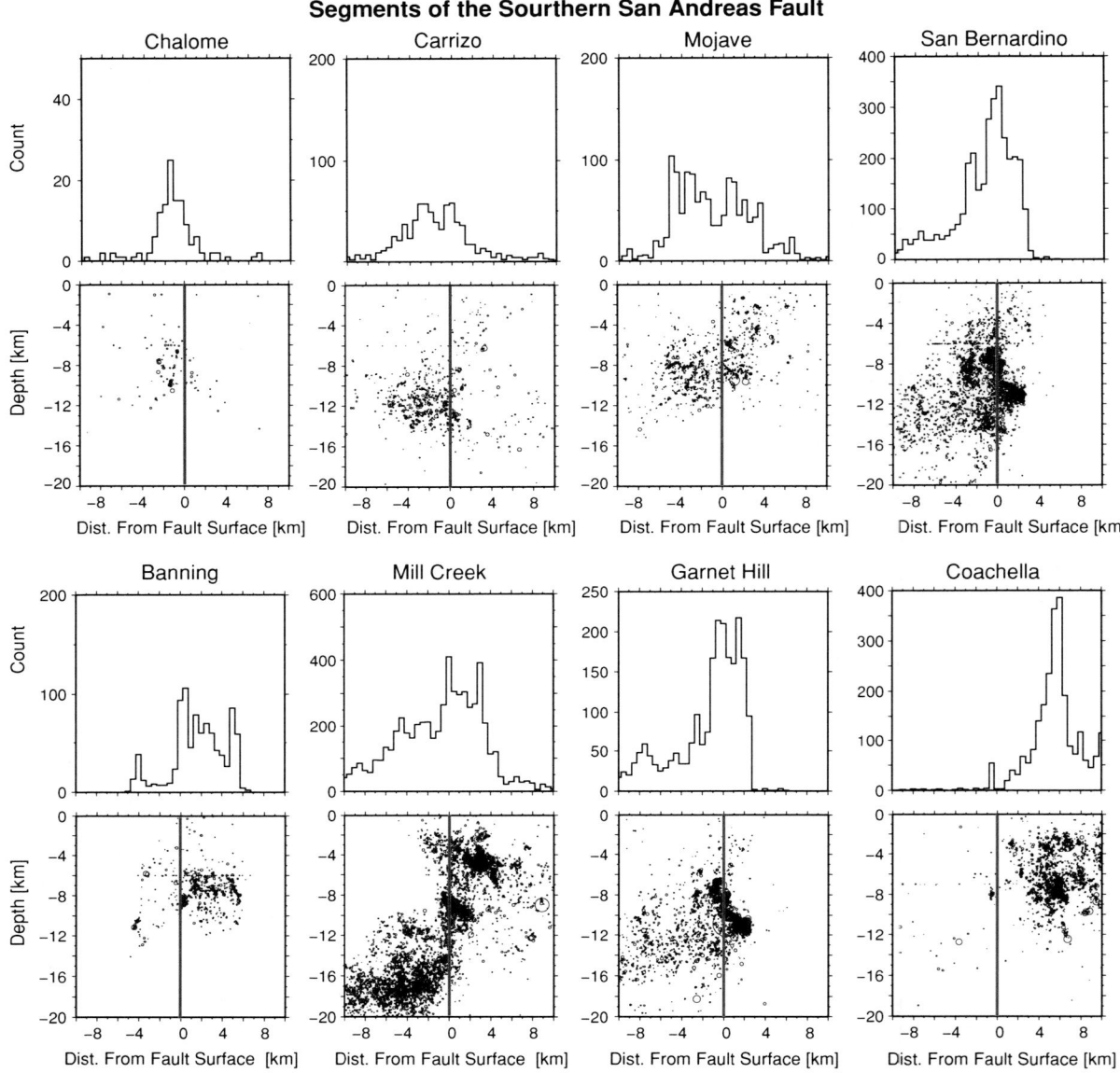

Figure 3
Histograms and depth sections for seismicity along segments of the San Andreas fault. The zero distance (*vertical reference line*) in depth sections is relative to the CFM surface and is the PSZ. The "Distance from CFM fault surface" is measured between each hypocenter and the nearest CFM triangular element of a fault surface. Thus, the PSZs for dipping faults are also vertical

although this decrease is probably not statistically significant. The presence of seismic quiescence near the core of the PSZs could indicate the possible presence of a very thin, almost not resolvable, zone of weakness or that the entire PSZ is locked and not slipping.

3.1.2 San Jacinto Fault

The San Jacinto fault zone exhibits the highest level of damage zone seismicity when compared to all other faults in southern California (Fig. 4). The corresponding depth distributions of the seismicity

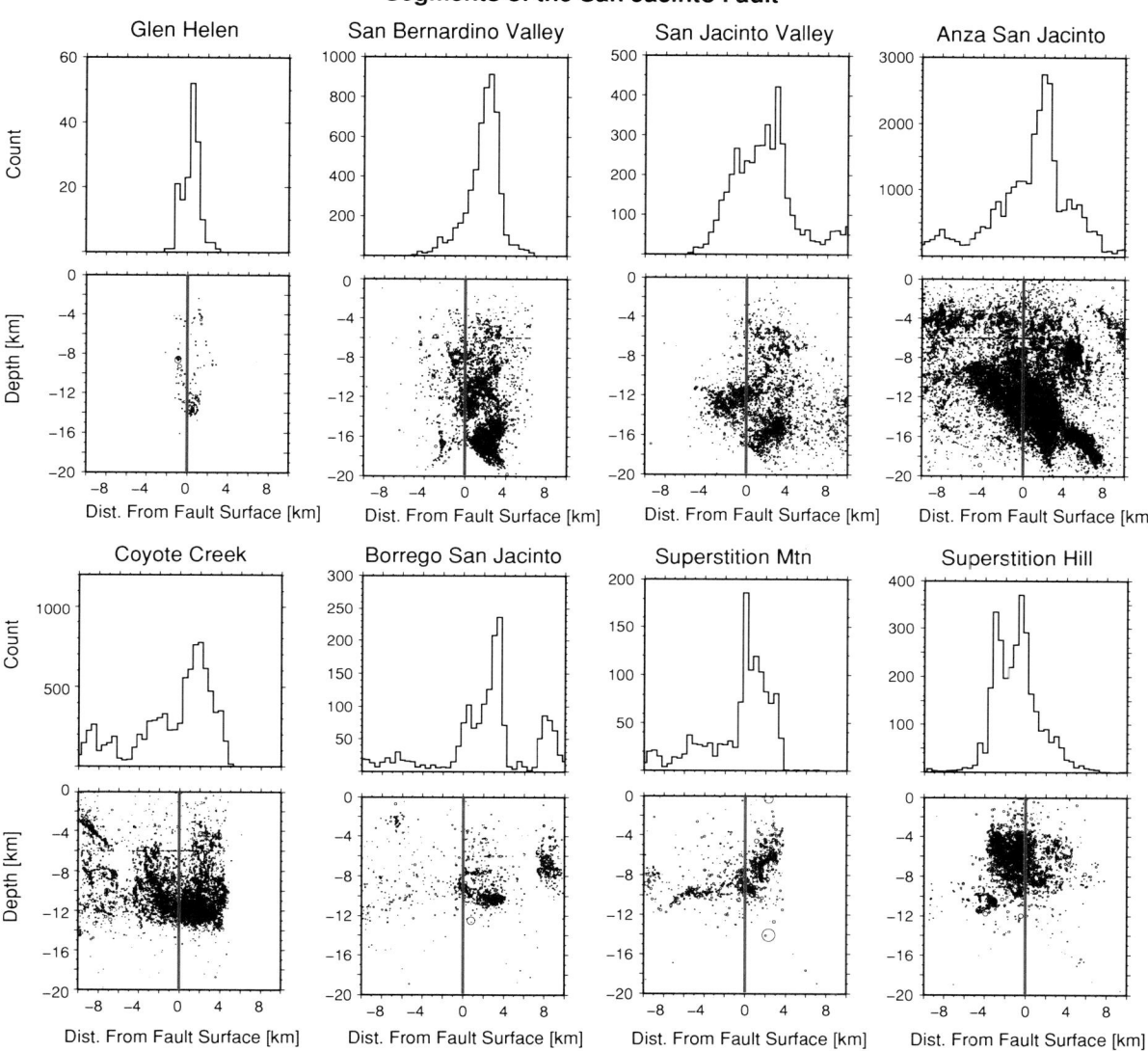

Figure 4
Histograms and depth sections for seismicity along segments of the San Jacinto fault. The zero distance (*vertical reference line*) in depth sections is relative to the CFM surface and is the PSZ

also illustrate complex distributions and the absence of clustering near the PSZs. Often the density and depth distribution patterns are different on either side of the PSZs. The histograms of hypocentral distances for the San Jacinto fault segments exhibit different shapes and other complexity in part caused by their *en-echelon* juxtaposition with other CFM segments. In several cases this seismicity is clustered in extensional step-overs, between the *en-echelon* fault segments. The north segments have distinct peaked distributions that are offset from the PSZs. In some cases the dip of the PSZ may not be correct and thus the seismicity is artificially offset. One of these segments, the seismic damage zone of the Anza San Jacinto segment differs from other damage zones because it has a considerably higher rate of seismicity than adjacent segments. It also has a wide damage zone with three to five times the productivity associated with it when compared to the adjacent segments.

The southern segments, Coyote Creek, Borrego San Jacinto, Superstition Mountain, and Superstition Hill all exhibit changes in density of seismicity across the PSZs, with some of the seismicity clustered near the brittle-ductile transition zone. They all exhibit 4–6 km shallower seismicity than observed to the north (Fig. 4). The shallower seismicity is consistent with higher heat flow in this region.

3.1.3 Elsinore Fault

The seismicity distributions next to the PSZs of the Elsinore fault segments are diffuse (Fig. 5). The seismic damage zones around the north Elsinore PSZs exhibit similar seismicity rates as the San Andreas fault segments. The Whittier, Chino, Glen Ivy, and Temecula Elsinore segments exhibit low level of activity, with different seismicity rates on each side of the PSZs.

The Julian, Earthquake Valley, Coyote Mountain, and Laguna Salada segments have higher levels of seismicity, although the maximum depth of the seismicity becomes shallower to the south. Similar to some of the PSZs of the San Andreas fault, the Earthquake Valley segment apparently exhibits seismic quiescence around the PSZ in the depth range of 4–12 km. However, this could be an artifact because the PSZ associated seismicity and focal mechanisms suggest that it changes dip along strike (C. Nicholson, written communication, 2008). The seismic damage zones of the Chino and Earthquake Valley segments form asymmetric truncated distributions because the seismicity is assigned to other adjacent segments. The Julian and Coyote Mountain segments exhibit somewhat peaked histogram distributions near the PSZs. The corresponding depth distributions show no obvious features related to the PSZs except for the absence of shallow seismicity in the Coyote Mountain section as well as a higher seismicity rate on the east side. Thus, the Whittier, Chino, Temecula, and Laguna Salada PSZs are juxtaposing two crustal blocks with different background seismicity rates. Although the Laguna Salada segment to the south has poorly defined seismicity, it shows a clear increase in seismicity from east to west.

3.1.4 The 1992 Landers Sequence

The fault-normal depth distributions of the 1992 $M_w7.3$ Landers seismicity recorded before and after the main shock are very different (Fig. 6). The 1981–1991 background seismicity preceding the main shock is a factor of 30 lower even though we combine the seismicity for all the Landers PSZs. It is distributed around the fault out to distances of ±10 km as we observe for the other strike-slip faults. In contrast, the aftershocks around all of the Landers PSZs show peaked distributions with a kurtosis of ~ 10, which is larger than the kurtosis of ~ 0 for the pre-main-shock background seismicity. The aftershock distributions are clearly centered on the PSZ, with an average width of ± 2 km.

The Johnson Valley, Homestead Valley, or Eureka Peak faults display peaked distributions with similar shapes. However, in some of the histograms and depth sections the seismicity is truncated by nearby fault segments, making artificial abrupt terminations to the distributions. The northernmost Camp rock segment, where the mainshock fault rupture terminated, has the lowest level of activity. Overall the aftershock depth distributions are symmetric, exhibit the highest level of seismicity at the PSZs, and decay with distance away from the PSZs.

3.1.5 Summary

The background seismicity and aftershocks form very different spatial patterns. The background seismicity appears to be driven by localized heterogeneous crustal shear and the availability of small slip surfaces near the PSZs. The density of these small slip surfaces decreases with distance away from the PSZs. In contrast, the aftershocks are clearly centered at the PSZ of the main shock, and probably driven by the heterogeneous stress field left behind by the main shock.

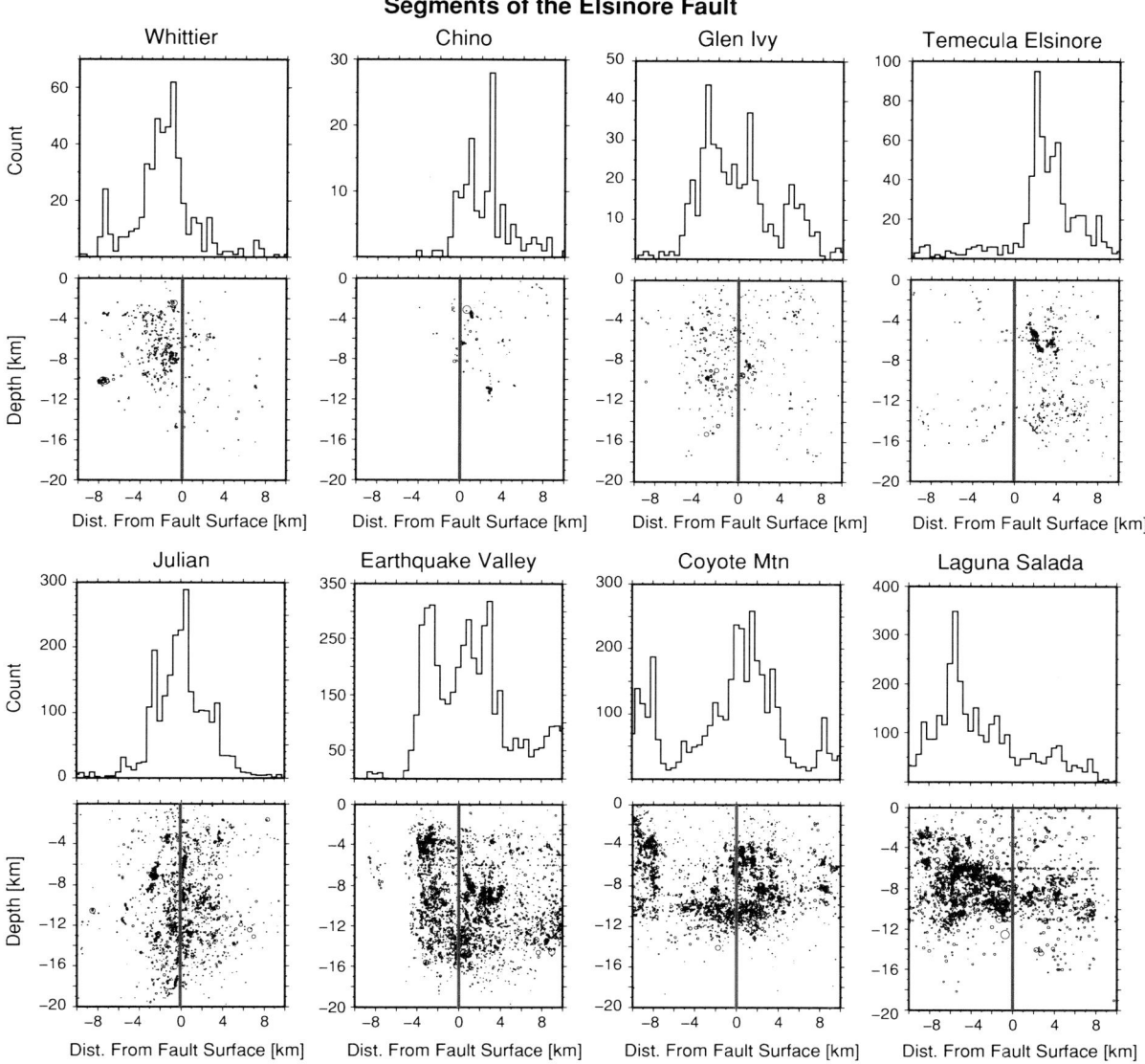

Figure 5
Histograms and depth sections for seismicity along segments of the Elsinore fault. The zero distance (*vertical reference line*) in depth sections is relative to the CFM surface and is the PSZ

3.2. Decay of Seismicity with Distance Away from PSZs

To explore the spatial relationship between the seismicity and the PSZs we have determined the distance decay of the fault normal density of seismicity. To search for possible differences in distance decay, we analyzed the decay rate of interseismic background seismicity near high sliprate strike-slip faults, and the 1992 Landers aftershocks. We also compared the distance decay of the five groups of PSZs.

The fault normal density of the interseismic seismicity, next to three major strike-slip faults, shows a constant rate of seismicity within a ±2.5-km-wide

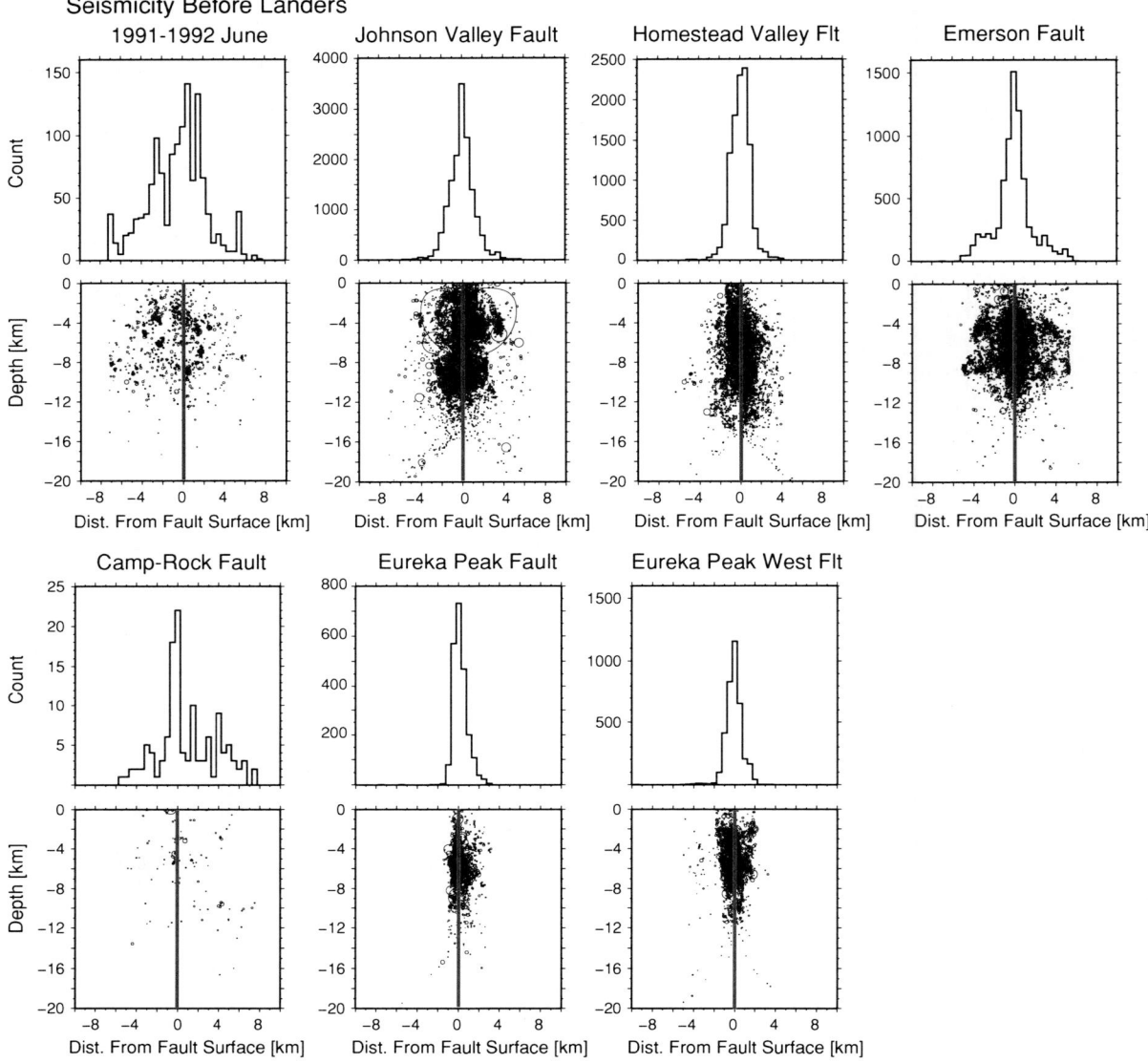

Figure 6
Histograms and depth sections for seismicity along segments that ruptured in the 1992 Landers sequence. The zero distance (*vertical reference line*) in depth sections is relative to the CFM surface and is the PSZ

fault zone (Fig. 7). The presence of the fault zone is consistent with geological features that form adjacent to major PSZs. Outside of the fault zone, the background seismicity rate decays at a rate ranging from $10^{-1.28}$ to $10^{-1.79}$. The distance decay of seismicity could reflect a possible decrease in permeability and porosity, which affect effective strength, or levels of tectonic stress as well as a decrease in availability of small slip surfaces.

The PSZs of the 1992 Landers earthquake exhibit similar constant rate of activity within a ±2-km-wide fault zone. Outside of the fault zone, the distance decay of pre-seismicity and aftershocks associated with the 1992 Landers surface rupture exhibits a

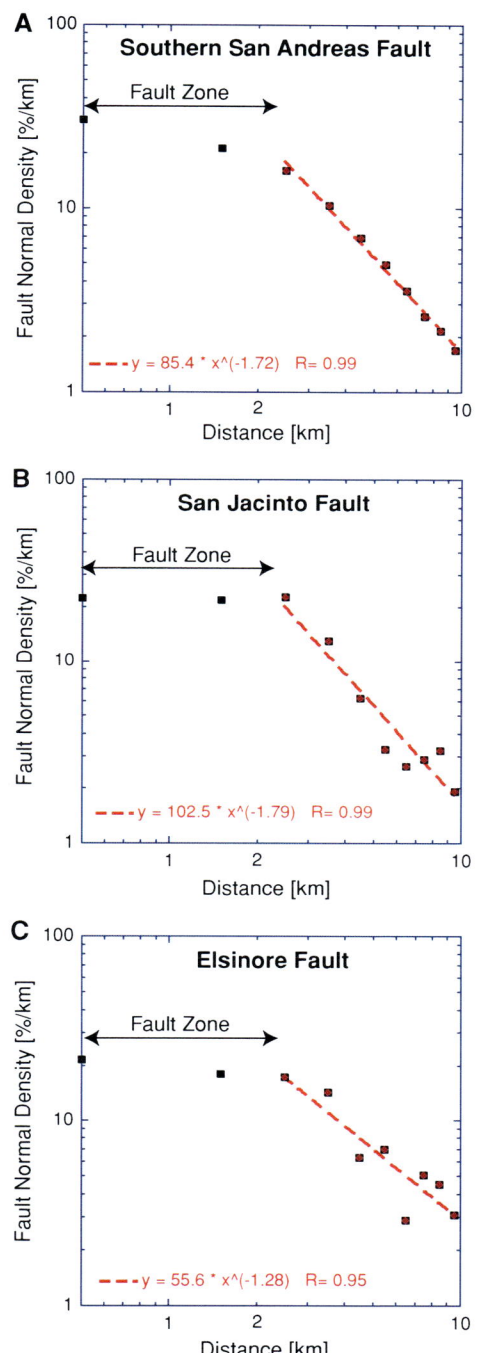

Figure 7
Decay of seismicity with distance for three high slip-rate faults, **a** San Andreas fault, **b** San Jacinto fault, and **c** Elsinore fault

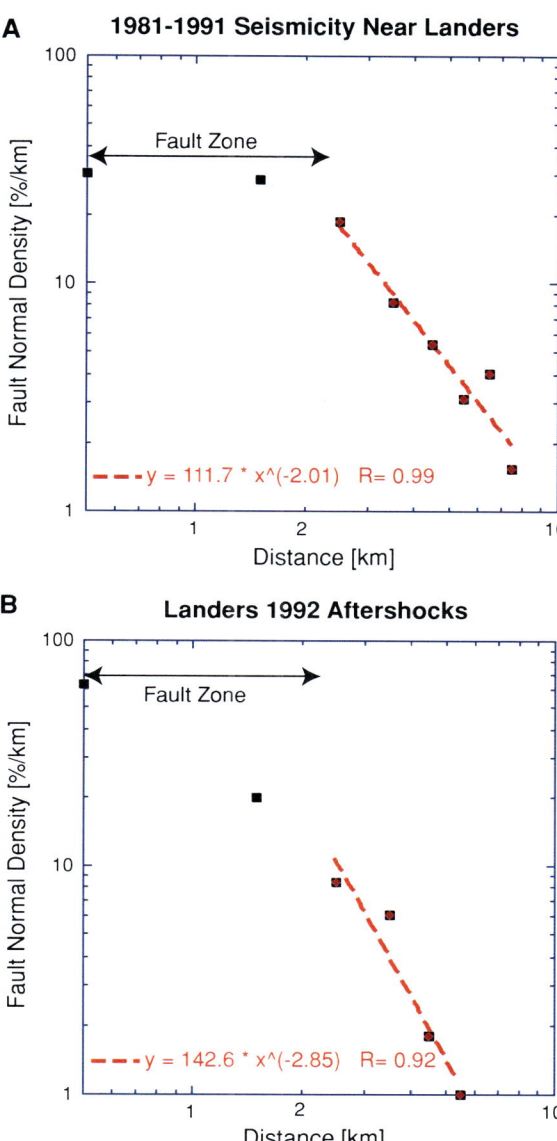

Figure 8
Decay of Landers seismicity with distance from the PSZs. **a** Distance decay of pre-main-shock seismicity; **b** Distance decay of aftershocks

sharp fall-off (Fig. 8). The preseismicity decay rate of $10^{-2.0}$ is somewhat larger than what we observed for major strike-slip faults during their interseismic period. The 1992 Landers aftershocks decay much faster with distance or as $10^{-2.8}$. The exponent of the power-law decay with distance for the 1992 Landers aftershocks is about twice as large as reported for smaller earthquakes (FELZER and BRODSKY, 2006). This difference may in part be explained because FELZER and BRODSKY (2006) analyzed a very selected data set. They analyzed all available main shocks in selected magnitude ranges, as well as shorter time and larger spatial scales, which would have included both sequences close to and distant from PSZs.

The decay of seismicity with distance away from the PSZs appears to be a stationary feature of the seismicity. However, the aftershocks decay faster with distance than the interseismic seismicity adjacent to high and low slip rate faults (Fig. 9a). The high rate of decay of the aftershocks is consistent with the transient nature of aftershocks and non-elastic fracturing of the region surrounding the PSZs of the main shock. The similar distance decay rates for high and low slip-rate faults suggest that the geological moment rate does not affect the decay rate significantly.

The fourth and fifth fault groups that are defined by "seismicity" and by "unconstrained seismicity" exhibit complex distance decay (Fig. 9b). In particular for the group of faults with unconstrained seismicity, the distance decay is not present, reflecting the lack of constraints on the seismicity and possible incorrect association between seismicity and the PSZs. The group of faults that are defined by seismicity decays more irregularly than aftershock zones, in part because they are small data sets of earthquake swarms that exhibit behavior that is in between the behavior of background seismicity and aftershocks.

The distance decay patterns show that each PSZ is surrounded by an approximately ±2-km-wide weak zone with a constant rate of seismicity. At greater distances, to about 10 km, the seismicity decays to a low background level. Thus, a core fault zone surrounds the PSZs and accommodates the aftershocks and some fraction of the background seismicity. A damage zone containing mostly small slip surfaces and gradually decaying with distance, accommodates elevated seismicity to ~10 km distance. One of many possible explanations for the presence of the damage zone is a wide zone of strain softening surrounding the PSZs. Alternatively, continuous slip below the brittle-ductile transition could load the slip surfaces of small earthquakes within the damage zone, resulting in elevated background seismicity.

3.3. Seismicity Characteristics Associated with each Fault Group

We have compared the seismicity parameters with the geological parameters of the PSZs (Fig. 10). The geological parameters describing each PSZ are the slip-rate and the geologic moment rate. The 'slip-rate' multiplied by 'fault area' is equivalent to geologic moment rate, and thus can be considered a

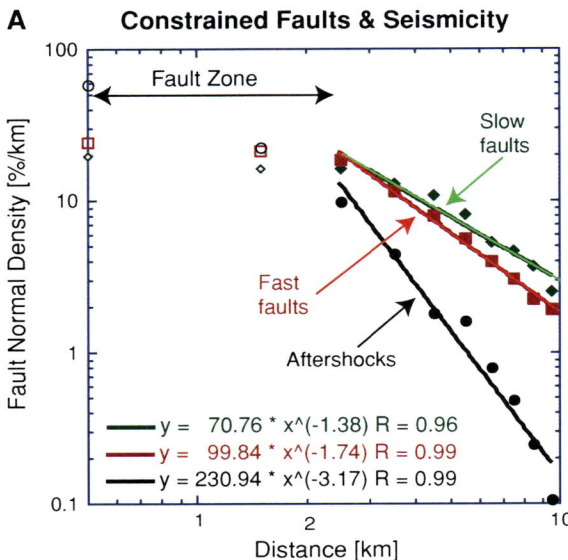

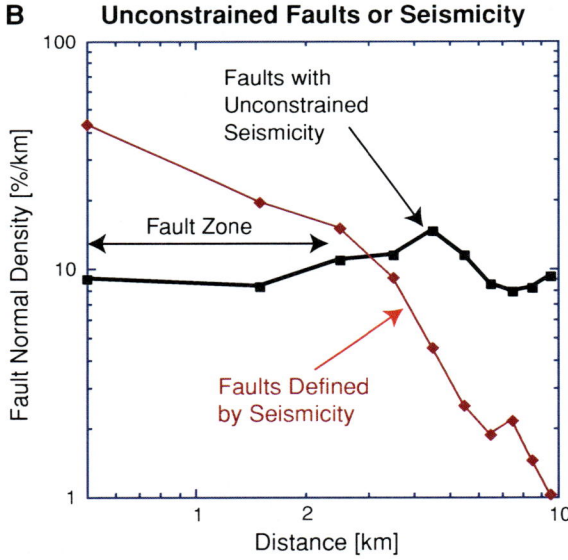

Figure 9
Fault normal density of hypocenters per kilometer plotted versus distance for the different groups of PSZs. a High slip-rate, low slip-rate, and aftershock groups; b Faults defined only by seismicity, and faults in which the seismicity in the region is not constrained caused by lack of seismic network coverage

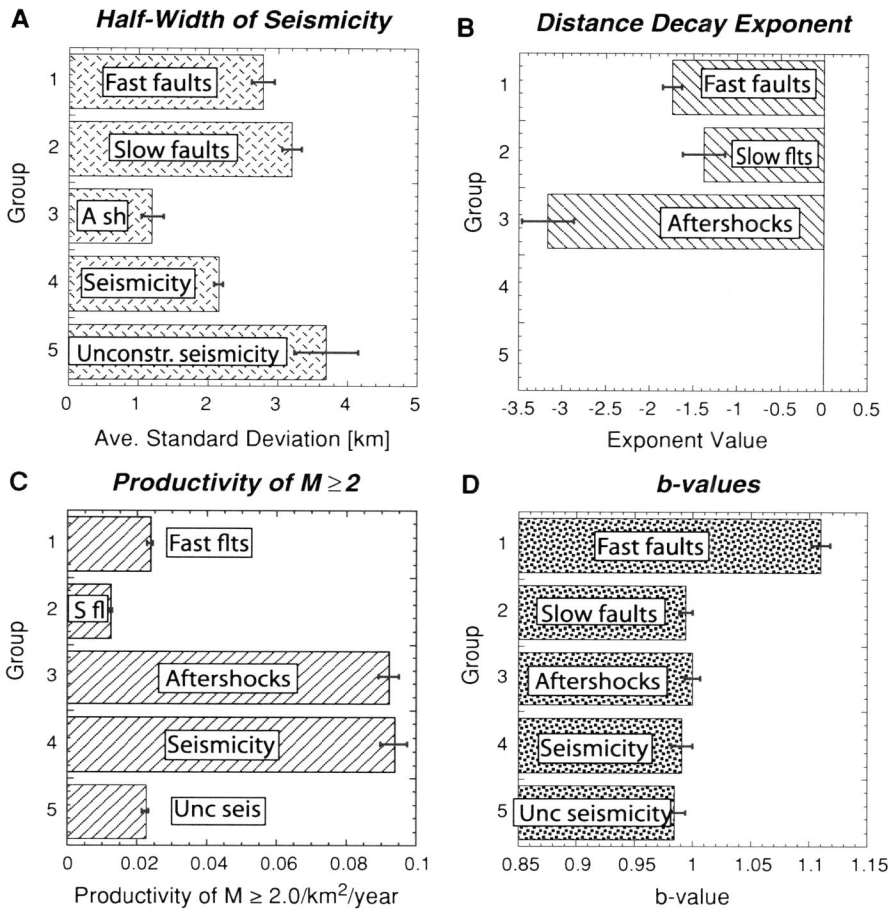

Figure 10
Bar graphs of seismicity parameters for the five fault groups of PSZs. The PSZs are divided into five groups as discussed in the text. *Error bars* of ±1σ are included. **a** Half-width of the histogram distributions of hypocentral distances; the "half-width" is calculated as the statistical "average deviation" or the statistical width of each histogram; "A sh"—aftershock fault group. **b** Distance decay parameter; this parameter is not available for PSZs that are defined by seismicity or are located near the edges or outside the network reporting area; **c** Seismicity productivity of $M \geq 2.0$ per area and year for each of the groups; "s fl"—slow faults group, "Unc seis" unconstrained seismicity fault group; and **d** b-value for each of the groups. "Unc seismicity" unconstrained seismicity fault group

proxy for the long-term tectonic strain loading along a particular CFM fault segment.

The seismicity parameters of each of the five PSZ groups are the standard deviation (the halfwidth of each seismicity distribution clustered around the PSZs), the distance decay, the productivity [derived from the a-value as ($10^{**}(a\text{-value} -2.0*b\text{-value})$/area)] and b-value, which quantifies the relative rate of large and small earthquakes. The productivity is the rate of $M \geq 2$ events per area and per year. Other geometrical distribution parameters such as skewness and kurtosis are not easily interpreted and do not exhibit simple relationships with the parameters of the PSZs. The uncertainty in the half-width of seismicity was determined by calculating the difference in the half-width for the full data set and half the data set. Similarly, the uncertainty in the distance decay exponent was determined by removing one data value from the regression calculation at a time. The b-value uncertainty estimate is approximately $b/\sqrt{N}$ for large N where N is the number of earthquakes with magnitude larger than the magnitude of completeness (Utsu, 2003). The productivity uncertainty was determined from the b-value uncertainty by estimating the change in productivity from the minimum and maximum b-value slopes.

The seismicity distributions for the five different fault groups have different half-widths and range from 1 km for aftershocks to ~4 km for unconstrained seismicity (Fig. 10a). The aftershock-defined and seismicity-defined faults have the narrowest distributions. The fast and slow slip-rate faults along with unconstrained seismicity faults have the broadest distributions. The distance decay rate is more rapid for aftershock-defined faults than for fast and slow slip-rate faults with interseismic seismicity (Fig. 10b). Thus aftershocks, and the interseismic background seismicity behave differently. This difference in behavior could be interpreted as being caused by the heterogeneous strain-field in the immediate vicinity of the PSZs which was left behind by the main shock.

The productivity is considerably higher for the aftershock-defined and seismicity-defined fault groups (Fig. 10c). The fast slip-rate, slow slip-rate, and unconstrained seismicity faults have lower productivity. In part, this result is expected because aftershock sequences are much more productive and constitute more than half of the southern California earthquake catalog. As a group, the high slip-rate faults exhibit the largest b-value (Fig. 10d). The low productivity and high b-value of high slip-rate faults is in agreement with the absence of moderate-sized events within their seismic zones. In particular, there is a lack of main shock-aftershock sequences in the intermediate magnitude range from M 5 to M 7.

There is an inverse relationship between the half-width of the fault groups and their productivity (Fig. 11). The aftershock-defined and seismicity-defined segments have very narrow and high producing distributions. The other three groups of faults that are in essence in their interseismic period have broader distributions with lower productivity. This observation is consistent with the main-shock rupture providing most of the heterogeneous driving strain field for the aftershocks. During the interseismic period all the faults seem to behave similarly.

The characteristic time and space clustering features of aftershock distributions suggest that the background seismicity within the ±10-km-wide seismic damage zone is not aftershocks, and is not accommodating seismic slip on the corresponding PSZ. Because the aftershock distributions do not

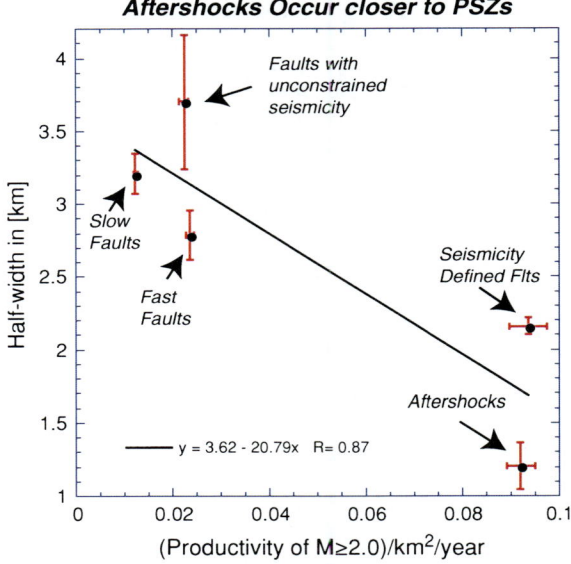

Figure 11
Half-width of seismicity distributions plotted versus productivity. Data points representing the five fault groups (Table 1) are labeled. Error bars of ±1σ are included. On average, the aftershock segments and the seismicity defined segments are narrower and exhibit higher productivity than faults in the interseismic stage

diffuse away from the PSZs and maintain their initial spatial distribution (HELMSTETTER et al., 2003), it is easy to compare their spatial patterns to the background seismicity. Using the halfwidth versus productivity relations, we can separate the aftershock distributions from the background seismicity distributions. These results for aftershocks are consistent with the clustering models of ZALIAPIN et al. (2007) who showed that aftershocks form a statistically distinct clustered spatial group from background seismicity.

The high slip-rate faults are the most important faults because they are responsible for most of the earthquake hazards. The slip-rate by itself gives instantaneous deformation rate while the slip-rate multiplied by fault area is a proxy for the long-term strain release rate. For the high slip-rate faults, both the productivity and b-value show variations with slip-rate and geologic moment rate (Fig. 12). The uncertainties in geological slip rates are from FRANKEL et al. (2002). The three most productive fault segments in southern California are the Anza San Jacinto segment, Imperial fault, and the San Andreas Mill Creek fault segment. The high productivity of the first two fault segments includes seismicity along the whole length of the faults. The Mill Creek segment of the San Andreas fault that includes a number of earthquakes at depth below Banning Pass is not an obvious high seismicity producer. In contrast, the three segments of the San Andreas fault: Chalome, Carrizo, and Mojave, exhibit extremely low productivity which can be attributed both to large cumulative offset and associated smoothing.

In Figs. 12c, d, the b-value plots display no clear trends, although the SAF segments to the north tend to have b-values on the high side. Small events are less common within the SAF damage zones, and some of the heterogeneity within the seismic damage zone could have been removed through high cumulative offset. The combined high b-value and low productivity for the SAF segments are in agreement with the observed absence of moderate-sized events adjacent to these faults.

4. Discussion

The clusters of background seismicity near the PSZs may be related to the cumulative offset or slip-rate of the faults. If some faults were older with large cumulative offset or weaker than others, we would expect that the corresponding seismicity distributions

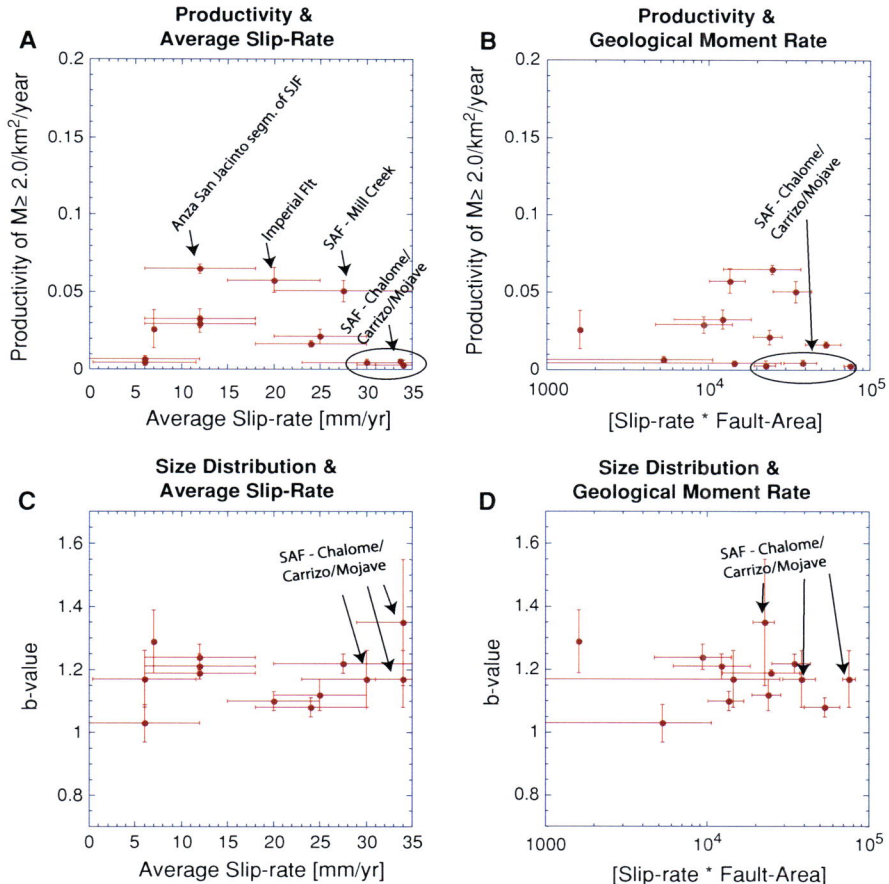

Figure 12
Comparison of fault parameters and seismicity parameters for high slip-rate fault segments. *Error bars* of $\pm 1\sigma$ are included. **a, b** show productivity and b-value versus average sliprate. Productivity is the rate of $M \geq 2.0$ events normalized per area and per year. High productivity and San Andreas fault segments are labeled. **c** shows productivity normalized by area and year, and **d** shows b-value versus geologic moment rate that is represented by (Slip* Fault Area). San Andreas fault segments that exhibit low productivity and high b-values are also labeled

could have different geometrical shapes or degree of clustering. For instance, the San Andreas fault is old and if its PSZs were exceptionally weak, we would expect that the background seismicity could be concentrated within or close to the PSZs, similar to what is observed along the creeping section in central California (PROVOST and HOUSTON, 2001). We do not observe such clustering next to the PSZs of the southern San Andreas fault. Similarly, only the three fastest moving PSZs of the San Andreas fault (Chalome, Carrizo, and Mojave) have unusually low productivity. Thus, the nearfault background seismicity appears to be mostly controlled by factors other than cumulative offset or slip-rate.

One possible explanation for the seismic damage zones is that they form as part of the fault to accommodate bends and other geometrical irregularities (Fig. 13). As the fault accumulates more offset, beyond its initial formation, the widths of the inner ±2-km-wide fault zone and the outer seismic damage zone do not change significantly. However, the rate of background seismicity within the seismic damage zone, or seismic productivity, may remain high during the initial offset and associated smoothing of the secondary heterogeneous fault networks. In particular, the seismic damage zone of the Anza San Jacinto segment of the San Jacinto fault accommodates a considerably higher level of seismicity than any other fault segments.

The productivity of a seismicity distribution may also depend on where within the seismic cycle the fault segment happens to be. If the fault just had a main shock, the seismic damage zone may be dominated by aftershocks such as the 1992 Landers PSZs. If the fault is late in the seismic cycle, it may be in a state of seismic quiescence. Similarly, external effects such as triggering by nearby main shocks or regional stress release caused by other earthquakes may influence the seismicity.

The relative strength of faults and the crust play an important role in understanding crustal strength (HARDEBECK and MICHAEL, 2006). The strength of faults is often presumed to vary with slip-rate, because fast moving faults slip a longer distance and more quickly remove geometrical irregularities and become weaker. Also, fast moving faults may be weaker because they have less time to heal. The features of damage zone seismicity thus may reflect some combination of the overall crustal strength in the region rather than the strength or slip-rates of individual faults. HARDEBECK and MICHAEL (2006) proposed a model of "*all major active faults being weak*" which explains our results of lack of special seismicity features for high slip-rate faults. The properties of the seismic damage zones of the San Andreas and San Jacinto faults are very similar to the other faults, suggesting similar fault strength.

The damage zone seismicity can also be affected by the crustal structure and heat flow as modeled by BEN-ZION and LYAKHOVSKY (2006). The more complex histograms that are observed to the south as compared to the north along the San Andreas, San Jacinto, and Elsinore faults suggest a different explanation for the seismicity than the influence of the PSZs. The tectonic difference between the northern and southern parts of our study region is the presence of the extensional tectonics and high heat flow to the southeast. The damage zones of fault segments located to the north of the Salton Trough and in areas of low to moderate heat flow, have a lower rate of seismicity and are spatially concentrated along only a few PSZs. The damage zones of fault segments located in the higher heat flow areas of the Salton Trough have a shallower brittleductile

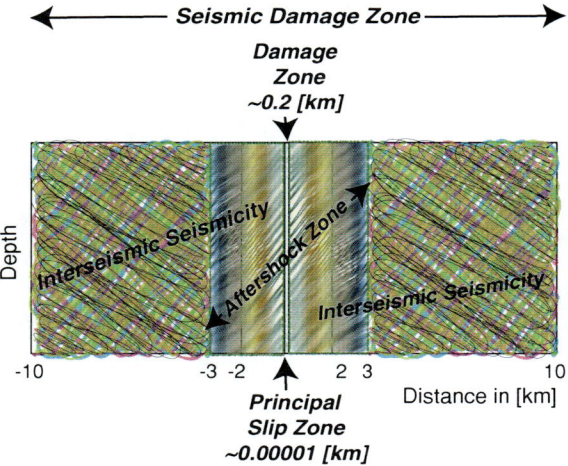

Figure 13
Cartoon cross section of fault structure illustrating the very narrow (~1 to 10 mm) PSZ, the ~200-m-wide damage zone determined from field mapping, and the ~10-km-wide seismic damage zone

transition (NAZARETH and HAUKSSON, 2004). They are also more numerous and have higher seismicity rates, although their width remains overall very similar as observed in other parts of southern California. This suggests that the seismic damage zones are of similar width but less productive where the seismogenic zone is thick.

The intensity of aftershocks along the PSZs could be caused by coseismic stress variability (MARSAN, 2006). Smooth static stress models predict a seismicity shadow along the aftershock zone because the main shock released most of the available shear stress. MARSAN (2006) suggested that the onset of the seismicity shadow is delayed following the main shock. We observe such seismicity shadows along some of the fault segments that are in the interseismic period such as the low productivity segments of the southern San Andreas fault and the Earthquake Valley segment of the Elsinore fault. Thus these segments may have released all the stress heterogeneity associated with the last main shock.

We have also compared our results with theoretical rupture models of SHAW (2004) who proposed that event sizes scaled with segment length. Similarly, SHAW (2006) showed preferential epicenter clustering of small events at the end of the major fault segments in his model. We do not detect similar behavior of seismicity near late Quaternary faults in southern California. There are no obvious concentrations of events near segment ends and seismicity rates do not simply scale with fault length.

Previous studies have attempted to relate the seismicity rate with cumulative fault offset. WESNOUSKY (1990) inferred that seismicity rate adjacent to late Quaternary faults in southern California is controlled by cumulative fault offset. His Figs. 4 and 5 suggests that normalized productivity (a-value/area/slip-rate) decreases with cumulative offset. We see only minor hints of this effect when we plot productivity versus slip-rate, which is a proxy for cumulative offset. The normalization by area is the correct procedure but because area has high variability, it tends to dominate the small variations in the a-value. We obtain a slope of ~ 1.09, which suggests at most a 9% effect on the scaling of a-value with cumulative offset, which is significantly smaller than that which WESNOUSKY (1990) implied.

It is beyond the goals of this study to thoroughly analyze the scaling relations between the different populations of events that occur within fault zones. The background seismicity manifests selfsimilar scaling when major events along the PSZs are not included. The PSZs may evolve with time and interact with the background seismicity. ANDO and YAMASHITA (2007) suggested that the self-similar scaling between small and large earthquakes breaks down because the major earthquakes can form fault branches through strong nonlinear interactions.

Other studies have also discovered spatial and temporal heterogeneities in the crustal processes in southern California, which in part explains the heterogeneity in the seismicity distributions. SPOTILA et al. (2007) who modeled the long-term vertical crustal deformation around the San Andreas fault showed that fault convergence in the near-field of the fault does not completely accommodate oblique plate motion. They inferred that vertical deformation along the San Andreas fault was influenced by relative slip partitioning as well as other factors such as surface processes, crustal anisotropy, and strain-weakening. They pointed out that heterogeneous deformation may be maintained through a positive-feedback effect of strain-softening. WOESSNER and HAUKSSON (2006) documented effects of strain-softening following the 1992 Landers earthquake. They found a small signal suggesting that the strain rate within the damage zones is higher the closer the seismicity is to the PSZs. The presence of the PSZs, ± 2-km-wide fault zones, and the ± 10-km-wide seismic damage zones also agrees with a mode of deformation that includes positive-feedback strain-softening.

5. Conclusions

The majority of small earthquakes do not occur on the same principal slip surfaces (PSZs) of late Quaternary faults in southern California as the major earthquakes. The background seismicity only exhibits weak clustering surrounding the different PSZs, forming ± 10-km-wide seismic damage zones. In contrast, aftershocks are clustered around the PSZs and decay away both in time and space. The 3-D geometrical shapes and productivities of these zones are similar, although they may be influenced by many factors, including the slip rates and the geological moment rates as well as the elapsed time since the last major earthquake on a PSZ. One possible interpretation is that the geometry of seismic damage zones develops early in the history of a fault, or alternatively the strengths of the near-fault crustal materials are very similar. For high slip-rate faults, the productivity of background seismicity of the damage zones is low in regions with thick crust. Because small and major earthquakes occur on spatially separated surfaces, their source physics may be different, with only large earthquakes being able to nucleate on the PSZs.

Acknowledgments

This research was supported by the US Geological Survey Grant 08HQGR0030 to Caltech, and by the Southern California Earthquake Center. SCEC is funded by NSF Cooperative Agreement EAR-0529922 and USGS Cooperative Agreement 07HQAG0008. We thank L. Jones and P. Shearer for feedback and discussions; and PAGEOPH reviewers and editors for detailed comments. Most figures were done using GMT (WESSEL and SMITH 1998). We thank A. Plesch for doing the distance measurements. SCEC contribution number 1214. Contribution number 10,007, Seismological Laboratory, Division of Geological and Planetary Sciences, California Institute of Technology, Pasadena.

REFERENCES

ANDO, R. and YAMASHITA, T. (2007), *Effects of mesoscopic-scale fault structure on dynamic earthquake ruptures: Dynamic formation of geometrical complexity of earthquake faults*, J. Geophys. Res. *112*, B09303, doi:10.1029/2006JB004612.

BEN-ZION, Y. and LYAKHOVSKY, V. (2006), *Analysis of aftershocks in a lithospheric model with seismogenic zone governed by damage rheology*, Geophys. J. Int. *165*, 197–210 doi: 10.1111/j.1365-246X.2006.02878.x.

CHAMBON, G., SCHMITTBUHL, J., CORFDIR, A., ORELLANA, N., DIRAISON, M. and GÉRAUD, Y. (2006), *The thickness of faults: From laboratory experiments to field-scale observations*, Tectonophysics *426*, 77–94.

FELZER, K.R. and BRODSKY, E.E. (2006), *Decay of aftershock density with distance indicates triggering by dynamic stress*, Nature *441*, 735–738.

FRANKEL, A.D. et al. (2002), *Documentation for the 2002 Update of the National Seismic Hazard Maps*, Tech. Rep. Open-File Report 02-420, US Geological Survey.

HARDEBECK, J.L. and MICHAEL, A. (2006), *Spatial and temporal stress inversion*, J. Geophys. Res *111*, B11310, doi: 10.1029/2005JB004144.

HELMSTETTER, A., OUILLON, G. and SORNETTE, D. (2003), *Are aftershocks of large California earthquakes diffusing?*, J. Geophys. Res. *108*, doi:10:1029/2003JB002503.

JENNINGS, C.W. (1994), *Fault activity map of California and adjacent areas*: California Department of Conservation, Division of Mines and Geology, Geologic Data Map No. 6, scale 1:750,000.

LIN, G., SHEARER, P.M. and HAUKSSON, E. (2007), *Applying a three-dimensional velocity model, waveform cross correlation, and cluster analysis to locate southern California seismicity from 1981 to 2005*, J. Geophys. Res. *112*, B12309, doi:10.1029/2007JB004986.

LIU, J., SIEH, K. and HAUKSSON, E. (2003), *A structural interpretation of the aftershock "cloud" of the 1992 $M_w7.3$ Landers Earthquake*, Bull. Seismol. Soc. Am. *93*, 3, 1333–1344.

MARSAN, D. (2006), *Can coseismic stress variability suppress seismicity shadows? Insights from a rate-and-state friction model*, J. Geophys. Res. *111*, B06305, doi:10.1029/2005JB004060.

NAZARETH, J.J. and HAUKSSON, E. (2004), *The seismogenic thickness of the southern California crust*, Bull. Seismol. Soc. Am. *94*, 940–960.

PLESCH, A., SHAW, J.H., BENSON, C., BRYANT, W.A., CARENA, S., COOKE, M., DOLAN, J., FUIS, G., GATH, E., GRANT, L., HAUKSSON, E., JORDAN, T., KAMERLING, M., LEGG, M., LINDVALL, S., MAGISTRALE, H., NICHOLSON, C., NIEMI, N., OSKIN, M., PERRY, S., PLANANSKY, G., ROCKWELL, T., SHEARER, P., SORLIEN, C., SÜSS, M.P., SUPPE, J., TREIMAN, J., and YEATS, R. (2007), *Community fault model (CFM) for southern California*, Bull. Seismol. Soc. Am., Dec., *97*, 1793–1802.

PRESS, H.W., TEUKOLSKY. S.A., VETTERLING, W.T. and FLANNERY, B.P., *Numerical Recipes in Fortran 77, The Art of Scientific Computing*, 2nd Edition, Vol. 1 (Cambridge University Press, New York, NY (1997)) 1447 pp.

PROVOST, A.-S. and HOUSTON, H. (2001), *Orientation of the stress field surrounding the creeping section of the San Andreas Fault: Evidence for a narrow mechanically weak fault zone*, J. Geophys. Res., *106*, B6, 11,373–11,386.

SAGY, A., BRODSKY, E.E. and AXEN, G.J. (2007), *Evolution of fault-surface roughness with slip*, Geol. *35*, 283–286.

SIEH, K., JONES, L.M., HAUKSSON, E., HUDNUT, K., EBERHART-PHILLIPS, D., HEATON, T.H., HOUGH, S., HUTTON, K., KANAMORI, H., LILJE, A., LINDVALL, S., MCGILL, S.F., MORI, J., RUBIN, C., SPOTILLA, J.A., STOCK, J., THIO, H.K., TREIMAN, J., WERNICKE, B. and ZACHARIASEN, J. (1993), *Near-field investigations of the Landers earthquake sequence April–July 1992*, Science *260*, 171–175.

SHAW, B.E. (2004), *Variation of large elastodynamic earthquakes on complex fault systems*, G. Res. Lett., *31*, L18609, doi:10.1029/2004GL019943.

SHAW, B.E. (2006), *Initiation propagation and termination of elastodynamic ruptures associated with segmentation of faults and shaking hazard*, J. Geophys. Res. *111*, B08302, doi:10.1029/2005JB004093.

SIBSON, R.H. (2003), *Thickness of the seismic slip zone*, Bull. Seism. Soc. Am. *93*, 3. 1169–1178; doi:10.1785/0120020061.

SPOTILA, J., NIEMI, A., BRADY, R., HOUSE, M., BUSCHER, J. and OSKIN, M. (2007), *Long-term continental deformation associated with transpressive plate motion: The San Andreas fault*, Geology 35, 11, 967–970; doi:10.1130/G23816A.1;

UTSU, T., *Statistical Features of Seismicity, Int'l. Handbook of Earthquake and Engineering Seismology V. 81B: Centennial publication of the Intl'. Assn. of Seism. and Physics of the Earth's Interior* (P. Jennings, H. Kanamori, and W. Lee, eds), pp. 719–732 (2003).

WESNOUSKY, S.G. (1990), *Seismicity as a function of cumulative geologic offset: Some observations from southern California*, Bull. Seism. Soc. Am. *80*, 1374–1381.

WESSEL, P. and SMITH, W.H.F. (1998), *New version of the generic mapping tools released*, EOS 79, 579.

WESSON, R.L., BAKUN, W.H. and PERKINS, D.M. (2003), *Association of earthquakes and faults in the San Francisco Bay Area using Bayesian inference*, Bull. Seismol. Soc. Am. *93*, 1306–1322.

WIEMER, S. (2001), *A software package to analyze seismicity: ZMAP*, Seis. Res. Lett., 373–382.

WOESSNER, J. and HAUKSSON, E. (2006), *Associating Southern California seismicity with Late Quaternary Faults: Implications for Seismicity Parameters (abstract)*, Southern California Earthquake Center Annual Meeting, Palm Springs, CA.

WOOD, H.O. (1916), *The earthquake problem in the Western United States*, Bull. Seismol. Soc. Am. *VI*, 196–217.

ZALIAPIN, I., GABRIELOV, A., KEILIS-BOROK, V. and WONG, H. (2007), *Aftershock identification, arXiv:0712.1303v1 [physics.geo-ph]*.

(Received August 5, 2008, revised March 11, 2009, accepted May 24, 2009, Published online March 18, 2010)

Earthquake Source Zones in Northeast India: Seismic Tomography, Fractal Dimension and b Value Mapping

PANKAJ M. BHATTACHARYA,[1] J. R. KAYAL,[2] SAURABH BARUAH,[3] and S. S. AREFIEV[4]

Abstract—We have imaged earthquake source zones beneath the northeast India region by seismic tomography, fractal dimension and b value mapping. 3D P-wave velocity (Vp) structure is imaged by the Local Earthquake Tomography (LET) method. High precision P-wave (3,494) and S-wave (3,064) travel times of 980 selected earthquakes, $m_d \geq 2.5$, are used. The events were recorded by 77 temporary/permanent seismic stations in the region during 1993–1999. By the LET method simultaneous inversion is made for precise location of the events as well as for 3D seismic imaging of the velocity structure. Fractal dimension and seismic b value has been estimated using the 980 LET relocated epicenters. A prominent northwest–southeast low Vp structure is imaged between the Shillong Plateau and Mikir hills; that reflects the Kopili fault. At the fault end, a high-Vp structure is imaged at a depth of 40 km; this is inferred to be the source zone for high seismic activity along this fault. A similar high Vp seismic source zone is imaged beneath the Shillong Plateau at 30 km depth. Both of the source zones have high fractal dimension, from 1.80 to 1.90, indicating that most of the earthquake associated fractures are approaching a 2D space. The spatial fractal dimension variation map has revealed the seismogenic structures and the crustal heterogeneities in the region. The seismic b value in northeast India is found to vary from 0.6 to 1.0. Higher b value contours are obtained along the Kopili fault (~ 1.0), and in the Shillong Plateau (~ 0.9) The correlation coefficient between the fractal dimension and b value is found to be 0.79, indicating that the correlation is positive and significant. To the south of Shillong Plateau, a low Vp structure is interpreted as thick (~ 20 km) sediments in the Bengal basin, with almost no seismic activity in the basin.

Key words: Microearthquake, fault plane solutions, seismotectonics, seismic tomography, fractal dimension, b value.

---

[1] Geological Survey of India, 27J N Road, Kolkata 700016, India.
[2] School of Oceanographic Studies, Jadavpur University, Kolkata 700032, India. E-mail: jr_kayal@yahoo.com
[3] North East Institute of Science and Technology, Jorhat, India.
[4] Institute of Physics of the Earth, Russian Academy of Sciences, Moscow, Russia.

1. Introduction

The northeast India region under study, Lat: 22°–29°N and Long: 89°–98°E, displays a complex geological setting (Fig. 1). The region is tectonically dissected into several mosaics by deep-rooted faults/thrusts along which episodic block/thrust/strike-slip movements are reported (NANDY, 1980). Two profound Tertiary mobile belts encircle the region; the east–west Himalayan fold belt to the north, caused by collision and continued north–south convergence along the Himalayan arc since Eocene (e.g., SEEBER *et al.*, 1981), and the north–south Indo-Burma fold belt to the east, which was caused by the east–west convergence along the Burmese arc since early Tertiary (e.g., NANDY 1980, 2001). The convergence tectonics to the north as well as to the east resulted in several thrusts and other faults in the region.

The region produced two great earthquakes ($\sim M_s$ 8.7) (RICHTER 1958), one in 1897 in the Shillong Plateau and the other in 1950 on the Assam–Tibet border at the Assam syntaxis zone; the meeting zone of the Himalayan arc and the Burmese arc (Fig. 1). As many as 17 large earthquakes $7.0 \geq M < 8.0$ occurred in the region during the last 100+ years since the 1897 great Shillong earthquake (KAYAL, 1996). The shallower (depth ≤ 20 km) earthquakes in the western Himalayan arc are attributed to collision tectonics and are correlated with the known regional thrusts, the Main Boundary Thrust (MBT) and the basement thrust or the plane of detachment (e.g., KAYAL, 2001). The plane of detachment is defined as the interface between the gently dipping Indian shield and the Himalayan sedimentary wedge (SEEBER *et al.*, 1981). In the northeast Arunachal Himalaya, microearthquake investigations,

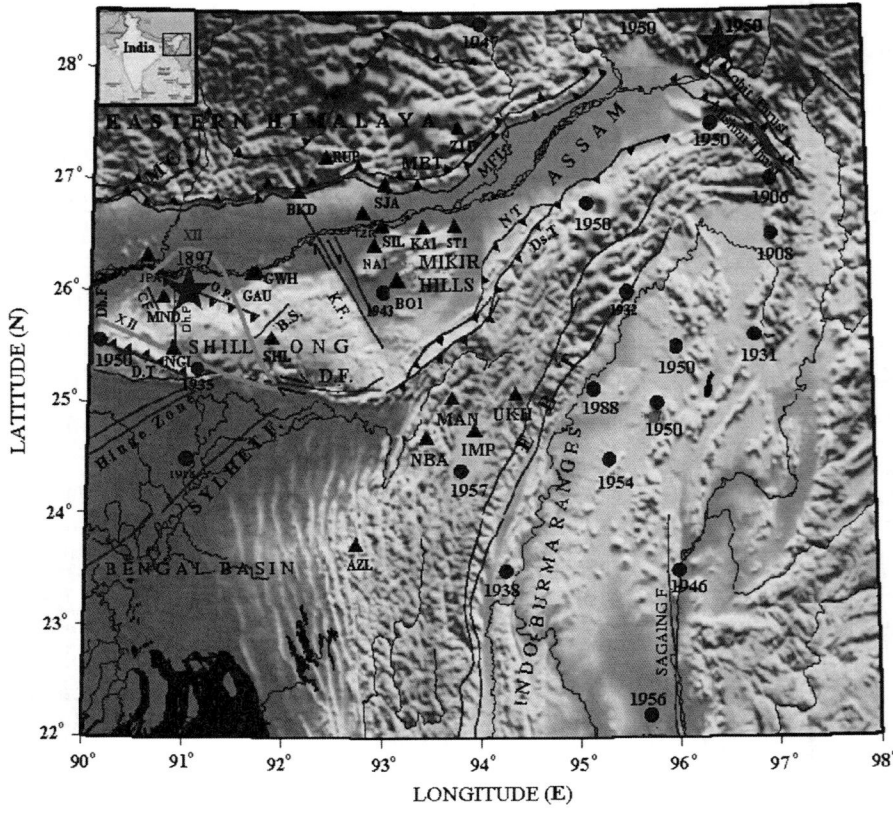

Figure 1
Tectonic map of the study region showing epicentres of the large and great earthquakes (KAYAL et al., 2006); *circles* indicate large earthquakes $M \geq 7.0$ and the *stars* the two great earthquakes $M > 8.0$. The permanent seismic stations are shown by *solid triangles*. MCT Main Central Thrust, MBT Main Boundary Thrust, D.F. Dauki Fault, D.T. Dapsi Thrust, K.L. Kopili Lineament. *Inset* key map of the study area

however, revealed that the earthquakes are deeper, to a depth of 80 km; transverse tectonics is suggested in this part of the Himalaya (KAYAL et al., 1993; KAYAL, 2001). The earthquakes in the Burmese arc to the east, on the other hand, result from subduction tectonics; normal, thrust and strike-slip faulting earthquakes are observed from the surface down to about 200 km (e.g., KUMAR and RAO, 1995; KAYAL, 1996). The Assam syntaxis zone is also seismically active and produced the great 1950 Assam–Tibet earthquake (M_s 8.7) (TANDON 1954). The Shillong Plateau-Assam valley area is bounded by the MBT to the north and by the Dauki fault to the south; the gigantic east–west Dauki fault separates the Plateau from the Bengal basin to its south (Fig. 1). The Shillong Plateau was the source area of the 1897 great earthquake M_S 8.7 (OLDHAM, 1899). In the Plateau area, the earthquakes are mostly confined within a depth of 35 km; oblique reverse faulting is reported (KAYAL and DE, 1991; KAYAL et al., 2006). To the east of the Shillong Plateau lies the Mikir massif, which is separated from the Shillong massif by the northwest–southeast Kopili lineament (Fig. 1). The Kopili lineament is defined as the Kopili fault due to its intense seismic activity; transverse tectonics are reported along this fault (KAYAL et al., 2006). In this study we have analyzed approximately 1,000 selected earthquakes $m_d \geq 2.5$ in northeast India to image the seismic source zones at depth by tomography and by mapping the fractal dimension and b value of the epicenter distribution.

2. Seismotectonics of the Region

CHEN and MOLNAR (1990) re-examined source parameters of earthquakes (shallower than 100 km) that occurred in the region, and presented 17 reliable fault plane solutions (Fig. 2). In the Shillong Plateau and its adjoining area the solutions (events 1–6) show reverse faulting. KAYAL and DE (1991), based on microearthquake surveys, also presented similar solutions for four cluster of events (A–D) in the Plateau area. All these solutions show N–S compression. BILHAM and ENGLAND (2001) suggested a pop-up tectonics for the Shillong Plateau. In the Indo-Burma ranges, CHEN and MOLNAR (1990) obtained ten solutions and these solutions show pure thrust to mixture of reverse and strike-slip faulting with a NNE–SSW compressional stress (Fig. 2); subduction tectonics and or dragging of the dipping Indian lithosphere is suggested below the Indo-Burma ranges (KAYAL, 1996; LE DIAN et al., 1984).

In Arunachal Himalaya, in the northeastern Himalayan collision zone, three composite fault-plane solutions were reported from temporary microearthquake surveys (KAYAL et al., 1993), two for the groups E and F earthquakes (depth 15–40 km), and one for the group G earthquakes (depth 50–80 km) (Fig. 2). These solutions show reverse faulting with strike-slip components. These earthquakes did not occur on the so-called plane of detachment; a transverse tectonics was suggested (KAYAL et al., 1993). The eastern syntaxis zone, the meeting zone of the Himalayan arc and the Burmese arc, was the source area for the 1950 great earthquake. CHEN and MOLNAR (1990) determined a thrust solution using the first-motion data of the 1950 great event (solution X). BEN-MENAHAM et al. (1974), on the other hand, obtained a right lateral strike-slip mechanism for this event (solution Y) (Fig. 2). A detailed review of the seismotectonics of the northeast India region is given by KAYAL (2008).

3. Data Source

The database of the present study is obtained from the earthquake catalogs, seismological bulletins (1993–1999) published by the National Geophysical Research Institute-Hyderabad (NGRI-H) and by the Regional Research Laboratory-Jorhat (RRL-J). These bulletins incorporate the events recorded by the temporary and permanent networks; 77 digital and analog seismic stations were in operation in time and space during the reported period 1993–1999. Out of the 77 stations, phase data were available from only 59 stations. These 59 stations are shown in Fig. 3a. Coordinated Universal Time (UTC) was maintained for the analog/digital seismic stations. The overall timing accuracy ±0.1 s was maintained for the analog seismic stations (KAYAL, 1996). The input data from the digital seismic stations were GPS time-based. The three component digital seismograms provided higher precision P-wave (±0.01 s) and S-wave (±0.05 s) arrival times. The total data set was fairly good for simultaneous inversion for seismic imaging.

The selected events are of medium magnitude (m_d 2.5 ≤ 5.5). A total of 3,190 events, during the period of 7 years (1993–1999), were reported in the bulletin. From this huge data set, about 1940 events with reliable P- and S-arrival times were used to make preliminary estimates of hypocentral parameters using the HYPO71 program (LEE and LAHR, 1975) and the 1D velocity model (Table 1) of DE and KAYAL (1990). Locations of these 1940 events are shown in Fig. 3b. The location errors were examined, but only 980 events with root-means-square (RMS) error lower than 1.0 s were selected for simultaneous inversion. The average RMS of the subsampled 980 events located by HYPO71 was 0.56 s.

4. Data Analysis

4.1. Simultaneous Inversion

The selected 980 events were then relocated using the Local Earthquake Tomography (LET) method of THURBER (1983). In this method, along with the high-precision locations, the heterogeneous 3D velocity structure is modeled by simultaneous inversion. The high-precision epicenters reduced the average RMS from 0.56 to 0.06 s. Epicenters of the 980 relocated events are shown in Fig. 3b. The earthquakes are relocated with an average precision of ±2 km in horizontal direction, and ±2 km in depth. The coupled problem was solved with 980 events.

THURBER'S (1983) LET method incorporates the parameter separation method of PAVILIS and BOOKER

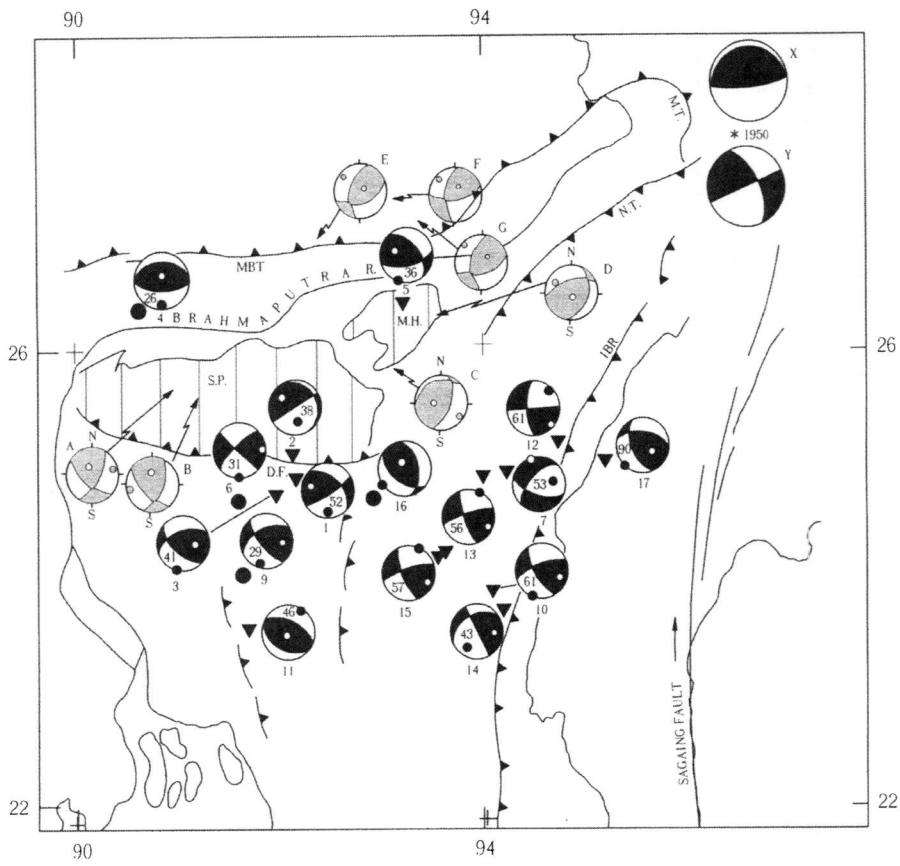

Figure 2
Focal mechanisms of earthquakes in NE India (fault plane solutions 1–16 and X from CHEN and MOLNAR, 1990, and Y from BEN-MENAHAM et al., 1974; A–G from KAYAL, 2001). The *numbers* inside the darker mechanisms indicate the focal depth of the earthquakes are reference numbers of the solutions given by CHEN and MOLNAR (1990). The *small solid circles* inside the focal mechanisms are the *P* and *open circles* the *T* axes

(1980) to simultaneously estimate velocities along a 3D flexible grid. Resolution of the data set and efficacy of the tomographic inversion are strongly dictated by model parameterization of the 3D inversion. The number of rays passing near each grid intersection, which controls the resolution at that node, arises from the station coverage, earthquake distribution and node spacing. We have used the maximum number of stations available in our database. Increasing the node spacing may improve the resolution, nonetheless it smooths velocity anomalies over a large volume, making a correlation with tectonic units difficult. Conversely, inverting for small anomalies by reducing the node spacing causes a considerable decrease in resolution. We tried to maintain at least 300 rays passing near most nodes by inverting a large volume with a relatively coarse grid spacing of 100 km and with a fine grid spacing of 50 km. Decreasing the damping factor increases resolution although at the expense of an increased standard error. Hence, we choose an optimal value of the damping parameter that yields low data variance, low solution variance and low standard error with a relatively good average resolution. The values of the resolution matrix varied from 0.60 to 0.90. Hence we tried to optimize various parameters, which consist of grid parameterization, selection of initial 1D velocity model, suitable damping parameter and the number of iterations allowed to reach a convergent and consistent solution.

The grid configuration with an origin at 26°N and 93°E was set up in the study area, about

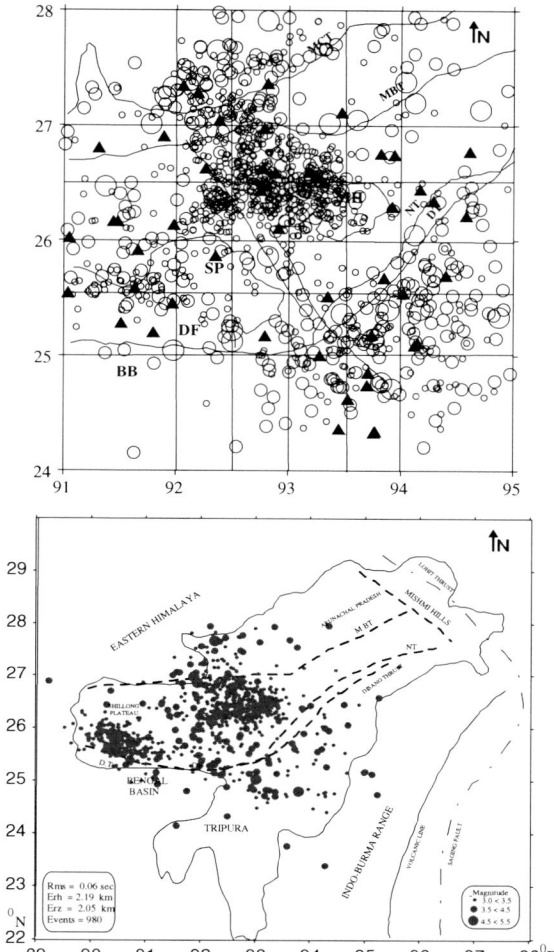

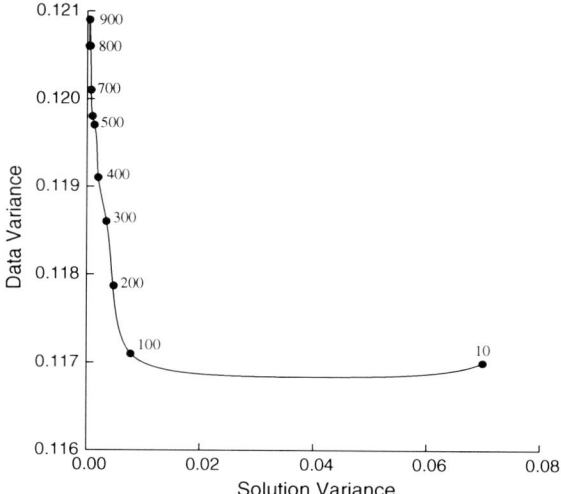

Figure 4
Emperical determination of damping parameter for velocity inversions (BHATTACHARYA et al., 2008)

Figure 3
a Map showing the epicentres of the 1940 events located by the HYPO 71 method, and **b** relocated 980 events by the LET method (see text)

Table 1

(1D velocity model: DE and KAYAL 1990)

Depth (km)	Vp (km/s)
0	5.55
20	6.52
41	8.10
46	8.57

250 × 250 km, in the east–west and north–south directions, respectively (Fig. 3a). A damping value was obtained as suggested by EBERHART-PHILLIPS (1986). Damping was selected by running a series of single-iteration inversions with a large range of damping values, and the data variance versus the solution variance was examined for these runs (EBERHART-PHILLIPS, 1986, 1993; ZHAO et al., 1992). A large range of damping values (1–1,000) was explored. A damping value of 100 was chosen and used throughout the inversion procedure (Fig. 4). This damping value provided a reasonable constraint in the resultant velocity anomalies, while constraining the effect of the noise in the data (BHATTACHARYA et al., 2008). The 3D tomographic inversion is made using the 1D inverted model as the starting model. The 1D inverted model (Table 2) is obtained by the LET method. The seismic images, depth slices of heterogeneous Vp structures beneath the region thus obtained, are shown in Fig. 5.

4.2. Fractal Dimension Mapping

Though major surface traces of the faults are generally well mapped, a significant fraction of regional seismicity occurs on secondary and sometimes on hidden structures (HANKSSON, 1990; JONES et al., 1990). The fractal dimension provides a measure of the degree of fractal clustering of points in the space. TOSI (1998) illustrated that possible values of fractal dimension (D) are bound to range between 0 and 2, which is dependent on the

Table 2

(1D velocity model estimated by the LET method)

Depth (km)	Vp (km/s)
0	5.56
10	6.10
20	6.45
30	6.90
40	7.60
50	8.40

dimension of the embedding space. Interpretation of such limit values is that a set with $D \sim 0$ has all events clustered into one point. At the other end of the scale, $D \sim 2$ indicates that the events are randomly or homogeneously distributed over a 2D embedding space. IDZIAK and TEPER (1996) suggested that the $D \sim 2$ is an evidence of multiple external forces which act on the rock mass. Multiple tectonic stresses, from the Himalayan arc and the Burmese arc in this region are reported by several authors (e.g., CHEN and MOLNAR, 1990; KAYAL, 1996; KUMAR and RAO, 1995). Hence the evaluation of fractal dimension is of significant importance in such cases. Spatial resolution of fractal dimension mapping is limited by the location precision of epicenters (WYSS et al., 2004). Hence only these high precision 980 epicenters, relocated by the simultaneous inversion, have been used for estimation of fractal dimension.

The fractal dimension was estimated using the correlation integral method of KAGAN and KNOPOFF (1980), which measures the correlation dimension

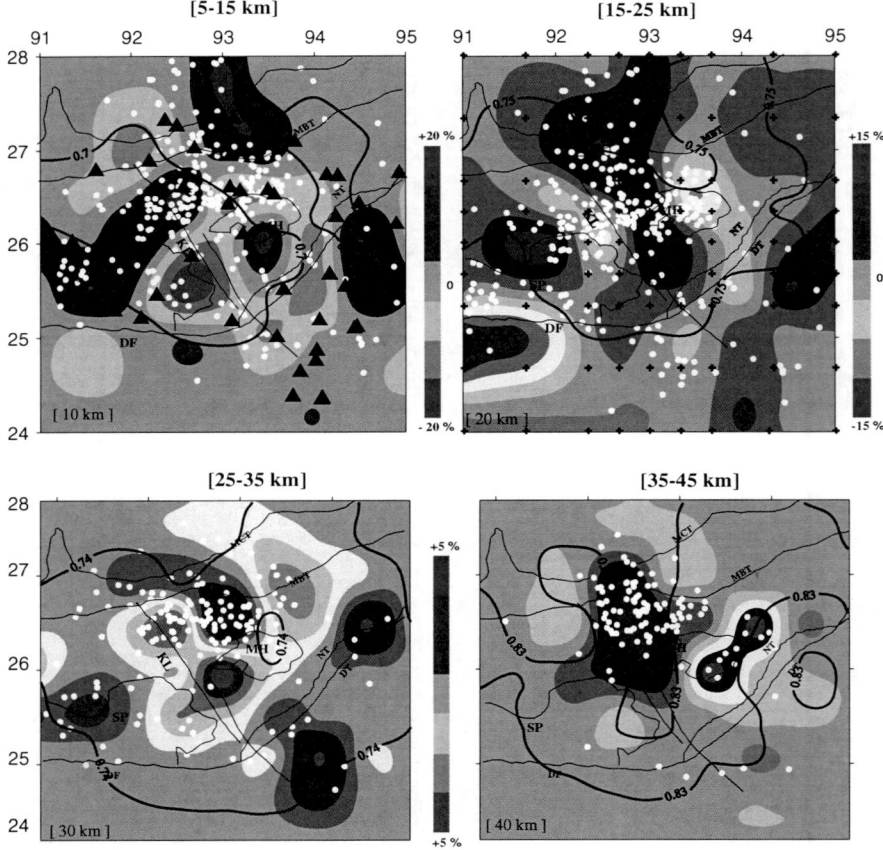

Figure 5
Distribution of Vp perturbations at different depths. The depth of each layer is shown at the *bottom*, and the depth range of the earthquakes is shown at the *top*. *Solid triangles* are seismic stations. The *black crosses* show grid point distributions. The *thick lines* show the threshold value of the resolution matrix, and events (*white dots*) within the specified depth ranges are shown. *Curvilinear lines* show the tectonic features as indicated in Fig. 3. The perturbation scale is shown (BHATTACHARYA et al., 2008)

D_2. The correlation integral method is widely applied in seismology, especially to spatial distributions of earthquake epicenters. This technique is preferred to the box-counting method because of its greater reliability and sensitivity to small changes in clustering properties (KAGAN and KNOPOFF, 1980; HIRATA, 1989). The correlation integral is related to the standard correlation function as given by KAGAN and KNOPOFF (1980):

$$C_r \sim r^{D_2},$$

where D_2 is a fractal dimension, more strictly the correlation dimension. GRASSBERGER and PROCACCIA (1983) introduced a practical algorithm for the measure of the correlation dimension, commonly referred to as the Grassberger–Procaccia algorithm, GPA. By plotting C_r against r on a double logarithmic coordinate, we can practically obtain the fractal dimension D_2 from the slope of the graph. The distance r between two events, θ_1, ϕ_1 and θ_2, ϕ_2, is calculated by using a spherical triangle as given by HIRATA (1989):

$$r = \cos^{-1}(\cos\theta_1\cos\theta_2 + \sin\theta_1\sin\theta_2\cos(\phi_1 - \phi_2)).$$

Examples of a few C_r versus r plots are shown in Fig. 6.

KAGAN (2007) reviewed various methods for determining fractal dimension of earthquake epicenters and hypocenters, paying special attention to the problem of errors, biases and systematic effects. They have shown that any value of correlation dimension can be obtained if the errors and inhomogeneities in observational data as well as deficiencies in data processing are not properly considered. In the practical calculation, the fractal dimension analysis is based on a power law and is turned into a linear law after logarithmic transformation. Therefore, sufficient data points are the key for a reliable estimate of fractal dimension based on the ensuing linear regression (XU and BURTON, 1999). SMITH (1988) suggested the minimum number of points or events required for a reliable calculation of a correlation dimension as:

$$N_{\min} \geq \{R(2-Q)/r(1-Q)\}^{\mu},$$

where Q is a quality factor and $0 < Q < 1$. $R = r_{\max}/r_{\min}$, where r is a scale to calculate C_r, and μ is the

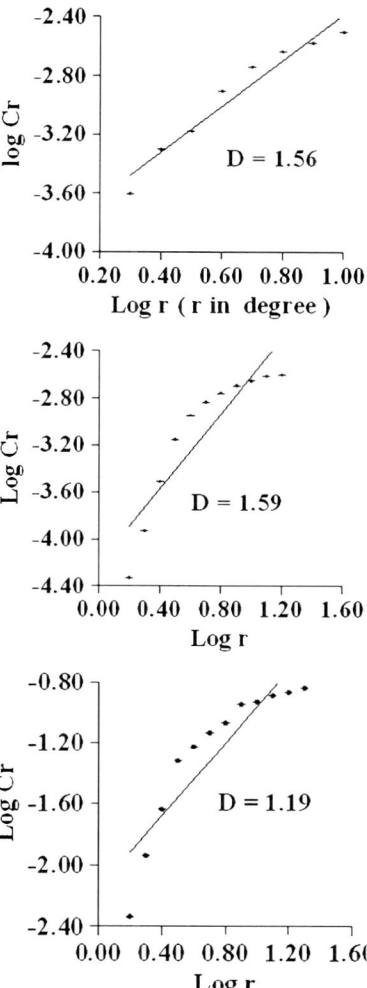

Figure 6

a, b c Three examples showing the plot of log C_r versus log r (see text). The slope of the best-fit line estimates the fractal dimension. *Error bars* are shown

greatest integer less than the obtained fractal dimension of the set. The smallest topological dimension in which the distribution of epicenters embeds is 2, therefore the value of D_2 will be less than 2, hence $\mu = 1$. If $Q = 0.95$ and $r = 4$, we will have $N_{\min} > 42$.

In order to spatially map the fractal dimension, the study area was gridded such that each grid contained a minimum of 42 points. Each grid was overlapped both in X and Y. This exercise generated 43 grids. Taking the center of the grid as a plotting point, contour maps of estimated fractal dimension

D_2 with a contour interval of 0.1 are prepared. In order to map the fractal dimension as a function of depth, two depth ranges are selected on the basis of hypocenter distribution. These maps are shown in Fig. 7a and b. Figure 7a is the fractal dimension contour map of events lying in the depth range between 0 and 20 km, and Fig. 7b is the fractal dimension contour map of events in the depth range from 20 to 40 km. A fractal dimension contour map for all the events is shown in Fig. 8a. A contour interval of 0.1 has been selected on the basis of error estimation. The sampling error is estimated using the sampling distribution theory. For this contour map, the error is found to vary from 0.04 to 0.06. Hence the contour interval of 0.1 is within the permissible limit. The value of 'test statistics' or 'Z-score' is found to lie between 1.72 and 1.93, which is in the critical region $-1.96 \leq Z \geq 1.96$ (GHOSH and SAHA, 2002). This falls in 5% level of significance, hence the level of confidence is 95%.

4.3. b Value Mapping

One of the most analyzed and discussed topics in statistical seismology concerns variations of b value of the Gutenberg–Richter (G–R) log-linear relation. Numerous methods have been proposed in the literature for computation of the b value (e.g., UTSU, 1965; AKI, 1965; PAGE, 1968; BENDER, 1983). We have estimated b value by the maximum likelihood method (UTSU, 1965; AKI, 1965) because it is reported to be a more appropriate way to compute a better estimation of b value, since it is inversely proportional to the mean magnitude as follows:

$$b = \frac{\log_{10} e}{M - M_0},$$

where M is the average magnitude of the events exceeding the threshold magnitude M_0 and $\log_{10} e = 0.4343$. A stable estimation of the b value by this method, however, requires at least 50 events (UTSU, 1965), and our data set satisfies this condition.

The frequency–magnitude relation should be examined carefully as the self-similarity may break into the following three stages: smaller events ($M < 3.0$), medium events ($3.0 < M < M_{\text{saturate}}$) and larger events ($M > M_{\text{saturate}}$). The smaller events may give lower b value because of a shortage of smaller events recorded in the catalogs, while bigger events may give higher b value because of the saturation of the magnitude (SCHOLTZ, 1990). We have, however, estimated b values for the medium events $3.0 < M < 5.5$ and we believe that self-similarity is maintained in this magnitude range. A b value contour map for the entire set of events is shown in Fig. 8b. To examine the correlation between the fractal dimension and b value, the fractal dimension is plotted against the b value for 43 subsets and is shown in Fig. 9.

5. Results and Discussion

A visual examination of the seismic activity and major geological features reveals clustering of earthquakes in four distinct areas; these are: (1) Shillong Plateau, (2) Mikir hills and lower Assam valley, (3) Arunachal Himalaya and (4) the Indo-Burma ranges (Fig. 3). The location errors obtained by using the 3D model are much improved for the selected 980 events; the average RMS is reduced from 0.56 to 0.06 s, the average epicenter error is reduced from 5.58 to 2.19 km and the average focal depth error from 3.03 to 2.05 km (Table 3). The seismic images of Vp at different depths show strong heterogeneity in velocity structure at all the depth slices (Fig. 5). The seismic activity is mostly concentrated in the high Vp zones, indicating that high velocity zones are the stress accumulators in the heterogeneous medium. The seismic activity is relatively sparse in the low Vp zones. Lateral heterogeneities in velocity structures at different depths are well reflected in the images. A Restore Resolution Test (RRT) given by ZHAO et al. (1992) was also performed using the results of observed tomography to compute the theoretical arrival times. The 3D velocity image (Fig. 5) is used as our initial model for the Vp structures. Then we inverted the synthetic data by using the same algorithm to synthesize the restored images of the actual result by adding random Gaussian noise of 0.30 s as picking accuracy for P-wave arrival times. We found that the input anomalies are well recoverable in the study area for the selected grid setup, which we used in the 3D

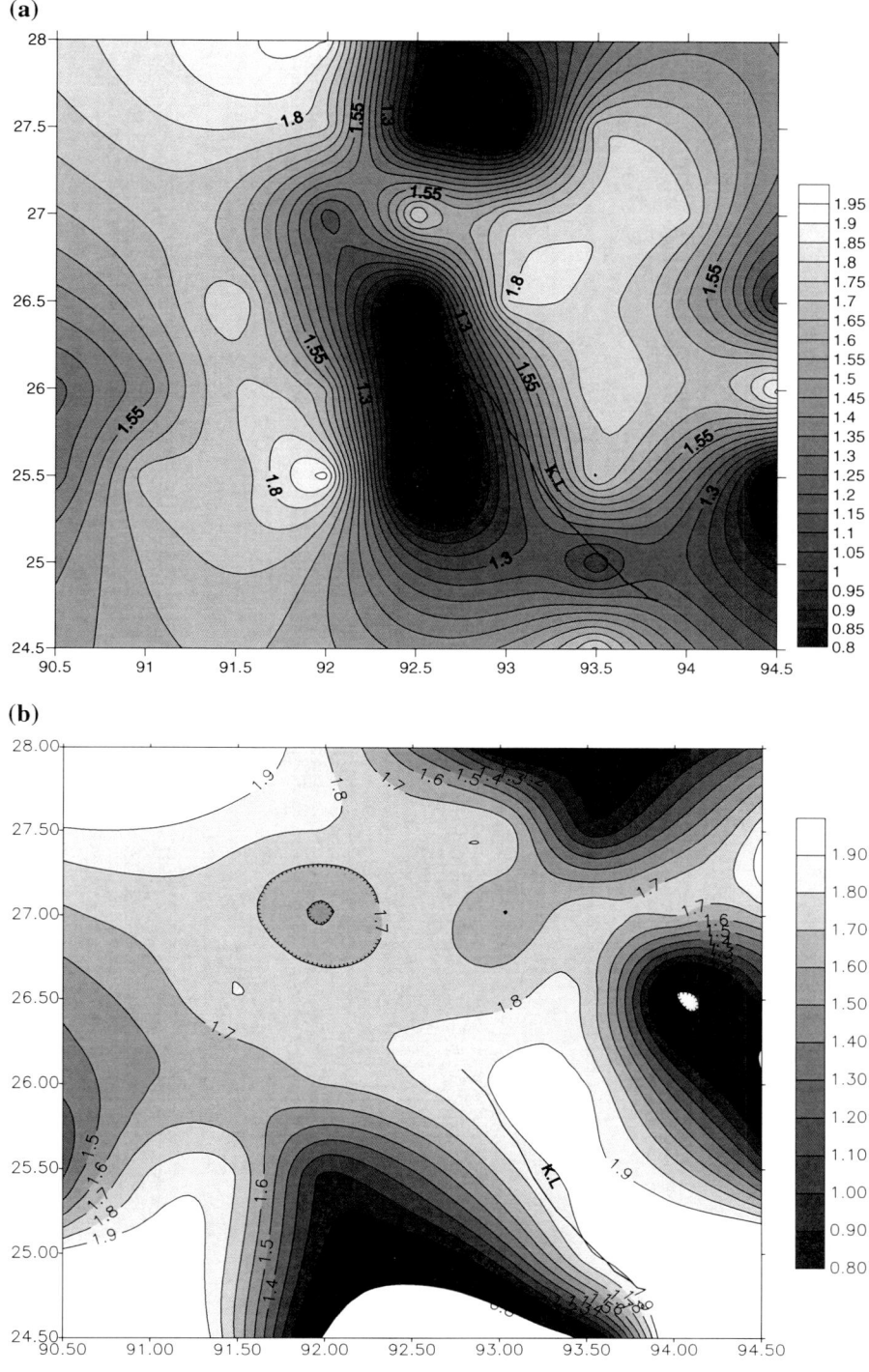

Figure 7
Fractal dimension contour maps: **a** events 0–20 km depth, and **b** events 20–40 km depth

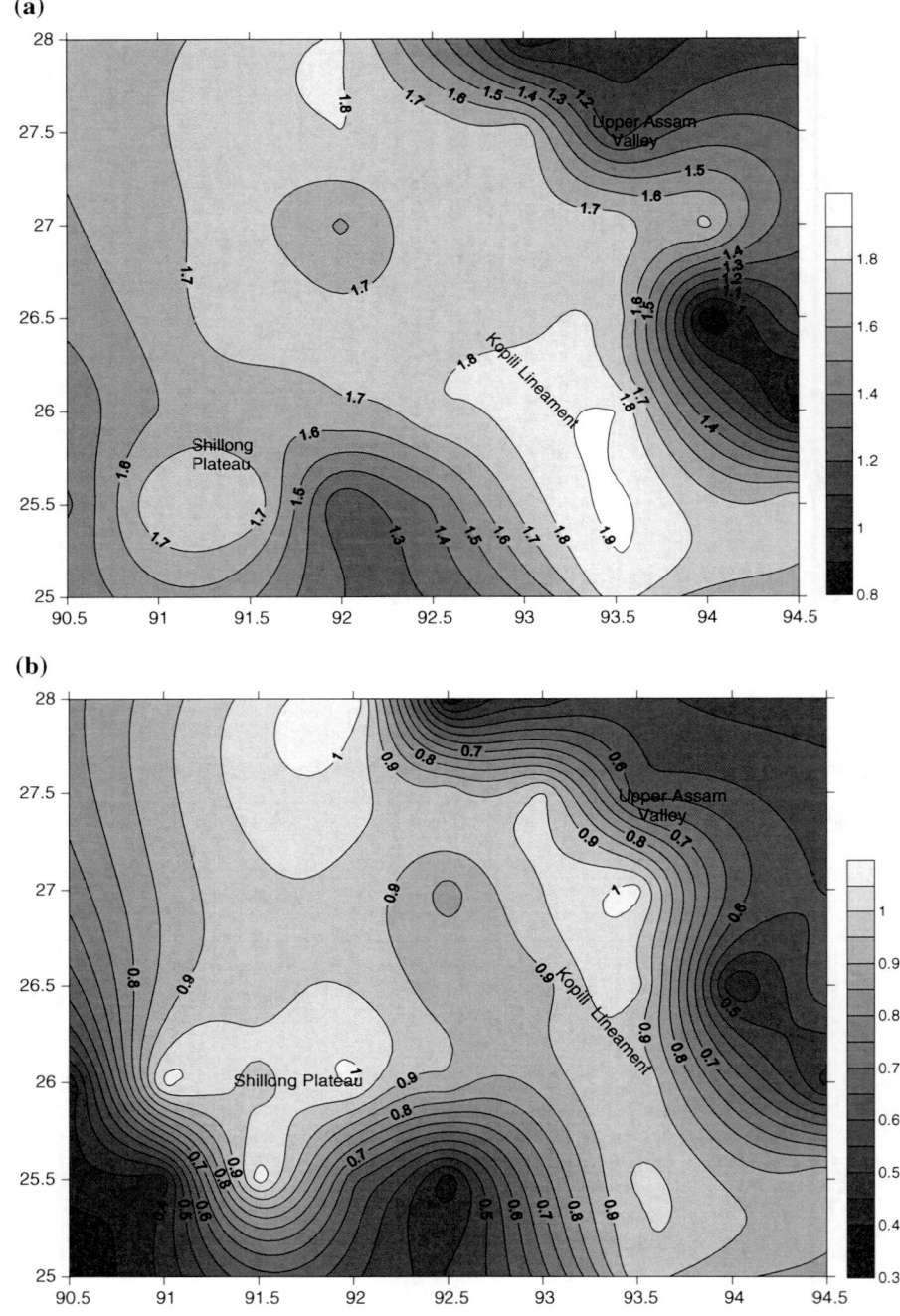

Figure 8
a Fractal dimension contour map of all the events. b b-value contour map of all the events (BHATTACHARYA et al., 2002)

inversion. Comparison of the actual Vp (Fig. 10a) images are shown with the restored Vp (Fig. 10b) and these are in good agreement; the reconstruction of the model is good at all depths for Vp.

In the uppermost crust, the depth slice at 10 km is characterized by low-Vp as well as by high-Vp zones. The Shillong Plateau is distinguished as a high velocity structure, and the Bengal basin as a low Vp

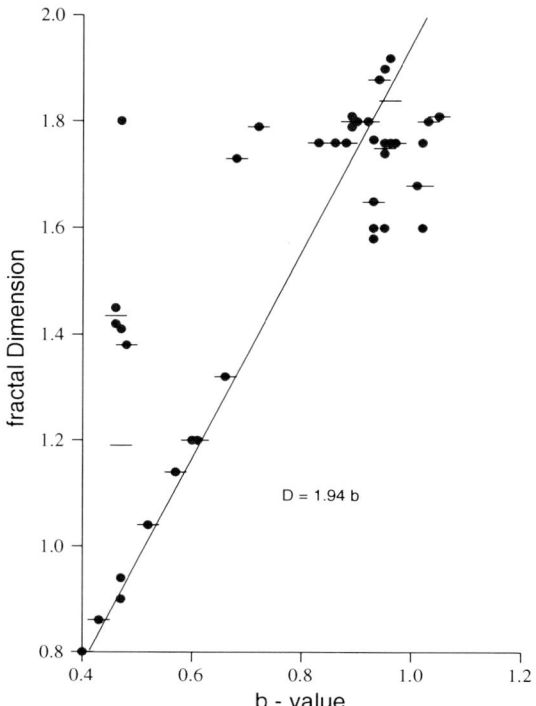

Figure 9
Plot showing the relation between fractal dimension and *b* value.
Error bars are shown on the slope

structure. The Mikir hills and Indo-Burma ranges are also demarcated as high Vp structures. In the mid crust at a depth of 20 km, the Shillong Plateau/Mikir hills and Indo-Burma ranges are well identified as high-Vp zones. A prominent northwest–southeast trending low-Vp zone is well imaged in the lower crust (depth 20–30 km), between the Shillong Plateau and the Mikir hills (Fig. 5). This low Vp structure can be correlated with the Kopili fault shear zone (Fig. 1). The northwest–southeast trending low Vp and high Vp structures at 20–30 km depth slices are conspicuous; higher seismic activity is prominent in the high Vp zones and in its surroundings. The thick Bengal basin sediments are well reflected down to 20 km, the low Vp structure at 20-km-depth slice is very prominent, and little seismic activity is recorded. It may be noted that the low Vp structure is no longer visible at the 30 km depth slice below the Bengal basin. This observation supports the reported sediment thickness of the basin ∼20 km (NANDY 1980, 2001).

The 40 km depth slice shows a near north–south trending high Vp zone, which is very prominent below the Mikir Hills/Assam valley (Fig. 5). The events are mostly concentrated in this high Vp zone; this is interpreted to be the base of the Kopili fault. The geologically mapped Kopili lineament is possibly a surface expression of the deep seated seismogenic structure, the Kopili fault. We infer that this high Vp zone at 40 km depth is the base of the seismogenic zone beneath the surface trace of the Kopili lineament. The stress is accumulated in the high velocity zone at the 'fault end'; the fault system is ∼300 km long and ∼50 km wide. It is noted that there is no such seismic source zone beneath the Shillong Plateau at 40 km depth (Fig. 5); the seismic source zone below the Plateau is confined within 35 km. This indicates a rehological change of the crust at the Moho depth; Moho depth beneath the Plateau is ∼35 km (MITRA et al., 2005).

The fractal dimension values for the entire region vary between 0.80 and 1.90. This observation suggests that the faults are spatially distributed in the entire region, and the entire region is seismically active. The fractal dimension at 0–20 km depth indicates a low value (∼1.0) along the Kopili fault (Fig. 7a) and a high value (∼1.9) at 20–40 km depth (Fig. 7b). The fractal dimension contour map for the entire data set shows a higher trend (∼1.8) along the Kopili fault (Fig. 8a). This is indicative of the fact that Kopili fault is a deep-rooted fault. The highest

Table 3

Earthquake location quality

Earthquake location Errors (average)	HYPO71 Using the 1D velocity (980 events)	SIMULPS Using 3D velocity model (980 events)
RMS (s)	0.49	0.06
ERH (km)	5.58	2.19
ERZ (km)	3.03	2.03

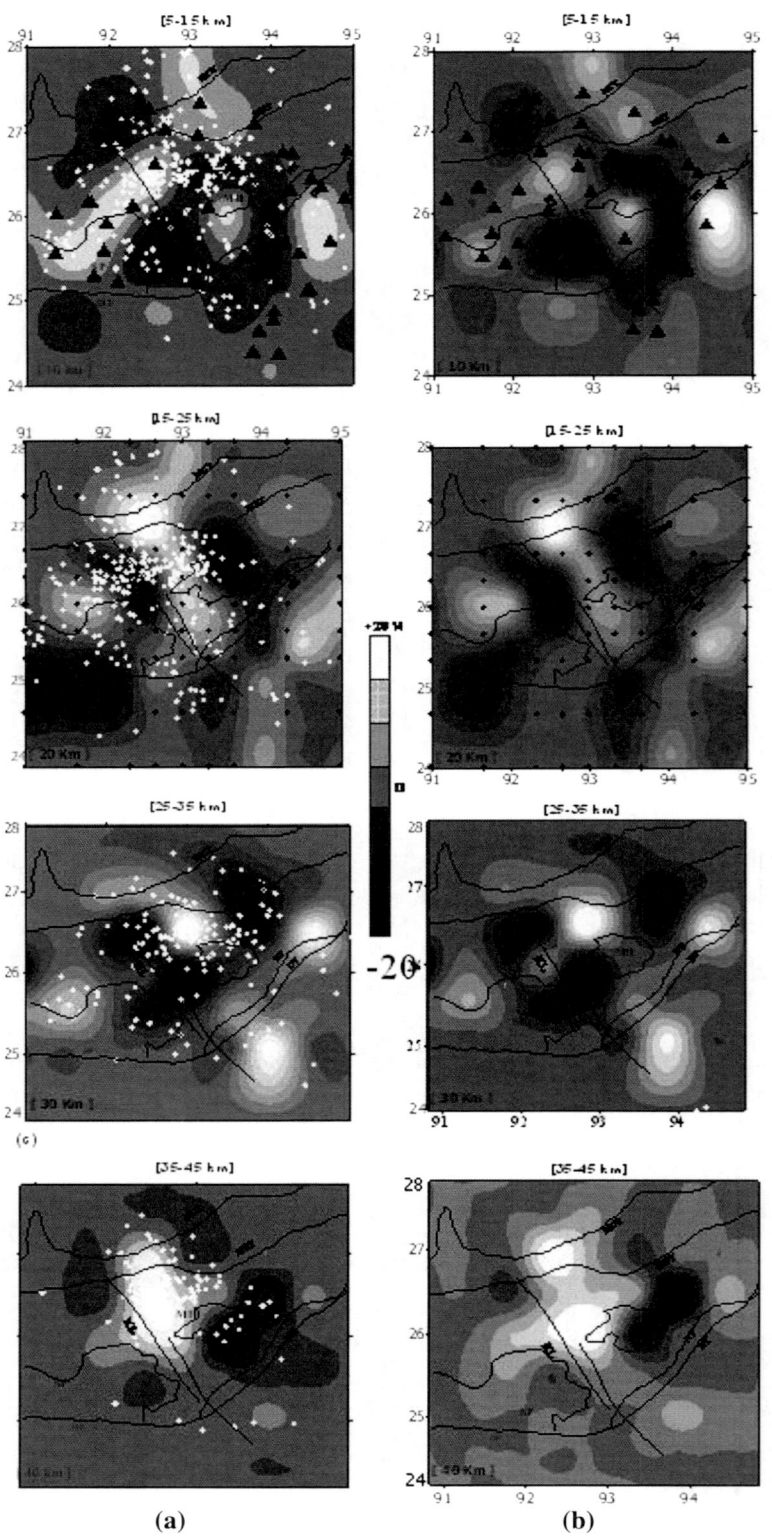

(a)　(b)

Figure 10
a Distribution of Vp perturbations at different depths. **b** Distribution of Vp perturbations at different depths after conducting restoring resolution test. The depth of each layer is shown at the *bottom*. The depth ranges of the earthquakes are shown at the *top* of each map. *Solid triangles* are the permanent and temporary seismic stations as in Fig. 3. The *black solid circles* show grid point distributions. The *thick lines* show the threshold value of the resolution matrix, and events (*white dots*) at different depths are shown. *Curvilinear lines* show the tectonic features as indicated in Fig. 3. The perturbation scale is shown right to the images (BHATTACHARYA et al., 2008)

fractal dimension of the order of 1.90 along this fault indicates heterogeneity and multiple/complex tectonic stresses in this fault zone; these stresses are from the Himalayan arc as well as from the Burmese arc, and the earthquakes mostly occur by strike slip-faulting (KAYAL et al., 2006). The b value map (Fig. 8b) clearly depicts a spatial variation of earthquake frequency in the region. Higher b value contours along the Kopili fault ($b \sim 1.0$) and in the Shillong Plateau ($b \sim 0.9$) reflect the two main seismic sources beneath the area. The NW–SE trend of higher b value along the Kopili fault extends from the Mikir Hills to Arunachal Himalaya across the MBT. The Kopili fault is suggested to be an active fault, transverse to the Himalayan trend (KAYAL et al., 2006; BHATTACHARYA et al., 2008).

The Shillong Plateau activity, on the other hand, is concentrated within the Plateau region; the fractal dimension contours show a higher trend 1.45–1.75, which is nearly circular (Fig. 8b). This value is, however, lower in comparison to the fractal dimension (~ 1.90) along the ~ 300 km long Kopili fault, which is more complex due to shear stress. The comparatively lower value in the Shillong Plateau indicates a relative lack of fracturing in the Shillong massif. In other words, compared to the Kopili fault, the Shillong Plateau is characterized as relatively unfractured and strong. The Plateau earthquakes are explained by 'pop-up' tectonics (BILHAM and ENGLAND, 2001), rather than shear stress. KAYAL et al. (2006) supported the pop-up tectonics of the Plateau between the Dapsi thrust and the Brahmaputra fault, and reported reverse faulting earthquakes beneath the Plateau. The circular trend of the fractal dimension contours could represent block uplift or pop-up of the Plateau unlike the linear trend along the Kopili fault. The higher fractal dimension along the deep rooted long Kopili fault is of considerable interest in this study, and may be a potential seismic zone for an impending large earthquake in the region.

We obtained a relation between D and b as: $D = 1.94\,b$ (Fig. 9). This supports the idea put forward by LEGRAND (2002) that for intermediate magnitude earthquakes the relation $D = 2b$ is fairly well satisfied. The correlation coefficient is found to be 0.79, which indicates that the correlation is positive and significant (HIRATA, 1989). As this region has a history of great and moderate magnitude earthquakes, this correlation implies that due to its complicated tectonic setting, the region experiences stress accumulation and the energy releases in the long-term period, forming locally fractured zones of high seismic activity. The NW–SE trending low Vp and high Vp zones at 20–30 km depth slices below the Kopili fault are conspicuous; the high velocity zone is seismically more active (Fig. 5). Higher b value and higher fractal dimension are also observed in this fault zone (Figs. 7, 8). Similar high Vp, high b value and high fractal dimensions are observed below the western part of the Shillong Plateau, and this part of the Plateau is also seismically very active too.

Acknowledgment

This research was carried out under an ILTP scheme, Indo-Russian project. It has been a part of Ph. D. thesis of PMB. JRK presented this paper at the Evison Symposium, Victoria University of Wellington, New Zealand, February, 2008.

REFERENCES

AKI, K. (1965), *Maximum likelihood estimate of b in the formula log N = a − b M and its confidence limits*, Bull Earthquake Res Inst Tokyo Univ *43*, 237–239.

BENDER, B. (1983), *Maximum likelihood estimation of b values for magnitude grouped data*, Bull Seism Soc Am *73*, 831–851.

BHATTACHARYA, P. M., MAJUMDAR, R. K., and KAYAL, J. R. (2002), *Fractal dimension and b-value mapping in northeast India*, Curr Sci *82* (*12*), 1486–1491.

BHATTACHARYA, P. M., MUKHOPADHYAY, S., MAJUMDAR, R. K., and KAYAL J. R. (2008), *3D seismic structure of the northeast India region and its implications for local and regional tectonics*, J Asian Earth Sci *33*, 25–41.

BILHAM, R. and ENGLAND, P. (2001), *Plateau 'pop-up' in the great 1897 Assam earthquake*, Nature *410*, 806–809.

CHEN, W. P. and MOLNAR, P. (1990), *Source parameters of earthquakes and intraplate deformation beneath the Shillong Plateau and northern Indo-Burma ranges*, J Geophys Res 95, 12,527–12,552.

DE REENA, and KAYAL, J. R. (1990), *Crustal P-wave velocity and velocity–ratio study in northeast India by a microearthquake survey*, Pure Appl Geophys 134, 93–108.

EBERHART-PHILLIPS, D. (1986), *Three-dimensional velocity structure in northern California coast ranges from inversion of local earthquake arrival times*, Bull Seism Soc Am 76, 1025–1052.

EBERHART-PHILLIPS, D., *Local earthquake tomography: Earthquake source regions*. In Seismic Tomography: Theory and Practice (eds. H. M. Iyer and K. Hirahara) (Chapman and Hall, London, 1993) pp. 613–643.

GHOSH, R.K. and SAHA S., *Mathematics and Statistics Advanced Level* (New Central Book Agency Publishers, India, 2002) 202 pp.

GRASSBERGER, P. and PROCACCIA, I. (1983) *Characterisation of strange attractors*, Phys Rev Lett 50, 346–349.

HANKSSON, E. (1990), *Earthquakes, faulting, and stress in the Los Angels basin*, J Geophys Res 95 (15), 365–394.

HIRATA, T. (1989), *A correlation between the b value and the fractal dimension of earthquakes*, J Geophys Res 94, 7507–7514.

IDZIAK, A. and TEPER, T. (1996), *Fractal dimension of faults network in the upper Silesian coal basin Poland: Preliminary studies*, Pure Appl Geophys 147, 239–247.

JONES, L. M., SIEH, K. E., HANKSON, E., and HUTTON, L. K. (1990), *The December 3, 1998 Pasadena earthquake: Evidence for strike-slip motion on the Raymond fault*, Bull Seism Soc Am 80, 474–482.

KAGAN, Y.Y. (2007), *Earthquake spatial distribution: The correlation dimension*, Geophys J Int 168, 1175–1194. doi: 10.1111/j.1365-246X.2006.03251.

KAGAN, Y.Y. and KNOPOFF, L. (1980), *Spatial distribution of earthquakes: The two-point correlation function*, Geophys J R Astron Soc 62, 303–320.

KAYAL, J.R. (1996), *Earthquake source processes in northeast India: A review*, J Him Geol 17, 53–69.

KAYAL, J.R. (2001), *Microearthquake activity in some parts of the Himalaya and the tectonic model*, Tectonophysics 339, 331–351.

KAYAL, J. R., *Microearthquake Seismology and Seismotectonics of South Asia* (Springer, Heidelberg, 2008) 503 pp.

KAYAL, J. R. and DE, R. (1991), *Microseismicity and tectonics in northeast India*, Bull Seism Soc Am 81, 131–138.

KAYAL, J. R., DE, R., and CHAKRABORTY, P. (1993), *Microearthquakes at the Main Boundary Thrust in eastern Himalaya and the present day tectonic model*, Tectonophysics 218, 375–381.

KAYAL, J. R., AREFIEV, S.S., BARUAH, S., HAZARIKA, D., GOGOI, N., KUMAR, A., CHOWDHURY, S.N., and KALITA, S. (2006), *Shillong Plateau earthquakes in northeast India region: Complex tectonic model*, Curr Sci 91(1), 109–114.

KUMAR, M.R. and RAO, N.P., (1995), *Significant trends related to slab seismicity and tectonics in the Burmese Arc region from Havard CMT solutions*, Phys Earth Planet Int 90, 75–80.

LE DIAN, A. Y., TAPPONNIER, P., and MOLNAR, P. (1984), *Active faulting and tectonics of Burma and surrounding regions*, J Geophys Res 89, 453–472.

LEE, W. H. K. and LAHR, J.C. (1975), *HYPO71 (revised): A computer program for determining hypocenter, magnitude and first motion pattern of local earthquakes*, US Geol Surv Open file Rep., Vol 75–311, pp 1–116.

LEGRAND, D. (2002), *Fractal dimension of small, intermediate and large earthquakes*, Bull Seism Soc Am 92, 3318–3320.

MITRA, S., PREISTLY, K., BHATTACHARYA, A. K., and GAUR, V. K. (2005), *Crustal structure and earthquake focal depths beneath northeastern India and southern Tibet*, Geophys J Int 160, 227–248. doi:10.1111/j.1365-246X.2004.02470.

NANDY, D. R. (1980), *Tectonic Pattern in NE India*, Ind J Earth Sci 7, 103–107.

NANDY, D. R., *Geodynamics of Northeastern India and the Adjoining Region* (ACB publication, New Delhi, 2001) 209 pp.

OLDHAM, R. D. (1899), *Report on the great earthquake of the 12th June 1897*, Geol Surv India Mem Vol 29, Reprinted, 1981, Geological Survey of India, Calcutta, 379 pp.

PAGE, R. (1968), *Aftershocks and micro-aftershocks*, Bull Seism Soc Am 58, 1131–1606.

PAVILIS, G. L. and BOOKER, J.R. (1980), *The mixed discrete–continuous inverse problem: Application to simultaneous determination of earthquake hypocenters and velocity structure*, J Geophys Res 88, 8226–8236.

RICHTER, C.F., *Elementary Seismology* (W.H. Freeman, San Francisco, 1958) 768 pp.

SCHOLTZ, C. H., *Mechanics of Earthquakes and Faulting* (Cambridge University Press, Cambridge, 1990) 290 pp.

SEEBER, L., ARMBUSTER, J.G., and QUITTMEYER, R. (1981), *Seismicity and continental subduction in the Himalayan Arc*. In Zagros, Hindu Kush Himalaya Geodynamic Evolution, Geodyn Ser. (eds. H. K. Gupta and F.M. Delany), 3, 215–242.

SMITH, L. A. (1988), *Intrinsic limits on dimension calculations*, Phys Lett 133, 283–288.

TANDON, A. N. (1954), *A study of Assam earthquake of August 1950 and its aftershocks*, Indian J Meteorol Geophys 5, 95–137.

THURBER, C. H. (1983), *Earthquake location and three dimensional crustal structure in the Coyote Lake area, Central California*, J Geophys Res 88, B10, 8226–8236.

TOSI, P. (1998), *Seismogenic structure behaviour revealed by spatial clustering of seismicity in the Umbria-Marche region, (Central Italy)*, Ann. De Geofisica 41(2), 215–224.

UTSU, T. (1965), *A method for determining the value of b in a formula $\log N = a - bM$ showing the magnitude–frequency relation for earthquakes*, Geophys Bull Hokkaido Univ 13, 99–103.

WYSS, M., SAMMIS, C. G., NADEAU, R.M., and WEIMER, S. (2004), *Fractal dimension and b value on creeping and locked patches of the San Andreas fault near Parkfield California*, Bull Seism Soc Am 94(2), 410–421.

XU, Y. and BURTON, P. (1999), *Spatial fractal evolutions and hierarchies for microearthquakes in central Greece*, Pure Appl Geophys 154, 73–99.

ZHAO, D., HASEGAWA, A., and HORIUCHI, S. (1992), *Tomographic imaging of P and S wave velocity structure beneath northeastern Japan*, J Geophys Res 97, 19909–19928.

(Received May 14, 2008, revised February 4, 2009, accepted May 24, 2009, Published online March 13, 2010)

Pure Appl. Geophys. 167 (2010), 1013–1048
© 2010 Birkhäuser / Springer Basel AG
DOI 10.1007/s00024-010-0085-1

Pure and Applied Geophysics

Seismic Hazard Evaluation in Western Turkey as Revealed by Stress Transfer and Time-dependent Probability Calculations

P. M. Paradisopoulou,[1] E. E. Papadimitriou,[1] V. G. Karakostas,[1] T. Taymaz,[2] A. Kilias,[3] and S. Yolsal[2]

Abstract—Western Turkey has a long history of destructive earthquakes that are responsible for the death of thousands of people and which caused devastating damage to the existing infrastructures, and cultural and historical monuments. The recent earthquakes of Izmit (Kocaeli) on 17 August, 1999 ($M_w = 7.4$) and Düzce ($M_w = 7.2$) on 12 November, 1999, which occurred in the neighboring fault segments along the North Anatolian Fault (NAF), were catastrophic ones for the Marmara region and surroundings in NW Turkey. Stress transfer between the two adjacent fault segments successfully explained the temporal proximity of these events. Similar evidence is also provided from recent studies dealing with successive strong events occurrence along the NAF and parts of the Aegean Sea; in that changes in the stress field due to the coseismic displacement of the stronger events influence the occurrence of the next events of comparable size by advancing their occurrence time and delimiting their occurrence place. In the present study the evolution of the stress field since the beginning of the twentieth century in the territory of the eastern Aegean Sea and western Turkey is examined, in an attempt to test whether the history of cumulative changes in stress can explain the spatial and temporal occurrence patterns of large earthquakes in this area. Coulomb stress changes are calculated assuming that earthquakes can be modeled as static dislocations in elastic half space, taking into account both the coseismic slip in large ($M \geq 6.5$) earthquakes and the slow tectonic stress buildup along the major fault segments. The stress change calculations were performed for strike-slip and normal faults. In each stage of the evolutionary model the stress field is calculated according to the strike, dip, and rake angles of the next large event, whose triggering is inspected, and the possible sites for future strong earthquakes can be assessed. A new insight on the evaluation of future seismic hazards is given by translating the calculated stress changes into earthquake probability using an earthquake nucleation constitutive relation, which includes permanent and transient effects of the sudden stress changes.

Key words: Stress transfer, earthquake probabilities, seismic hazard.

1. Introduction

Many destructive earthquakes occurred in the territory of western Turkey and the adjacent eastern part of the Aegean Sea, some of them close both in time and space. Observations on temporal and spatial clustering of strong events have led several authors to highlight the importance of fault interactions on the basis of physical models. Earthquake triggering or delay due to changes in stress was recognized more than a decade ago (e.g., Harris, 1998 and references therein) and is worked out in assessing earthquake occurrence and future seismic hazard in a certain area. Stein (1999), reviewing the role of stress transfer, emphasized the earthquake interaction as a fundamental feature of seismicity that promises a deeper understanding of the earthquake occurrence and a better description of the seismic hazard, when stress transfer is incorporated into probability models (Stein et al., 1997; Toda et al., 1998). In association with physical fault models and fault properties, such models were more developed and statistically assessed (Parsons, 2004, 2005; Hardebeck, 2004; among others).

The first goal of the present study is to investigate how the stress changes caused by the strong earthquakes of $M \geq 6.5$ that occurred during the instrumental era, that is since the beginning of the twentieth century in the area of eastern Aegean Sea

---

[1] Department of Geophysics, School of Geology, Aristotle University of Thessaloniki, 54124 Thessaloniki, Greece. E-mail: popi.paradis@gmail.com; ritsa@geo.auth.gr; vkarak@geo.auth.gr

[2] Department of Geophysics, The Faculty of Mines, İstanbul Technical University, Maslak, 34469 Istanbul, Turkey. E-mail: taymaz@itu.edu.tr; yolsalse@itu.edu.tr

[3] Department of Geology, School of Geology, Aristotle University of Thessaloniki, 54124 Thessaloniki, Greece. E-mail: kilias@geo.auth.gr

and western Turkey (Fig. 1), influence future occurrences. This is attempted by the application of the stress evolutionary model (DENG and SYKES, 1997) according to which the long-term tectonic loading on the major regional faults is added to the coseismic slips of the strong events. As a second step the static stress changes on specific faults that have accumulated to date will be incorporated into probabilistic models, in an attempt to assess the seismic hazard in the study area. The first relevant investigation along the North Anatolian Fault was compiled by STEIN et al. (1997) who found that 9 out of 10 earthquakes with $M \geq 6.7$ were triggered by previous events, and estimated stress-based probabilities. Investigation of stress transfer in northwestern Turkey and the North Aegean Sea was performed by NALBANT et al. (1998) by the calculation of the static stress changes due to the coseismic slips of $M \geq 6.0$ events, whereas PAPADIMITRIOU and SYKES (2001) applied the stress evolutionary model in the Northern Aegean Sea for strike-slip faulting. HUBERT-FERRARI et al. (2000) calculated the stress field that resulted from the coseismic slips of events of M 6 or greater since 1700 and the secular interseismic stress changes to show that the 1999 events were anticipated. For the southeastern Aegean area, part of which coincides partially with our study area, the evolutionary model satisfactorily explained the clustering of strong ($M \geq 6.5$) normal faulting earthquakes (PAPADIMITRIOU et al., 2005). In all these studies possible future occurrences are suggested, which will be discussed in the following sections along with the results obtained in the present study.

This study differs from the previously mentioned ones, as regards calculations of static stress changes, in that it aims to integrate the stress evolution history of the entire territory of western Turkey and its adjacent Aegean Sea area. This integration concerns both the areal coverage and the involvement of different faulting types and the continuous tectonic loading, since the above-mentioned studies dealt either with parts of our study area, or a single faulting type and coseismic slips only. The variability of the stress change calculations from friction coefficient, dip and rake angles were assessed following PARSONS (2005). The perspective of the stress field evolution calculations is the identification of active fault segments that are currently in stress enhanced areas; a first step for the seismic hazard assessment. We start with the stationary and conditional probability models estimation of the probability of occurrence in the next 30 years of an earthquake with $M \geq 6.5$ on known fault segments of the study area. Thereafter the accumulated stress changes due to coseismic slips of the modeled events were incorporated into the estimation of earthquake probability. Change in the probability on a given fault is calculated from the change in seismicity rate, which is computed taking into account both permanent and transient effects.

2. Seismotectonic Setting

The complexity of the plate interactions and associated crustal deformation in the eastern Mediterranean region is reflected in many destructive earthquakes that have occurred throughout its recorded history, and many of them are rather well documented and studied. The region features complex tectonics because it relates to the interaction of Eurasian, Arabian, and African lithospheric plates (Fig. 1). The subduction of the eastern Mediterranean oceanic lithosphere, the frontal part of the northward moving African lithosphere, along the Hellenic Arc is a key feature that influences the active deformation of the region, causing an extension of the continental crust in the overlying Aegean province (PAPAZACHOS and COMNINAKIS 1969, 1971; PAPAZACHOS et al., 1998).

North and East Anatolian faults represent the lateral movement of Turkey toward the west (MCKENZIE, 1970). This motion is transferred into the Aegean in a southwesterly direction. It has been suggested that the Aegean Sea and much of Anatolia should be considered as two separate microplates observed from geodetic information combined with the seismological data (TAYMAZ et al., 1991a; JACKSON, 1994; PAPAZACHOS, 1999; MCCLUSKY et al., 2000; NYST and THATCHER, 2004). The southern boundary of the south Aegean plate is defined by low angle thrust faults that are located along the Hellenic Arc (PAPAZACHOS et al., 1984, 1998; TAYMAZ, 1990, 1996). Interplay between dynamic effects of the relative motions of adjoining plates thus controls the

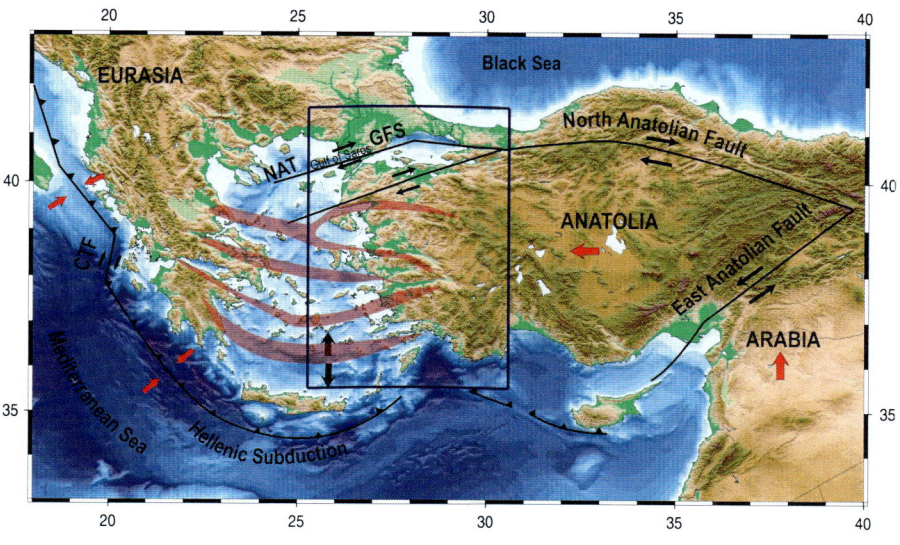

Figure 1
Summary sketch map of the active tectonic boundaries in the eastern Mediterranean Sea region. Large *arrows* show relative motions of plates with respect to Eurasia. The main extensional structures are shaded in *red*. *NAT* North Aegean Trough, *CTF* Cephalonia Transform Fault (after PAPAZACHOS *et al.*, 1998; ARMIJO *et al.*, 1999; MCCLUSKY *et al.*, 2000)

large-scale crustal deformation and the associated earthquake activity in the study area.

Our study area is one of the most seismically active and deforming regions in the world bounded on the north by the NAF and the Ganos Fault System (GFS). The NAF is one of the longest active right lateral strike-slip fault systems, of about 1,500 km in length extending from eastern Turkey, through the Marmara Sea where it bifurcates into two branches. Along its northern branch, the Ganos Fault System constitutes the most significant tectonic element controlling the tectonic evolution of the area. The western termination of this fault system is in the Gulf of Saros, which is a neotectonic basin with ENE trending depression placed at the northeastern part of the Aegean Sea, where the North Aegean Trough (NAT) is developed. The dextral strike-slip motion of NAF is translated into the Aegean where it additionally accommodates the rapid N–S extension of the backarc Aegean region. Western Turkey is located in the boundary area between these regions and it has been under an N–S extension since Late Oligocene (SAUNDERS *et al.*, 1998). The major neotectonic features of western Turkey are the E–W trending grabens (e.g., Gediz, Küçük Menderes, Büyük Menderes) and their basin bounding active normal faults (Fig. 2) as well as other less prominent structures with NE–SW trending basins (e.g., SEYITOĞLU and SCOTT, 1991; TAYMAZ and PRICE, 1992; WESTAWAY, 1993; BOZKURT, 2001, 2003).

3. Methodology

The stress evolutionary model that is applied in the present study was proposed by DENG and SYKES (1997) and originally tested in southern California. Cumulative stress changes are assumed to arise from the following two sources: Tectonic loading generated by plate motions and coseismic displacements on faults associated with earthquakes. Interseismic stress accumulation between strong events is modeled by introducing "virtual negative displacements" along the major regional faults using the best available information on their long-term slip rates. These virtual dislocations are imposed on the faults with sense of slip opposite to the observed slip. The magnitude of this virtual slip is incremented in time according to the long-term rate of the fault. This is equivalent to the constant positive slip extending from the bottom of the seismogenic layer to infinite depth. Hence, tectonically-induced stress builds up in

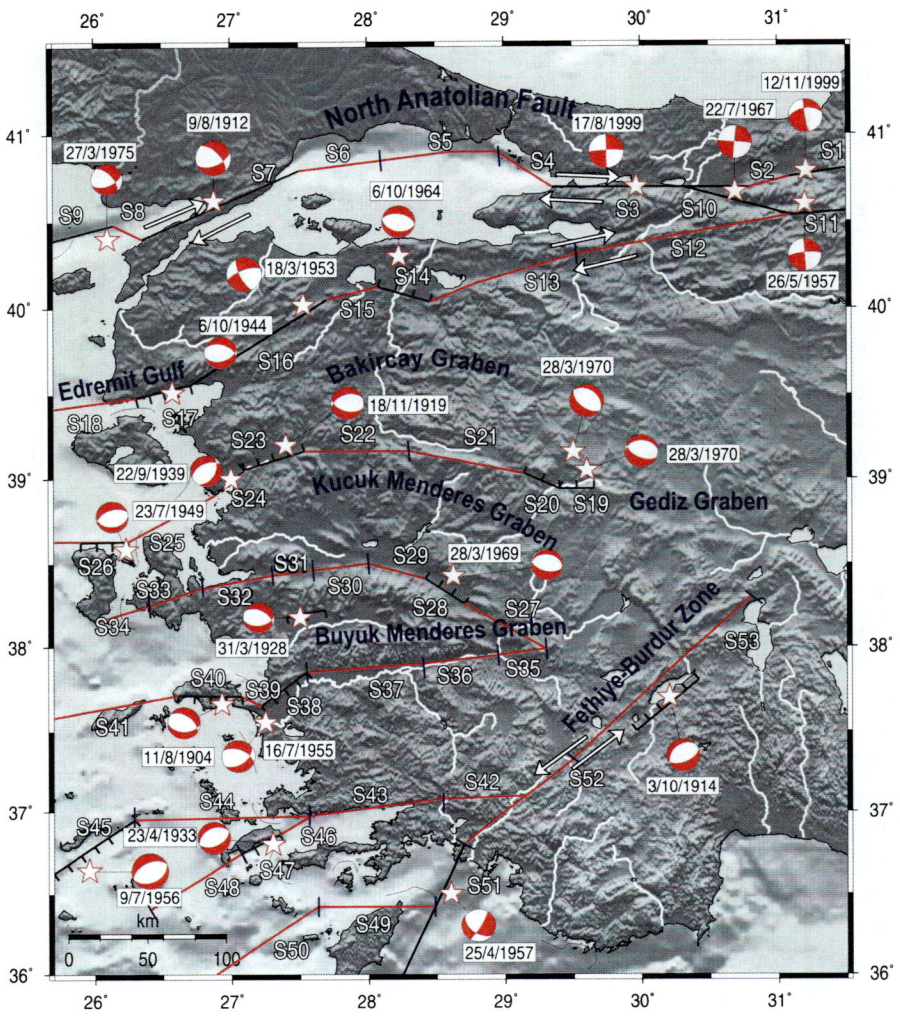

Figure 2
Simplified map of active faulting in the area of western Turkey. The code names of the fault segments are shown next to each segment. The segments that are associated with earthquakes of $M \geq 6.5$ that occurred since 1900, are shown in *black*. The fault plane solutions of $M \geq 6.5$ events are shown as lower hemisphere equal area projection whereas their epicenters are denoted by stars, linked with a light line with the beach balls. The occurrence day of each event (month/date, year) is given on top of the focal spheres

the vicinity of faults during interseismic periods. All computed interseismic stress accumulation is associated with the deformation caused by the time-dependent virtual displacement on major faults extending from the free surface to the seismogenic depth. Stress build-up is released wholly or in part during the next strong earthquake, with positive real displacements on given fault segments. Changes in stress associated with strong earthquakes are calculated for coseismic displacements on the ruptured fault segment and by adding the changes in the components of the stress tensor together as they occur in time.

Stress changes associated with both the virtual dislocations and actual earthquake displacements are calculated using a dislocation model of a planar fault surface, Σ, embedded in an elastic half space (OKADA, 1992). Earthquakes occur when stress exceeds the strength of the fault. The closeness to the failure was quantified by using the change in Coulomb failure function (ΔCFF). It depends on both changes in shear stress, $\Delta\tau$, and normal stress, $\Delta\sigma$, and in the presence of pore fluid it takes the form:

$$\Delta CFF = \Delta\tau + \mu(\Delta\sigma + \Delta p), \quad (1)$$

where $\Delta\tau$ is the shear–stress change (computed in the slip direction), $\Delta\sigma$ is the fault–normal stress change (positive for extension), Δp is the pore pressure change within the fault, and μ is the friction coefficient, which ranges between 0.6 and 0.8 (HARRIS, 1998 and references therein). Throughout this study we ignore the time-dependent changes in pore fluid pressure and consider only the undrained case (BEELER et al., 2000), meaning that Δp depends on the fault–normal stress whereas the fluid mass content per unit volume remains constant. Induced changes in pore pressure resulting from a change in stress under undrained conditions, according to RICE and CLEARY (1976) are calculated from:

$$\Delta p = -B\frac{\Delta\sigma_{kk}}{3}. \quad (2)$$

B is the Skempton's coefficient, where $0 \leq B < 1$, and $\Delta\sigma_{kk}$ indicates summation over the diagonal elements of the stress tensor. The Skempton's coefficient, B, denotes the relative proportion of fault–normal stress and change in pore pressure as it is assumed in Coulomb stress analysis (see KING et al., 1994; HARRIS, 1998, and references therein). If the air fills the pores then B is nearly zero, whereas if water fills the pores, it is typically between 0.5 and 1.0 for fluid-saturated rock and close to 1.0 for fluid-saturated soil. Sparse experimental determinations of B for rocks indicate a range from 0.5 to 0.9 for granites, sandstones, and marbles (RICE and CLEARLY, 1976). We assume a $B = 0.5$ and $\mu = 0.75$ (as in ROBINSON and McGINTY, 2000; among others). If in the fault zone $\Delta\sigma_{11} = \Delta\sigma_{22} = \Delta\sigma_{33}$, so that $\Delta\sigma_{kk}/3 = \Delta\sigma$, then the apparent coefficient of friction is defined as $\mu' = \mu(1 - B)$. The above selected values for B and μ result to a value of apparent coefficient of friction equal to 0.4, which is widely used in studies of Coulomb stress modeling. We will investigate the effects of different values of Skempton's coefficient, B, namely equal to 0.2 and 0.9, which are the extreme values expressing the percentage of water filling the pores.

In Eq. 2 $\Delta\sigma_{kk}$ is the summation of the stress normal components, which, along with $\Delta\tau$ are calculated according to the fault plane solution of the next earthquake in the sequence of events, whose triggering is inspected. $\Delta\tau$ is positive for increasing shear stress in the direction of the relative slip on the observing fault while $\Delta\sigma$ is positive for tensional normal stress. When compressional normal stress on a fault plane decreases, the static friction across the fault plane also decreases. A positive value of ΔCFF for a particular fault denotes movement of that fault towards the failure (that is, likelihood that it will rupture in an earthquake is increased). The shear modulus and Poisson's ratio are fixed at 3.3×10^5 bar and 0.25, respectively.

Earthquakes nucleating on active fault surfaces are often approximated with rectangles dipping within the brittle layer of the Earth's crust. Fault planes are adequately described by the use of geometrical parameters such as the length, L, and the width, w, of the fault zone, and the fault plane solution. To calculate the rupture parameters that are necessary for the model application we use empirical relationships when field observations or relevant information from previous investigations are not available. These relationships are taken from PAPAZACHOS et al. (2004) who collected worldwide data and proposed scaling laws for different seismotectonic environments, according to which fault length, (L, in km), and coseismic displacement, (u, in cm), can be calculated as a function of the earthquake magnitude. For the dip slip faults the following scaling laws are derived:

$$\log L = 0.50M - 1.86, \quad (3)$$

$$\log u = 0.72M - 2.82, \quad (4)$$

whereas for strike-slip faults the respective equations are:

$$\log L = 0.59M - 2.30, \quad (5)$$

$$\log u = 0.68M - 2.59. \quad (6)$$

Estimates from Eqs. 3–6 and the respective relations proposed by WELLS and COPPERSMITH (1994) were found to be in a good agreement.

When the seismic moment (M_o) of an earthquake is known the coseismic displacement, $\bar{u}$, is calculated from the following equation:

$$M_o = \mu \cdot \bar{u} \cdot S = \mu \cdot \bar{u} \cdot L \cdot w, \qquad (7)$$

where μ is taken equal to 3.3×10^5 bar, as above, and S is the fault surface ($S = L\,w$). Fault width, in km, was estimated from the dip angle of the fault and the distance measured down-dip from the surface to the upper and lower edges of the rectangular dislocation plane, respectively.

For the evolutionary model application, it is necessary to define the seismogenic layer where the distribution of the coseismic slip is considered. In our study area it is known that the majority of the foci of the crustal earthquakes are located in the depth range of 3–15 km, which defines the brittle part of the crust. Considering all the above information combined with previous investigations for the study area (PAPADIMITRIOU and SYKES, 2001), the seismogenic layer in our calculations is taken to be in this range (3–15 km) for all the strong events ($M \geq 6.5$) modeled.

In addition to the fault geometry parameters, knowledge of the fault plane solutions is essential because the variation of these parameters affects the shape of the calculated stress field. Information on the events' fault plane solutions and moment magnitudes was collected from several studies (Global CMT solutions; MCKENZIE, 1972; EYIDOĞAN and JACKSON, 1985; EYIDOĞAN, 1988; TAYMAZ et al., 1991b; TAYMAZ and PRICE, 1992; KIRATZI and LOUVARI, 2001; BARKA et al., 2002; PAPAZACHOS and PAPAZACHOU, 2003). The fault planes are defined with the use of available field surveys, surface ruptures and in the absence of visible tectonic features of their corresponding faults, the ones consistent with the regional stress field are chosen. Magnitudes that are provided from PAPAZACHOS and PAPAZACHOU (2003) are equivalent moment magnitudes, M_w^* (PAPAZACHOS et al., 1997).

4. Long-term Slip Rate Constraints on Major Faults

The incorporation of the tectonic loading in the evolutionary stress field calculations requires the identification of the position and geometry of the major faults in the study area, along with the definition of the long-term slip rates on them. It is possible to estimate the slip rates on the existing faults by using the relative motions between GPS stations straddling them. Such information is available mostly from ARMIJO et al. (2003) who decomposed the present-day GPS velocity into two superposed velocity fields, associated with corresponding sets of slip rates on the major structures, and REILINGER et al. (2006) who presented a new GPS-derived velocity field including data from 1988 to 2005, updating the results by MCCLUSKY et al. (2000; 2003). These observations deal with the zone of interaction of the Arabian, African (Nubian, Somalian) and Eurasian plates, adjacent parts of Zagros and central Iran, Turkey and the Aegean/Peloponnesus relative to Eurasia.

Active faults in western Turkey, Aegean and the Greek mainland, are closely related with the plate boundary processes, namely, the westward propagation of the North Anatolian Fault System and the Aegean extension related to subduction processes. GPS measurements have shown that large regions of the lithosphere move coherently while deformation is mostly localized on a small number of structures that extend to the base of the lithosphere. During the interseismic period, a fault is locked at seismogenic depths in the brittle schizosphere. In the lower part of the crust, in the plastosphere, the fault is continuously creeping, loading its locked upper part. Block boundaries have been determined from geologically active faults, which account for the present-day block motions and regional deformation, seismicity and historic earthquakes.

ARMIJO et al. (2003) incorporated both the geodetic and the geological constraints and provide a robust description of the present-day deformation of the Anatolian–Aegean region. They use for their model localized deformation zones, which are represented by dislocation elements and extended from the base of the lithosphere to the locking depth at the base of the seismogenic layer. Some representative values of slip rates on elements at the area of NAF are 12–20 mm/year at the northern strand of NAF, 12 or 2–6 mm/year at the southern part of NAF. Other recent GPS data indicate rates of about 15–25 mm/year (REILINGER et al., 1997; MCCLUSKY et al., 2000, 2003) or 24 ± 1 mm/year (REILINGER et al., 2006).

The latter authors used a simple, kinematic block model, including elastic strain accumulations on the block-bounding faults, to quantify relative block motions and to determine present-day rates of the strain accumulation on the block bounding faults.

Normal faulting in the area of western Turkey is related with backarc extension of the Aegean. Stretching appears localized in a few regularly spaced rift zones in the Aegean which taper out into Anatolia and Greece (FLERIT et al., 2004). These rift zones are flanked by active faults, some of them being associated with the strong events modelled in this study, or with historical events. The extension rates of these fault segments range between 2 and 6 mm/year for the central part of western Turkey, whereas at the southern part reach values up to 8 mm/year. The transtensional character for some of these faults up to the central western part is modelled by a strike-slip component of the order of 1–2 mm/year (ARMIJO et al., 2003; REILINGER et al., 2006).

The defined major active fault segments and their long-term slip rates with their code names are shown in Fig. 2 along with the fault plane solutions of the modeled strong events ($M \geq 6.5$), as lower-hemisphere equal-area projections. Information on the names of the fault segments, their code names, geographical position, geometry (strike and dip) and their average long-term slip rates is given in Table 1. All values of slip rates represent about 60% of the slip rates provided by ARMIJO et al. (2003), FLERIT et al. (2004) and REILINGER et al. (2006) in order to account only for the seismically released strain energy. This constraint of the maximum possible accuracy slip rate for each fault segment will promote better estimates of earthquake hazard.

As regards the value of 60% of seismic coupling coefficient concerns, our choice was based on previous relevant investigations. According to AMBRASEYS and JACKSON (1990) a significant proportion (as much as 60%) of the strain may be aseismic. JACKSON et al. (1994) concluded that seismicity can account for at most 50% of the deformation in the Aegean area. KING et al. (1994), comparing plate rates to seismic moment release rates at the area of California and Nevada, found that the relative plate motion occurred about 60% seismically and 40% aseismically. For the area of NAF in particular KING et al. (2001) found that the rate of moment release accounts for about 60% of the relative plate motion. DAVIES et al. (1997) found that the seismic expression of strain for Greece accounts for only 20–50% of the geodetically determined strain. AYHAN et al. (2001) comparing GPS and seismic shear strain rates discovered that about 70% of GPS shear strain is accounted for by coseismic strain release. BIRD and KAGAN (2004) showed that continental transform faults (like NAF) have a 74% seismic coupling.

5. Calculation of the Evolutionary Stress Field

Stress changes, i.e., values of ΔCFF, are computed for the faulting types present in the study area, that is, right-lateral strike-slip faults oriented almost E–W or NE–SW and normal faults with almost E–W or ENE–WSW strike directions. At each stage of the evolutionary model, ΔCFF is calculated for a specific faulting type, that of the next inspected event. Information pertinent to the fault plane solutions of the events included in the calculations is given in Table 2 along with the rupture dimensions, length, L, and width, w, and the along strike, SS, and along dip direction, DS, coseismic slip, u, components.

Figures 3a–v are snapshots of ΔCFF at a depth of 10 km, chosen to be several kilometers above the locking depth (15 km) in the evolutionary model, since the nucleation depth is not known for most of the events. This is in agreement with KING et al. (1994) who found that seismic slip peaks at mid-depths in the seismogenic zone. In these figures, blue regions denote negative changes in Coulomb stress models and are called stress shadows (HARRIS and SIMPSON, 1993, 1996). Yellow to red areas are characterized as stress bright zones, representing positive values of ΔCFF. Pure green area indicates no significant change in CFF. Shadow zones and bright zones are specific to the strike, dip and rake angles of the fault that experiences the ΔCFF. We will present that, in each stage of the stress evolution calculations, the strong events are located inside the stress enhanced areas. The same applies for moderate events with faulting similar to the type for which the stress calculations were performed.

Table 1

Information on the fault segments on which tectonic loading is considered for the evolutionary stress field calculations

Segment number	Code name	Fault segment start		Fault segment end		Strike (°)	Dip (°)	Length (km)	Width (km)	Faulting type	SS mm/year	DS mm/year
		Latitude °N	Longitude °E	Latitude °N	Longitude °E							
S1	Duzce	40.75	31.08	40.83	31.73	262	53	56	0–19	RS	−7.2	0.0
S2	Karadere	40.69	30.79	40.75	31.08	255	55	26	0–18	RS	−7.2	0.0
S3	Izmit	40.70	29.35	40.69	30.79	268	84	120	0–15	RS	−10.8	0.6
S4	Cinarcik	40.70	29.35	40.90	28.95	308	88	40.4	0–15	RS	−10.2	4.8
S5	Marmaras 1	40.90	28.95	40.85	28.10	263	84	75	0–15	RS	−11.4	0.0
S6	Marmaras 2	40.85	28.10	40.79	27.50	263	78	51	0–16	RS	−11.4	0.0
S7	Ganos	40.40	26.35	40.79	27.50	68	55	110	0–18	RS	−12	0.6
S8	Saros 1	40.48	26.15	40.40	26.35	297	60	20	0–17	RS	−12	3.0
S9	Saros 2	40.39	25.70	40.48	26.15	68	55	40	0–18	RS	−12	3.0
S10	Mudurnu	40.53	31.10	40.70	30.30	275	88	80	0–15	RS	−3.6	1.8
S11	Abant	40.53	31.10	40.60	32.20	265	78	90	0–16	RS	−7.2	0.0
S12	Iznik	40.53	31.10	40.31	29.53	260	78	135.5	0–16	RS	−3	0.6
S13	Bursa 1	40.31	29.53	40.04	28.46	248	78	95.7	0–16	RS	−3	0.6
S14	Manyas	40.12	28.08	40.04	28.46	280	50	35	0–19	N	−0.6	3.0
S15	Yenice 1	40.05	27.70	40.12	28.08	256	70	33.2	0–16	RS	−1.2	0.0
S16	Yenice 2	40.05	27.70	39.56	26.70	250	70	101	0–16	RS	−1.8	0.0
S17	Edremit1	39.48	26.31	39.56	26.70	74	46	35	0–21	N	−1.2	2.4
S18	Edremit2	39.48	26.31	39.41	25.60	82	45	61.3	0–21	N	−1.2	2.4
S19	Gediz1	38.94	29.4	38.94	29.65	270	35	24	0–26	N	0.0	1.6
S20	Gediz2	39.03	29.14	38.94	29.40	308	35	24	0–26	N	0.0	1.6
S21	Bergama1	39.16	28.30	39.03	29.14	282	35	74	0–26	N	0.0	1.6
S22	Bergama2	39.16	27.54	39.16	28.30	271	35	66	0–26	N	−0.6	3.0
S23	Soma	39.05	27.10	39.16	27.54	253	45	43	0–21	N	−0.6	3.0
S24	Dikilli	39.05	27.10	38.90	26.88	211	45	26	0–21	N	−0.6	3.0
S25	Elaia	38.90	26.88	38.63	26.25	241	45	62.8	0–21	N	−1.2	1.8
S26	Chios	38.63	26.25	38.63	25.70	84	36	47.6	0–26	N	0.0	3.6
S27	Philadelphia	38.26	28.70	38.00	29.30	328	45	60.36	0–21	N	−0.6	2.4
S28	Alasehir	38.41	28.42	38.26	28.70	281	34	25	0–27	N	−0.6	2.4
S29	Sardeis	38.41	28.42	38.50	28.00	286	35	38	0–26	N	−0.6	2.4
S30	Turgutlu	38.50	28.00	38.46	27.60	263	45	35	0–21	N	0	3.6
S31	Kemalpasa	38.46	27.6	38.43	27.30	263	45	26	0–21	N	0	3.6
S32	Izmir	38.40	27.00	38.15	26.00	260	45	47	0–21	N	0	3.6
S33	Cezme	38.25	26.40	38.35	26.77	252	45	34	0–21	N	0	3.6
S34	Urla	38.35	26.77	38.43	27.30	252	45	37	0–21	N	0	3.6
S35	Denizli	37.85	27.55	37.95	28.95	83	45	31	0–21	N	−0.6	4.2
S36	Nazili	37.85	27.55	37.90	28.40	84	45	48	0–21	N	−0.6	4.2
S37	Aydin	37.90	28.40	37.95	28.95	86	45	75	0–21	N	−0.6	4.2
S38	Kusadaci 1	37.65	27.22	37.85	27.55	55	51	36	0–19	N	−0.6	4.2
S39	Kusadaci 2	37.71	27.08	37.65	27.22	117	45	16	0–21	N	0	0.6
S40	Samos	37.71	26.61	37.71	27.08	91	45	46	0–21	N	0	3.0
S41	Ikaria	37.71	26.61	37.58	25.70	80	45	80	0–21	N	0	2.4
S42	Mugla	37.08	28.53	37.10	29.14	88	45	54	0–21	N	0	3.0
S43	Marmaris	36.97	27.55	37.08	28.53	83	45	88	0–21	N	0	3.0
S44	Kalymnos	36.95	26.30	36.97	27.55	90	45	111	0–21	N	0	3.0
S45	Amorgos	36.41	25.30	36.95	26.30	65	40	100	0–23	N	0	2.4
S46	Bodrum	36.98	27.60	36.86	27.35	59	50	26	0–19	N	0	3.0
S47	Kos	36.74	27.07	36.86	27.35	65	50	28	0–19	N	0	3.0
S48	Astypalaia	36.74	27.07	36.40	26.40	58	50	71	0–19	N	0	3.0
S49	Symi	35.90	26.71	36.42	27.63	91	45	91	0–21	N	0	0.6
S50	Tilos	36.80	28.70	35.99	28.25	56	50	100	0–19	N	0	0.6
S51	Rodos	36.42	27.63	36.42	28.65	30	80	98	0–15	LS	0	1.8
S52	Fethiye	37.32	29.48	36.80	28.70	50	80	90	0–15	LS	0	1.8

Table 1 continued

Segment number	Code name	Fault segment start		Fault segment end		Strike (°)	Dip (°)	Length (km)	Width (km)	Faulting type	SS mm/year	DS mm/year
		Latitude °N	Longitude °E	Latitude °N	Longitude °E							
S53	Burdur	38.30	30.80	37.32	29.48	50	80	158	0–15	LS	0	1.8

Some of these segments are associated with the strong ($M \geq 6.5$) modeled events or with known historical events. The first two columns give the number (as it appears in Fig. 2) and the code name of each fault segment. The next four columns give the geographical coordinates of the segment edges. The fifth and sixth columns give the strike and dip angles, respectively, for each segment, whereas the seventh and eighth columns give the corresponding fault length and width. The faulting type of each segment is indicated in the 11th column (*RS* right–lateral strike-slip; *N* normal; *LS* left-lateral strike-slip). The last two columns give the annual slip rate assigned in each segment (*SS* strike-slip component, negative for dextral motion; *DS* dip-slip component: positive for normal faulting)

The fault segment associated with the occurrence of each incoming earthquake is shown in Fig. 3 by a black color and the faults that already failed are in white, where the changes in stress are presented for the whole study area. In Fig. 3a, c–i, k–l, o, and q–s the calculations of ΔCFF were performed for normal or oblique normal faulting type. The remaining figures show the evolutionary stress field calculated for dextral strike-slip faulting. Initial values of ΔCFF are assumed to be zero everywhere on each fault plane just before the Samos earthquake of 1904, which is the first strong event in our data sample.

Figure 3a, displays the coseismic stress changes associated with the 1904 Samos event, which created a shadow zone for the normal faulting type. Bright zones are observed to the east and west. We expect these stress changes to affect the occurrence of future events. The 1912 Ganos earthquake occurred between the Gulf of Saros and the Sea of Marmara at the western part of the North Anatolian Fault. Figure 3b shows the state of stress before its occurrence with respect to the 1904 baseline. The event is inside an area of positive static stress changes due to the virtual model of stress accumulation. The 1914 Burdur earthquake is located inside at the borders of a region of positive ΔCFF (Fig. 3c). The 1919, Soma earthquake is located inside a bright zone (Fig. 3d), when the evolutionary stress field is calculated just before its occurrence and according to its fault plane solution. An extended shadow zone covers the areas to the north and south of the 1919 rupture, due to the coseismic stress changes of the 1904 and 1912 earthquakes. The stress evolutionary model successfully explains the location of the 1928 Torbali and the 1933 Kos events since the causative faults are inside bright zones (Fig. 3e, f, respectively). The epicenter and part of the causative fault of the 1939 Dikili event are located inside the stress enhanced area (Fig. 3g), partly created from the 1919 occurrence. The shadow zone at the north of the study area is eliminated over time as stress accumulates from 1939 to 1944, thus creating a bright zone inside which the 1944 Ayvacik event is located (Fig. 3h). Figure 3i depicts the stress state before the occurrence of the 1949 Chios earthquake, with respect to the 1904 baseline, which reveals that the rupture zone is located in a region of positive ΔCFF.

The state of ΔCFF before the 1953 Yenice earthquake is shown in Fig. 3j. The rupture is located in a region of positive ΔCFF, although, as we will show later, a part of the surface plane is inside positive stress changes. Figure 3k shows the accumulated Coulomb stress just before the 1955 Agathonisi earthquake calculated according to its fault plane solution. The causative fault is located inside a region of positive ΔCFF, to the east of the 1904 rupture, which probably hastened the 1955 occurrence. Figure 3l shows the state of stress before the 1956 Amorgos large event, with the associated fault seated in a stress enhanced area.

The fault of the 1957 Rhodos earthquake is located at the borders of stress bright zone and stress shadow (Fig. 3m). The stress field shown in Fig. 3n is the result of the accumulated stress changes (since 1904) calculated according to the fault plane solution of the 1957 Abant event. Its rupture zone was located inside a large region of positive ΔCFF. We note here the positive effect of the large 1944 Bolu–Gerede

Table 2

Rupture models for earthquakes with M ≥ 6.5 that occurred in the study area since the beginning of the twentieth century

Origin time epicenter					M_w^*	M_o (×10²⁶ dym cm)	Fault plane solution			L (km)	w (km)	SS (m)	DS (m)
Year	Date	Time	Latitude (°N)	Longitude (°E)			Strike (°)	Dip (°)	Rake (°)				
1904	Aug. 11	06:08:30	37.66	26.93	6.8 (4)		91	45	−115	46	17	−0.50	1.08
1912	Aug. 9	01:29:00	40.62	26.88	7.4 (2)	6.0 (2)	68	55	−145 (2)	110	14.6	−1.02	0.72
1914	Oct. 3	22:07:00	37.70	30.20	7.0 (5)	4.4 (4)	230	35	−105	52	21	−0.43	1.60
1919	Nov 18	21:54:50	39.20	27.40	6.9 (2)	1.38 (2)	253	45	−115 (2)	43	17	−0.26	0.57
1928	Mar. 31	00:29:47	38.18	27.50	6.5 (4)		83	45	−94	25	17	−0.05	0.72
1933	Apr. 23	05:57:37	36.80	27.30	6.6 (4)		65	50	−90 (6)	28	15.6	0.18	0.84
1939	Sep. 22	00:36:32	39.00	27.00	6.6 (4)		211	45	−115	26	17	−0.36	0.78
1944	Oct. 6	02:34:41	39.51	26.57	6.8 (2)		74	46	−114 (2)	35	17	0.41	1.12
1949	July 23	15:03:30	38.58	26.23	6.7 (4)	1.85 (6)	84	36	−80	31	20.4	0.17	1.01
1953	Mar. 18	19:06:16	40.02	27.53	7.2 (2)	8.7 (5)	250	70	−160	60	13	−3.55	1.29
1955	July 16	07:07:10	37.55	27.15	6.9 (4)		55	51	−133 (8)	38	15.4	−0.81	0.87
1956	July 9	03:11:40	36.30	25.70	7.7 (4)		65	40	−90 (9)	75	18.7	0.00	5.30
1957	Apr. 25	02:25:42	36.50	28.60	7.2 (5)	4.4 (5)	30	80	−41	67	12.2	1.01	0.88
1957	May 26	06:33:30	40.60	31.00 (1)	7.0 (2)	6.76 (2)	265	78	179 (2)	40	12.2	−1.47	−0.03
1964	Oct. 6	14:31:23	40.30	28.23 (2)	6.9		280	50	−90 (2)	35	16.7	−0.12	1.41
1967	July 22	16:56:58	40.67	30.69 (2)	7.2		275	88	−178 (2)	80	12	−2.02	0.07
1969	Mar. 28	01:48:29	38.42	28.6 (3)	6.5 (3)	0.625 (3)	313	34	−90 (3)	25	21	0.00	0.61
1970	Mar. 28	21:02:23	39.055	29.60 (3)	7.1 (3)	1.09 (3)	308	35	−90 (3)	24	21	0.00	1.60
1970	Mar. 28	23:00:00	39.16	29.50 (3)	7.1 (3)	3.06 (3)	270	35	−110 (3)	24	21	−0.82	2.25
1975	Mar. 27	05:15:08	40.40	26.10	6.6 (7)	0.64 (7)	68	55	−145 (2)	40	14.7	−0.70	0.49
1999	Aug. 17	00:01:37	40.76	29.97	7.4 (10)	1.31 (10)	268	84	180 (10)	35	12	−2.76	0.00
							260	87	164 (10)	20	12	−1.92	−0.55
							265	87	164 (10)	26	12	−3.36	−0.96
							271	87	164 (10)	35	12	−1.83	−0.52
1999	Nov. 12	20:00:00	40.79	31.21	7.2 (7)	6.56 (7)	262	53	−177 (11)	56	15	−2.06	−0.14

First five columns give information on the occurrence time and the epicentral coordinates of each event. The next two columns give the magnitude and seismic moment when available. The eighth through tenth columns give the strike, dip and rake angles of the fault plane. Eleventh and 12th columns give the fault length, L, and width, w. SS and DS in the last two columns, respectively, give the strike-slip component (negative for dextral faulting) and the dip-slip component (positive for normal faulting). Note that for the 1999 Izmit main shock a multi segmented fault is considered. Numbers in parentheses indicate references

References: 1. AMBRASEYS (2001); 2. TAYMAZ et al. (1991b); 3. EYIDOĞAN and JACKSON, 1985; 4. PAPAZACHOS and PAPAZACHOU (2003); 5. PACHECO and SYKES (1992); 6. EYIDOĞAN, 1988; 7. Global CMT determination; 8. MCKENZIE (1972); 9. SHIROKOVA (1972); 10. BARKA et al. (2002); 11. KIRATZI and LOUVARI (2001)

earthquake (M 7.3), which occurred outside and to the east of our study area, as being along strike, created a wide stress enhanced zone encompassing the ruptures of 1957, 1967 and the two 1999 events (NALBANT et al., 1998). In Fig. 3o the accumulated Coulomb stress changes just before the 1964 Manyas earthquake of 1964 are shown for faulting in agreement with its focal mechanism. The activated fault is situated in a stress-enhanced area. Figure 3p shows the accumulated stress changes calculated according to the fault plane solution and just before the occurrence of the 1967 Mudurnu earthquake which is probably inhibited by the 1957 Abant earthquake by the stress transfer between the two adjacent fault segments. Figure 3q shows the accumulated Coulomb stress changes just before the occurrence of the 1969 Alaşehir rupture, which is located in a stress enhanced area.

Figures 3r and s show the accumulated Coulomb stress just before the 1970 Gediz main shock and its major aftershocks, respectively. The activated faults are located inside stress-enhanced areas. The later one, occurring just 2 h after the first event, is most probably triggered by the first. Figure 3t shows the accumulated Coulomb stress changes just before the 1975 Saros event, occurring on a site where the 1912

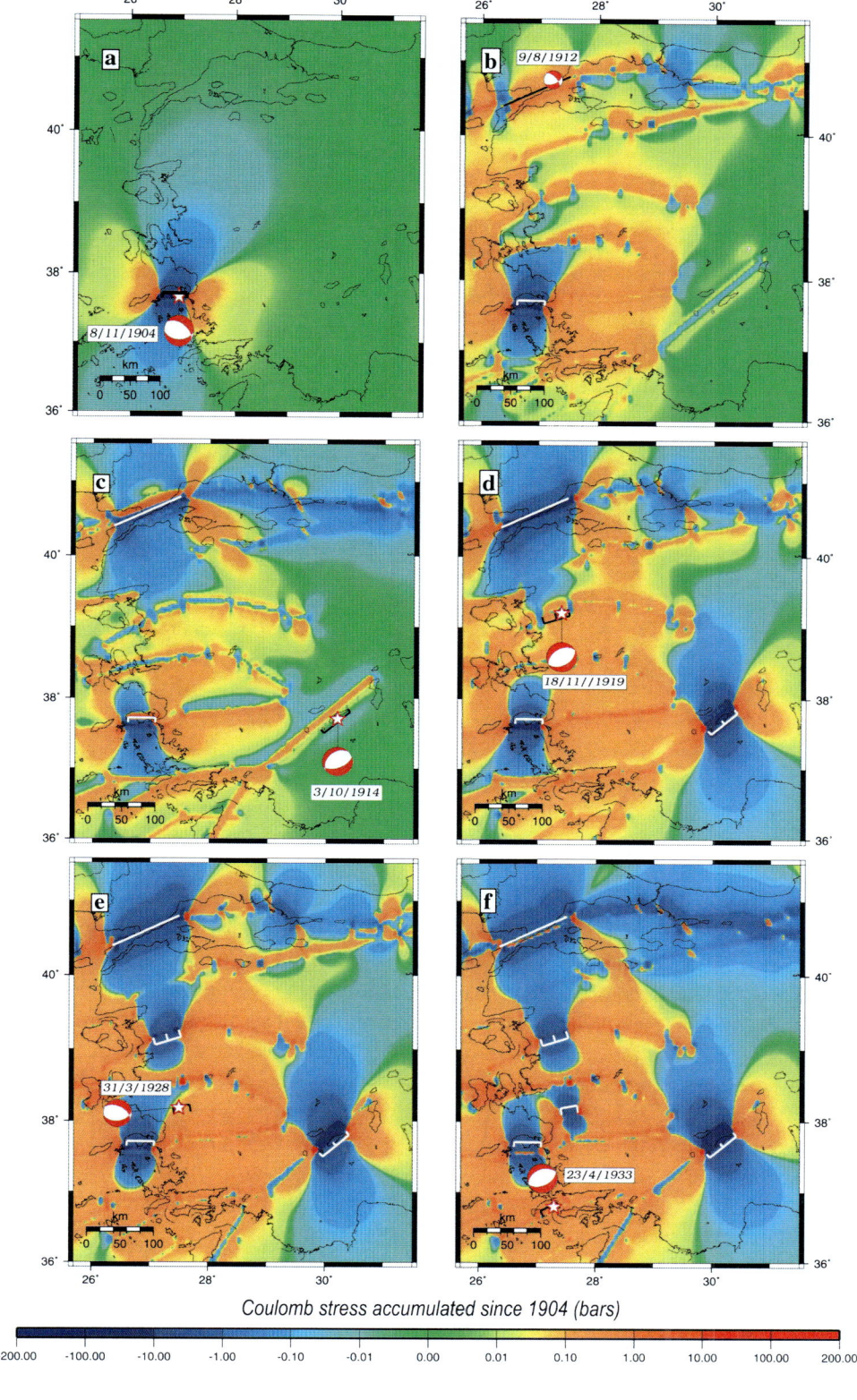

Coulomb stress accumulated since 1904 (bars)

Figure 3

Stress evolution in western Turkey and eastern Aegean Sea since 1904. Coulomb stress is calculated for dextral strike-slip and normal faults at a depth of 10 km. The stress pattern is calculated for the faulting type of the next strong event in the sample. The colour scale in the bottom shows the stress changes in *bars*. Fault plane solutions are denoted as in Fig. 2. Stars denote epicenters of earthquakes linked with a thin line with the beach balls. The fault segment associated with the occurrence of each event is shown by *black color*, while the segments that already failed are shown in white. **a** Coseismic Coulomb stress changes associated with the 1904 event. **b** Stress evolution until just before the 1912 Ganos event. Coseismic stress changes associated with the 1904 earthquake and tectonic loading on the fault segments since then are included. **c** State of stress before the 1914 Burdur earthquake. **d** ΔCFF before the 1919 Soma event. **e** State of stress just before the 1928 Torbali event. **f** Stress evolution before of the occurrence of the 1933 Kos event. **g** Coulomb stress changes before the 1939 Dikili earthquake. **h** Stress evolution until the 1944 Ayvacik main shock. **i** State of stress just before the 1949 Chios island event. **j** ΔCFF up to the 1953 Yenice earthquake. **k** Stress evolution before the 1955 Agathonisi earthquake. **l** State of stress just before the 1956 Amorgos main event. **m** Coulomb stress changes until the 1957 Rhodes earthquake. **n** Stress evolution just until the 1957 Abant earthquake. **o** State of stress before the 1964 Manyas earthquake. **p** Stress evolution until just before the 1967 Mudurnu earthquake. **q** ΔCFF just before the 1969 earthquake in Alaşehir. **r** State of stress before the first 1970 Gediz event, and **s** before the second 1970 Gediz event. **t** State of stress before the 1975 earthquake in Saros. **u** ΔCFF before the 1999 Izmit earthquake. **v** ΔCFF before the 1999 Düzce earthquake

Ganos earthquake has accumulated positive Coulomb stress changes. Figure 3u shows the accumulated stress changes calculated just before the occurrence of the 1999 Izmit (Kocaeli) large main shock, which is here shown as one fault segment, although a multi-segmented source is considered for its modeling. After the 1967 earthquake a bright zone had been created in this part of NAF; a branch of which includes the rupture zone of the 1999 event. An extended bright zone appeared at the eastern area of Kocaeli where the next Düzce earthquake occurred (Fig. 3v). This is in agreement with PARSONS et al. (2000), HUBERT-FERRARI et al. (2000) and PAPADIMITRIOU et al. (2001) who found that the spatial distribution of Coulomb Stress changes caused by the Izmit (Kocaeli) earthquake showed an extended stress enhanced zone comprising the rupture area of the Düzce earthquake. The bright zone in Fig. 3v that now encompasses the fault associated with the 1999 Düzce earthquake is evidently due to the stress changes caused by the Izmit coseismic slip, thus evidencing its possible triggering by the previous strong event.

We extended our calculations of the evolutionary stress field to 2008, whereas after 1999 no strong event ($M \geq 6.5$) has occurred in our study area. Figure 4 depicts the evolved stress state from 1904 to the present and includes the addition of coseismic stress changes associated with the occurrence of the 1999 Düzce earthquake and the stress accumulation caused by 105 years of tectonic loading. The stress field is inverted for three faulting types (dextral strike-slip, normal and sinistral strike-slip) and different values of the Skempton's coefficient ($B = 0.2$, $B = 0.5$ and $B = 0.9$), for testing the effect of the pore fluid on the stress changes calculations. The lower and maximum values assigned here are the extreme values that this coefficient can take, expressing the lower and maximum filling of the pores, respectively. Figures 4a–c shows the evolved stress change for a typical normal slip fault for the area (strike $= 275°$, dip $= 45°$, rake $= -90°$) along with the focal mechanisms of smaller magnitudes ($M < 6.5$) earthquakes since 1999. The same holds for Fig. 4d–f but for dextral strike-slip faulting (strike $= 90°$, dip $= 87°$, rake $= -178°$) and Fig. 4g–i for left-lateral strike-slip faulting (strike $= 30°$, dip $= 80°$, rake $= -41°$). Information regarding the fault plane solutions depicted in Fig. 4 is given in Table 3. A lower value of Skempton's coefficient ($B = 0.2$) is used in Fig. 4a, d and g where the stress pattern looks very similar to that depicted in Figs. 4b and 5b, which was calculated for $B = 0.5$. Figure 4c, f and i show the stress pattern calculated for a higher Skempton's coefficient ($B = 0.9$), in which the resulting pattern also remains almost unaffected. This shows that the value of the Skempton's coefficient selected ($B = 0.5$) is suitable for our calculations. Most of the smaller events plotted in Fig. 4 are located inside bright zones or in the borders between bright and shadow zones.

In order to investigate the effect of the evolutionary stress field on the incoming ruptures, the cumulative stress changes were calculated onto the rupture plane of each event just before its occurrence and are shown in Fig. 5. Although precise hypocenter location or details on rupture initiation are not available for the majority of the modeled events, and

Stress Transfer and Time-dependent Probability

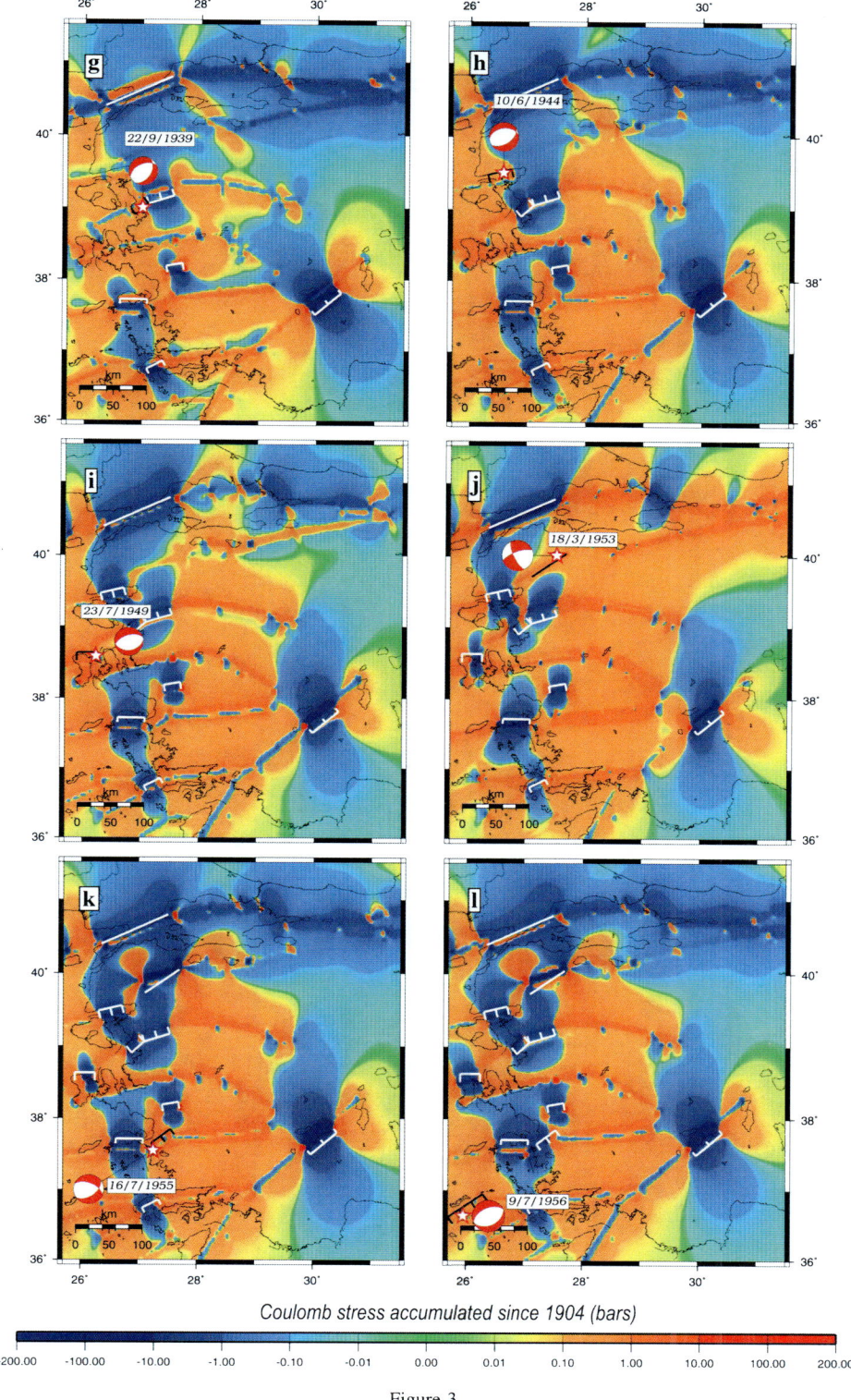

Figure 3
continued

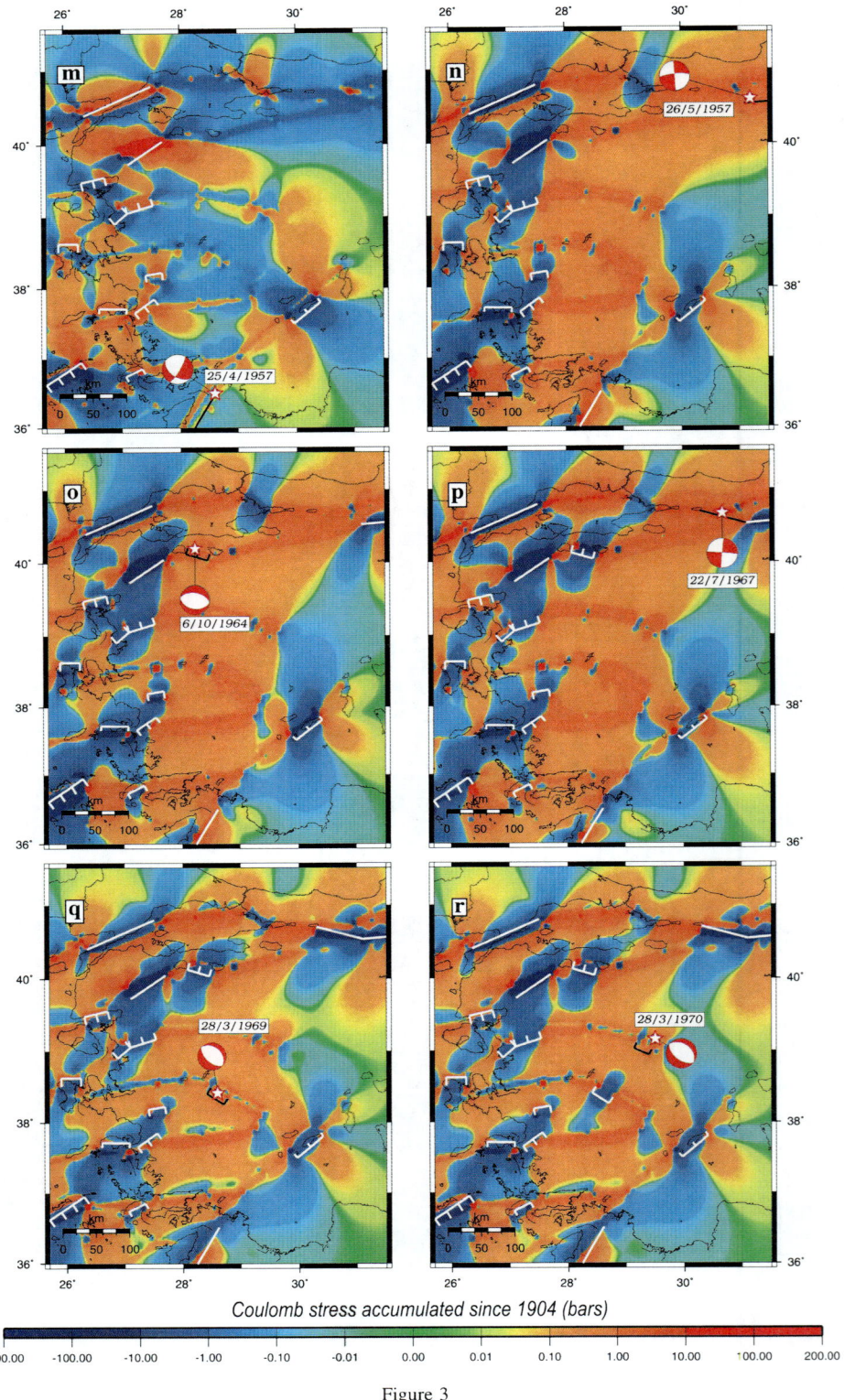

Figure 3
continued

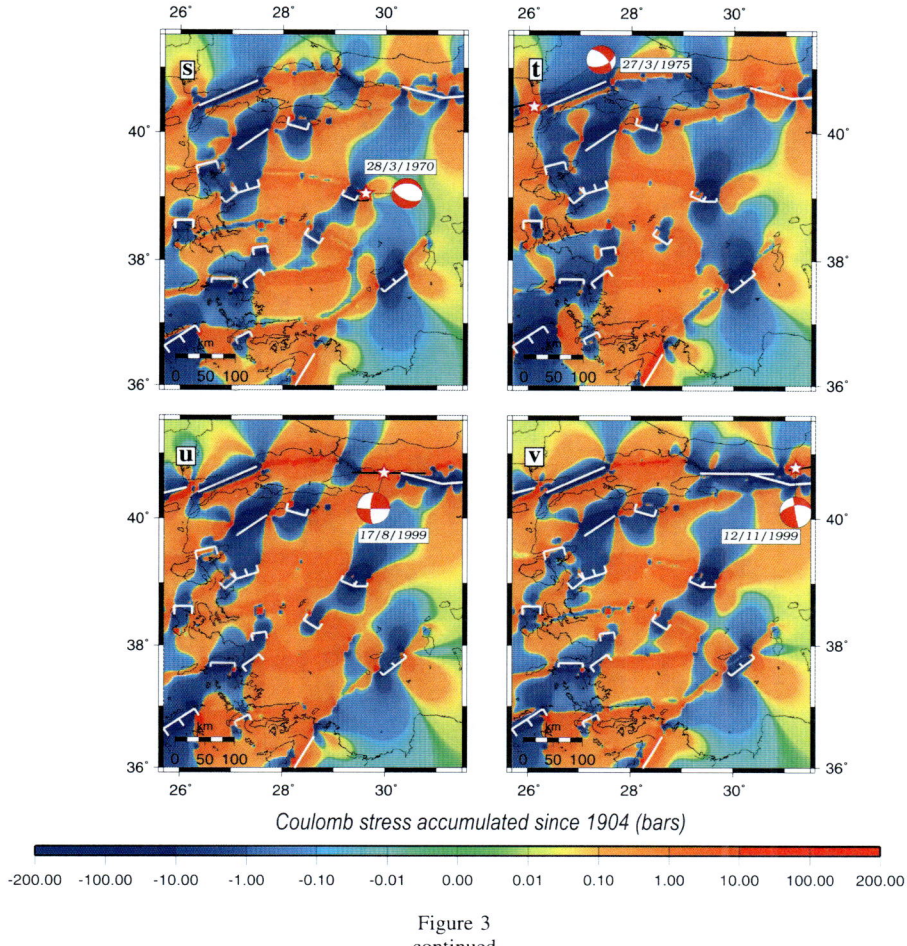

Figure 3 continued

in this case it is not feasible to correlate them with the position of the maximum ΔCFF, it is evident that in most cases the assumed rupture surfaces are inside stress enhanced areas. Examining the rupture plane of the normal faulting 1914 Burdur event (Fig. 5b), although it is influenced by tectonic loading on the nearby strike-slip Fetiye–Burdur fault, its occurrence is not inhibited since it is partially enhanced by positive stress changes. Even the rupture plane of the 1953 Yenice main shock, which is of strike-slip faulting in a parallel branch with the one associated with the 1912 Ganos event, is partially inhibited by this previous occurrence (Fig. 5i). It is thus worthy of note here, that the evolutionary model is adequate to explain the sequential occurrence of the strong events.

A quantitative evaluation of the calculation is given in Table 4, where the percentage of the rupture plane with positive or even larger of 0.1 bars, stress changes values is given (4th and 5th columns of the Table). Given the uncertainty in knowing the nucleation depth, this percentage is also estimated at depths of 6, 8, 10 and 12 km (last four columns of the Table), which are considered the most presumable depths for crustal events in the study area.

6. Influence of the Skempton's Coefficient, and Rake and Dip Angles on the ΔCFF Calculations

In order to investigate to which extend the uncertainties involved in the fault parameters (dip

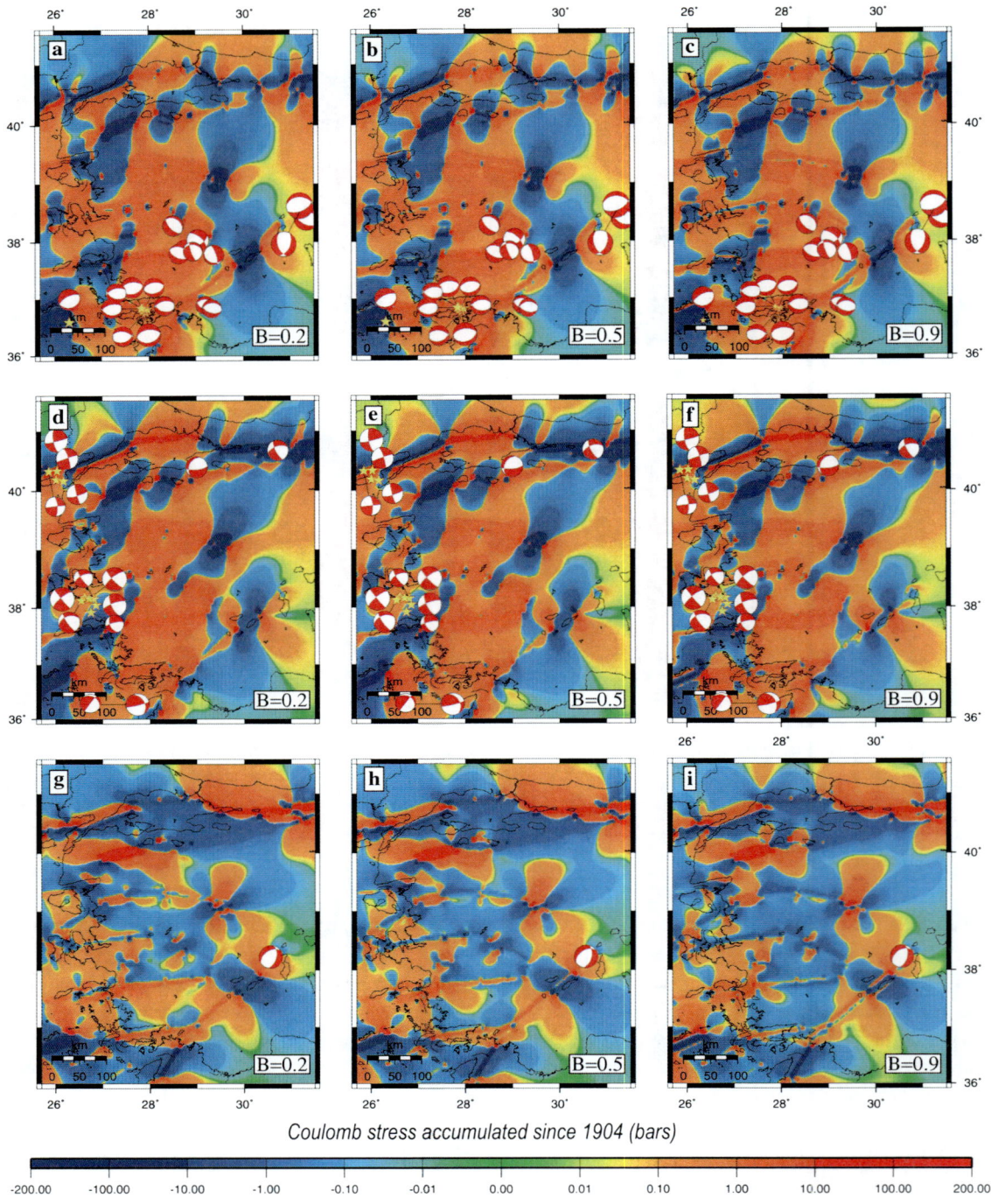

Figure 4
a Coulomb stress evolution until 2008 calculated for normal faulting representative (strike = 275°, dip = 45°, rake = 90°) of the area at a depth of 10.0 km for Skempton's coefficient equal to 0.2. The fault plane solutions of the events with $M \leq 6.5$ that occurred during 2000–2008 and are associated with normal faulting are also shown as lower-hemisphere equal area projections. b Same as a with $B = 0.5$. c Same as a with $B = 0.9$. d Same as in a but for dextral strike-slip faulting representative for the area (strike = 90°, dip = 87°, rake = 178°). e Same as in d but for $B = 0.5$. f Same as in d but for $B = 0.9$. g Same as in a but for sinistral strike slip faulting representative (strike = 30°, dip = 80°, rake = −41°) of the area. h Same as in g but for $B = 0.5$. i Same as in g but for $B = 0.9$

Table 3

Information on the available fault plane solutions for earthquakes that occurred in the study area from 1999 to present (Global-CMT determination)

Origin time		Epicenter		M	Depth (km)	Focal mechanism		
Year	Date	Longitude (°E)	Latitude (°N)			Strike (°)	Dip (°)	Rake (°)
2000	21 Apr.	29.39	37.78	5.4	15.0	110	23	−139
2000	23 Aug.	30.72	40.68	5.3	15.3	253	57	−160
2000	15 Dec.	31.35	38.40	6.0	15.0	285	41	−100
2002	3 Feb.	31.21	38.62	6.4	15.0	269	37	−71
2002	3 Feb.	30.56	38.23	5.8	15.0	236	45	−58
2002	3 Feb.	31.22	38.52	5.3	15.0	76	43	−70
2003	4 Oct.	26.86	38.05	5.7	15.0	155	70	−15
2003	17 Apr.	26.75	37.92	5.2	15.0	156	50	−15
2003	6 July	26.02	40.19	5.7	15.0	169	771	7
2003	6 July	26.17	40.17	5.2	15.0	73	77	173
2003	9 July	25.86	40.33	4.8	18.0	356	71	3
2003	23 July	28.77	37.88	5.3	15.0	97	31	−111
2003	26 July	29.05	38.03	5.4	15.0	60	57	−147
2004	15 June	26.04	40.34	5.2	12.0	342	78	5
2004	3 Aug.	27.93	36.77	5.2	12.0	74	38	−97
2004	4 Aug.	27.88	36.80	5.5	12.0	75	40	−95
2004	4 Aug.	27.97	36.82	5.2	12.0	71	42	−111
2004	4 Aug.	27.91	36.81	5.3	12.0	75	41	−94
2004	20 Dec.	28.33	36.88	5.3	12.0	105	45	−69
2005	10 Jan.	27.87	36.84	5.4	15.1	110	45	−63
2005	11 Jan	27.84	36.84	5.0	12.2	100	33	−69
2005	17 Oct.	26.82	38.15	5.5	15.2	242	61	−166
2005	17 Oct	26.62	38.18	5.8	12.0	231	76	−177
2005	17 Oct	26.54	38.12	5.2	17.8	250	42	−161
2005	20 Oct.	26.72	38.16	5.8	12.9	231	73	−169
2006	5 June.	28.65	37.80	4.8	21.7	295	34	−88
2006	24 Oct.	29.00	40.40	5.0	14.3	205	32	−144
2007	23 Jan.	28.52	38.28	4.9	13.2	301	28	−103
2007	30 Mar.	30.91	37.92	4.5	13.5	158	45	−129
2007	10 Apr.	30.87	37.96	5.1	14.6	161	50	−122
2007	31 Aug.	26.32	36.59	5.2	15.7	71	25	−83
2007	29 Oct.	29.21	36.89	5.3	12.0	275	37	−107
2007	16 Nov.	29.34	36.83	5.1	13.0	263	38	−108
2008	25 Apr.	28.94	37.84	5.0	12.2	276	28	−151

and rake angles) influence the calculated stress pattern, two earthquakes were selected and the correlation between calculated stress changes and different values of Skempton's coefficient, rake and dip angle of the fault are tested following the technique of PARSONS (2005). We have chosen for this purpose a dip-slip and a dextral strike-slip event for the sake of comparison, and performed the calculations at a depth of 10 km.

The values of static stress changes at the hypocenter of the 1969 Alaşehir normal faulting earthquake ($M_w = 6.5$) as a function of assumed different values of the Skempton's coefficient ($0.2 \leq B \leq 0.9$) and varying the rake angles (ranging between $-70°$ and $-110°$ to keep the normal character of faulting) are shown in Fig. 6a. For a constant value of the rake angle variation, the selection of the Skempton's coefficient value, B, can cause differences in the calculated static stresses up to 0.2 bar (20% variation). Almost the same difference is found (0.16 bar) for dip angle variation when the calculations are performed for a constant value of B (12–13% variation). Keeping the rake value constant and equal to $-90°$, different dip angle values ranging from 30° to 60° were tested (considering that the typical mean value for crustal normal faults is 45°

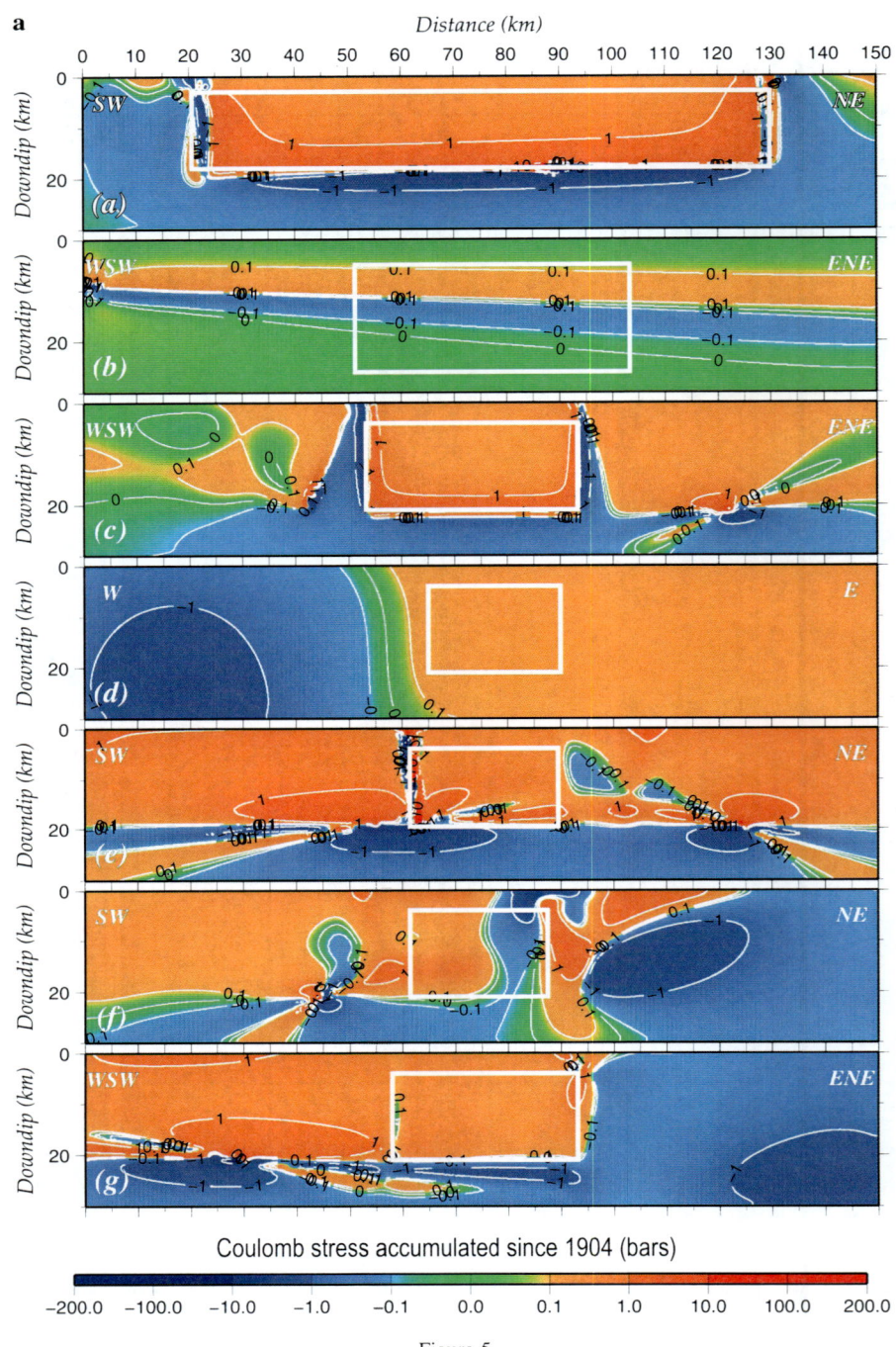

Figure 5
Accumulated static stress changes associated with the tectonic loading on the major faults and the coseismic slip of the earthquakes taken into account in the stress evolutionary model, resolved onto the rupture plane of the next strong event. *Contour lines* are accompanied with corresponding values of stress changes in *bars*. *Rectangles* denote the rupture areas, considered as rectangular surfaces with two edges parallel to the Earth's surface, for: **a** the 1912 Ganos main shock, **b** the 1914 Burdur earthquake, **c** the 1919 Soma main event, **d** the 1928 Torbali main shock, **e** the 1933 Kos earthquake, **f** the 1939 Dikili earthquake, **g** the 1944 Ayvacik event, **h** the 1949 Chios main shock, **i** the 1953 Yenice earthquake, **j** the 1955 Agathonisi event, **k** the 1956 Amorgos main shock, **l** the 1957 Rhodes earthquake, **m** the 1957 Abant earthquake, **n** the 1964 Manyas event, **o** the 1967 Mudurnu main shock, **p** the 1969 Alasehir earthquake, **q** the 1970 first Gediz event, **r** the 1970 second Gediz event, **s** the 1975 Saros main shock, **t** the 1999 Izmit main shock and **u** the 1999 Düzce event

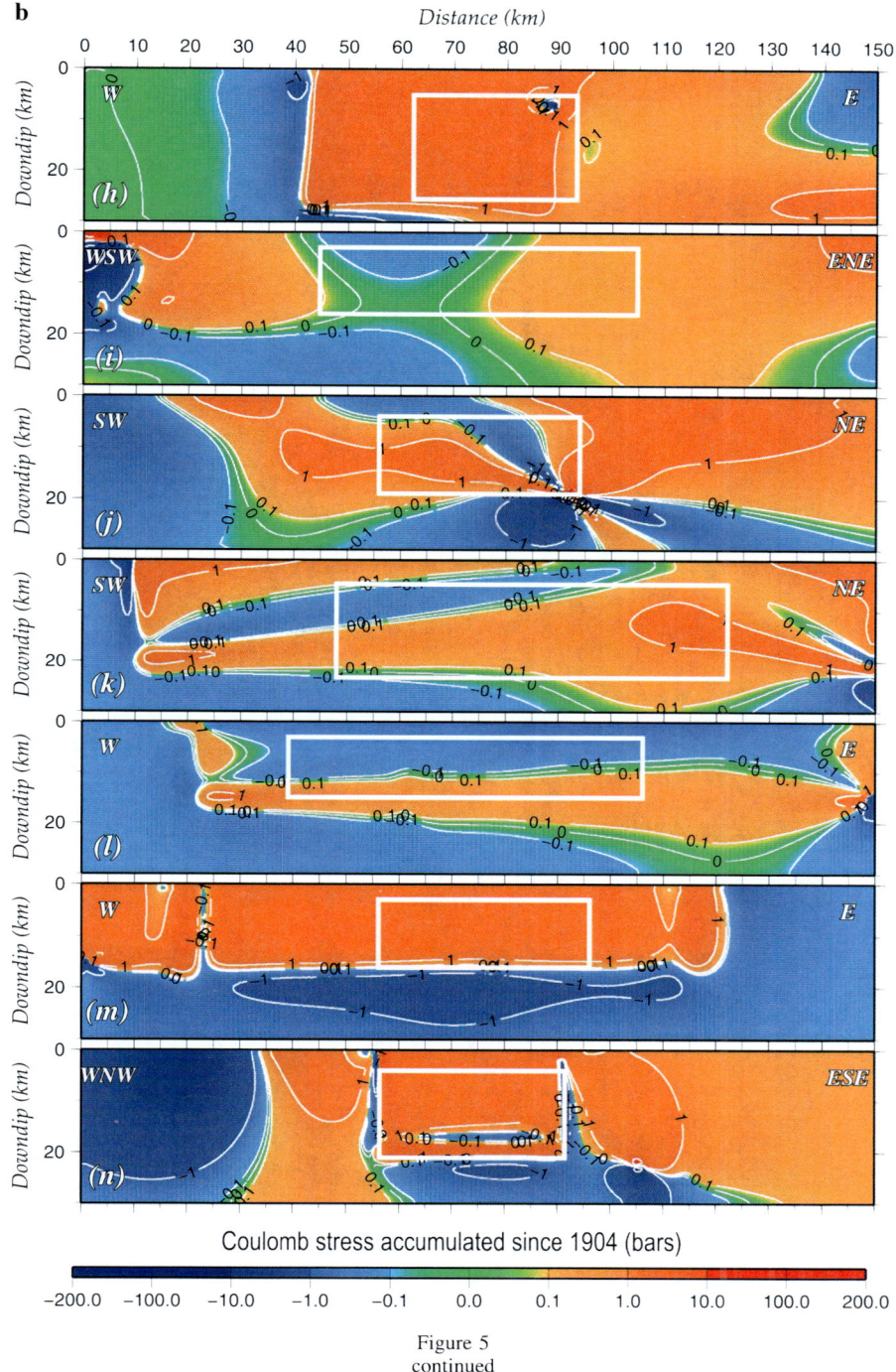

Figure 5 continued

and that the 1969 earthquake occurred on a fault with dip angle 34°, see Table 2). The stress changes were found to vary up to 0.7 bar (38–60% variation), whereas a change of up to 0.2 (11–12% variation) bar was found to depend on the variation of friction coefficient (Fig. 6b).

The same procedure was followed for the 1999 Düzce earthquake ($M_w = 7.2$) of dextral strike-slip

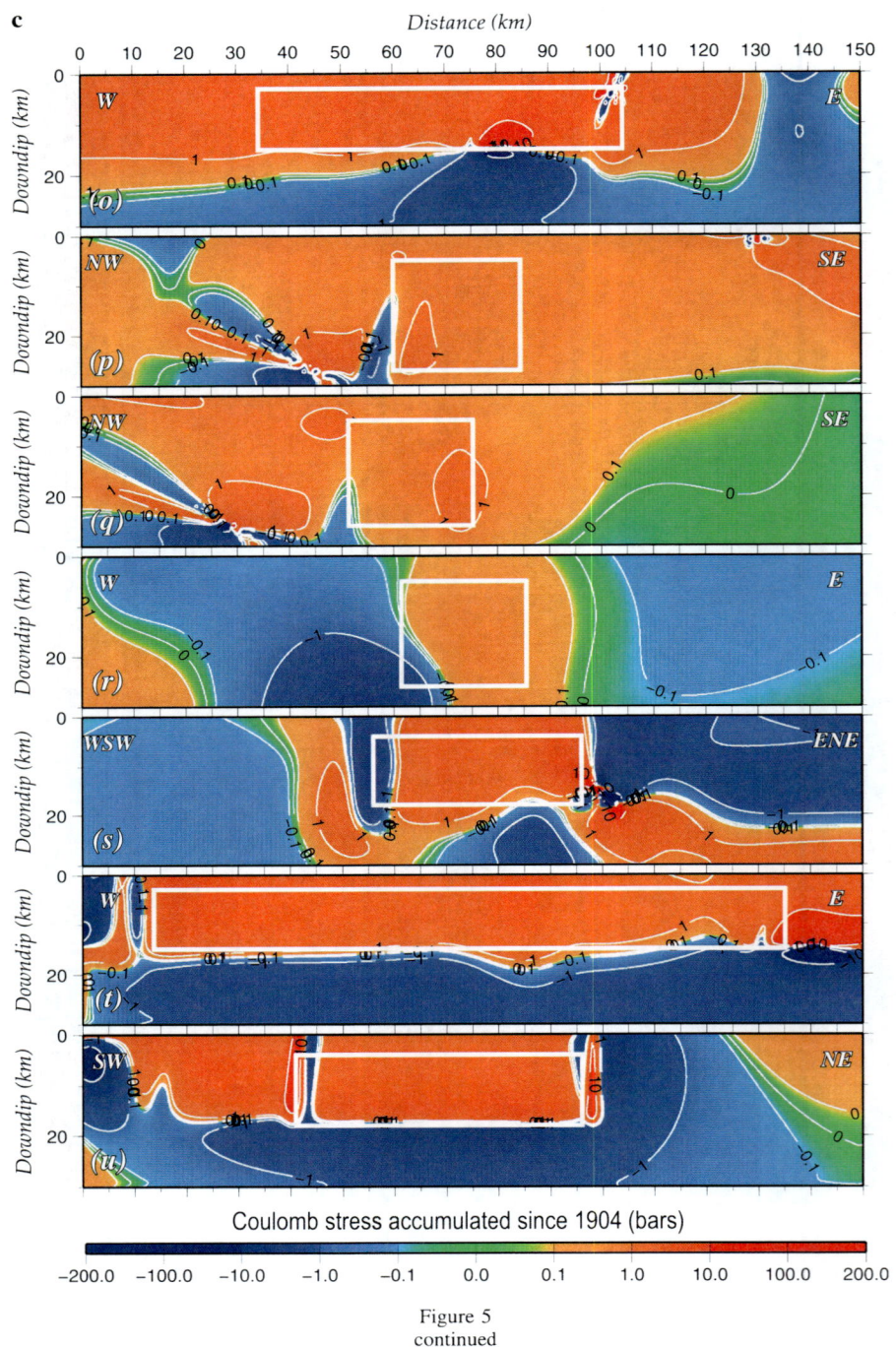

Figure 5
continued

faulting. We selected this event, which also has been tested by PARSONS (2005), for the reasons stated by the above author and for the sake of comparison. The values of Skempton's coefficient are varied from 0.2 to 0.9 and the ones of the rake angle from −140° to −180° keeping the sense of the dextral strike slip faulting. Differences reaching 0.55 bar were found (5% variation) in the static stress changes depending on Skempton's coefficient, whereas these changes reached up to 1.5 bar (12–13% variation) for the rake

Table 4

Coulomb stress change values calculated onto the rupture planes of the modelled strong events

Occurrence Year	ΔCFF minimum (bars)	ΔCFF maximum (bars)	ΔCFF mean (bars)	Percentage ΔCFF > 0.0 bars	Percentage ΔCFF > 0.1 bars	Percentage ΔCFF > 0.1 h = 6 km	Percentage ΔCFF > 0.1 h = 8 km	Percentage ΔCFF > 0.1 h = 10 km	Percentage ΔCFF > 0.1 h = 12 km
1912	−256.25	95.96	0.92	94.78	94.67	96.33	97.25	97.25	97.25
1914	−0.50	0.28	−0.01	62.59	26.05	100.00	0.00	0.00	0.00
1919	0.26	3.17	0.75	100.00	100.00	100.00	100.00	100.00	100.00
1928	0.15	0.36	0.27	100.00	100.00	100.00	100.00	100.00	100.00
1933	−35.49	48.93	1.20	93.08	91.07	53.85	57.69	96.43	71.43
1939	−1.90	3.02	0.15	66.24	56.41	100.00	100.00	65.38	69.23
1944	−0.29	2.03	0.47	95.40	93.65	97.14	97.14	97.14	100.00
1949	−11.10	5.47	1.21	99.08	99.08	100.00	100.00	100.00	100.00
1953	−0.26	0.22	0.05	60.71	42.26	38.33	43.33	45.00	48.33
1955	−3.11	47.36	0.74	73.36	70.07	63.16	71.05	78.95	86.84
1956	−0.42	1.36	0.40	80.65	74.32	54.05	83.78	100.00	100.00
1957a	−0.58	1.16	−0.02	38.81	32.49	0.00	0.00	1.49	85.07
1957b	−0.32	3.21	2.23	94.82	93.93	100.00	100.00	100.00	100.00
1964	−4.33	6.41	1.72	92.38	91.43	97.14	97.14	97.14	22.86
1967	−92.47	44.56	3.73	97.47	97.36	97.14	97.14	98.57	100.00
1969	0.11	1.20	0.85	100.00	100.00	100.00	100.00	100.00	100.00
1970a	−0.13	1.29	0.76	98.67	96.78	95.83	91.67	95.83	91.67
1970b	−2.45	0.88	0.39	88.07	84.66	90.00	87.50	87.50	75.00
1975	−4.45	11.69	2.41	89.33	88.50	90.00	90.00	90.00	92.50
1999a	−3.11	21.90	3.92	97.48	97.10	100.00	100.00	100.00	99.18
1999b	−4.47	9.65	4.57	87.76	87.39	92.73	92.73	96.36	96.36

The first column gives the occurrence year of each event (multiple occurrences in the same year are signified as a and b, according to the occurrence date and time). The second, third and fourth columns give the minimum, maximum and mean values, respectively, of the calculated ΔCFF in bars. The fifth and sixth columns give the percentage of the rupture plane onto which ΔCFF exceeded the values of 0.0 and 0.1 bars, respectively. The last four columns give the percentage of ΔCFF ≥0.1 bar, at depths of 6, 8, 10 and 12 km, respectively

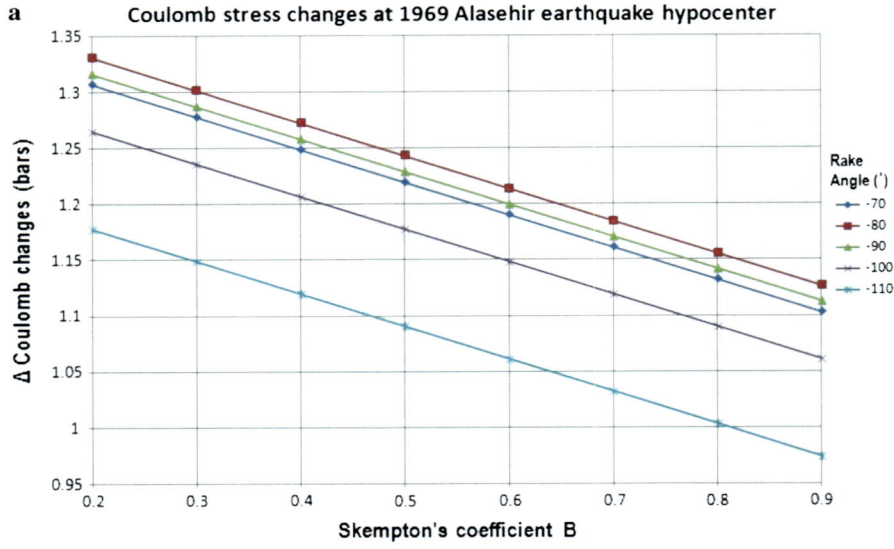

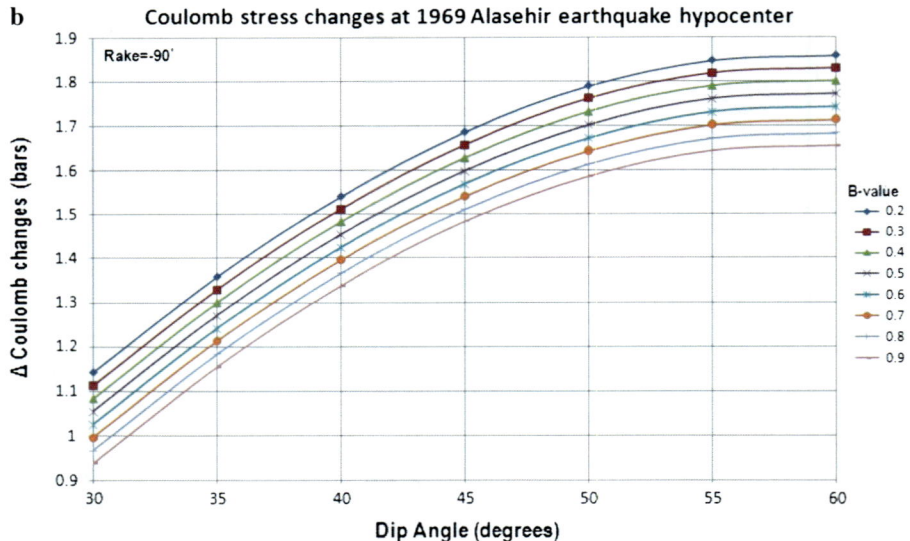

Figure 6
Changes in stress calculated at the 1969 Alaşehir hypocenter, **a** versus different values of the Skempton's coefficient (0.2–0.9) and values of rake ranging from −70° to −110°, and **b** versus different values of dip angle (30°–90°) and values of Skempton's coefficient ranging from 0.2 to 0.9

angle variance (Fig. 7a). Keeping the rake angle constant and equal to −177°, it was found that if the fault dip at Düzce hypocenter is allowed to vary from 40° to 90° the ΔCFF values may vary up to 2.1 bar (17–20% variation), whereas values up to 0.7 bar resulted from the variation of the Skempton's coefficient (7% variation) (Fig. 7b). PARSONS (2005) found 20–80 and 40–50% changes for rake and dip angle variation, respectively, and 20–50% when examining the range of the values coefficient of friction.

The definition of the dip angle of the fault plane seems to play the most important role in the variations of the calculated static stress changes, since for both cases investigated these variations registered the larger values. The influence of the rake angle is still important, however, because the absolute differences found, of the order of 0.1–1.5 bars, are significant

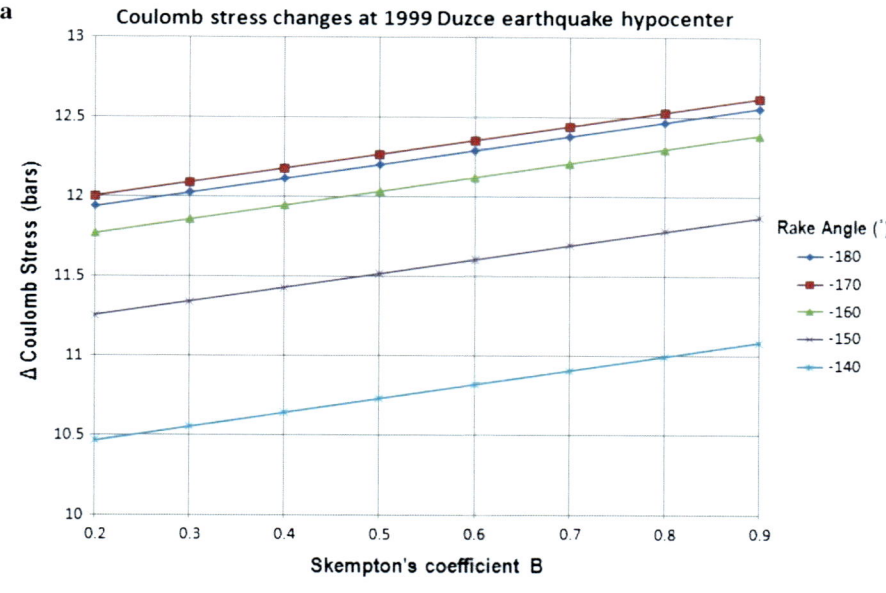

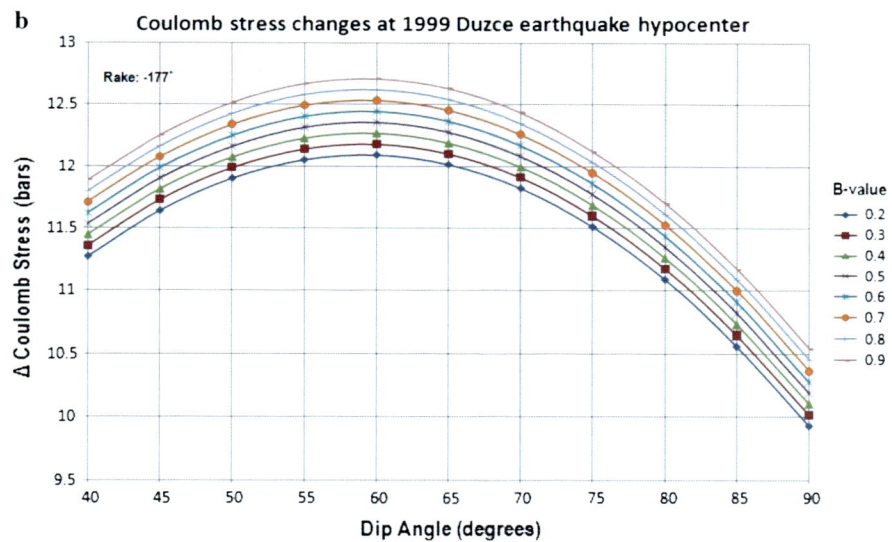

Figure 7
Changes in stress calculated at the Düzce hypocenter, **a** as a function of different values of friction coefficient from 0.1 to 0.8 and fault rakes ranging from −140° to −180°, **b** as a function of different values of Skempton's coefficient from 0.2 to 0.9 and dip angle from 40° to 90°

when triggering is inspected or when these values are incorporated in probability calculations.

7. Stress Transfer and Earthquake Probabilities

An attempt is made in this section to estimate probabilities for the occurrence of future strong ($M \geq 6.5$) events on the fault segments associated with events of $M \geq 6.5$ that occurred either during the instrumental period or during the past centuries and for which available information exists. For a probabilistic earthquake forecast in a region under the influence of past events it is considered that the stress transfer might hasten or delay an upcoming earthquake. Calculations of time dependent probability are a means of expressing variability in an earthquake renewal process. For this purpose we followed the methodology of STEIN et al. (1997), TODA et al. (1998) and PARSONS (2004, 2005) who consider both permanent and transient effects of the stress changes on earthquake probabilities.

Two models for the estimation of earthquake probabilities are generally in use, the stationary Poisson model and the conditional probability model (CORNELL et al., 1968; HAGIWARA, 1974). We first present results (Table 5) from the simple Poisson model for comparative purposes. This model is one that treats earthquakes as random in time (t) about an average interevent time (T_r) as:

$$P(t \leq T \leq t + \Delta t) = 1 - e^{-\Delta t/T_r}. \quad (8)$$

The values of T_r and the corresponding calculated probabilities are given in Table 5.

With the conditional probability model, probability can increase with time to represent increasing stress on a fault segment toward an uncertain stress threshold. A time-dependent probability (in any time interval ($t, t + \Delta t$)) is calculated by a probability density function $f(t)$ as:

$$P(t \leq T \leq t + \Delta t) = \int_t^{t+\Delta t} f(t)\,\mathrm{d}t, \quad (9)$$

where P is the probability that an earthquake will occur at some time T in some interval ($t, t + \Delta t$). Two commonly applied probability density functions, $f(t)$, the lognormal distribution (e.g., NISHENKO and BULAND, 1987):

$$f(t, \alpha, \beta) = \frac{1}{\beta t \sqrt{2\pi}} \exp\left[\frac{-\left(\ln \frac{t}{\alpha}\right)^2}{2\beta^2}\right], \quad (10)$$

where $\beta^2 = \ln\left(\frac{s_t^2}{T_r^2} + 1\right)$, $\alpha = \ln[T_r \exp(-0.5\beta^2)]$, T_r is the average interevent time, s_t: 'the standard deviation of interevent time, and the Brownian Passage Time (KAGAN and KNOPOFF 1987; MATTHEWS et al., 2002):

$$f(t, T_r, a) = \sqrt{\frac{T_r^2}{2\pi \alpha^2 t^3}} \exp\left[\frac{-(t - T_r)^2}{2T_r \alpha^2 t}\right], \quad (11)$$

where T_r is the mean interevent time and α is the aperiodicity (coefficient of variation), have characteristics that qualitatively mimic earthquake renewal. The lognormal distribution is assumed here and the mean interevent time (T_r) with the corresponding standard deviation (s_t) were estimated. For this estimation historical information is mainly taken from PAPAZACHOS and PAPAZACHOU (2003), AMBRASEYS and JACKSON (2000) and AMBRASEYS (2002), whereas for some of the fault segments, and especially along the NAF, additional information on the corresponding recurrence times was taken from paleoseismological investigations. Historical information is thus enhanced, especially in cases of large events that seemed to have broken multiple segments, as the 1509 and 1766 events, or in cases of clustering, such as the three large earthquakes in 1343, 1344 and 1354, which according to ROCKWELL et al. (2001) comprise a mini-sequence rupturing much of the NAF.

In addition to the 1999 earthquake the Duzce fault (our S1 segment) is associated with the 967 and 1,878 earthquakes, with no obvious correlation with a third palaeoearthquake (1,495–1,700), result in an average recurrence time of 330–370 years (PANTOSTI et al., 2008). The 1719 and August 1999 earthquakes both appear to have ruptured the Izmit (S3) segment with an observed interevent time of 280 years and a calculated one of 288 years (PARSONS, 2004). The 1719 rupture is also supported by PONDARD et al. (2007) and KLINGER et al. (2003). The 1556, 1754 and 1894 earthquakes are associated with the Cinarcik (S4) segment, with an observed mean frequency of 170 years and a modeled ~250 year interevent time (PARSONS, 2004). The 1754 and 1894 ruptures on Cinarcik are compatible with the most plausible scenario of rupturing by PONDARD et al. (2007). The 1509 and the May 1766 earthquakes appear to have broken the same fault segment according to PARSONS (2004), our fault segment N. Marmara (S5), giving an observed mean interevent time of 257 years and a calculated one of 270 years. The May 1766 rupture agrees with PONDARD et al. (2007) scenario, while KLINGER et al. (2003) assigned this event to the Izmit segment. Observations on the Ganos fault (542, 824, 1354, 1509, 1766 and 1912) support an average return period of about 275 years (ROCKWELL et al., 2001). PARSONS (2004) gives a calculated mean interevent time of 207 years. The most conservative interpretation of the trench stratigraphy and faulting evidence suggests that at least one palaeoearthquake

Table 5

Estimated 30 year probabilities on the fault segments of the study area

Segment Code Name	Tr (years)	Elapsed Time (years)	Probability before the stress step		Minimum ΔCFF (bars)	Maximum ΔCFF (bars)	Average ΔCFF (bars)	Stressing rate $\dot{\tau}$ (bars/yr)	Conditional probability after the stress step (permanent effect)			Conditional probability after the stress step (permanent + transient effect)		
			Poisson	Conditional					Minimum $P_{mod}(30)$	Maximum $P_{mod}(30)$	Average $P_{mod}(30)$	Minimum $P_{mod}(30)$	Maximum $P_{mod}(30)$	Average $P_{mod}(30)$
S1	516	9	0.0564817	8.43E−08	−24.5782	3.4056	−12.2939	0.06994	1.28E−13	2.00E−07	1.44E−10	0	0.0005081	0
S2														
S3	280	9	0.1016027	4.13E−05	−52.3904	−13.9618	−46.7524	0.17983	6.94E−10	3.40E−06	2.43E−09	0	0	0
S4	281	114	0.1012601	0.0756	0.7583	20.7883	4.562	0.2146	0.0817	0.2664	0.1148	0.0867	0.5178	0.1541
S5	141	242	0.1916547	0.3815	−0.4677	−0.3041	−0.3479	0.2583	0.3689	0.3733	0.3721	0.3691	0.3734	0.3722
S6	149	354	0.1823675	0.1096	−32.6356	5.6021	−0.6235	0.18927	0.0002266	0.1878	0.1018	0.0003309	0.1796	0.1024
S7	279	96	0.1019476	0.0536	−10.2436	−0.3658	−6.6026	0.09318	0.0013	0.0489	0.0058	7.50E−11	0.0443	7.42E−05
S8														
S9	116	33	0.2278842	0.1389	−5.4096	4.275	−3.6749	0.14403	0.0506	0.2352	0.0729	0.0005075	0.4812	0.0079
S10	274	41	0.103708	0.0043	−46.1886	−19.8572	−33.8926	0.06677	7.99E−15	5.66E−08	1.26E−11	0	0	0
S11	500	51	0.0582355	0.0001469	−84.5332	−16.0104	−29.6677	0.12638	6.66E−16	1.78E−06	2.37E−08	0	0	0
S12	495.7	153	0.0587255	0.0185	−58.6503	−1.1137	−6.3017	0.04397	0	0.0095	0.0001166	0	0.0052	9.82E−10
S13	500	153	0.0582355	0.0179	−8.0555	4.8023	0.9243	0.06091	0.000175	0.0586	0.0247	1.98E−08	0.1594	0.0326
S14	641	44	0.0457236	7.29E−06	−33.7195	−0.9043	−20.8138	0.03791	0	3.88E−06	4.11E−15	0	6.64E−08	0
S15	1013	452	0.0291808	0.0218	3.8029	73.2066	11.447	0.01973	0.1454	0.9356	0.3571	0.2529	0.9987	0.7043
S16	265	55	0.1070347	0.014	−22.4072	−0.035	−11.8283	0.02232	0	0.0135	1.69E−11	0	0.0123	0
S17	500	64	0.0582355	0.0004609	−36.6035	−12.1088	−14.9083	0.03219	0	2.60E−10	8.03E−12	0	0	0
S18	137	199	0.1966607	0.3908	−0.2629	5.6945	0.1151	0.02748	0.3242	0.977	0.4203	0.3235	0.9652	0.4205
S19	500	148	0.0582355	0.0161	−204.0237	−20.7238	−53.5348	0.03041	0	2.68E−08	0	0	0	0
S20	500	38	0.0582355	3.33E−05	−230.4551	−4.2825	−64.5768	0.0313	0	2.88E−06	0	0	0	0
S21														
S22														
S23	311.5	89	0.0918159	0.0282	−25.5562	−9.0319	−10.6422	0.03526	1.99E−08	0.0002332	9.37E−05	0	0	0
S24	500	69	0.0582355	0.000668	−52.8621	−13.1149	−19.3354	0.04078	5.00E−15	1.47E−06	6.97E−08	0	0	0
S25														
S26	560	59	0.0521618	0.0001244	−139.1693	−12.0385	−22.8667	0.03593	0	2.09E−07	5.05E−10	0	0	0
S27														
S28	500	39	0.0582355	3.79E−05	−41.0145	−7.6932	−13.0689	0.02836	0	2.28E−07	5.16E−09	0	0	0
S29	500	163	0.0582355	0.0217	−0.0461	0.6387	0.0933	0.02605	0.0211	0.0306	0.0228	0.0204	0.0447	0.0243
S30	500	128	0.0582355	0.0098	−0.8781	−0.0961	−0.4526	0.04347	0.0072	0.0095	0.0084	0.0037	0.0089	0.0061
S31	500	128	0.0582355	0.0098	−1.0768	−0.9031	−1.0313	0.04788	0.0069	0.0073	0.007	0.0033	0.004	0.0034
S32	425.5	128	0.0680772	0.0211	−1.0753	−0.4059	−0.8155	0.0399	0.0138	0.018	0.0153	0.0062	0.0139	0.0087
S33	500	125	0.0582355	0.009	−1.8369	−0.2005	−0.5217	0.04387	0.0046	0.0084	0.0075	0.0008247	0.0073	0.0051
S34	500	127	0.0582355	0.0095	−3.5359	−1.4241	−2.7912	0.04287	0.0023	0.0056	0.0032	1.80E−05	0.0016	0.0001312

Table 5 continued

Segment Code Name	T_r (years)	Elapsed Time (years)	Probability before the stress step		Minimum ΔCFF (bars)	Maximum ΔCFF (bars)	Average ΔCFF (bars)	Stressing rate $\dot{\tau}$ (bars/yr)	Conditional probability after the stress step (permanent effect)			Conditional probability after the stress step (permanent + transient effect)		
			Poisson	Conditional					Minimum $P_{mod}(30)$	Maximum $P_{mod}(30)$	Average $P_{mod}(30)$	Minimum $P_{mod}(30)$	Maximum $P_{mod}(30)$	Average $P_{mod}(30)$
S35	500	357	0.0582355	0.0923	0.0126	0.0315	0.0234	0.05287	0.0926	0.0931	0.0929	0.0926	0.0932	0.093
S36	500	306	0.0582355	0.078	−0.0538	0.0115	−0.0313	0.04618	0.0767	0.0783	0.0773	0.0764	0.0784	0.0771
S37	246	109	0.1148085	0.1041	−0.1418	4.2329	0.2102	0.04254	0.0992	0.3001	0.1116	0.0941	0.5298	0.1199
S38	500	53	0.0582355	0.0001785	−48.5545	−16.3565	−22.0164	0.04809	4.16E−13	2.72E−07	2.58E−08	0	0	0
S39														
S40	500	104	0.0582355	0.0044	−36.9901	−15.4043	−17.7095	0.03264	8.59E−13	4.77E−07	1.16E−07	0	0	0
S41														
S42														
S43	500	139	0.0582355	0.0131	−0.0831	0.4846	0.0602	0.02981	0.0126	0.0166	0.0135	0.0117	0.0237	0.0142
S44														
S45	500	52	0.0582355	0.0001621	−16.9506	−0.61	−8.4607	0.02325	1.03E−10	0.0001071	1.53E−07	0	3.46E−06	0
S46														
S47	400	75	0.0723	0.0043	−44.8297	−13.0816	−17.0323	0.03916	6.68E−13	7.36E−06	9.79E−07	0	0	0
S48														
S49														
S50														
S51	500	51	0.0582355	0.0001469	−9.339	3.5846	−4.0189	0.02571	1.39E−10	0.0007856	5.46E−07	0	0.164	0
S52	500	1590	0.0582355	0.1123	−0.4279	0.2749	−0.0141	0.02575	0.0879	0.1301	0.1114	0.0882	0.1298	0.1114
S53	500	94	0.0582355	0.0028	−28.934	10.6422	−1.8285	0.02561	5.97E−13	0.0376	0.0008495	0	0.8131	8.81E−07

The three-first columns of the table give the code name of each fault segment, the recurrence time, T_r, and the time elapsed since the occurrence of the last strong ($M \geq 6.5$) earthquake on the corresponding segment, respectively. The fourth and fifth columns give the Poissonian and Conditional probabilities. The next three columns give the minimum, maximum and average value of the accumulated Coulomb stress changes onto each fault segment, which are due to the coseismic slips of the modeled events. The ninth column gives the stressing rate of each segment in bars/year. The tenth through twelfth columns give the maximum, minimum and average probability of occurrence for the next 30 years modified by the permanent effect of ΔCFF. The last three columns give the corresponding probabilities modified by both the permanent and transient effect of ΔCFF. Calculations were not performed for the fault segments where there is no information available for the occurrence of events with $M \geq 6.5$

(most probably two) occurred after A.D. 1693 on the 1967 Mudurnu (S11) segment (PALYVOS et al., 2007). For the Yenice (S17) fault KURCER et al. (2008) estimated a recurrence interval of 660 ± 160 years for large morphotectonic earthquakes, creating linear surface ruptures.

The paleoseismological observations were combined with information from the historical catalogs mentioned above, for events of $6.5 \leq M \leq 7.0$. In cases where only one or two events were reported for a particular fault segment, interevent times equal to 500 years and $\alpha = 0.5$ were assumed. This later value is in accordance with previous investigations in the area (STEIN et al., 1997; ERDIK et al., 2004; PARSONS, 2004).

The incorporation of calculated stress changes in conditional probabilities calculations needs the treatment of a stress change as an advance or delay in the earthquake cycle. A sudden stress change should be equivalent to a sudden shift in the time, T', to the next earthquake. The 'life clock' of the fault of interest can be estimated as:

$$T' = \frac{\Delta CFF}{\dot{\tau}}, \quad (12)$$

where ΔCFF is the stress change due to the coseismic stress changes by the nearby events and $\dot{\tau}$ is the tectonic stressing rate (in bars/year). Therefore, for the calculation of the conditional probability for the fault of interest an adjusted time by the clock change is taken into account:

$$P_c(t_1) = \frac{\int_{t_1}^{t_1+\Delta t} f_t(t+T)\,dt}{\int_{t_1}^{\infty} f_t(t+T)\,dt}. \quad (13)$$

Stressing rates were calculated for each fault segment from the yearly slip rate and the use of the same dislocation program as for static stress change calculations. The $\dot{\tau}$ values are displayed in Table 5 and are in agreement with those from STEIN et al. (1997) who estimated a value of 0.15 bar/year along most of the NAF system. The ΔCFF value on each fault segment was achieved by extending the calculations of the accumulated static stress changes due to the coseismic slip of the modeled events up to 2009. Since uncertainties are involved in these estimations and because stress change is spatially variable, we considered three different values, i.e. the minimum, maximum and average calculated ΔCFF values (Table 5), and consequently three different clock change values.

The next step was to estimate the rate-state transient effect that describes an expected enhanced rate of earthquake nucleation resulting from a stress increase and which can be expressed as a probability. For a stress decrease the rate of nucleation declines and eventually recovers. The time-dependent seismicity rate $R(t)$ after a stress perturbation is equal to (DIETERICH, 1994):

$$R(t) = \frac{r}{[\exp(-\Delta CFF/A\sigma) - 1]\exp[-t/t_\alpha] + 1}, \quad (14)$$

where r is the steady state seismicity rate, ΔCFF is the stress step, σ is normal stress, A is a fault constitutive constant, t_α is the observed aftershock duration. The transient change in the expected earthquake rate $R(t)$ after a stress step can be related to the probability of an earthquake of a given size over the time interval Δt (we use 30 years for these computations) through a non stationary Poisson process as (DIETERICH and KILGORE, 1996):

$$P(t, \Delta t) = 1 - \exp\left[-\int_t^{\Delta t} R(t)\,dt\right]$$
$$= 1 - \exp[-N(t)], \quad (15)$$

where $N(t)$ is the expected number of earthquakes in the interval Δt and is equal to:

$$N(t) = r_p\left\{\Delta t + t_\alpha \ln\left[\frac{1 + [\exp(-\frac{\Delta CFF}{A\sigma}) - 1]\exp[\frac{-(\Delta t)}{t_s}]}{\exp(\frac{-\Delta CFF}{A\sigma})}\right]\right\} \quad (16)$$

where r_p is the expected rate of earthquakes and is equal to (TODA et al., 1998):

$$r_p = -\frac{1}{\Delta t}\ln[1 - P_c]. \quad (17)$$

Note that the transient effect disappears if $\Delta CFF = 0$, that is $N = r_p \cdot \Delta t$. We set the aftershock duration equal to 10% of the minimum interevent time, according to DIETERICH (1994). Thus, for the area of the North Anatolian fault $t_\alpha = 25$ year, considering a minimum return period of 250 years. For the same area a regional aftershock decay time for $M \geq 6.7$ earthquakes was found to be ~ 35 years by PARSONS et al. (2000). As this duration is inversely proportional to the presumed fault stressing rate (DIETERICH, 1994) a value of $t_\alpha = 50$ year was set for the remainder of our study area. This is also in accordance with the longer observed interevent times (~ 500 years). Knowing the parameters t_α and ΔCFF and using the equation (DIETERICH, 1994):

$$t_\alpha = \frac{A\sigma}{\dot{\tau}} \qquad (18)$$

we calculated the $A\sigma$. In summary, the net probability of events rupturing each fault segment combines both the permanent and transient effects of a stress step. Net probability is obtained by first computing the permanent effect of a stress change on the conditional probability using the approach of Eq. 13. Then the expected rate of earthquakes, r_p, for the permanent effect is obtained using Eq. 17 to evaluate Eqs. 15, 16 for the net probability. The conditional probability after the stress step (minimum, maximum and average) for the permanent effect and both the transient and permanent effects is displayed in Table 5.

The affect of the stress step in the probability estimates becomes more evident in the cases in which the fault segment has recently failed (the cases of Izmit and Duzce segments) and where a fault segment is located along strike with a previously failed segment, resulting in the positive static stress changes on the first segments. In these cases the differences between the probability estimates before and after the stress step are significant and must be included in any assessment for the future seismic hazard. As can be observed from Table 5, the fault segments adjacent to previous ruptures (segments S4, S5, S9, S15, S18, S37, S53) exhibit high estimates of time dependent probabilities, which are appreciably larger than the estimates before the stress step was considered.

8. Discussion

The present study is an effort to interpret the occurrences of strong ($M \geq 6.5$) earthquakes in the area of western Turkey and the eastern Aegean Sea and to evaluate the future seismic hazard. The methodology applied is based on a model assuming the fault interaction that led to the triggering of one event by previous ones and explains the probable mechanism of their occurrence in space and time. The stress interactions of 22 strong earthquakes ($M \geq 6.5$) that occurred since 1904 in the study area have been investigated by calculating Coulomb Stress changes (ΔCFF). We constructed a model of the evolution of stress for the time interval of 1904–2008 in order to examine if the history of cumulative changes in stress can explain the spatial and temporal occurrence of strong ($M \geq 6.5$) earthquakes in the region. Tectonic stress loading is simulated by introducing a negative virtual slip on major fault segments. From this study and from previous investigations, it has become clear that changes in Coulomb stress are associated with areas where future events are likely to occur. Thus, regions of increased stress must be considered as subject to greater hazard than anywhere else.

When considering the accumulated stress changes our calculations indicate that the Coulomb stress evolution model can successfully explain the location of strong earthquakes in the study area. Stress loading on the eastern Aegean, North Anatolian and the rest of western Turkey fault segments transform the stress acting on strike-slip and normal faults. The model satisfies our expectations in explaining the locations of the vast majority of the modeled events that are located in stress-enhanced regions, meaning that each earthquake seems to encourage the failure in the adjacent regions. For example, the calculated static stress changes following the 1912 Ganos ($M_w = 7.4$) earthquake encouraged the failure at the site of the 1975 Saros event ($M_w = 6.6$), as these two events are associated with nearby fault segments (S7 and S9, respectively) of the same dextral strike-slip faulting type. Similar evidence is presented from the 1957 Abant ($M_w = 7.0$) for the 1967 Mudurnu ($M_w = 7.2$) event (S11 and S10 segments, respectively), from the 1999 Izmit ($M_w = 7.4$) for the 1999 Düzce ($M_w = 7.2$) event (S3 and S1 segments, respectively), from the 1944 Ayvacik ($M_w = 6.8$) oblique faulting earthquake for the 1953 Yenice ($M_w = 7.2$) dextral strike-slip event (S17 and S16 segments, respectively), from the 1919 Soma ($M_w = 6.9$) for the 1939

Dikili ($M_w = 6.6$) earthquake, from the 1970 Gediz doublet (both of $M_w = 7.1$ on adjacent segments S20 and S19) and from the 1904 Samos ($M_w = 6.8$) for the 1955 Agathonisi ($M_w = 6.9$) earthquakes (S40 and S38 segments, respectively). It became evident that 12 out of 22 strong events occurred in nearby or adjacent fault segments, which means that the occurrence time of the subsequent ones most probably advanced, since the respective causative faults received positive values of static stress changes due to the coseismic slip of the preceding earthquakes.

The choice of the pore pressure model significantly influences the calculations of Coulomb stress changes caused by a shear dislocation in an elastic isotropic half-space (BEELER et al., 2000; COCCO and RICE, 2002). For this reason we performed our calculations by considering different values of normal stress components instead of choosing a value of apparent coefficient of friction and equality among these components. We investigated the effect of the fluid pore pressure in the modelling by considering different values for Skempton's coefficient ($B = 0.2$, $B = 0.5$ and $B = 0.9$). Differences are observed on a small scale and in particular close to the tips of the faults that failed. This investigation was accomplished for different faulting types in an attempt to examine if the current state of stress as derived from our evolutionary model, explains the location of the smaller events that occurred after 1999, when the last strong earthquake occurred. The results are encouraging because the majority of these events are located inside stress enhanced areas.

It is important to determine the hazardous segments that might generate an impeding earthquake. According to these results the fault segments of North Sea of Marmara (segments S4, S5 and S6), the smaller fault segment in Saros Gulf (S8), the Bursa segments (S12 and S13) and the Yenice1 segment (S15), on the NAF branches, have received positive static stress changes from the failure of adjacent fault segments in addition to the continuous tectonic loading. Some of the probable sites found in this study, namely the segments along the northern part of Marmara Sea and Saros Gulf, were also identified by previous investigations (STEIN et al., 1997; NALBANT et al., 1998; PAPADIMITRIOU and SYKES, 2001) as contestant regions for the occurrence of a future strong earthquake. Several of the normal fault segments in the central part of the study area are currently in stress enhanced areas. It is worth noting here that although a large area is presented as continuously loading, we have to focus our attention only at the sites of the active faults.

With respect to probability estimations, the first remarkable result is that for most of the segments the renewal model based on the lognormal distribution predicts conditional probabilities of failure of these segments for the next 30 years, that are differentiated from those based on the Poisson model ranging between 3 and 23% (Table 5). These results agree with previous investigations, especially at the Marmara and Izmit region (STEIN et al., 1997; PARSONS, 2004), although the first authors found larger values with the Poisson model. For the conditional probabilities estimations we used mean interevent time equal to 500 years for some fault segments, because reliable historical information was not available. In the cases where the Poissonian probabilities are larger than the conditional ones for some fault segments, the time elapsed since the last event of $M \geq 6.5$ is shorter than the estimated mean interevent time (see Table 5). The uncertainties involved in these estimates concern the mean interevent time and the corresponding standard deviation, the aftershock duration (t_α) and the value of $A\sigma$ (we calculate the $A\sigma$ value using Eq. 18 considering that t_α is known from previous investigations).

The ΔCFF values, calculated according to the faulting type of each fault segment, where incorporated into the probability estimates, as the permanent stress effects and both the permanent and transient effects. For this purpose we considered minimum, maximum and average values of ΔCFF, as well as the minimum, maximum and average values of clock advanced or delay (Eq. 12). It is interesting to note that these values affect the estimated probabilities, by increasing them in comparison with Poissonian and conditional probability estimates when positive values of ΔCFF were found on a certain fault segment, or decreasing them in the cases of negative corresponding ΔCFF values. For example, in the Izmit fault segment, the ΔCFF effect decreases substantially the conditional probabilities (by a factor of 10^{-5}).

The stress transfer between adjacent fault segments considerably influences the probability estimates. For certain fault segments the differences between Poissonian and conditional estimates before the stress step are significantly different than those incorporated the stress step, and worthy of mention for future seismic hazard assessment. For the fault segments along the north Marmara Sea (S4 and S5) the Poissonian probabilities are found equal to 10% and 19%, respectively, while the corresponding time dependent ones are equal to 52 and 37%. The opposite but also significant consequence of the ΔCFF effect is observed for the Izmit (S3) and Duzce (S1) segments last ruptured in 1999, yielding a 30-year Poisson probability of 5% and 10%, respectively, whereas the time-dependent probabilities on these segments are $\sim 0\%$. These findings agree with PARSONS (2004). The ΔCFF effect resulted in high probability estimates for normal fault segments being along strike with previous ruptures, in the central and southern part of the study area.

Acknowledgments

The paper greatly benefited from insightful comments of two anonymous reviewers and the editorial assistance of Martha Savage. The stress tensors were calculated using a program written by J. Deng (DENG and SYKES, 1997), based on the DIS3D code of S. Dunbar, which later improved (ERIKSON, 1986) and the expressions of G. Converse. Paradisopoulou P. M. wishes to express her sincere and profound thanks to Professor Stanislaw Lasocki and Assistant Professors Janusz Mirek and Beata Orlecka–Sikora from the Faculty of Geology, Geophysics and Environmental Protection, AGH University of Science and Technology, Krakow Poland, for many useful suggestions, ideas and assistance for the preparation of the useful application for our calculations. Part of this undertaking entailed a 2-month visit of the first author to AGH University in the frame of PENED grand. Critical reading of the manuscript by Rodolfo Console, from Istituto Nazionale Geofisica e Vulcanologia, Roma, Italy, is greatly appreciated. The GMT system (WESSEL and SMITH, 1998) was used to plot the figures. The first author is a grantee of the 03ED375 research project implemented within the framework of the "Reinforcement Programme of Human Research Manpower" (PENED) and co-financed by the National and Community Funds (25% from the Greek Ministry of Development-General Secretariat of Research and Technology and 75% from E.U.-European Social Fund) (Grant No. 03EΔ815 61585 23-09-05). Geophysics Department contribution 742.

Appendix: Events ($M \geq 6.5$) Included in the Stress Evolutionary Model

1904, Samos earthquake ($M_w = 6.8$): A rupture length equal to 46 km (PAPAZACHOS and PAPAZACHOU, 2003) and an average displacement of 1.2 m estimated from relation (4) were considered for this oblique normal faulting event (strike = 91°, dip = 45°, rake = −115°) for calculating the Coulomb stress changes due to its coseismic displacement.

1912, Ganos (Mürefte) earthquake ($M_w = 7.4$): This earthquake occurred between the Gulf of Saros and the Sea of Marmara at the western part of the North Anatolian Fault. The main earthquake was followed by two aftershocks, the first one ($M = 6.2$) on August 10 and the second ($M = 6.7$) on September 13 at the SE of the main shock (PAPAZACHOS and PAPAZACHOU, 2003). Maps, reports and photographs taken just after the earthquake are available (MACOVEI, 1912; MIHAILOVIC, 1927, 1933). Surface expressions on the 50-km-long strike-slip fault were observed with ENE direction linking the Marmara and the Saros fault systems (ATES and TABBAN, 1976; BARKA 1992). The surface rupture pattern was complex with a substantial right-lateral strike-slip component (up to 3 m) (AMBRASEYS and FINKEL, 1987). NALBANT et al. (1998) modeled this event with a rupture length of 90 km extended the rupture seen on land by 15 km to the east and 25 km to the west. PAPADIMITRIOU and SYKES (2001) use 110 km length and 3.32 m slip derived from scaling laws. In the present study a rupture equal to 116 km and an average slip of 2.8 m were estimated from Eqs. 5 to 6, respectively.

1914, Burdur earthquake ($M_w = 7.0$): This event occurred near the Burdur Lake and is associated with a 52-km-long normal fault (strike = 230°, dip = 35°, rake = −105°) dipping to NW (PAPAZACHOS and PAPAZACHOU, 2003), with a calculated mean displacement of 1.66 m (Eq. 4).

1919, Soma earthquake ($M_w = 6.9$): This earthquake occurred at Bakircay Graben (strike = 253°, dip = 45°, rake = −115°, TAYMAZ et al., 1991b) on a segment adjacent to the 1939 rupture. A 43-km-fault length was estimated from relation (3) and a mean displacement of 0.63 m from the event's scalar moment.

1928, Torbali earthquake ($M_w = 6.5$): This earthquake caused considerable damage in Torbali, Izmir and Kücük Menderes Graben (PAPAZACHOS and PAPAZACHOU, 2003). It is associated with a normal fault (strike = 83°, dip = 45°, rake = −94°) with 25-km length and a mean displacement of 0.72 m calculated by using Eqs. 3 and 4.

1933, Kos earthquake ($M_w = 6.6$): This earthquake is associated with a fault segment almost parallel to the south coastline of the Kos Island (strike = 65°, dip = 50°, rake = −90°). An average displacement of 0.85 m and a fault length equal to 28 km were estimated using Eqs. 4 and 5, respectively.

1939, Dikili earthquake ($M_w = 6.6$): The Dikili earthquake occurred near the coastal Aegean area south of the Edremit Gulf. The isoseismal maps indicate that this event was located at the western extremity of the Bakircay Graben, a normal fault zone (ARPAT and BINGÖL, 1969; WESTAWAY, 1990). The event is associated with a NE–SW trending normal faulting dipping to the north (strike = 211°, dip = 45°, rake = −115°). An estimated coseismic displacement of 0.85 m and a rupture length of 26 km were assigned for this event, from Eqs. 4 and 3, respectively.

1944, Ayvacık earthquake ($M_w = 6.8$): The earthquake occurred near the Edremit Gulf, where the southern branch of the North Anatolian Fault reaches the Aegean Sea through the Edremit Gulf (strike = 74°, dip = 46°, rake = −114° after TAYMAZ et al., 1991b). A fault length equal to 35 km and an average displacement equal to 1.4 m were estimated from Eqs. 3 and 4, respectively.

1949, Chios earthquake ($M_w = 6.7$): This earthquake occurred to the north of Chios Island, associated with normal faulting (strike = 84°, dip = 36°, rake = −80°) with an estimated fault length of 31 km, from scaling law (3), and a mean coseismic displacement equal to 1.03 m (Eq. 4).

1953, Yenice earthquake ($M_w = 7.2$): The Yenice earthquake occurred between the Sea of Marmara to the north and the Edremit Gulf to the south. The rupture took place at the southern branch of NAF over 60 km (PINAR, 1952; AMBRASEYS, 1970). The earthquake focal mechanism parameters (MCKENZIE, 1972; TAYMAZ et al., 1991a) indicate pure southwest–northeast trending right-lateral strike-slip faulting (strike = 250°, dip = 70°, rake = −160°). The slip reaches 3.5 m in the eastern part and diminishes to 1.5 m at both ends (KETIN and ROESLI, 1953; AMBRASEYS, 1970). NALBANT et al. (1998) modeled this event using the observed slip distribution and the geometry (length of 60 km) of the mapped surface rupture. Based on the above information the fault length is taken equal to 60 km and the mean displacement, derived from the event's scalar moment, is equal to 3.78 m.

1955, Agathonisi earthquake ($M_w = 6.9$): The earthquake occurred in the Büyük Menderes graben, near Agathonisi Island. The focal mechanism (MCKENZIE, 1972) shows NE–SW normal faulting (strike = 55°, dip = 51°, rake = −113°). We have modeled this event using a 38-km-fault length with a mean displacement of 1.19 m (Eqs. 3, 4, respectively).

1956, Amorgos earthquake ($M_w = 7.7$): This is the strongest event that occurred in the backarc Aegean area during the instrumental era. It occurred on an ENE-trending normal fault that is seated parallel to the Island's southern coastline and was followed by a strong event in an adjacent fault to its southwest, which most probably was triggered by the first occurrence. Its fault plane solution (strike = 65°, dip = 40°, rake = −90°) was determined by SHIROKOVA (1972). A fault length of 75 km, in accordance with the submarine topography, and a mean displacement of 5.30 m, were estimated for this large earthquake from Eqs. 3 and 4, respectively.

1957, Rhodes earthquake ($M_w = 7.2$): A preshock ($M = 6.8$) took place before the main earthquake ($M = 7.2$) near the Rhodes Island and many aftershocks followed, from which the largest one registered magnitude $M = 6.1$ (PAPAZACHOS and

PAPAZACHOU, 2003). The main shock is associated with a left-lateral strike-slip faulting with a NE–SW strike direction (strike = 30°, dip = 80°, rake = −41°). The fault length is estimated equal to 67 km from Eq. 5 and the mean displacement equal to 1.34 m from Eq. 6.

1957, Abant earthquake ($M_w = 7.0$): The Abant event occurred on the North Anatolian fault at the eastern part of the study area. The 40-km-long surface faulting was mapped by AMBRASEYS (1970). The focal mechanism (MCKENZIE, 1972; TAYMAZ et al., 1991a) indicates strike-slip faulting (strike = 265°, dip = 78°, rake = 179°). The slip is not well constrained, being measured at only two localities (1.4 and 1.6 m). Taking into account the morphology, a fault length of 40 km was estimated from Eq. 5 in accordance with the morphology, and a mean displacement of 1.47 was calculated from (6), in good agreement with the reported values.

1964, Manyas earthquake ($M_w = 6.9$): The Manyas earthquake occurred at the south of the Sea of Marmara in the southern branch of NAF between the Lakes Manyas and Uluabat. The focal mechanism (strike = 280°, dip = 45°, rake = −90°) (TAYMAZ et al., 1991b) indicates a WNW–ESE normal faulting although strike-slip faulting prevails in this part of our study area. The 40-km surface normal faulting (NALBANT et al., 1998) (en echelon surface rupture and fissuring over a wide zone) was interpreted as resulting from the right-lateral strike-slip motion (ERENTÖZ and KURTMAN, 1965; KETIN, 1966). A 35-km-long WNW–ESE normal fault dipping to the north is considered here with a mean displacement of 1.4 m (Eq. 6).

1967, Mudurnu earthquake ($M_w = 7.2$): The Mudurnu earthquake occurred on the NAF at the easternmost part of the Sea of Marmara, Mudurnu Valley, and extended towards the west. Its fault plane solution (strike = 275°, dip = 45°, rake = −178°) based on teleseismic body-waveform (P- and SH-) inversion (TAYMAZ et al., 1991a) shows a E–W dextral strike-slip faulting mechanism. A large aftershock (July 30, 1967, $m_b = 5.6$) occurred at its western extremity with NW–SE striking and normal fault plane solution (STEWART and KANAMORI, 1982; MCKENZIE, 1972; JACKSON and MCKENZIE, 1984). This illustrates the change on the NAF in this area between strike–slip motion to the east and normal and strike–slip motion on several branches to the west. NALBANT et al. (1998) used detailed maps of the 80-km-long surface rupture and the fault slip distribution for the event modeling (AMBRASEYS and ZATOPEK, 1969; GÜÇLÜ, 1969) which is greatest, 2.5 m, in the east and decreases steadily to the west. A fault length of 80 km is taken and a calculated mean displacement from Eq. 6, equal to 2.02 m.

1969, Alaşehir earthquake ($M_w = 6.6$): The Alaşehir earthquake occurred in the Gediz River Valley, associated with about 30–36 km of surface rupture and extending from NW through Alaşehir to SE (AMBRASEYS and TCHALENKO, 1972). The strike of the surface varied from N85°W in the NW to N50°W in the SE (KETIN and ABDUSSSELAMOĞLU 1969). Displacements at the surface measured an average of about 20 cm. The fault plane solution of the main earthquake shows a normal faulting with a dip of 32°NNE and a strike of N79°W, consistent with the strike observed at the NW end of surface ruptures (EYIDOĞAN and JACKSON, 1985; BRAUNMILLER and NABELEK, 1996). We model this event using the reported fault plane solution of EYIDOĞAN and JACKSON (1985) (strike = 281°, dip = 34°, rake = −90°) as a normal fault with a length of 25 km, estimated from the Eqs. 3, and a mean displacement of 0.61 m, from Eq. 4.

1970, Gediz earthquakes ($M_w = 7.1$): About 45 km of complicated surface normal faulting was associated with this earthquake, trending both NNW–SSE and E–W down thrown to the east and north (AMBRASEYS and TCHALENKO, 1972). The aftershock sequence defined a 40-km-wide, 200-km-long, E–W zone (AMBRASEYS and TCHALENKO, 1972). The observed seismograms show complexity and were modeled using three main subevents (EYIDOĞAN and JACKSON, 1985). The first subevent occurred on a 15-km-long NNW–SSE segment with a mean displacement of 1.6 m and a dip of 35°. The second subevent, of the same magnitude (M_w 7.1), triggered by the first shock and ruptured about the 24-km-long E–W segment with a mean displacement of 2.4 m and a dip pf 35°. The third subevent, much smaller in magnitude (M 5.7), occurred on a ∼15° dipping fault extending the second fault segment from 12.5 to 17.5 km depth (EYIDOĞAN and JACKSON, 1985). We modeled this event as comprising the two major subevents.

1975, Saros earthquake ($M_w = 6.6$): The Saros segment is located in the prolongation of the Ganos (Gaziköy) fault zone at the western part of the North Anatolian Fault, where the 1975 earthquake occurred. It is an oblique right-lateral strike-slip fault as the focal mechanism indicates (strike = 68°, dip = 55°, rake = −145°) (TAYMAZ et al., 1991a) with ENE–WSW strike consistent with the orientation of NAF at this particular location. The rupture length is taken equal to 40 km and the mean displacement equal to 0.86 m from Eqs. 5 and 6, respectively.

1999 Izmit (Kocaeli) earthquake ($M_w = 7.4$): The Izmit earthquake, one of the most destructive earthquakes in Turkey, occurred at the western part of NAF. About 115 km of surface strike-slip faulting was associated with its occurrence, trending E–W from Sapanca–Akyazi at the east to Hersek Delta to the west (BARKA et al., 2002). A rupture constituted from four segments (with lengths equal to 35, 20, 26 and 35 km, going from west to east) is considered for modeling this event, according to BARKA et al. (2002). Details of the geometry and coseismic slip of each segment are given in Table 2.

1999 Düzce earthquake ($M_w = 7.2$): This event occurred in Bolü basin, in the adjacent fault segment associated with the previous Izmit earthquake and with in <3 months afterwards. The fault length is about 40–56 km long (KIRATZI and LOUVARI, 2001; AKYÜZ et al., 2002; AYDIN and KALAFAT, 2002) and the focal mechanism indicates a right-lateral strike-slip faulting with E–W strike and dip to the north (strike = 262°, dip = 53°, rake = −177°). We model this event according to KIRATZI and LOUVARI (2001), who suggested a fault length of 56 km and a mean displacement of 2.60 m.

REFERENCES

AKYÜZ, H. S., HARTLEB, R., BARKA, A., ALTUNEL, E., SUNAL, G., MEYER, B., and ARMIJO, R. (2002), *Surface rupture and slip distribution of the November 1999 Düzce earthquake (M = 7.1), North Anatolian Fault, Bolu, Turkey*, Bull. Seismol. Soc. Am. 92, 61–66.

AMBRASEYS, N. N. (1970), *Some characteristic features of the North Anatolian fault zone*, Tectonophysics 9, 143–165.

AMBRASEYS, N. N. (2001), *Reassesment of earthquakes, 1900–1999, in the Eastern Mediterranean and the Middle East*, Geophys. J. Int. 145, 471–485.

AMBRASEYS, N. N. (2002), *The seismic activity of the Marmara Sea region over the last 2000 years*, Bull. Seismol. Soc. Am. 92, 1–18.

AMBRASEYS, N. N. and ZATOPEK, A. (1969), *The Mudurnu Valley, west Anatolia, Turkey, earthquake of 22 July, 1967*, Bull. Seismol. Soc. Am. 59, 521–589.

AMBRASEYS, N. N. and TCHALENKO, J. S. (1972), *Seismotectonic aspects of the Gediz, Turkey, earthquake of March 1970*, Geophys. J. R. Astron. Soc. 30, 229–252.

AMBRASEYS, N. N. and FINKEL, C. (1987), *The Saros–Marmara earthquake of 9 August, 1912*, Earthq. Eng. Struct. Dyn. 15, 189–211.

AMBRASEYS, N. N. and JACKSON, J. A. (1990), *Seismicity and associated strain of central Greece between 1890 and 1988*, Geophys. J. Int. 101, 663–708.

AMBRASEYS, N. N. and JACKSON, J. A. (2000), *Seismicity of Marmara (Turkey) since 1500*, Geophys. J. Int. 141, F1–F6.

ARMIJO, R., MEYER, B., HUBERT, A., and BARKA, A. (1999), *Propagation of the North Anatolian fault into the north Aegean: Timing and kinematics*, Geology 27, 267–270.

ARMIJO, R., FLERIT, F., KING, G. and MEYER, B. (2003), *Linear elastic fracture mechanics explains the past and present evolution of the Aegean*, Earth Plan. Sci. Lett. 217, 85–95.

ARPAT, E. and BINGÖL, E. (1969), *The rift system of western Turkey, through on its development*, Bull. Miner. Res. Expl. I. 73, 1–9.

ATES, R. and TABBAN, A. (1976), *A preliminary report on August 9, 1912 Mürefte–Sarköy earthquake (in Turkish)*, Earthquake Res. Inst. Ankara, Turkey.

AYDIN, A. and KALAFAT, D. (2002), *Surface ruptures of the August 17 and November 12, 1999, Izmit, and Duzce Earthquakes in NW Anatolia*, Turkey: Their tectonic and kinematic significance and the associated damage, Bull. Seismol. Soc. Am. 92, 95–106.

AYHAN, M. E., BURGMANN, R., MCCLUSKY S., LENK, O., AKTUG, B., HERECE, E., and REILINGER, R. E. (2001), *Kinematics of the $M_w = 7.2$, 12 November, 1999, Duzce, Turkey Earthquake*, Geophys. Res. Lett. 28, 367–370.

BARKA, A. A. (1992), *The North Anatolian fault*, Ann. Tectonicae 6, 164–195.

BARKA, A. A., AKYÜZ, H. S., ALTUNEL, E., SUNAL, G., ÇAKIR, Z., DIKBAS, A., YELI, B., ARMIJO, R., MEYER, B., J. B. DE CHABALIER, ROKWELL, T., DOLAN, J. R., HARTLEB, R., DAWSON, T., CHRISTOFFERSON, S., TUCKER, A., FUMAL, T., LANGRIDGE, R., STENNER, H., LETTIS, W., BACHHUBER, J., and PAGE, W. (2002), *The surface rupture and slip distribution of the 17 August, 1999 Izmit earthquake, M = 7.4, North Anatolian Fault*, Bull. Seismol. Soc. Am. 92, 43–60.

BEELER, N. M., SIMPSON, R. W., HICKMAN, S. H., and LOCKNER, D. A. (2000), *Pore fluid pressure, apparent friction, and Coulomb failure*, J. Geophys. Res. 105, 25533–25542.

BIRD, P. and KAGAN, Y. Y. (2004), *Plate–tectonic analysis of shallow seismicity: Apparent boundary width, beta, corner magnitude, coupled lithosphere thickness, and coupling in seven tectonic settings*, Bull. Seismol. Soc. Am. 94, 2380–2399.

Bozkurt, E. (2001), Neotectonics of Turkey–a synthesis, Geodynamica Acta 14, 3–30.

Bozkurt, E. (2003), Origin of the NE–trending basins in western Turkey, Geodynamica Acta 16, 61–81.

BRAUNMILLER, J. and NABELEK, J. (1996), *Geometry of continental normal faults: Seismological constraints*, J. Geophys. Res. 101, 3045–3052.

COCCO, M. and RICE, J. R. (2002), *Pore pressure and poroelasticity effects in Coulomb stress analysis of earthquake interactions*, J. Geophys. Res. *107*, doi:10.1029/2000JB000138.

CORNELL, C. A., WU, S.-C., WINTERSTEIN, S. R., DIETRICH, J. H., and SIMPSON, R. W. (1968), *Seismic hazard induced by mechanically interactive fault segments*, Bull. Seismol. Soc. Am. *83*, 436–449.

DAVIES, R., ENGLAND, P., PARSONS, B., BILLIRIS, H., PARADISSIS, D., and VEIS, G. (1997), *Geodetic strain of Greece in the interval 1892–1992*, J. Geophys. Res. *102*, 24,571–24,588.

DENG, J. and SYKES, L. (1997), *Evolution of the stress field in Southern California and triggering of moderate size earthquakes: A 200-year perspective*, J. Geophys. Res. *102*, 9859–9886.

DIETERICH, J. H. (1994), *A constitutive law for rate of earthquake production and its application to earthquake clustering*, J. Geophys. Res. *99*, 2601–2618.

DIETERICH, J. H. and KILGORE, B. (1996), *Implications of fault constitutive properties for earthquake prediction*, Proc. Natl. Acad. Sci. USA *93*, 3787–3794.

ERDIK, M., DEMIRCIOGLU, M., SESETYAN, K., DURUKAL, E., and SIYAHI, B. (2004), *Earthquake hazard in Marmara region, Turkey*, Soil Dyn. Earth. Eng. *24*, 605–631.

ERENTÖZ, C. and KURTMAN, F. (1965), *A report on the 1964 Manyas earthquake (in Turkish)*, Bull. Miner. Res. Expl. Inst. *63*, 1–5.

ERIKSON, L. (1986), *User's Manual for DIS3D: A three-dimensional dislocation program with applications to faulting in the Earth*, Masters Thesis, Stanford Univ., Stanford, California, pp. 167.

EYIDOĞAN, H. (1988), *Rates of crustal deformation in western Turkey as deduced from major earthquakes*, Tectonophysics *148*, 83–92.

EYIDOĞAN, H. and JACKSON, J. (1985), *A seismological study of normal faulting in the Demirci, Alasehir and Gediz earthquakes of 1969–70 in western Turkey: Implications for the nature and geometry of deformation in the continental crust*, Geophys. J. R. Astron. Soc. *81*, 569–607.

FLERIT, F., ARMIJO, R., KING. G., and MEYER, B. (2004), *The mechanical interaction between the propagating North Anatolian fault and the backarc extension in the Aegean*, Earth Planet. Sci. Lett. *224*, 347–362.

GÜÇLÜ, U., *Section 1: Field investigation (in Turkish)*, in *Investigations on July 22, 1967 Mudurnu Valley earthquake* (ed. Ergin K.) (Publ. 27, Istanbul Technical University, Earth Physics Institute, Istanbul, Turkey, 1969) pp. 1–27.

HAGIWARA, Y. (1974), *Probability of earthquake occurrence as obtained from a Weibull distribution analysis of crustal strain*, Tectonophysics *23*, 313–318.

HARDEBECK, J. L. (2004), *Stress triggering and earthquake probability estimates*, J. Geophys. Res. *109*, B04310, doi:10.1029/2003JB002437.

HARRIS, R. (1998), *Introduction to special section: Stress triggers, stress shadows, and implications for seismic hazard*, J. Geophys. Res. *103*, 251–254.

HARRIS, R. and SIMPSON, R. (1993), *In the shadow of 1857: An evaluation of the static stress changes generated by the M 8 Ft. Tejon, California, earthquake*, EOS, Trans. Am. Geophys. Un. *74*(43), 427.

HARRIS, R. and SIMPSON, R. (1996), *In the shadow of 1857: The effect of the great Ft Tejon earthquake on subsequent earthquakes in Southern California*, Geophys. Res. Lett. *23*, 229–232.

HUBERT-FERRARI, A., BARKA, A., JAQUES, E., NALBANT, S. S., MEYER, B., ARMIJO, R., TAPPONNIER, P., and KING G. C. P. (2000), *Seismic hazard in the Marmara Sea region following the 17 August, 1999 Izmit earthquake*, Nature *404*, 269–273.

JACKSON, J. (1994), *Active tectonics of the Aegean region*, Ann. Rev. Earth Planet. Sci. *22*, 239–271.

JACKSON, J. and MCKENZIE, D. P. (1984), *Active tectonics of the Alpine–Himalayan Belt between western Turkey and Pakistan*, Geophys. J. R. Astron. Soc. *77*, 185–246.

JACKSON, J., HAINS, A. J., and HOLT, W. E. (1994), *A comparison of satellite laser ranging and seismicity data in the Aegean region*, Geophys. Res. Lett. *21*, 2849–2852.

KAGAN, Y. Y. and L. KNOPOFF (1987), *Random stress and earthquake statistics: Time dependence*, Geophys. J. R. Astron. Soc. *88*, 723–731.

KETIN, İ. (1966), *Tension cracks occurred on the surface during the October 6, 1964 Manyas earthquake (in Turkish)*, Bull. Turk. Geol. Union *10*, 44–51.

KETIN, İ. and ROESLI, T. (1953), *Macroseismic research of the Northwest Anatolian earthquake of the 18 March 1953 (in German)*, Eclogae Geol. Helv. *46*, 187–208.

KETIN, İ. and ABDUSSSELAMOĞLU, S. (1969), *Macroseismic observations on March 23, 1969 Demirci and March 28, 1969 Alasehir–Sarigöl earthquakes (in Turkish)*, Min. Mag. Istanbul Univ. *4*(5), 21–26.

KING, G. C. P., OPPENHEIMER, D., and AMELUNG, F. (1994), *Block versus continuum deformation in the western United States*, Earth Planet. Sc. Lett. *128*, 55–64.

KING, G. C. P., HUBERT-FERRARI, A., NALBANT, S. S., MEYER, B., ARMIJO, R., and BOWMAN, D. (2001), *Coulomb interactions and the 17 August, 1999 Izmit, Turkey earthquake*, Earth and Planet. Sci., C. R. Acad. Sci. Paris *333*, 557–569.

KIRATZI, A.A. and LOUVARI, E. (2001), *Source parameters of the Izmit–Bolu 1999 (Turkey) earthquake sequences from teleseismic data*, Annali di Geofisica *44*, 33–47.

KLINGER, Y., SIEH, K., ALTUNEL, E., AKOGLU, A., BARKA, A., DAWSON, T., GONZALEZ, T., MELTZNER, A., and ROCKWELL, T. (2003), *Paleoseismic evidence of characteristic slip on the western segment of the North Anatolian Fault, Turkey*, Bull. Seismol. Soc. Am. *93*, 2317–2332.

KURCER, A., CHATZIPETROS, A., TUTKUN, S. Z., PAVLIDES, S., ATES, O., and VALKANIOTIS, S. (2008), *The Yenice–Gonen active fault (NW Turkey): Active tectonics and palaeoseismology*, Tectonophysics *453*, 263–275.

MACOVEI, G. (1912), *About the Sea of Marmara earthquake of the 9 August, 1912 (in French)*, Bull. Sect. Acad. Roumanie, Bucarest *1* (1), pp. 1–10.

MATTHEWS, M. V., ELLSWORTH, W. L., and REASENBERG, P. A. (2002), *A Brownian model for recurrent earthquakes*, Bull. Seismol. Soc. Am. *92*, 2233–2250.

MCCLUSKY, S., BALASSANIAN, S., BARKA, A., DEMIR, C., ERGINTAV, S., GEORGIEV, I., GURKAN, O., HAMBURGER, M., HURST, K., KAHLE, H., KASTENS, K., KEKELIDZE, G., KING, R., KOTZEV, V., LENK, O., MAHMOUD, S., MISHIN, A., NADARIYA, M., OUZOUNIS, A., PARADISSIS, D., PETER, Y., PRILEPIN, M., REILINGER, R., SANLI, I., SEEGER, H., TEALEB, A., TOKSOZ, M.N., and VEIS, G. (2000), *Global positioning system constraints on plate kinematics and dynamics in the eastern Mediterranean and Caucasus*, J. Geophys. Res. *105*, 5695–5719.

MCCLUSKY, S., REILINGER, R., MAHMOUD, S., BEN-SARI, D., and TEALEB, A. (2003), *GPS constraints on Africa (Nubia) and Arabia plate motions*, Geophys. J. Intern. *155*, 126–138.

McKenzie, D. P. (1970), *The plate tectonics of the Mediterranean region*, Nature 226, 271–299.

McKenzie, D. P. (1972), *Active tectonics of the Mediterranean region*, Geophys. J. R. Astron. Soc. 30, 109–185.

Mihailovic, J. (1927), *The large seismic disasters around the Sea of Marmara (in French)*, Inst. Seismol. De l' Univ. de Belgrade, Belgrade, Yugoslavia.

Mihailovic, J. (1933), *The seismicity of the Sea of Marmara and Asia minor (in French)*, Monogr. Trav. Sci. Inst. Seismol., 2B.

Nalbant, S. S., Hubert, A., and King, G. C. P. (1998), *Stress coupling between earthquakes in northwest Turkey and the north Aegean Sea*, J. Geophys. Res. 103, 469–486.

Nishenko, S. P. and Buland R. (1987), *A generic recurrence interval distribution for earthquake forecasting*, Bull. Seismol. Soc. Am. 77, 1382–1399.

Nyst, M. and Thatcher, W., (2004), *New constraints on the active tectonic deformation of the Aegean*, J. Geophys. Res. 109, B11406, doi:10.1029/2003JB002830.

Okada, Y. (1992), *Internal deformation due to shear and tensile faults in a half-space*, Bull. Seismol. Soc. Am. 82, 1018–1040.

Pacheco, J. F. and Sykes, L. R. (1992), *Seismic moment catalogue of large shallow earthquakes, 1900 to 1989*, Bull. Seismol. Soc. Am. 82, 1306–1349.

Palyvos, N., Pantosti, D., Zabci, C., and D' Addezio, G. (2007), *Paleoseismological evidence of recent earthquakes on the 1967 Mudurnu valley earthquake segment of the North Anatolian Fault zone*, Bull. Seismol. Soc. Am. 97, 1646–1661.

Pantosti, D., Pucci, S., Palyvos, N., De Martini, P. M., D' Addezio, G., Collins, P. E. F., and Zabci, C. (2008), *Paleo-earthquakes of the Düzce fault (North Anatolian Fault Zone): Insights for large surface faulting earthquake recurrence*, J. Geophys. Res. 113, doi:10.1029/2006JB004679.

Papadimitriou, E. E. and Sykes, L. R. (2001), *Evolution of the stress field in the northern Aegean Sea (Greece)*, Geophys. J. Int. 146, 747–759.

Papadimitriou, E.E., Karakostas, V., and Papazachos, B. C. (2001), *Rupture zones in the area of the 17.08.99 Izmit (NW Turkey) large earthquake (M_w 7.4) and stress changes caused by its generation*, J. Seismol. 5, 269–276.

Papadimitriou, E. E., Sourlas, G., and Karakostas, V. G. (2005), Seismicity variations in southern Aegean, Greece, before and after the large (M_w 7.7) 1956 Amorgos earthquake due to evolving stress, Pure Appl. Geophys. 162, 783–804.

Papazachos, B. C. and Comninakis, P. E. (1969), *Geophysical features of the Greek Island Arc and Eastern Mediterranean Ridge*, Com. Ren. Séances Conf. Reunie Madrid, 16, 74–75.

Papazachos, B. C. and Comninakis, P. E. (1971), *Geophysical and tectonic features of the Aegean arc*, J. Geophys. Res. 76, 8517–8533.

Papazachos, B. C. and Papazachou, C. (2003), *The earthquakes of Greece*, Ziti publications, Thessaloniki, pp. 289.

Papazachos, B. C., Kiratzi, A. A., Hatzidimitriou, P. M., and Rocca, A. Ch. (1984), *Seismic faults in the Aegean area*, Tectonophysics 106, 71–85.

Papazachos, B. C., Kiratzi, A. A., and Karakostas, B. G. (1997), *Toward a homogeneous moment magnitude determination in Greece and surrounding area*, Bull. Seismol. Soc. Am. 87, 474–483.

Papazachos, B. C., Papadimitriou, E. E., Kiratzi A. A., Papazachos, C. B., and Louvari, E. K. (1998), *Fault plane solutions in the Aegean Sea and the surrounding area and their tectonic implications*, Bull. Geof. Teor. Appl. 39, 199–218.

Papazachos, B. C., Scordilis, E. M., Panagiotopoulos, D. G., Papazachos, C. B., and Karakaisis, G. F. (2004), *Global Relations between seismic fault parameters and moment magnitude of Earthquakes*, Bull. Geol. Soc. Greece XXXVI, 1482–1489.

Papazachos, C. (1999), *Seismological and GPS evidence for the Aegean Anatolia interaction*, Geophys. Res. Lett. 17, 2653–2656.

Parsons, T. (2004), *Recalculated probability of $M \geq 7$ earthquakes beneath the Sea of Marmara, Turkey*, J. Geophys. Res. 109, doi:10.1029/2003JB002667.

Parsons, T. (2005), *Significance of stress transfer in time–dependent earthquake probability calculations*, J. Geophys Res. 110, doi:10.1029/2004JB003190.

Parsons, T., Toda, S., Stein, R. S., Barka, A., and Dieterich, J. H. (2000), *Heightened odds of large earthquakes near Istanbul: An interaction-based probability calculation*, Science 288, 661–665.

Pinar, N. (1952), *The Yenice earthquake of the 18 March, 1953 and the fracture line of Yenice–Gönen*, Rev. Fac. Sci. Univ. Istanbul, Ser. A., 18, 131–141.

Pondard, N., Armijo, R., King, G. C. P., Meyer, B., and Flerit, F. (2007), *Fault interactions in the Sea of Marmara pull–apart (North Anatolian Fault): Earthquake clustering and propagating earthquake sequences*, Geophys. J. Int. 171, 1185–1197.

Reilinger, R. E., McClusky, S. C., Oral M. B., King, R. W., King, R. W., and Toksöz, M. N. (1997), *Global Positioning System measurements of present–day crustal movements in the Arabia–Africa–Eurasia plate collision zone*, J. Geophys Res. 102, 983–999.

Reilinger, R., McClusky, S., Vernant, P., Lawrence, S., Ergintav, S., Cakmak, R., Ozener, H., Kadirov, F., Guliev, I., Stepanyan, R., Nadariya, M., Hahubia, G., Mahmoud, S., Sakr, K., ArRajehi, A., Paradissis, D., Al-Aydrus, A., Prilepin, M., Guseva, T., Evren, E., Dmitrotsa, A., Filikov, S. V., Gomez, F., Al-Ghazzi R., and Karam, G. (2006), *GPS constraints on continental deformation in the Africa–Arabia–Eurasia continental collision zone and implications for the dynamics of plate interactions*, J. Geophys. Res. 111, doi:10.1029/2005JB004051.

Rice, J. R. and Clearly, M. P. (1976), *Some basic stress diffusion solutions fro fluid-saturated elastic porous media with compressible constituents*, Rev. Geophys. 14, 227–241.

Robinson, R. and McGinty, P. J. (2000), *The enigma of the Arthur's Pass, New Zealand, earthquake. 2. The aftershock distribution and its relation to regional and induced stress fields*, J. Geophys. Res. 105, 16139–16150.

Rockwell, T., Barka, A., Dawson, T., Akyuz, S., and Thorup, K. (2001), *Paleoseismology of the Gazikoy-Saros segment of the North Anatolia fault, northwestern Turkey: Comparison of the historical and paleoseismic records, implications of regional seismic hazard, and models of earthquake recurrence*, J. Seismology 5, 433–448.

Saunders, P., Priestley, K., and Taymaz, T. (1998), *Variations in the crustal structure beneath western Turkey*, Geophys. J. Intern. 134, 373–389.

Seyitoğlu, G. and Scott, B. C. (1991), *Late Cenozoic crustal extension and basin formation in west Turkey*, Geol. Mag. 128, 155–166.

Shirokova, E. (1972), *Stress pattern and probable motion in the earthquake foci of the Asia-Mediterranean seismic belt, in elastic strain field of the Earth and mechanisms of earthquake sources*. In L. M. Balakina et al. (eds.), Nauka, Moscow, 8.

STEIN, R. S. (1999), *The role of stress transfer in earthquake occurrence*, Nature *402*, 605–609.

STEIN, R. S., BARKA, A. A., and DIETERICH, J. H. (1997), *Progressive failure on the North Anatolian fault since 1939 by earthquakes stress triggering*, Geophys. J. Int. *128*, 594–604.

STEWART, S. G. and KANAMORI, H. (1982), *Complexity of rupture in large strike–slip earthquakes in Turkey*, Phys. Earth Planet. Inter. *28*, 70–84.

TAYMAZ, T. (1990), *Earthquake source parameters in the Eastern Mediterranean Region*, Ph.D. Thesis, Darwin College, University of Cambridge, UK, 244 pp.

TAYMAZ, T. (1996), *S–P–wave travel-time residuals from earthquakes and lateral heterogeneity in the upper mantle beneath the Aegean and the Hellenic Trench near Crete*, Geophys. J. Int. *127*, 545–558.

TAYMAZ, T., and PRICE, S. (1992), *The 1971 May 12 Burdur earthquake sequence, SW Turkey: A synthesis of seismological and geological observations*, Geophys. J. Int. *108*, 589–603.

TAYMAZ, T., JACKSON, J. A., and MCKENZIE, D. P. (1991a), *Active tectonics of the North and Central Aegean Sea*, Geophys. J. Int. *106*, 433–490.

TAYMAZ, T., EYIDOĞAN, H., and JACKSON, J. (1991b), *Source parameters of large earthquakes in the east Anatolian fault zone (Turkey)*, Geophys. J. Int. *106*, 537–550.

TODA, S., STEIN, R. S., REASENBERG, P. A., DIETERICH, P. H., and YOSHIDA, A. (1998), *Stress transfer by the 1995 $M_w = 6.9$ Kobe, Japan, shock: Effect of aftershocks and future earthquakes probabilities*, J. Geophys. Res. *103*, 24,543–24,565.

WELLS, D. L. and COPPERSMITH, K. J. (1994), *New empirical relationships among magnitude, rupture length, rupture width, rupture area and surface displacement*, Bull. Seismol. Soc. Am. *84*, 972–1002.

WESSEL, P. and SMITH, W. H. F. (1998), *New improved version of the Generic Mapping Tools Released*, EOS Trans. AGU *79*, 579.

WESTAWAY, R. (1990), *Block rotation in Western Turkey, 1. Observational evidence*, J. Geophys. Res. *95*, 19857–19884.

WESTAWAY, R. (1993), *Neogene evolution of the Denizli region of western Turkey*, J. Struct. Geol. *15*, 37–53.

(Received August 21, 2008, revised April 24, 2009, accepted June 18, 2009, Published online March 10, 2010)

ical rate changes associated with major earth-

Correlation of Static Stress Changes and Earthquake Occurrence in the North Aegean Region

D. A. Rhoades,[1] E. E. Papadimitriou,[2] V. G. Karakostas,[2] R. Console,[3] and M. Murru[3]

Abstract—A systematic analysis is made of static Coulomb stress changes and earthquake occurrence in the area of the North Aegean Sea, Greece, in order to assess the prospect of using static stress changes to construct a regional earthquake likelihood model. The earthquake data set comprises all events of magnitude $M \geq 5.2$ which have occurred since 1964. This is compared to the evolving stress field due to constant tectonic loading and perturbations due to coseismic slip associated with major earthquakes ($M \geq 6.4$) over the same period. The stress was resolved for sixteen fault orientation classes, covering the observed focal mechanisms of all earthquakes in the region. Analysis using error diagrams shows that earthquake occurrence is better correlated with the constant tectonic loading component of the stress field than with the total stress field changes since 1964, and that little, if any, information on earthquake occurrence is lost if only the maximum of the tectonic loading over the fault orientation classes is considered. Moreover, the information on earthquake occurrence is actually increased by taking the maximum of the evolving stress field since 1964, and of its coseismic-slip component, over the fault orientation classes. The maximum, over fault orientation classes, of linear combinations of the tectonic loading and the evolving stress field is insignificantly better correlated with earthquake occurrence than the maximum of the tectonic loading by itself. A composite stress-change variable is constructed from ordering of the maximum tectonic loading component and the maximum coseismic-slip component, in order to optimize the correlation with earthquake occurrence. The results indicate that it would be difficult to construct a time-varying earthquake likelihood model from the evolving stress field that is more informative than a time-invariant model based on the constant tectonic loading.

Key words: Earthquake prediction, static stress changes, Greece.

1. Introduction

Coseismic stress changes in the vicinity of strong earthquakes suggest that perturbations of 0.1–1 bar may affect the occurrence of other earthquakes. Changes in the occurrence rate of local and regional seismicity (Toda and Stein, 2003; Toda et al., 2005; Mallman and Zoback, 2007), as well as observed clustering of strong earthquakes (Papadimitriou and Karakostas, 2003; Papadimitriou et al., 2004), suggest that failure on one fault may affect earthquake occurrence on another fault, with changes to the static stress field being an obvious physical mechanism (Stein et al., 1997). Detailed studies of stress changes and seismicity following the occurrence of major earthquakes provide a body of anecdotal evidence that the location of aftershocks, ensuing major events and other changes in seismicity patterns in the vicinity of a major earthquake can often be explained by changes in the static stress field resulting from coseismic slip associated with the major earthquake (e.g., King et al., 1994a; Deng and Sykes, 1997; Harris, 1998 and references therein; Robinson and McGinty, 2000; Papadimitriou and Sykes, 2001; Steacy et al., 2005 and references therein). Coseismic stress changes have been incorporated as an important component in time-dependent probabilistic hazard assessment models (Stein et al., 1997; Hardebeck, 2004; Michael, 2005; Parsons, 2005; among others), and poroelasticity effects and post-earthquake relaxation associated with coseismic stress transfer have been introduced to account for the spatiotemporal distribution of aftershocks (Cocco and Rice, 2002; Pollitz et al., 2006; Perfettini and Avouac, 2007; Savage, 2007).

A previous study, in a wider region of Greece, compared the evolving stress field and precursory scale

---

[1] GNS Science, Lower Hutt, New Zealand. E-mail: d.rhoades@gns.cri.nz
[2] Geophysics Department, University of Thessaloniki, 54124 Thessaloniki, Greece.
[3] Istituto Nazionale di Geofisica e Vulcanologia, Rome, Italy.

increase approaches to long-term seismogenesis (PAPADIMITRIOU et al., 2006). It was found that recent major earthquakes are largely consistent with both approaches, and also that the evolving stress field was already positive for the occurrence of a major earthquake before the onset of the precursory scale increase, i.e., a long time (years to decades) before the actual time of the earthquake. This is further anecdotal evidence that the evolving stress field can provide an explanation for temporal and spatial fluctuations in seismicity.

Here we attempt to advance these studies beyond the anecdotal stage by systematically comparing the evolving stress field and earthquake occurrences in an extended region over an extended period of time. The goal is to enable the use of static stress changes to construct a regional earthquake likelihood model (FIELD, 2007). For the evolving stress field calculations a purely elastic model is used that takes into account both the coseismic slip of the stronger events and the long-term tectonic loading on the major regional faults. Moreover, the stress field is calculated each time according to the faulting type of the target fault. This model has proved to be effective in predicting the locations of future earthquakes (e.g., DENG and SYKES, 1997; PAPADIMITRIOU and SYKES, 2001), while in many investigations tectonic loading is not included and assumptions are made about the directions and magnitudes of regional stresses. An intermediate step attempted here is to establish the level of correlation between static Coulomb stress changes and seismicity. The North Aegean Sea region in Greece is selected for this investigation because it has an adequate number of strong ($M \geq 6.4$) earthquakes which are included in the stress evolutionary model, whose coseismic slip is considered to perturb the evolving stress field, along with an adequate number of moderate ($M \geq 5.2$) events which are inspected for triggering. Our data sample starts at 1964, from which time the location of earthquakes became more accurate, and the determination of focal mechanisms is more reliable for the stronger events and available for many of the smaller magnitude ones.

2. Data and Methods

The North Aegean study region covers the latitude range 38.3°–40.5°N. and longitude range 23.5°–26.5°E. All earthquakes with $M \geq 5.2$ in the Aristotle University of Thessaloniki (AUTH) catalogue since 1964 (67 events) are included (see Table 3 of Appendix for a list), and all earthquakes with $M \geq 6.4$ (8 events) are considered as contributing to the stress field perturbations. The threshold of 6.4 is chosen because the coseismic slip of such events is sufficiently large to disturb the stress field. In addition, the fault plane solutions of these stronger events have been determined by waveform modeling, and the only other event with $M \geq 6.0$ in our catalogue is the one in 1965 with M 6.1 (see Table 3 of Appendix). The 67 earthquake locations, and available focal mechanisms for 27 events, are shown in Fig. 1. For 40 earthquakes the focal mechanism is unknown and must be inferred from that of nearby earthquakes, albeit with some uncertainty.

When searching for a potential correlation between static stress changes and seismicity changes, one approach is to calculate these changes for the nodal planes of the subset of shocks with known focal mechanisms (STEIN, 1999). Since the stress field depends on the fault orientation, it is necessary to calculate the stress field for a representative set of fault orientation classes which cover all the earthquakes in the catalogue. From a computational perspective, the number of classes should be as small as possible. The distribution of strike angles, dip angles and rake angles in the 27 known focal mechanisms is shown in Fig. 2. From these distributions, it was possible to divide the strike angles into five groups, the dip angles into three groups, and the rake angles into five groups. In this division the M 6.6 earthquake of 1967 March 4, of oblique normal faulting, formed a group of its own in both strike angle and rake angle. All the known focal mechanisms were found to be contained in only 15 of the 75 resulting possible classes for combinations of strike angle, dip angle and rake angle groups. However, a 16th class was included, in which no earthquakes in the current database fall, to allow for the possibility, however unlikely, of earthquakes occurring with very different focal mechanisms from those observed to date. Table 1 shows how the 15 classes were derived from combinations of ranges of strike angle, rake angle and dip angle. Where earthquakes are observed in a particular class, the restricted range of dip angles

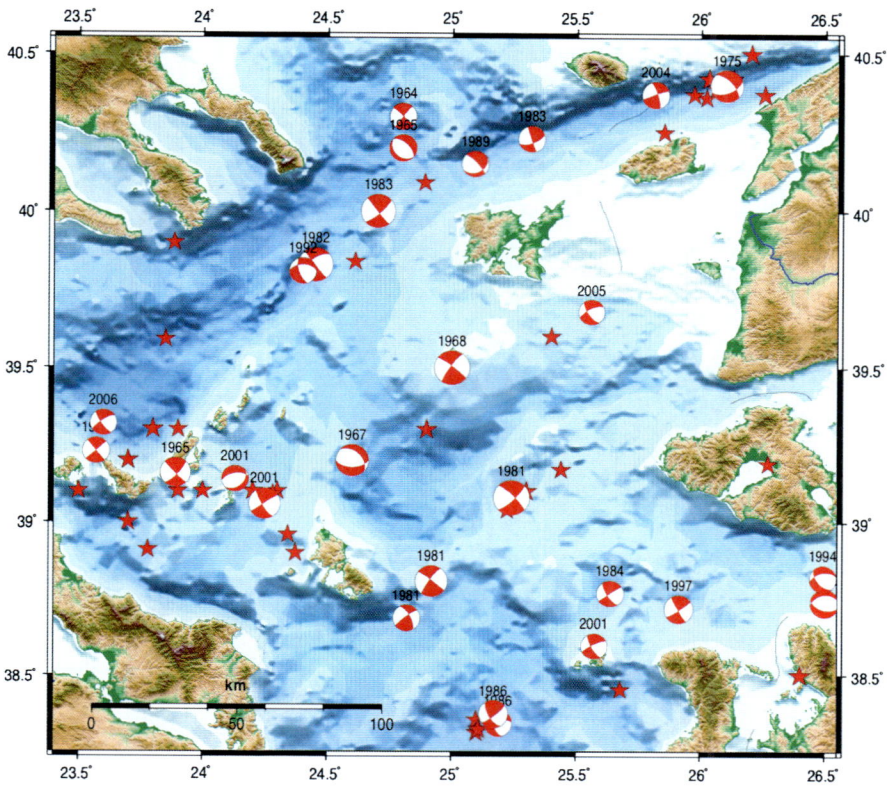

Figure 1
Map of the North Aegean study region, showing locations of 67 earthquakes with M ≥ 5.2 since 1964 and focal mechanisms where available

actually spanned by the earthquake focal mechanisms in the class is shown in the column corresponding to the broader range used to define the class, and is followed in parentheses by the number of earthquake focal mechanisms observed in the class.

Figure 3 shows how the so-defined classes were used to infer fault-orientation classes for the other 40 earthquakes in the catalogue. The inference is based on observed spatial clustering of the M ≥ 5.2 events and the similarity of the known fault plane solutions among neighboring events, although disagreements have been observed in some cases. These disagreements may be partly due to the limited amount of input information for the routine determination of focal mechanisms for the smaller and moderate events. In these cases the more representative faulting type, meaning the one that is more compatible with the orientation of the regional stress, is considered as the dominant faulting pattern. The rectangles in Fig. 3 each correspond to a fault orientation class, and earthquakes without well-defined focal mechanisms located in a given rectangle are assumed to belong to the same fault orientation class as the earthquakes with known focal mechanisms in the same rectangle. For each fault orientation class, the faulting type is represented by average values of the strike, rake and dip angles, as given in Table 2. There are only three isolated earthquakes which cannot be assigned to any fault orientation class. These are linked to locations with no historical or instrumental recordings of strong (M ≥ 6.0) events and the seismicity is sparse, consequently faults cannot easily be identified.

3. Stress Calculations

The evolving stress field is considered to have two main components—a constantly accumulating component due to tectonic loading on the major faults in

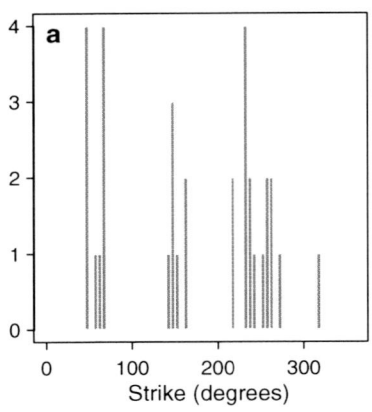 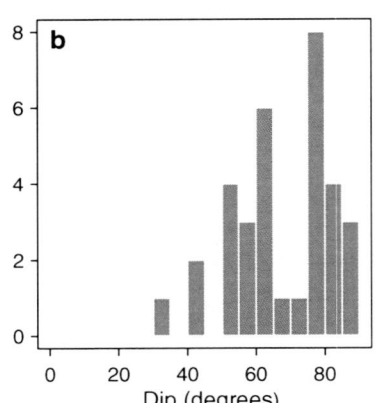 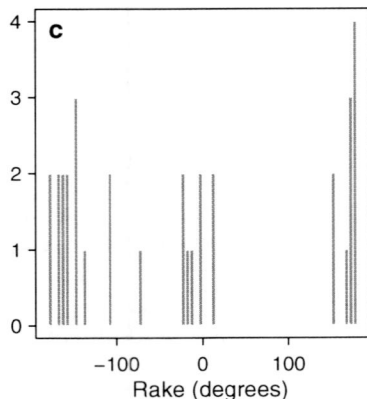

Figure 2
Histograms of **a** strike angle, **b** dip angle, and **c** rake angle, for earthquakes with determined fault plane solutions in the study region

Table 1
Fault orientation classes and number of earthquakes in each (in parentheses)

Strike angle range	Rake angle range	Dip angle range		
		30°–45°	50°–70°	70°–90°
45°–70°	−177° to −135°	34° (1)	55° (1)	77° (1)
45°–70°	−116°	37° (1)		
45°–70°	175°–177°		64° (1)	75°–83° (3)
130°–165°	−22° to 15°		59°–63° (2)	74°–76° (3)
215°–240°	−167° to −161°		62° (1)	89° (1)
215°–240°	153°–179°			79°–89° (4)
250°–275°	−156° to −108°	41° (1)	51°–68° (3)	
250°–275°	168°			85° (1)
313°	−56°	43° (1)		

the region, and a component consisting entirely of jumps due to coseismic slip accompanying the major earthquakes (DENG and SYKES, 1997). Interseismic stress accumulation between the strong events is modeled by "virtual negative displacements" along major faults in the entire region under study, using the best available information on long-term slip rates. These virtual dislocations are imposed on the faults with the sense of slip opposite to the observed slip. The magnitude is incremented according to the long-term slip rate of the fault. This virtual negative slip is equivalent to constant positive slip extending from the bottom of the seismogenic layer to infinite depth. Hence, tectonically induced stress builds up in the vicinity of faults during the time intervals between earthquakes. All computed interseismic stress accumulation is associated with the deformation caused by the time-dependent virtual displacement on major faults extending from the free surface up to the depth at which earthquakes and brittle behavior cease (∼15 km).

The major regional faults in our study area, which accommodate strain accumulation culminating in earthquake occurrence, are mainly submarine and therefore field information on their properties is sparse. Recent seismic activity for which hypocentral determinations are available is used to define these fracture lines, and their strike, dip and rake are defined according to the reliable fault solutions of the stronger (M ≥ 6.0) events associated with them (Fig. 4). It is possible to estimate slip rates for these faulting lines directly from the relative motions between GPS stations straddling them. Such information is available from MCCLUSKY et al. (2000) and REILINGER et al. (2006), who interpreted geodetic measurements of crustal motions. The latter authors

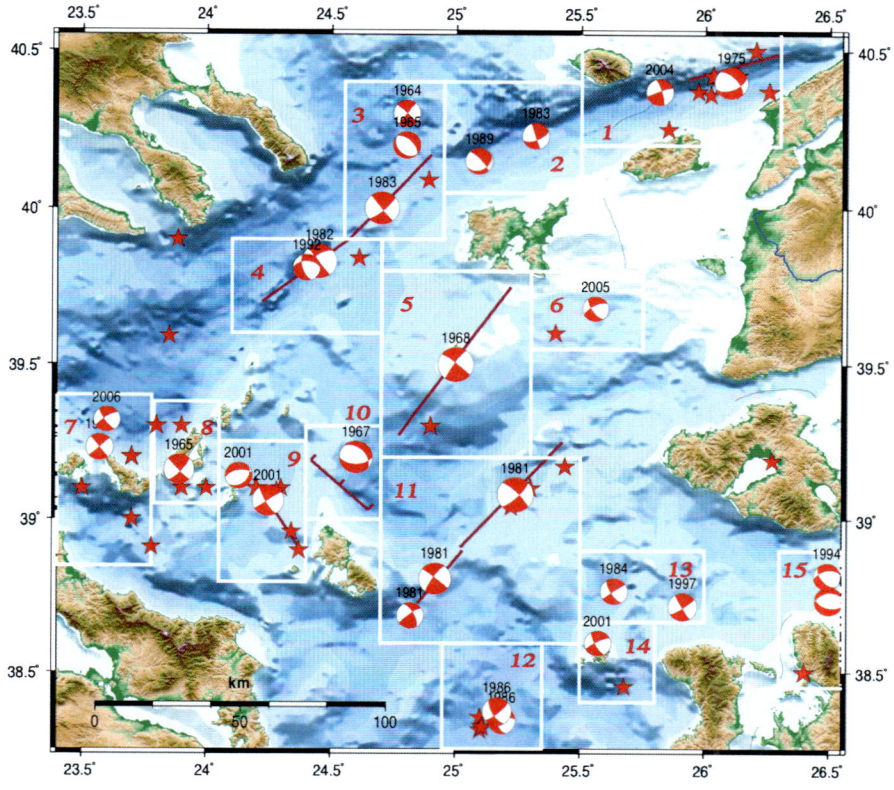

Figure 3
Map of the study region showing rectangles for grouping earthquakes into fault orientation classes

Table 2

Representative strike, dip and rake angles for fault orientation classes

Class number	Strike angle	Dip angle	Rake angle
1	65	55	−145
2	65	55	−165
3	50	76	177
4	233	62	−177
5	216	81	173
6	244	68	156
7	144	76	−15
8	44	75	175
9	148	76	−1
10	313	43	−56
11	47	77	−167
12	156	60	−5
13	60	85	−170
14	151	74	−12
15	260	50	−120
16	80	25	90

used a simple kinematic block model, including elastic strain accumulations on the block-bounding faults, to quantify relative block motions and to determine present-day rates of the strain accumulation on the block bounding faults. Based on the above, the long-term slip rate for each of the faulting lines is defined approximately, so that their sum is in accordance with the generally accepted motion. We assumed a total of 24 mm/year of right-lateral slip, placing a large part of this motion (12 mm/year) on the northern branch and distributing the rest along the four other parallel branches, reducing the amount of slip from north to south. For the left-lateral faults a total of 10 mm/year is assumed. The slip rate values we selected are also in agreement with ARMIJO *et al.* (2003) who incorporated both the geodetic and geological constraints, providing a description of the present day deformation of the Anatolian–Aegean region. They use in their model localized deformation zones which are represented by dislocation elements and extended from the base of the lithosphere to the locking depth at the base of the seismogenic layer. The values of slip rates we adopted are equal to 60% of the geodetically determined ones in order to account for the seismically released strain energy. This choice is based on previous investigations, for example JACKSON et al. (1994) who concluded that seismicity can account for at most 50% of the deformation in the Aegean area, and KING et al. (2001) who for the area of the North Anatolian Fault found that the rate of moment release accounts for about 60% of the relative plate motion. Nevertheless, more accurate long-term slip rates for each fault that contributes to the total plate motion will lead to better estimates.

Stress changes associated with both the virtual dislocations and actual earthquake displacements are calculated for an isotropic elastic half space (ERIKSON, 1986; OKADA, 1992) at a depth of 8 km. This depth, the choice of which is not very critical since the faults are almost vertical, was chosen to be several kilometers above the locking depth (15 km) in the evolutionary model. This is the mean of the centroid depths of the stronger events included in our evolutionary model and in agreement with KING et al. (1994b) who found that seismic slip peaks at mid-depths in the seismogenic layer, and thus deformation must be localized on the faults at these depths. The seismogenic layer in our calculations is taken to extend between 3 and 15 km, based on the centroid depths derived from waveform inversions (6–15 km, mostly) and the focal parameters of accurately relocated aftershocks (e.g., PAPSAZACHOS et al. 1984; ROCCA et al., 1985). The shear modulus and Poisson's ratio are fixed at 33 and 0.25 GPa, respectively. The selection of the value of the apparent coefficient of

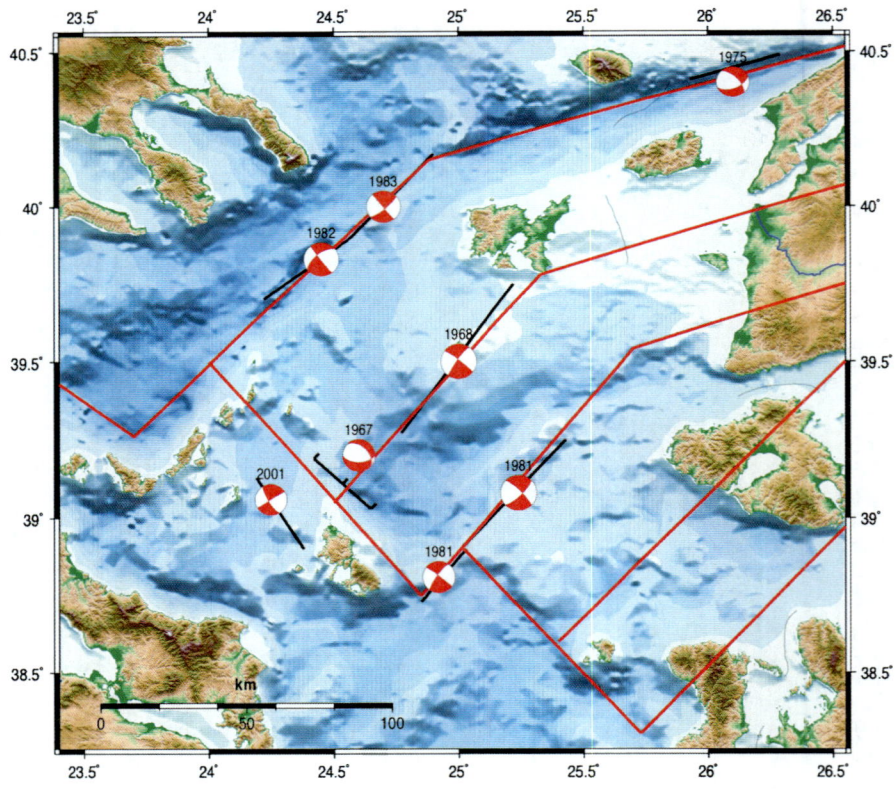

Figure 4
Map of the study area showing major earthquake focal mechanisms and associated faults, and major fracture lines on which the tectonic loading is assumed to accumulate

friction, μ', is based on previous results. A value of μ' equal to 0.4 was chosen and considered adequate throughout the calculations, as previous investigations and pertinent tests have revealed (KING et al., 1994a; PAPADIMITRIOU, 2002).

The annual Coulomb stress change in the absence of fault movement is calculated based on the slip rate, and is resolved for each of the 16 fault orientation classes. This tectonic component of the evolving stress field is illustrated in Fig. 5 for the faulting types given in Table 2. The stress field was calculated according to the faulting type assigned to each class, and must be viewed in the context of this specific style of fault slip, i.e., strike, dip and rake. This is because stress is a tensorial quantity which changes in space according to the observational plane and sense of slip. As can be seen, the spatial patterns for some of the orientation classes are quite similar, due to relatively slight differences between the faulting types that the classes represent. The jumps in the stress field due to coseismic slip accompanying the eight major ($M \geq 6.4$) earthquakes since 1964 are illustrated in Fig. 6, in which we show the Coulomb stress field change for the actual fault orientation of each earthquake. Combining these jumps with the tectonic component allowed us to calculate the total change in the Coulomb stress field due to tectonic loading on the major faults and the coseismic slip associated with major earthquakes from the beginning of the catalogue up to just before the occurrence time of any earthquake with $M \geq 5.2$. The evolving stress field is then calculated according to the faulting type assigned to the box inside which the earthquake is located.

All stress field components are calculated on a rectangular grid with 5 km steps. The grid cells are comparable in size to the source area of an earthquake of M 5.4, and larger than that for M 5.2 (WELLS and COPPERSMITH, 1994). However, the contributions to the stress field calculated here have only larger-scale features, so that the values at intermediate points, in particular at the epicenters of $M \geq 5.2$ earthquakes, can be well approximated by interpolation from the grid points. Therefore, the grid spacing used here is adequate for the purpose.

It is the actual stress field that affects earthquake occurrence. The change in the stress field over a period of time is not necessarily a good measure of the actual stress field at the end of the period, unless the stress field was uniform at the beginning of the period. The constant tectonic forcing component has been long contributing to the stress field, and therefore the large-scale features of the actual stress field at any time should resemble it in some ways, although the field is modified by every earthquake that occurs, and the effects of earthquakes that occurred prior to the beginning of the catalogue are unknown. The actual stress field at any time cannot be calculated from the available components. However if we were to attempt to construct something that would approximate it, there is no reason to begin the tectonic loading contribution only at the beginning of the catalogue. Equally, there is no reason to begin it at any other time, whether 10, 50 or 500 years prior to the start of the catalogue. In seeking to define a stress variable that is well correlated with earthquake occurrence, we need therefore to consider various combinations of the tectonic loading component and the coseismic slip component of the evolving stress field.

4. Correlation of Stress Changes and Earthquake Occurrence

In what follows, we denote the annual tectonic stress rate by R, the coseismic slip component of the evolving stress field by S, and the total evolving stress field since the beginning of 1964 by ESF. All of these variables are resolved for the 16 fault orientation classes, evaluated on a grid with 5-km spacing, and interpolated to intermediate values. S and ESF can be accumulated from 1964 up to any time of interest, and in particular up to the times of occurrence of $M \geq 5.2$ earthquakes.

An error diagram (MOLCHAN, 1990, 1991) is a useful tool for exploring the relation between earthquake occurrence and any scalar variable defined on the domain of possible times and locations of earthquake occurrence. In an error diagram, the x axis represents the proportion of space or space-time in which the scalar variable exceeds some value. The y axis represents the proportion of earthquakes that occur at times and locations when the scalar variable

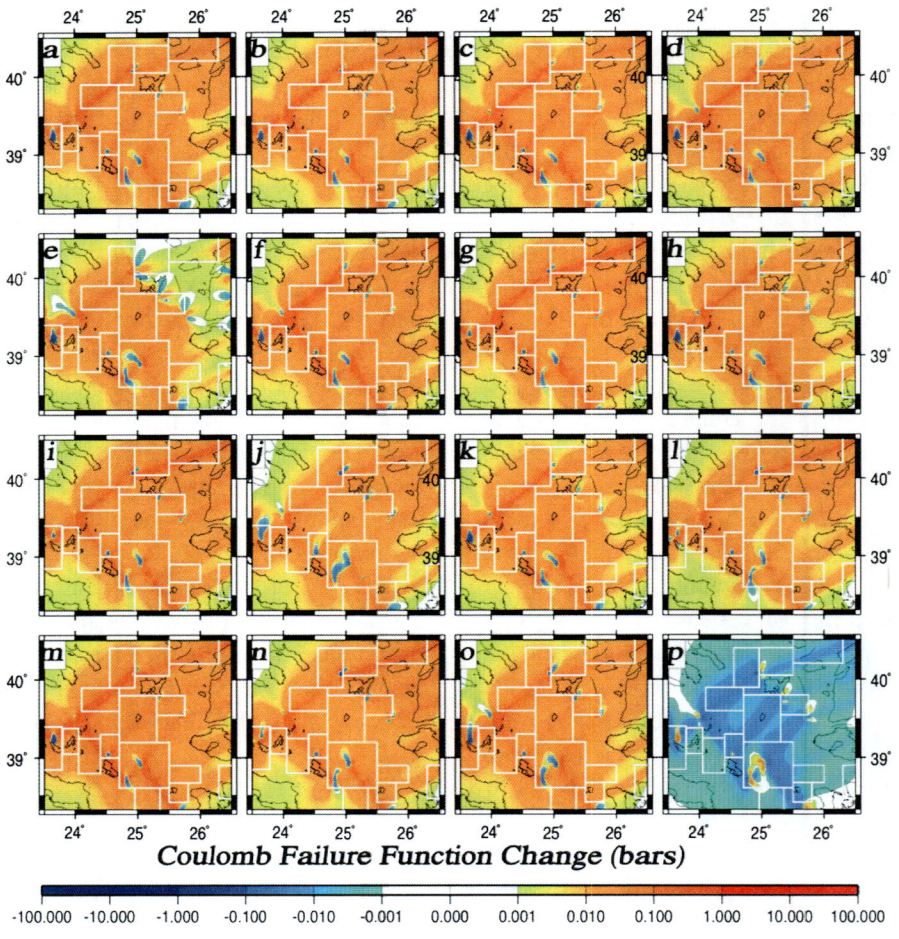

Figure 5
Annual Coulomb stress changes associated with the tectonic loading on the major regional faults. The stress pattern is calculated for each one of the 16 different faulting types (See Table 2). The *color scale* in the *bottom* gives the changes in stress in *bars*

does not exceed the same value. The error diagram is generated from a dense set covering the full range of possible values of the measured variable, with each point in the set contributing a point on the graph. The actual value of the scalar variable is unimportant; the error diagram is the same for any order-preserving transformation of its values (ZECHAR and JORDAN, 2008). If the strategy for declaring an earthquake alarm is that the scalar variable should exceed some value, then the corresponding point on the error diagram shows, on the x axis, the proportion of space time occupied by alarms and, on the y axis, the proportion of unpredicted earthquakes using this strategy. Two points on the error diagram are fixed, irrespective of the variable used: the point ($x = 0$, $y = 1$) where there are no alarms and therefore all earthquakes are unpredicted, and the point ($x = 1$, $y = 0$) where there are continuous alarms everywhere and therefore no earthquakes are unpredicted. Alarm strategies with no prediction skill are represented by the diagonal joining these two fixed points. If the error diagram lies close to this diagonal, there is little or no correlation between the scalar variable and earthquake occurrence. Skilful strategies are represented by points below the diagonal; if the error diagram lies predominantly below the diagonal, there is a positive correlation between the scalar variable and earthquake occurrence. If the error diagram lies above the diagonal, there is a negative correlation. The area above the error diagram curve has been called the area skill score (ASS) by ZECHAR and JORDAN (2008), and it is used here as a numerical index of the correlation. A value of ASS = ½ corresponds to no correlation between the scalar variable and earthquake occurrence, ASS = 1 to a perfect positive correlation and ASS = 0 to a perfect negative correlation.

Figure 7a shows the error diagram for the tectonic loading rate R, taking the fault orientation class into account. The proportion of space occupied is estimated from a synthetic earthquake catalogue with earthquakes distributed randomly according to a uniform distribution in time and space, and randomly assigned with equal probability to one of the 15 fault-orientation classes to which past earthquakes belong. The dotted lines show the 95% tolerance limits for alarm strategies with no skill, so the envelope between these limits is a zone of insignificant deviation from the diagonal (ZECHAR and JORDAN, 2008). The fact that the error diagram for R is outside and below this zone of insignificance shows that R is significantly correlated with earthquake occurrence. This correlation could be used to construct a time-invariant likelihood model for earthquake occurrence in the North Aegean Sea region. Note that R is dependent on the faulting model, which is itself derived in part from past earthquake occurrence. Therefore such a likelihood model would embody the hypothesis that earthquakes are likely to recur on faults where they have occurred in the past, because the faults represent chronic weak zones that re-rupture in preference to the rupture of unfaulted rock.

Figure 7b is a similarly constructed error diagram for ESF, except that now the stress varies with time as well as location and fault-orientation class. This shows a rather mixed picture and a lower ASS value than Fig. 7a. At the high end of the scale the graph lies below the zone of insignificance, showing that ESF is correlated with earthquake occurrence, however, at the low end of the scale (corresponding to low values of ESF), the graph is above the diagonal and touches the upper limit of the zone of insignificance, indicating a weak negative correlation with earthquake occurrence. These contrasting correlations indicate that very high and very low values of ESF are both associated with an increased likelihood of earthquake occurrence. The low values of ESF are actually quite strongly negative as seen in Fig. 8, which shows histograms of ESF values for the actual and random catalogues. The negative values at the low end of the distribution of ESF (Fig. 8a) are responsible for the excursion of the error diagram (Fig. 7b) above the diagonal. These negative values are probably due to unknown factors affecting the analysis, such as misclassification of earthquakes into fault-orientation classes or smaller scale changes in the stress field than are accounted for here.

On the matter of misclassification, several earthquakes could not be placed in a particular class, and actual fault plane solutions are available for less than half of the earthquakes in the catalogue. There is therefore some degree of uncertainty in the majority of the assignments of earthquakes to classes. Also, from a point of view of earthquake hazard, there is usually more interest in knowing the time and location of future earthquakes than the details of their fault orientation. The likelihood of an earthquake occurring at a given location is possibly more closely related to the maximum of the stress field over all classes at that location than to the value in any particular class. Therefore, there is interest in calculating the maximum of the stress field changes over all classes, and examining the associated error diagrams.

Figure 9 is the error diagram for the maximum of R over all 16 fault orientation classes, henceforth denoted max (R), superimposed on a 95% confidence band for the error diagram for R. The fact that the graph lies mostly inside and in some places slightly below the confidence band indicates that R provides no significant information about the fault-orientation of individual earthquakes as classified here. This conclusion is reinforced by a slightly higher value of ASS for max (R) than for R. Hence, in the remainder of our analyses, we consider only the maximum of stress changes over all fault-orientation classes, and address the question of whether we can construct a composite stress variable that is better correlated with earthquake occurrence than max (R). If so, such a variable could potentially be used to construct a time-varying model of earthquake occurrence in the region, which would be more informative than a time-invariant model, constructed from max (R).

It should be noted that the confidence bands on error diagrams in this paper account for sampling

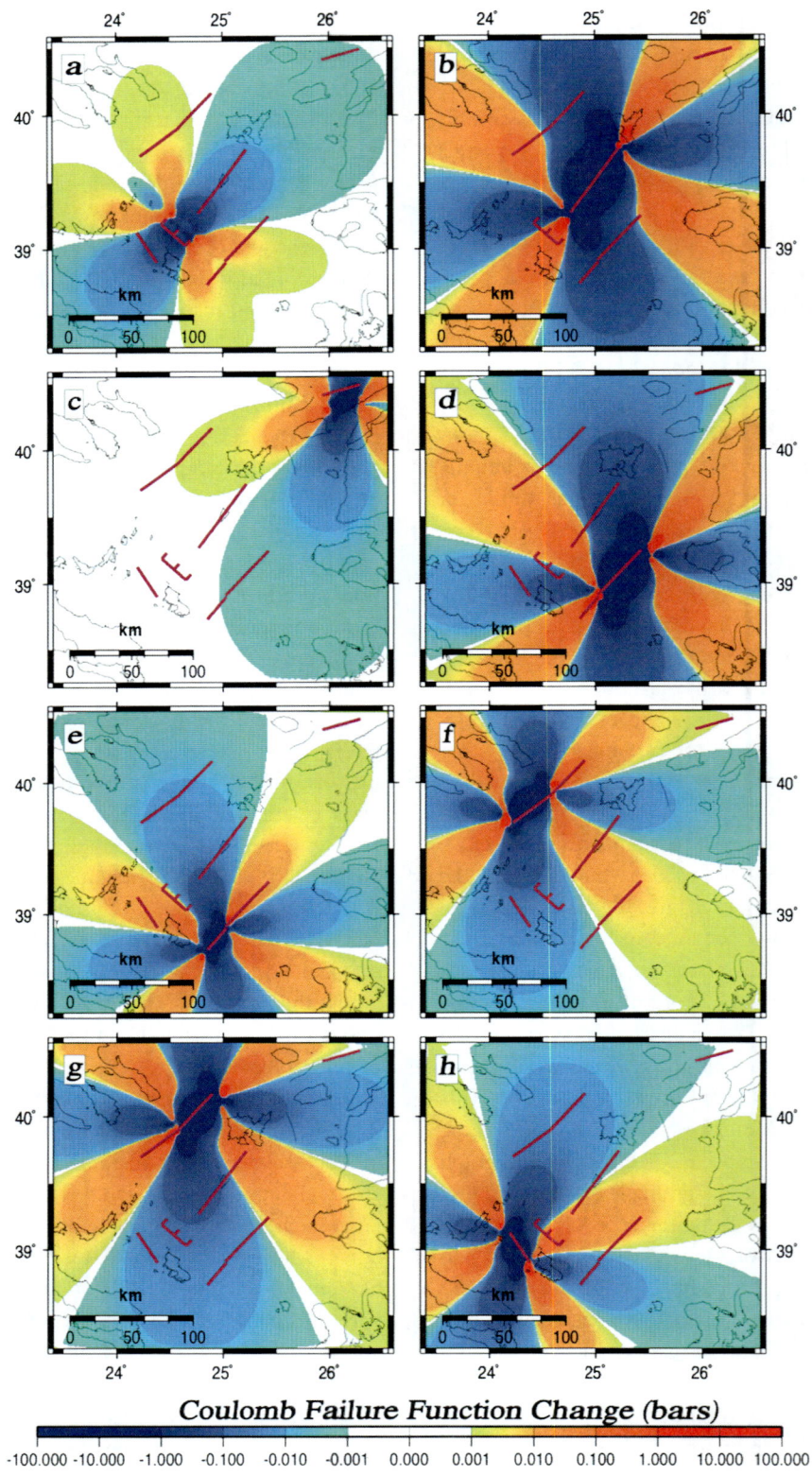

Figure 6
Coulomb stress changes associated with the coseismic slips for the eight major (M ≥ 6.4) earthquakes that occurred in the study area since 1964. The stress field is calculated according to the faulting type of the modeled event. The *color scale* in the *bottom* gives the changes in stress in *bars*. **a** 4 March 1967, M 6.3, Strike: 313, Dip: 43, Rake: −56; **b** 19 February 1968, M 7.1, Strike: 216, Dip: 81, Rake: 173; **c** 29 March 1975, M 6.6, Strike: 68, Dip: 55, Rake: −145; **d** 19 December 1981, M 7.2, Strike: 47, Dip: 77, Rake: −167; **e** 27 December 1981, M 6.5, Strike: 216, Dip: 79, Rake: 175; **f** 18 January 1982, M 7.0, Strike: 233, Dip: 62, Rake: −177; **g** 6 August 1983, M 6.6, Strike: 50, Dip: 76, Rake: 175; **h** 26 July 2001, M 6.4, Strike: 148, Dip: 76, Rake: −1

uncertainty only, and not for the uncertainties associated with the modeling of faults, calculations of stress and assignment of earthquakes to fault-orientation classes. The latter uncertainties are undoubtedly substantial, nonetheless no attempt is made here to formally estimate them.

There is no particular time at which the accumulation of stress in the evolving stress field can be assumed to begin. The present stress field is presumably affected by events in the arbitrarily distant past, including slow tectonic changes and sudden coseismic changes. We are unable to include the effects of coseismic changes prior to 1964, however we can include the effect of slow tectonic changes in the arbitrarily distant past, for as long a period as these can reasonably be assumed to be static. Therefore we considered variables constructed from the ESF since 1964 plus an arbitrary number of years of additional tectonic loading.

Figure 10 shows error diagrams for the variables max (ESF), max (ESF + 10 R), max (ESF + 30 R) and max (ESF + 100 R). In the latter three variables an extra 10, 30 and 100 years, respectively, of tectonic loading have been added to ESF. Figure 10a, when compared to Fig. 7b, shows that max (ESF) is better correlated with earthquake occurrence than ESF itself, and that the negative correlation seen for low values of ESF in Fig. 7b is no longer present, since the graph lies significantly below the diagonal for most of its length. However max (ESF) is not as well correlated with earthquakes as max (R), as can be seen by comparing Fig. 10a with Fig. 9. The error diagram for max (ESF + 10 R), shown in Fig. 10b, is much closer to that of max (R), and lies within the 95% confidence band of the latter for much of its length, although it lies partly below the band at the top end, indicating a better correlation with earthquake occurrence than max (R) in this range, and above the band for middle-range values. The error diagram for max (ESF + 30 R), shown in Fig. 10c, is closer again to that of max (R), and lies toward the

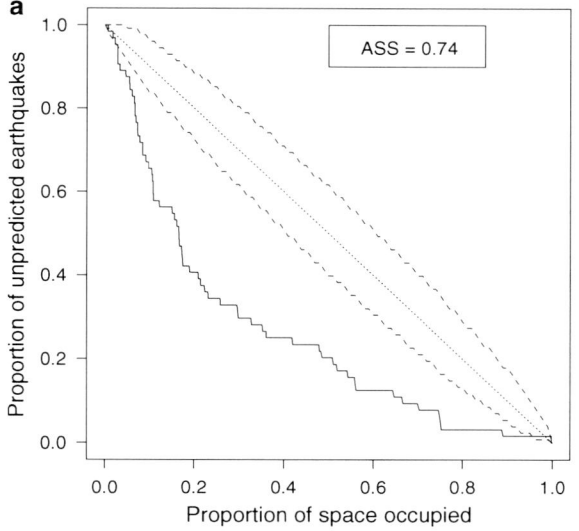
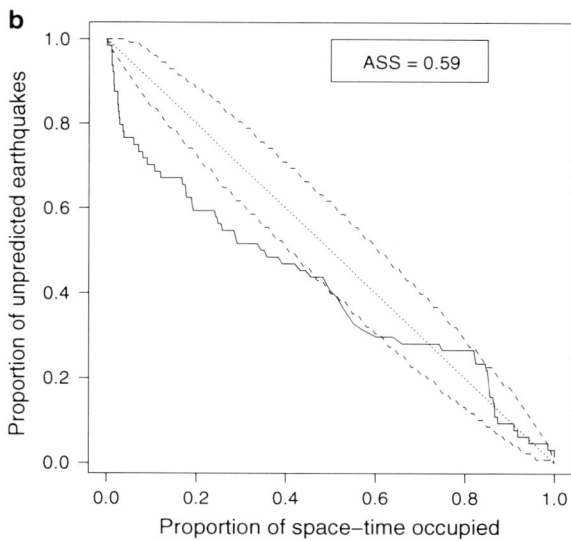

Figure 7
Error diagram for **a** annual tectonic Coulomb stress rate R and **b** evolving Coulomb stress field since 1964 (ESF) resolved into 15 fault orientation classes. For the purposes of computing the proportion of space-time occupied, all classes were given equal weighting. The dotted lines are 95% tolerance limits for alarm strategies with no skill. The area skill score (ASS) is also given

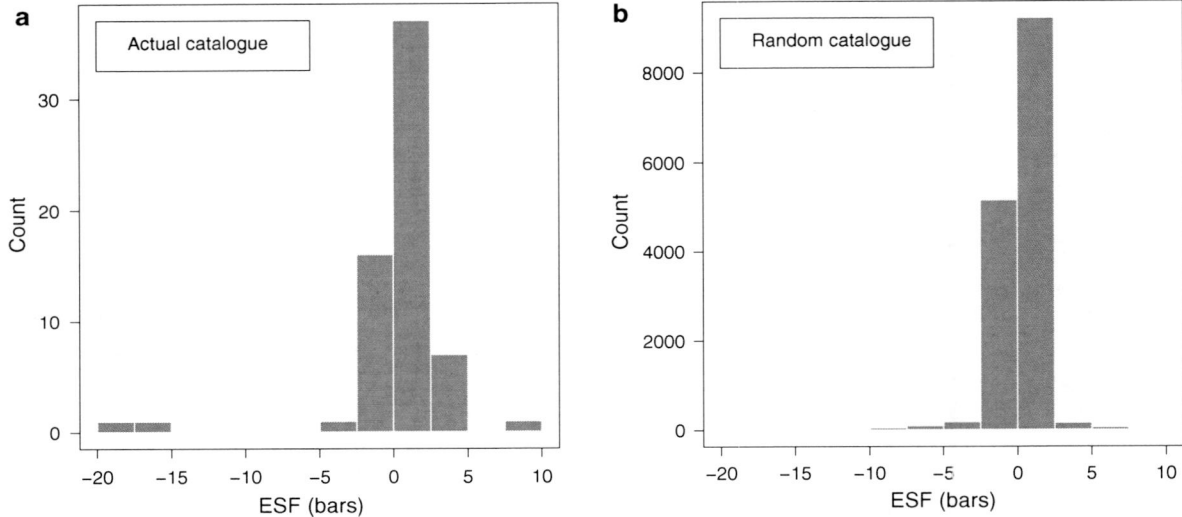

Figure 8
Histograms of **a** ESF values (*bars*) corresponding to the time of occurrence, location and fault orientation class of earthquakes in the catalogue. **b** ESF values corresponding to randomly chosen times, locations and 15 fault orientation classes

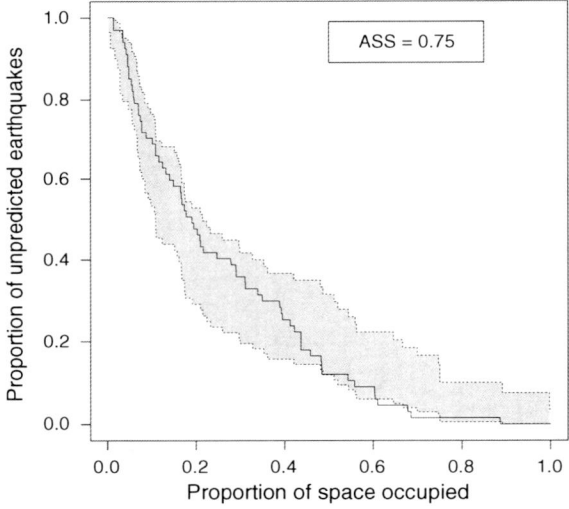

Figure 9
Error diagram for max (R), the maximum, over all 16 fault orientation classes, of the annual tectonic loading (R), and 95% confidence band of error diagram for R. The similarity of this diagram to that for R itself indicates that R contains little information on the fault orientation class of earthquakes

low end of the confidence band for max (R), though not outside of it, for a longer range at the top end. However, a section of the lower end lies above the confidence band. The diagram for max (ESF + 100 R) appears to be the best of all the error diagrams in Fig. 10, in that the ASS is highest, although no higher than that for max (R). Moreover the error diagram lies entirely within the 95% confidence band of that for max (R).

Increasing the tectonic loading beyond 100 years tends to shift the error diagram closer to that of max (R). It appears therefore that no variable of the form max (ESF + cR), where c is a positive constant, is better correlated with earthquake occurrence than max (R) itself. Therefore, it is necessary to consider other ways of defining composite statistics, which combine the earthquake-related information from the tectonic loading and coseismic-slip components of the evolving stress field. In so doing it is convenient to work with the raw variables S and R, which are independent, rather than with ESF, which is a mixture of the two. The error diagrams for S and max (S) are shown in Fig. 11. The graph for S, when compared with the zone of insignificance, shows that S is hardly correlated with earthquake occurrence, as confirmed by the ASS value of 0.47. The graph for max (S) shows a weak but marginally significant correlation with earthquake occurrence, with an ASS value of 0.57.

Contributing to this result is the fact that many of the smaller events, which can be considered as aftershocks of the main events, are located in stress

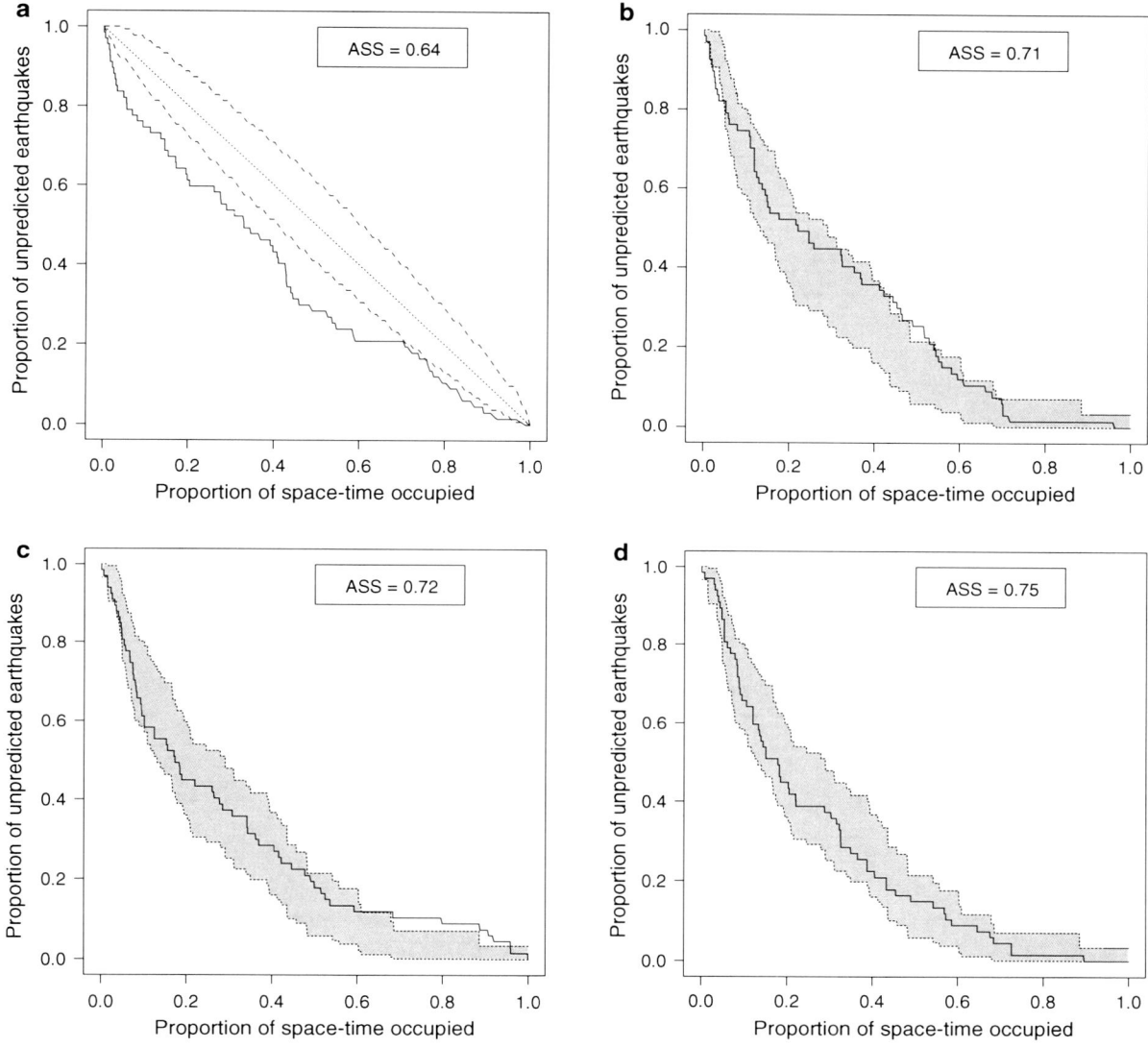

Figure 10
Error diagram for **a** max (ESF), **b** max (ESF + 10 R), **c** max (ESF + 30 R), and **d** max (ESF + 100 R). In **b–d**, the 95% confidence band of the error diagram for max (R) is also shown

shadows created by the coseismic slip of the main event. Either the present slip models of the main events are not detailed enough to predict their locations or the assignment of many of the minor events to fault orientation classes is in error.

Since it is the ordering of values that determines the error diagram, it is of interest to consider whether statistics based only on the ordering of values within components of the stress field may be more closely related to earthquake occurrence than statistics derived from linear combinations of the components. Therefore, as an alternative to the statistics of the form max (ESF + cR) discussed above, consider a composite statistic based on the ordering of values of max (S) and max (R), rather than the actual values. For a given value x of max (S), let s be the proportion of earthquakes in a random catalogue that have a lower value of max (S) than x. Likewise for a given value y of max (R), let r be the proportion of earthquakes in a random catalogue that have a lower value

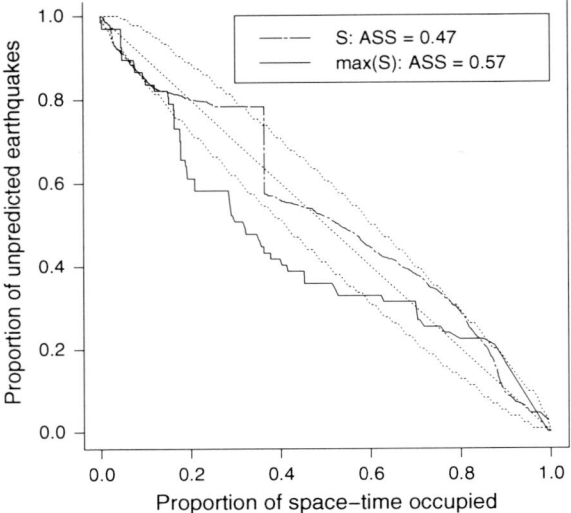

Figure 11
Error diagram for S, the coseismic contributions to the evolving stress field, and max (S), its maximum over the 16 fault orientation classes

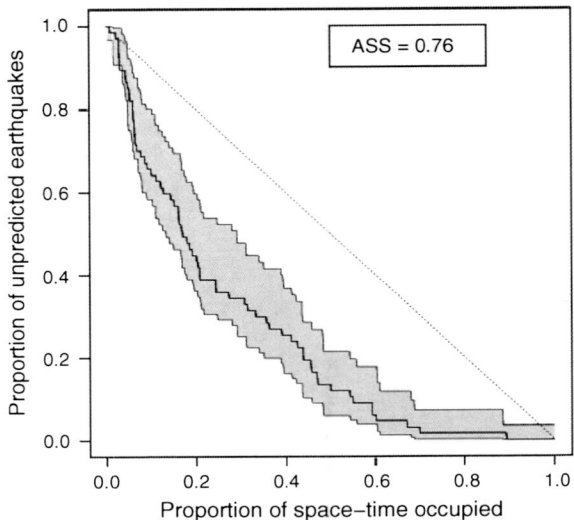

Figure 12
Error diagram for the composite statistic Q (27) (see text) compared to 95% confidence band for error diagram of max (R)

of max (R) than y. Then, if a point in space has values x and y for max (S) and max (R), respectively, we define the composite statistic

$$Q(c) = s + cr. \quad (1)$$

For $c = 0$, the ASS for $Q(c)$ is 0.57—the same as for max (S). As c is increased, the value of ASS increases until $c = 27$, then decreases gradually. For $c = 100$, the ASS is 0.75—the same as for max (R). Thus, the ASS for $Q(c)$ is maximized when $c = 27$, although the maximum value of ASS so attained exceeds that for max (R) by only 0.01. Figure 12 shows the error diagram for Q (27), compared to 95% confidence limits for the error diagram for max (R). Nowhere does the graph lie outside the confidence limits for max (R). However, near the top end it touches the lower limit. Neglecting the lack of statistical significance, we can examine the probability gain that could possibly be achieved from this statistic. Figure 13 shows the relative proportion of earthquakes predicted by Q (27) compared with that predicted by max (R) as a function of the proportion of space-time occupied. This ratio can be interpreted as a probability gain. The maximum gain of 3.5 applies to about 10% of predicted earthquakes using Q (27). Thus the advantage of using Q (27) rather than max (R) can be approximated to a probability

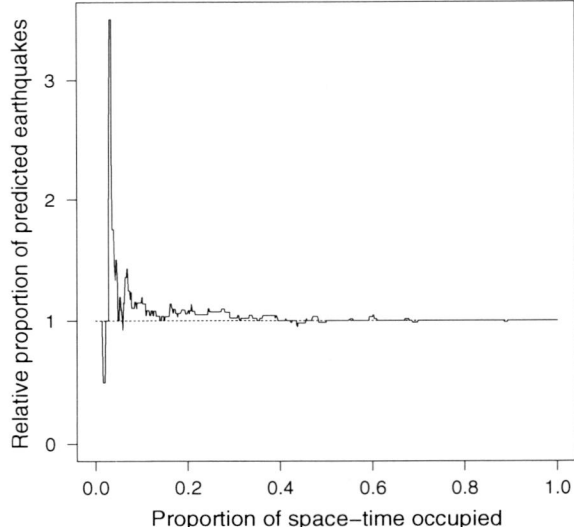

Figure 13
Proportion of earthquakes predicted by Q (27) relative to that predicted by max (R) as a function of the proportion of space-time occupied

gain of 3.5 for 10% of earthquakes and 1 for the remaining 90%. This would give a (geometric) mean probability gain per earthquake of 1.13, rather lower than existing models for long-range and short-range forecasting based only on the times, magnitudes and locations of previous earthquakes (CONSOLE et al.,

2006). Therefore, there is no indication from these data that changes in static stress could be used to produce a time-varying model of earthquake occurrence that would be significantly more informative than a time-invariant model, or as informative as existing time-varying models.

5. Conclusion

The available earthquake, fault and geodetic data have allowed the large-scale features of the coseismic-slip contribution to the evolving stress field since 1964 and the constant tectonic loading in the north Aegean Sea region to be evaluated. An analysis using error diagrams has shown that the constant tectonic stress loading and its maximum over all orientation classes are each well correlated with the location of $M \geq 5.2$ earthquakes in the region since 1964.

The maximum of the tectonic loading could be used to construct a static model of earthquake occurrence. The total evolving stress field since 1964 is less well correlated with earthquake occurrence than the tectonic loading. This agrees with KAGAN et al. (2005) who found that the most robust relationship is between the tectonic loading and the locations and mechanisms of earthquakes in southern California during 1850–2004, while the inclusion of the cumulative coseismic effects from past earthquakes did not significantly improve the correlation. Taking the maximum of the evolving stress field and that of its coseismic component over all fault orientation classes improves the correlation of these variables with earthquake occurrence. The maximum, over fault orientation classes, of linear combinations of the tectonic loading and the evolving stress field is insignificantly better correlated with earthquake occurrence than the maximum of the tectonic loading by itself. Contributing to this result is the fact that many aftershocks are located in apparent stress shadows created by the coseismic slip of the main events. This is consistent with PARSONS (2002), who found that only 61% of aftershocks could be associated with stress enhancements. It suggests that the actual stress changes resulting from the main events are more complex than those predicted by the present slip models and the assignment of many of the minor events to fault orientation classes may be in error.

However, the coseismic component of ESF contains information on the locations and times of occurrence of the larger earthquakes independent from the tectonic loading. An example has been given of a composite statistic constructed from the maximum of the tectonic loading and that of the coseismic-slip component of ESF that is slightly better correlated with earthquake occurrence than the maximum of the tectonic loading by itself. Such statistics may be useful in building time-varying earthquake likelihood models. However, with the current data, the probability gain over static models is likely to be quite small. When a larger data set becomes available, including focal mechanisms for more of the smaller earthquakes and covering a longer time-period, the coseismic-slip component of the evolving stress field is likely to provide more information toward prediction of time-varying earthquake occurrence.

Acknowledgements

Appreciation is extended to J. Ristau, R. Robinson, E. G. C. Smith, Editor M. K. Savage and an anonymous reviewer for many helpful suggestions. The stress tensors were calculated using the DIS3-D code of S. Dunbar, which was later improved by ERIKSON (1986) and the expressions of G. Converse. The GMT system (WESSEL and SMITH, 1998) was used to plot some of the figures. This research was supported by the Foundation for Research, Science and Technology under contract CO5X0402. Geophysics Department, AUTH, contribution number 741.

Appendix

See Table 3.

Table 3

Catalogue of earthquake origin times, locations, magnitudes, fault orientation classes and fault plane solutions (where available)

Date	Time (hours)	Lat.	Long.	Depth	M	Class	Strike	Dip	Rake	Reference
23 February 1964	2241	39.2	23.7	10	5.4	7				
11 April 1964	1600	40.3	24.8	33	5.5	3	220	89	179	McKenzie (1972)
29 April 1964	0421	39.2	23.7	20	5.6	7				
29 April 1964	1700	39.1	23.5	15	5.2	7				
9 March 1965	1757	39.16	23.89	7.0	6.1	8	44	75	175	Taymaz et al. (1991)
9 March 1965	1759	39.3	23.8	0.1	5.7	8				
9 March 1965	1837	39.3	23.9	33	5.2	8				
9 March 1965	1946	39.1	23.9	19	5.2	8				
13 March 1965	0408	39.1	24	11	5.3	8				
13 March 1965	0409	39	23.7	33	5.5	7				
23 August 1965	1408	40.5	26.2	33	5.6	1				
20 December 1965	0008	40.2	24.8	33	5.6	3	132	32	−90	McKenzie (1972)
4 March 1967	1758	39.2	24.6	10	6.6	10	313	43	−56	Taymaz et al. (1991)
19 February 1968	2245	39.5	25	15	7.1	5	216	81	173	Kiratzi et al. (1991)
20 February 1968	0221	39.6	25.4	8.0	5.2	6				
10 March 1968	0710	39.1	24.2	0.1	5.5	9				
24 April 1968	0818	39.3	24.9	20	5.5	5				
6 April 1969	0349	38.5	26.4	16	5.9	15				
17 March 1975	0511	40.36	26.02	15	5.3	1				
17 March 1975	0517	40.39	26.06	15	5.4	1				
17 March 1975	0535	40.38	26.1	16	5.8	1				
27 March 1975	0515	40.4	26.1	15	6.6	1	68	55	−145	Taymaz et al. (1991)
29 March 1975	0206	40.42	26.03	33	5.7	1				
14 June 1979	1144	38.74	26.5	8	5.9	15	262	41	−108	Taymaz et al. (1991)
12 November 1980	1604	39.1	24.3	0	5.3	9				
19 December 1981	1410	39	25.26	10	7.2	11	47	77	−167	Kiratzi et al. (1991)
21 December 1981	1413	39.17	25.43	10.5	5.2	11				
27 December 1981	1739	38.81	24.94	6	6.5	11	216	79	175	Taymaz et al. (1991)
29 December 1981	0800	38.7	24.84	15	5.4	11	235	81	153	Harvard CMT solutions
10 April 1982	0450	39.94	24.61	17.4	5.2	3				
18 January 1982	1927	39.78	24.5	7.0	7.0	4	233	62	−177	Taymaz et al. (1991)
18 January 1982	1931	39.44	24.61	35	5.6	10				
6 August 1983	1543	40	24.7	9	6.8	3	50	76	177	Kiratzi et al. (1991)
10 October 1983	1017	40.23	25.32	11	5.4	2	70	64	176	Louvari (2000)
6 May 1984	0912	38.77	25.64	9	5.4	13	237	89	−161	Louvari (2000)
29 July 1984	0158	40.37	25.97	15.9	5.2	1				
5 October 1984	2058	39.1	25.3	22.6	5.6	11				
25 March 1986	0141	38.34	25.19	15	5.5	12	163	59	−22	Louvari (2000)
29 March 1986	1836	38.37	25.17	14	5.8	12	149	63	15	Louvari (2000)
3 April 1986	2332	38.35	25.1	1	5.2	12				
3 June 1986	0616	38.31	25.1	6.7	5.3	12				
17 June 1986	1754	38.32	25.11	31.8	5.4	12				
6 August 1987	0621	39.19	26.27	13.4	5.2	U				
8 August 1987	2215	40.09	24.89	11.1	5.3	3				
27 August 1987	1646	38.91	23.78	6.3	5.2	7				
30 May 1988	1647	40.25	25.85	2.8	5.2	1				
19 March 1989	0536	39.23	23.57	15	5.4	7	320	90	0	Harvard CMT solutions
5 September 1989	0652	40.15	25.09	15	5.4	2	64	34	−159	Harvard CMT solutions
23 July 1992	2012	39.81	24.4	8	5.4	4	272	51	−148	Louvari (2000)
24 May 1994	0205	38.82	26.49	21.4	5.5	15	258	54	−135	Harvard CMT solutions
16 July 1997	1306	39.04	25.22	15	5.2	11				
14 November 1997	2138	38.72	25.91	10	5.8	13	58	83	175	Louvari (2000)
11 April 1998	0929	39.9	23.88	7	5.2	U				
22 August 2000	0335	39.59	23.85	11	5.2	U				
10 June 2001	1311	38.6	25.57	33.6	5.6	14	151	74	−12	Harvard CMT solutions

Table 3 continued

Date	Time (hours)	Lat.	Long.	Depth	M	Class	Strike	Dip	Rake	Reference
26 July 2001	0021	39.06	24.25	15	6.4	9	148	76	−1	Harvard CMT solutions
26 July 2001	0034	39.05	24.27	13.9	5.3	9				
26 July 2001	0206	38.96	24.34	14.6	5.2	9				
26 July 2001	0209	38.9	24.37	9.8	5.3	9				
30 July 2001	1524	39.14	24.13	15	5.4	9	259	58	−72	Harvard CMT solutions
29 October 2001	2021	39.09	24.28	10	5.4	9				
6 July 2003	1910	40.37	26.25	20	5.5	1				
6 July 2003	2010	40.42	26.13	17	5.2	1				
15 June 2004	1202	40.37	25.81	12	5.2	1	251	85	168	Harvard CMT solutions
22 November 2004	1913	38.45	25.68	20	5.2	14				
24 August 2005	0306	39.68	25.56	29	5.2	6	244	68	−156	Harvard CMT solutions
21 December 2006	1830	39.32	23.6	23	5.3	7	144	76	−15	Harvard CMT solutions

U: Orientation class unknown

REFERENCES

ARMIJO, R., FLERIT, F., KING, G., and MEYER, B. (2003), *Linear elastic fracture mechanics explains the past and present evolution of the Aegean*, Earth Planet. Sci. Lett. *217*, 85–95.

COCCO, M. and RICE, J. R. (2002), *Pore pressure and poroelasticity effects in Coulomb stress analysis of earthquake interactions*, J. Geophys. Res. *107*, doi: 10.1029/2000JB000138.

CONSOLE, R., RHOADES, D. A., MURRU, M., EVISON, F. F., PAPADIMITRIOU, E. E., and KARAKOSTAS, V. G. (2006), *Comparative performance of time-invariant, long-range and short-range forecasting models on the earthquake catalogue of Greece*, J. Geophys. Res. *111*, B09304, doi:10.1029/2005JB004113.

DENG, J. and SYKES, L. R. (1997), *Evolution of the stress field in southern California and triggering of moderate-size earthquakes: a 200-year perspective*, J. Geophys. Res. *102*, 9859–9886.

ERIKSON, L. (1986), *User's manual for DIS3D: a three-dimensional dislocation program with applications to faulting in the Earth*, Masters thesis, Stanford Univ., Stanford, CA, 167 pp.

FIELD, E. H. (2007), *Overview of the working group for the development of regional earthquake likelihood models*, Seism. Res. Lett. *78(1)*, 8–16.

HARDEBECK, J. L. (2004), *Stress triggering and earthquake probability estimates*, J. Geophys. Res. *109*, B04310, doi: 10.1029/2003JB002437.

HARRIS, R. A. (1998), *Introduction to special session: stress triggers, stress shadows, and implications for seismic hazard*, J. Geophys. Res. *103*, 24347–24358.

JACKSON, J., HAINS, A. J., and HOLT, W. E. (1994), *A comparison of satellite laser ranging and seismicity data in the Aegean region*, Geophys. Res. Lett. *21*, 2849–2852.

KAGAN, Y. Y., JACKSON, D. D., and LIU, Z. (2005), *Stress and earthquakes in southern California, 1850–2004*, J. Geophys. Res. *110*, doi:10.1029/2004JB003313.

KING, G. C. P., STEIN, R. S., and LIN, J. (1994a), *Static stress changes and triggering of earthquakes*, Bull. Seism. Soc. Am. *84*, 935–953.

KING, G., OPPENHEIMER, D., and AMELUNG, F. (1994b), *Block versus continuum deformation in the Western United States*, Earth Planet. Sci. Lett. *128*, 55–64.

KING, G. C. P., HUBERT-FERRARI, A., NALBANT, S. S., MEYER, B., ARMIJO, R., and BOWMAN, D. (2001), *Coulomb interactions and the 17 August, 1999 Izmit, Turkey Earthquake*, Earth Planet. Sci., C. R. Acad. Sci. Paris *333*, 557–569.

KIRATZI, A. A., WAGNER, G., and LANGSTON, C. (1991), *Source parameters of some large earthquakes in Northern Aegean determined by body waveform modeling*, Pure Appl. Geophys. *135*, 515–527.

LOUVARI, E. (2000), *A detailed seismotectonic study in the Aegean Sea and the surrounding area with emphasis on the information obtained from microearthquakes*, Ph.D. thesis, Aristotle Univ., Thessaloniki, Greece, 373 p.

MALLMAN, E. P. and ZOBACK, M. D. (2007), *Assessing elastic Coulomb stress transfer models using seismicity rates in southern California and southwestern Japan*, J. Geophys. Res. *112*, B03304, doi:10.1029/2005JB004076.

MCCLUSKY, S., BALASANIAN, S., BARKA, A., DEMIR, C., GEORGIEV, I., HAMBURGER, M., HURST, K., KAHLE, H., KASTENS, K., KEKELIDZE, G., KING, R., KOTSEV, V., LENK, O., MAHMOUD, S., MISHIN, A., NADARIYA, M., OUZOUNIS, A., PARADISIS, D., PETER, Y., PEILEPI, M., REILINGER, R., SANLI, I., SEEGER, H., TEALEB, A., TOKSOZ, M. N., and VEIS, G. (2000), *GPS constraints on crustal movements and deformations in the eastern Mediterranean (1988–1997): implications for plate dynamics*, J. Geophys. Res. *105*, 5695–5719.

MCKENZIE, D. (1972), *Active tectonics of the Mediterranean region*, Geophys. J. R. Astr. Soc. *30*, 109–185.

MICHAEL, A. J. (2005), *Viscoelasticity, postseismic slip, fault interactions, and the recurrence of large earthquakes*, Bull. Seism. Soc. Am. *95*, 1594–1603.

MOLCHAN, G. M. (1990), *Strategies in strong earthquake prediction*, Phys. Earth Planet. Inter. *61*, 84–98.

MOLCHAN, G. M. (1991), *Structure of optimal strategies of earthquake prediction*, Tectonophysics *193*, 267–276.

OKADA, Y. (1992), *Internal deformation due to shear and tensile faults in a half-space*, Bull. Seism. Soc. Am. *82*, 1018–1040.

PAPADIMITRIOU, E. E. (2002), *Mode of strong earthquake recurrence in Central Ionian Islands (Greece): possible triggering due to Coulomb stress changes generated by the occurrence of previous strong shocks*, Bull. Seismol. Soc. Am. *92*, 3293–3308.

PAPADIMITRIOU, E. E. and KARAKOSTAS, V. G. (2003), *Episodic occurrence of strong (M ≥ 6.2) earthquakes in Thessalia Area (Central Greece)*, Earth Planet. Sci. Lett. *215*, 395–409.

PAPADIMITRIOU, E. E. and SYKES, L. R. (2001), *Evolution of the stress field in the Northern Aegean Sea (Greece)*, Geophys. J. Int. *146*, 747–759.

PAPADIMITRIOU, E. E., EVISON, F. F., RHOADES, D. A., KARAKOSTAS, V. G., CONSOLE, R., and MURRU, M. (2006), *Long-term seismogenesis in Greece: comparison of the evolving stress field and precursory scale increase approaches*, J. Geophys. Res. *111*, B05318, doi:10.1029/2005JB003805.

PAPADIMITRIOU, E., WEN, X., KARAKOSTAS, V., and JIN, X. (2004), *Earthquake triggering along the Xianshuie fault zone of western Sichuan, China*, Pure Appl. Geophys. *161*, 1683–1707.

PAPAZACHOS, B. C., KIRATZI, A. A., VOIDOMATIS, PH. S., and PAPAIOANNOU, CH. A. (1984), *A Study of the December 1981–January 1982 seismic activity in Northern Aegean Sea*, Boll. Geof. Teor. Appl., *XXVI*, 101–113.

PARSONS, T. (2002), *Global Omori law decay of triggered earthquakes: large aftershocks outside the classical aftershock zone*, J. Geophys. Res. *107* (B9), 2199, doi:10.1029/2001JB000646.

PARSONS, T. (2005), *Significance of stress transfer in time-dependent earthquake probability calculations*, J. Geophys. Res. *110*, doi:10.1029/2004JB003190.

PERFETTINI, H. and AVOUAC, J.-P. (2007), *Modeling afterslip and aftershocks following the 1992 Landers earthquake*, J. Geophys. Res. *112*, doi:10.1029/2006JB004399.

POLLITZ, F. F., NYST, M., NISHIMURA, T., and THATCHER, W. (2006), *Inference of postseismic deformation mechanisms of the 1923 Kanto Earthquake*, J. Geophys. Res. *111*, B05408, doi: 10.1029/2005JB003901.

REILINGER, R., MCCLUSKY, S., VERNANT, PH., LAWRENCE, S., ERGIVTAV, S., CAKMAK, R., OZENER, H., KADIROV, F., GULIEV, I., STEPANYAN, R., NADARIYA, M., HAHUBIA, G., MAHMOUD, S., SAKR, K., ARRAJEHI, A., PARADISSIS, D., AL–AYDRUS, A., PRILEPIN, M., GUSEVA, T., EVREN, E., DMITROTSA, A., FILIKOV, S. V., GOMEZ, F., AL–GHAZZI, R., and KARAM, G. (2006), *GPS Constraints on continental deformation in the Africa–Arabia–Eurasia continental collision zone and implications for the dynamics of plate interactions*, J. Geophys. Res. *111*, B05411, doi:10.1029/2005JB004051.

ROBINSON, R. and MCGINTY, P. J. (2000), *The enigma of the Arthur's Pass, New Zealand, Earthquake 2: the aftershock distribution and its relation to regional and induced stress*, J. Geophys. Res. *105*, 16139–16150.

ROCCA, A. CH., KARAKAISIS, G. F., KARAKOSTAS, B. G., KIRATZI, A. A., SCORDILIS, E. M., and PAPAZACHOS, B. C. (1985), *Further evidence on the strike-slip faulting of the Northern Aegean Trough based on properties of the August–November 1983 seismic sequence*, Boll. Geof. Teor. Appl. *XXVII*, 101–109.

SAVAGE, J. C. (2007), *Postseismic relaxation associated with transient creep rheology*, J. Geophys. Res. *112*, B05412, doi: 10.1029/2006JB004688.

STEACY, S., GOMBER, J., and COCCO, M. (2005), *Introduction to special section: stress transfer, earthquake triggering, and time-dependent seismic hazard*, J. Geophys. Res. *110*, B05S01, doi: 10.1029/2005JB003692.

STEIN, R. (1999), *The role of stress transfer in earthquake occurrence*, Nature *402*, 605–609.

STEIN, R. S., BARKA, A. A., and DIETERICH, J. H. (1997), *Progressive failure on the North Anatolian Fault since 1939 by earthquakes stress triggering*, Geophys. J. Int. *128*, 594–604.

TAYMAZ, T., JACKSON, J., and MCKENZIE, D. (1991), *Active tectonics of the North and Central Aegean Sea*, Geophys. J. Int. *106*, 433–490.

TODA, S. and STEIN, R. (2003), *Toggling of seismicity by the 1997 Kagoshima Earthquake couplet: a demonstration of time-dependent stress transfer*, J. Geophys. Res. *108*, doi:10.1029/2003JB002527.

TODA, S., STEIN, R. S., RISHARDS–DINGER, K., and BOZKURT, S. B. (2005), *Forecasting the evolution of seismicity in Southern California: animations built on earthquake transfer*, J. Geophys. Res. *110*, B05S16, doi:10.1029/2004JB003415.

WELLS, D. L. and COPPERSMITH, K. J. (1994), *New Empirical relationships among magnitude, rupture length, rupture width, rupture area, and surface displacement*, Bull. Seism. Soc. Am. *84*, 974–1002.

WESSEL, P. and SMITH, W. H. F. (1998). *New improved version of the Generic mapping tools released*, EOS Trans. AGU *79*, 579.

ZECHAR, J. D. and JORDAN, T. H. (2008), *Testing alarm-based earthquake predictions*, Geophys. J. Int. *172*, 715–724.

(Received September 5, 2008, revised May 8, 2009, accepted June 5, 2009, Published online March 10, 2010)

Aftershock Sequences Modeled with 3-D Stress Heterogeneity and Rate-State Seismicity Equations: Implications for Crustal Stress Estimation

DEBORAH ELAINE SMITH[1] and JAMES H. DIETERICH[1]

Abstract—In this paper, we present a model for studying aftershock sequences that integrates Coulomb static stress change analysis, seismicity equations based on rate-state friction nucleation of earthquakes, slip of geometrically complex faults, and fractal-like, spatially heterogeneous models of crustal stress. In addition to modeling instantaneous aftershock seismicity rate patterns with initial clustering on the Coulomb stress increase areas and an approximately 1/t diffusion back to the pre-mainshock background seismicity, the simulations capture previously unmodeled effects. These include production of a significant number of aftershocks in the traditional Coulomb stress shadow zones and temporal changes in aftershock focal mechanism statistics. The occurrence of aftershock stress shadow zones arises from two sources. The first source is spatially heterogeneous initial crustal stress, and the second is slip on geometrically rough faults, which produces localized positive Coulomb stress changes within the traditional stress shadow zones. Temporal changes in simulated aftershock focal mechanisms result in inferred stress rotations that greatly exceed the true stress rotations due to the main shock, even for a moderately strong crust (mean stress 50 MPa) when stress is spatially heterogeneous. This arises from biased sampling of the crustal stress by the synthetic aftershocks due to the non-linear dependence of seismicity rates on stress changes. The model indicates that one cannot use focal mechanism inversion rotations to conclusively demonstrate low crustal strength (≤ 10 MPa); therefore, studies of crustal strength following a stress perturbation may significantly underestimate the mean crustal stress state for regions with spatially heterogeneous stress.

Key words: Stress heterogeneity, rate-state, fractal, aftershock, Coulomb stress, crustal strength.

1. Introduction

We investigate aftershock sequences using simulations that combine several features, namely: (1) Coulomb static stress change analysis, (2) seismicity equations based on rate-state friction nucleation of earthquakes from DIETERICH (1994) and DIETERICH et al. (2003), (3) slip on geometrically complex faults as in DIETERICH and SMITH (2009), and (4) spatially heterogeneous fault planes/slip directions based on a model of fractal-like spatially variable initial stress from SMITH (2006) and SMITH and HEATON (2010). Our goal is to investigate previously unmodeled effects of system heterogeneities on aftershock sequences. The resulting model provides a unified means to simulate the statistical characteristics of aftershock focal mechanisms, including inferred stress rotations, and to provide insights on the persistent low-level occurrence of aftershocks in the Coulomb stress "shadow zones" (regions where Coulomb stress decreases).

Coulomb static stress change failure analysis has been extensively used to study the spatial distribution of aftershocks for moderate to large earthquakes (DENG and SYKES, 1997a, b; HARDEBECK et al., 1998; HARRIS and SIMPSON, 1996; HARRIS et al., 1995; KING et al., 1994; OPPENHEIMER et al., 1988; REASENBERG and SIMPSON, 1992; STEIN et al., 1994). In general, the change of Coulomb stress due to fault slip in a mainshock works well in explaining aftershock patterns, but not perfectly. Depending upon the individual mainshock, the performance of Coulomb static stress triggering models can range from 50% correlation, which is no better than random noise, to a 95% correlation, and with many reports around the 85% correlation level (DENG and SYKES, 1997a, b; HARDEBECK et al., 1998).

Rate-state friction, as well as other mechanisms (such as viscoelastic relaxation), has been used to explain temporal changes in seismicity rate and

---

[1] University of California, Riverside, CA 92521, USA. E-mail: desmith144@gmail.com

migration of events (DIETERICH, 1994; POLLITZ and SACKS, 2002; STEIN, 1999; STEIN et al., 1997; TODA and STEIN, 2003; TODA et al., 1998). Our study employs the earthquake rate formulation of DIETERICH (1994) and DIETERICH et al. (2003), which is based on time- and stress-dependent earthquake nucleation on faults with rate- and state-dependent friction. The formulation explains temporal features of aftershocks, such as the Omori law decay in aftershock seismicity rate, as consequences of Coulomb stress changes; it provides a natural framework for investigation of the effects of heterogeneities on aftershock processes.

Natural systems, which are inherently complex, must be heterogeneous at some level. In our model, stresses drive the aftershock process and determine the orientations at which faults fail. Stress heterogeneity can arise through a variety of mechanisms, including propagation of fault slip through geometric complexities, rupture dynamics that creates highly non-uniform slip, and inhomogeneous elastic structure. A variety of observations indicate that stress and slip are spatially heterogeneous and possibly fractal in nature (ANDREWS, 1980, 1981; BEN-ZION and SAMMIS, 2003; HERRERO and BERNARD, 1994; LAVALLEE and ARCHULETA, 2003; MAI and BEROZA, 2002; MANIGHETTI et al., 2001, 2005). MCGILL and RUBIN (1999) in particular, observed extreme changes in slip over short distances for the Landers earthquake. Borehole studies of stress orientation provide additional evidence that stress can be quite heterogeneous (BARTON and ZOBACK, 1994; WILDE and STOCK, 1997). Studies also indicate that stress heterogeneity is wavelength dependent; namely, there is a greater stress uniformity at short scales than at long scales.

Faults in nature are not geometrically planar surfaces—faults have irregularities at all wavelengths and can be depicted approximately as random fractal topographies (POWER and TULLIS, 1991; SCHOLZ and AVILES, 1986). Geometric interactions from slip of faults with random fractal roughness generate complex, high amplitude stress patterns close to and along the fault (DIETERICH and SMITH, 2009). While these stress concentrations die off with distance, they may be the primary reason for the characteristic high density of aftershocks close to the fault in the traditional stress shadow zone. An intriguing observation derives from ZOBACK and BEROZA (1993), who reported scattered focal mechanism solutions for Loma Prieta aftershocks, including left-lateral orientations on fault planes parallel to the San Andreas. A plausible explanation is that the stress was highly heterogeneous after the earthquake with short wavelength pockets of high stress in random directions.

HELMSTETTER and SHAW (2006) modeled the effect of a heterogeneous shear stress change on a plane for aftershock rates in light of rate- and state-dependent friction. Using two different, heterogeneous stress formulations, they produced Omori law-like decay of aftershocks and found that stress shadows are difficult to see. In another study (HELMSTETTER and SHAW, 2009), they used a simple slider block system to examine afterslip and aftershocks for a fault obeying rate-state friction and found that stress heterogeneity, as opposed to frictional heterogeneity, could explain a variety of post-seismic phenomena.

In addition to heterogeneous stress changes at the time of a mainshock, we assume the initial stress is heterogeneous and produces heterogeneous fault plane orientations on which aftershocks occur. To generate a heterogeneous population of fault orientations (and slip vectors) for aftershocks, we use a representation of a heterogeneous stress field based on SMITH (2006) and SMITH and HEATON (2010). The spatially varying models of the full stress tensor allowed Smith and Heaton to estimate best fitting stochastic parameters for Southern California focal mechanism data. Also, the model indicates earthquake failures are preferred for faults that are optimally oriented with respect to stressing rate; hence, stress inversions of focal mechanism data may be biased as well.

The sample bias effect may bear directly on the use of stress inversions of aftershock focal mechanisms to determine crustal stress properties, such as crustal stress heterogeneity and crustal strength. An implicit assumption in these studies is that the Earth is a good random sampler of its stress state when generating earthquakes; therefore, changes in the stress inversion mean misfit angle, β, and rotations of the inferred maximum horizontal compressive stress, σ_H, from stress inversion of aftershock sequences are assumed to represent true changes in stress (HARDEBECK and HAUKSSON, 2001; HAUKSSON, 1994; PROVOST and HOUSTON, 2003; RATCHKOVSKI, 2003; WOESSNER, 2005). These studies used inferred σ_H rotations to constrain average crustal stress and often estimate $\leq$10 MPa. However, if seismicity is a biased sampler of conditions in the Earth, the inferred σ_H rotations could be larger than the "true" rotation of the total stress field, and the actual crustal stress could be much larger than 10 MPa. This may explain the discrepancy between estimates of crustal stress based on stress rotation and estimates of $\geq$80 MPa from independent measures of crustal strength, such as borehole breakouts (HICKMAN and ZOBACK, 2004; TOWNEND and ZOBACK, 2000, 2004; ZOBACK and TOWNEND, 2001; ZOBACK et al., 1993). Previous studies have proposed other potential sources of error in stress inversions (ARNOLD and TOWNEND, 2007; LUND and TOWNEND, 2007; TOWNEND, 2006; TOWNEND and ZOBACK, 2001; WALSH et al., 2008).

2. Rate- and State-dependent Friction

As with previous studies (STEIN, 1999; STEIN et al., 1997; TODA and STEIN, 2003; TODA et al., 1998), we use the seismicity rate formulation of DIETERICH (1994) to model aftershock rates. This formulation is based on rate- and state-dependent friction constitutive representation of laboratory observations, which can be written as

$$\tau = \sigma_n \left[\mu_0 + a \ln\left(\frac{\dot{\delta}}{\dot{\delta}_*}\right) + b \ln\left(\frac{\theta}{\theta_*}\right) \right], \quad (1)$$

where τ is shear stress, σ_n is normal stress, $\dot{\delta}$ is slip speed, and θ is a state variable that depends on sliding history and normal stress history. a, b, and μ_0 are coefficients determined by experiment, and $\dot{\delta}_*$ and θ_* are normalizing constants.

This earthquake rate formulation employs solutions for earthquake nucleation on faults with rate-state friction (DIETERICH, 1992), and it provides a way to represent seismicity. Earthquake rate is both time- and stress-dependent and can be written in terms of Coulomb stress (DIETERICH et al., 2003)

$$R = \frac{r}{\gamma \dot{S}_r} \quad (2)$$

and

$$d\gamma = \frac{1}{a\sigma_n}[dt - \gamma dS], \quad (3)$$

where R is earthquake rate in some magnitude interval, $S = \tau - \mu\sigma_n$ is a Coulomb stress, $\dot{S}_r$ and r are reference values of the stressing rate and steady-state earthquake rate, respectively, and γ is a state variable that evolves with time and stressing history. The equations also give the characteristic Omori aftershock decay law and predict that aftershock duration is proportional to mainshock recurrence time. Also see DIETERICH (2007) for a review and discussion of the rate-state formulation and applications to seismicity modeling.

3. A Model of 3-D Spatially Varying Stress Heterogeneity

To generate a system of temporally stationary heterogeneous fault planes/slip directions, we use the following model of spatially varying stress heterogeneity in 3-D (SMITH, 2006; SMITH and HEATON, 2010). Seismicity rates will be determined on these fault planes/slip directions using rate-state friction. SMITH (2006) is available online at http://etd.caltech.edu/etd/available/etd-05252006-191203/.

SMITH (2006) and SMITH and HEATON (2010) defined the initial stress as,

$$\boldsymbol{\sigma}^0(\mathbf{x}) = \boldsymbol{\sigma}^B + \boldsymbol{\sigma}^H(\mathbf{x}), \quad (4)$$

where $\boldsymbol{\sigma}^B$ is a spatially uniform background stress that is approximately equal to the spatial average of $\boldsymbol{\sigma}^0(\mathbf{x})$ for the entire grid. $\boldsymbol{\sigma}^H(\mathbf{x})$ is the full 3-D heterogeneous stress term with little to no spatial mean; i.e., $\boldsymbol{\sigma}^H(\mathbf{x}) \approx \boldsymbol{\sigma}^0(\mathbf{x}) - \bar{\boldsymbol{\sigma}}^0(\mathbf{x})$. This term, $\boldsymbol{\sigma}^H(\mathbf{x})$, is created by filtering Gaussian noise in 3-D and then added to $\boldsymbol{\sigma}^B$ to create $\boldsymbol{\sigma}^0(\mathbf{x})$. In generating the Gaussian noise, SMITH and HEATON (2010) prescribed the off-diagonal elements to have an expected mean/standard deviation of $(0, \sigma)$ and the diagonal elements to have an expected mean/standard deviation of $(0, \sqrt{2}\sigma)$. Then a 3-D filter is applied so the spatial amplitude spectrum of any component of stress along any line bisecting the model is described by a power law,

$$\tilde{\boldsymbol{\sigma}}^H(k_r) \sim k_r^{-\alpha}, \tag{5}$$

where k_r is wave number. The parameter, α, is a measure of the spatial correlation in the filtered heterogeneous stress term, $\boldsymbol{\sigma}^H(\mathbf{x})$. If $\alpha = 0.0$, there is no filtering, and as α increases, the spatial heterogeneity becomes increasingly smoother and correlated spatially.

Note, the only difference between the stress model of SMITH (2006) and the stress model of SMITH and HEATON (2010) arises from the methodology used to create $\boldsymbol{\sigma}^H(\mathbf{x})$. Instead of starting with normally distributed tensor components as described above, SMITH (2006) started with normally distributed principal stresses with a mean of zero and uniformly distributed random orientations based on quaternion mathematics. Wave number filtering is then applied to the three principal stresses and to the stress tensor orientation, represented by three angles ($\omega, [\theta, \phi]$), where ω is a total rotation angle about a rotation axis, $[\theta, \phi]$. Both methodologies produce similar seismicity statistics and biasing toward the stressing rate; however, for mathematical simplicity, we use the methodology of SMITH and HEATON (2010) for this paper in creating $\boldsymbol{\sigma}^H(\mathbf{x})$.

Once $\boldsymbol{\sigma}^H(\mathbf{x})$ has been filtered, its overall amplitude is set relative to the spatially uniform, $\boldsymbol{\sigma}^B$. This relative heterogeneity amplitude is described, using a second statistical parameter, HR (Heterogeneity Ratio), based on the deviatoric stresses, where

$$HR = \frac{\sqrt{\overline{[\boldsymbol{\sigma}'^H(\mathbf{x}) : \boldsymbol{\sigma}'^H(\mathbf{x})]}}}{\sqrt{\boldsymbol{\sigma}'^B : \boldsymbol{\sigma}'^B}}. \tag{6}$$

Note that

$$\boldsymbol{\sigma}'^B : \boldsymbol{\sigma}'^B = \left(\sigma_{11}'^B\right)^2 + \left(\sigma_{22}'^B\right)^2 + \left(\sigma_{33}'^B\right)^2 + 2\left(\sigma_{12}'^B\right)^2 + 2\left(\sigma_{23}'^B\right)^2 + 2\left(\sigma_{13}'^B\right)^2 \tag{7}$$

and

$$\overline{[\boldsymbol{\sigma}'^H(\mathbf{x}) : \boldsymbol{\sigma}'^H(\mathbf{x})]} = \frac{1}{N}\sum_{i=1}^{N} \boldsymbol{\sigma}'^H(x_i) : \boldsymbol{\sigma}'^H(x_i). \tag{8}$$

HR is analogous to a coefficient of correlation since it computes a quantity that is like the standard deviation of $\boldsymbol{\sigma}'^0(\mathbf{x})$ divided by its mean.

SMITH and HEATON (2010) and SMITH (2006) compared statistics of synthetic focal mechanisms from their 3-D spatially heterogeneous stress to the statistics of real focal mechanisms from HARDEBECK (2006) and Hardebeck's SCEC catalog (HARDEBECK and SHEARER, 2003) for Southern California to constrain α and HR. To create their synthetic focal mechanism catalogs for comparison with real focal mechanism data, Smith and Heaton added a stressing rate, $\dot{\boldsymbol{\sigma}}^T$, from far-field plate tectonics and used a plastic failure criterion to determine when points fail within the simulation space. They varied the two statistical parameters, α and HR, to create suites of synthetic focal mechanisms' catalogs with different stochastic properties. SMITH and HEATON (2010) undertook a five-parameter grid search (α, HR, ε_{FM}, ε_{hypo}, L), which included the two statistical parameters, α and HR, two simulated measurement error parameters, focal mechanism angular uncertainty (ε_{FM}) and location error (ε_{hypo}), and the outer-scale, L, to find which set of parameters best reproduces real focal mechanism statistics. Specifically, they calculated the average angular difference between pairs of focal mechanisms as a function of distance for each set of (α, HR, ε_{FM}, ε_{hypo}, L) and compared their results to HARDEBECK (2006), with a best fit in the range of ($\alpha = 0.7$–0.8, $HR = 2.25$–2.5) (SMITH and HEATON, 2010). SMITH (2006), using a less rigorous inversion technique and the slightly different stress model, found comparable results. Smith and Heaton also found their inverted parameters to be consistent with mean misfit angle, β, statistics. Applying the stress inversion program "slick" (MICHAEL, 1984, 1987) to their synthetic focal mechanisms for ($\alpha = 0.8$, $HR = 2.375$) and to Hardebeck's A and B quality focal mechanism data for Southern California (HARDEBECK and SHEARER, 2003), they found the mean misfit angle statistics between their simulated data and Southern California data to be compatible.

Figure 1 shows a 1-D cross section of filtered synthetic stress using ($\alpha = 0.7$, $HR = 2.5$), which are the heterogeneous stress parameters for the models in this paper. In Fig. 1, all the components of $\boldsymbol{\sigma}^B$ equal zero except $\sigma_{12}^B \neq 0$. A random $\boldsymbol{\sigma}^H(x)$ is filtered with $\alpha = 0.7$, then added to $\boldsymbol{\sigma}^B$, where the relative amplitudes are specified by $HR = 2.5$ to create $\boldsymbol{\sigma}^0(x)$. $\boldsymbol{\sigma}^0(x)$ is scaled so that $200 \text{ MPa} \geq \sigma_{12}^0(x) \geq$

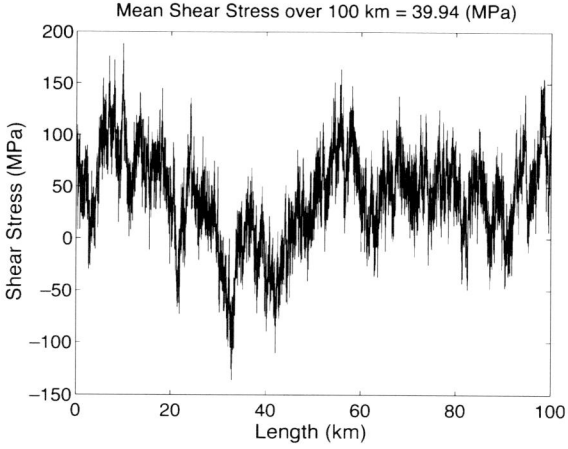

Figure 1
This is one realization of heterogeneous shear stress with parameters ($\alpha = 0.7$, $HR = 2.5$) and max shear stress about 200 MPa. Wave number filtering with $\alpha = 0.7$ produces this model of stress with greater spatial correlation at short distances than at long distances. Consequently, if one averages over the entire length of 100 km, the mean shear stress is approximately 40 MPa; however, if one were to average over an asperity, considerably higher mean shear stresses can be obtained (ELBANNA and HEATON, 2010; SMITH, 2006; SMITH and HEATON, 2010)

-200 MPa, and then $\sigma_{12}^0(x)$ is plotted. This model of stress heterogeneity produces great spatial variability in shear stress over the scale of 50–100 km; however, over the scale of 1–5 km, the stress is relatively uniform. This arises from the wave number filtering of $\boldsymbol{\sigma}^H(x)$, with $\alpha = 0.7$.

4. Methodology

In creating our synthetic aftershock sequences, we utilize: (1) the above 3-D heterogeneous stress model (SMITH, 2006; SMITH and HEATON, 2010) to define our failure planes/slip directions, (2) a spatially uniform far-field stressing rate and a spatially variable stress change from slip on a geometrically complex fault to create our stressing history, (3) rate-state seismicity equations (DIETERICH, 1994; DIETERICH et al., 2003) to evolve the seismicity rates on these failure planes/slip directions, given the stressing history, and (4) a random number generator to produce synthetic failures, assuming the earthquakes are a Poissonian process with spatially and temporally varying seismicity rates.

We are not aiming to delineate precise aftershock behavior, nor do we claim to know stress heterogeneity exactly. Instead, our goal is to demonstrate a general effect on aftershock sequences when pre-existing stress heterogeneity is included; hence, the parameters ($\alpha = 0.7$, $HR = 2.5$) are a reasonable place to start in creating the initial stress, $\boldsymbol{\sigma}^0(\mathbf{x})$. For all simulations, a deviatoric amplitude of $(\sigma_1 - \sigma_3)/2 = 50$ MPa is used for $\boldsymbol{\sigma}^B$, and the exact eigenvector orientations/relative eigenvalue sizes are selected a priori at the beginning of the simulation.

Then the outliers of $\boldsymbol{\sigma}^0(\mathbf{x})$ are clipped so that the maximum deviatoric amplitudes are in the range of granitic rock yield strength (SCHOLZ, 2002). The off-diagonal components are given a min/max value of ± 200 MPa, and the diagonal components are given a min/max value of $\pm 200\sqrt{2}$ MPa since the original heterogeneous stress, $\boldsymbol{\sigma}^H(\mathbf{x})$, is generated using a normal distribution with standard deviation σ for off-diagonal components and $\sqrt{2}\sigma$ for diagonal components.

A Coulomb failure criterion is then applied to the initial heterogeneous stress field, $\boldsymbol{\sigma}^0(\mathbf{x})$, to create two possible failure planes/slip directions at each point in the 3-D grid. The two possible failure planes are planes rotated $\pm\theta$ from the most compressive principal stress axis for $\boldsymbol{\sigma}^0(\mathbf{x})$, where $\theta = \frac{\pi}{4} - \frac{\tan^{-1}(\mu)}{2}$, where $\mu = 0.4$. A coefficient of friction slightly less than 0.6 is used partially because low coefficients of friction tend to best fit the Coulomb static stress analysis (REASENBERG and SIMPSON, 1992). Slip directions on the failure planes lie in the σ_1, σ_3 plane to produce optimal Coulomb failures. We label the two sets of failure planes/slip directions by the normal vectors to the planes and by the slip vectors, $(\mathbf{n}^{RL}, \mathbf{l}^{RL})$ for right-lateral mechanisms and $(\mathbf{n}^{LL}, \mathbf{l}^{LL})$ for left-lateral mechanisms.

Even though the total stress will change with time, any changes are treated as perturbations to the initial stress, $\boldsymbol{\sigma}^0(\mathbf{x})$; hence, $(\mathbf{n}^{RL}, \mathbf{l}^{RL})$ and $(\mathbf{n}^{LL}, \mathbf{l}^{LL})$ are stationary in time. The equation for total stress (SMITH, 2006; SMITH and HEATON, 2010) is

$$\boldsymbol{\sigma}(\mathbf{x}, t) = \boldsymbol{\sigma}^0(\mathbf{x}) + \dot{\boldsymbol{\sigma}}^T(t - t_0) + \Delta\boldsymbol{\sigma}^F(\mathbf{x}), \quad (9)$$

where $\dot{\boldsymbol{\sigma}}^T$ is the far-field stressing rate from plate-tectonics, t_0 is the time since the mainshock, and $\Delta\boldsymbol{\sigma}^F(\mathbf{x})$ is the static stress change from the mainshock.

$\dot{\boldsymbol{\sigma}}^T$ and $\Delta\boldsymbol{\sigma}^F(\mathbf{x})$ are treated as perturbations since their magnitudes are much smaller than the spatially heterogeneous initial stress, $\boldsymbol{\sigma}^0(\mathbf{x})$, in our simulations.

We now apply a stressing history defined by the background tectonic stressing rate, $\dot{\boldsymbol{\sigma}}^T$, and the 3-D static stress change, $\Delta\boldsymbol{\sigma}^F(\mathbf{x})$, calculated from Okada's equations for slip on a dislocation (OKADA, 1992), onto this population of failure planes/slip directions ($\mathbf{n}^{RL}$, $\mathbf{l}^{RL}$) and ($\mathbf{n}^{LL}$, $\mathbf{l}^{LL}$). In turn, this stressing history resolved onto ($\mathbf{n}^{RL}$, $\mathbf{l}^{RL}$) and ($\mathbf{n}^{LL}$, $\mathbf{l}^{LL}$) can be used as input for the rate-state earthquake rate equations from DIETERICH (1994) and DIETERICH et al. (2003) to update the seismicity rates at every point in the grid throughout the aftershock period. When we use Eqs. 2 and 3 in this paper, we set $a\sigma_n = 0.2$ MPa.

To implement Eqs. 2 and 3 to evolve the seismicity rates on the pre-existing planes/slip directions, it is necessary to first set the initial value γ_0 for each fault surface/slip direction in the model. We assume a steady-state condition wherein seismicity rate is constant. This requires that $\gamma_0 = \frac{1}{\dot{S}_r}$, where $\dot{S}_r = \dot{\tau}^T - \mu\dot{\sigma}_n^T$ is the resolved Coulomb stress rate for tectonic loading on the failure plane in the specified slip direction. $\dot{\boldsymbol{\sigma}}^T$ is resolved into both sets of failure orientations, ($\mathbf{n}^{RL}$, $\mathbf{l}^{RL}$) and ($\mathbf{n}^{LL}$, $\mathbf{l}^{LL}$), because when $\mu = 0.4$, the two planes at each point form an angle $< 90°$ and will not have the same resolved Coulomb stress rates, $\dot{S}_r$. Generally, $\dot{S}_r$ will have different values at each grid point because the tectonic stressing rate will not be optimally aligned with the heterogeneous array of failure plane orientations. To initialize the system for background seismicity prior to a main shock, we use only those failure orientations/slip directions with positive $\dot{S}_r$. Equation 2 can now be rewritten as

$$R(t) = r\frac{\gamma_0}{\gamma(t)}. \qquad (10)$$

The change in γ due to a static stress change, $\Delta\boldsymbol{\sigma}^F(\mathbf{x})$, at the time of the main shock is given by the solution to Eq. 3 for a step in stress

$$\gamma_1 = \gamma_0 \exp\left(-\frac{\Delta S^F}{a\sigma_n}\right), \qquad (11)$$

where ΔS^F is the Coulomb stress change from $\Delta\boldsymbol{\sigma}^F(\mathbf{x})$ resolved into ($\mathbf{n}^{RL}$, $\mathbf{l}^{RL}$) and ($\mathbf{n}^{LL}$, $\mathbf{l}^{LL}$). The evolution of γ with time following the stress step is given by the solution to Eq. 3 for a constant stress rate, $\dot{S}_r$,

$$\gamma_2(t) = \left(\gamma_1 - \frac{1}{\dot{S}_r}\right)\exp\left(-\frac{t}{t_a}\right) + \frac{1}{\dot{S}_r}, \qquad (12)$$

where $t_a = \frac{a\sigma_n}{\dot{S}_r}$ is the aftershock duration. In modeling aftershock sequences, previous studies (DIETERICH, 1994; DIETERICH, 2007; TODA et al., 1998) typically found values of t_a in the range of 2–10 years. The values $\gamma_2(t)$ for the two possible failure planes/slip directions at each point in the grid can then be used with Eq. 10 to calculate the time evolution of seismicity rate, R, at each point.

Last, to generate the synthetic aftershock catalogs, we use a random non-stationary Poissonian process with the seismicity rate, R, to sample the failure planes/slip directions ($\mathbf{n}^{RL}$, $\mathbf{l}^{RL}$) and ($\mathbf{n}^{LL}$, $\mathbf{l}^{LL}$). To simulate measurement uncertainty seen in real focal mechanism data, a random normal noise is added to the focal mechanisms orientations with a mean angular spread of 12° to simulate fairly high quality focal mechanisms, following the procedure of SMITH (2006) and SMITH and HEATON (2010).

5. Overview of Results

In the following, we present results for synthetic seismicity with spatially uniform stressing at a constant rate for aftershocks resulting from spatially uniform static stress changes and aftershocks resulting from spatially variable static stress changes from slip on a finite, geometrically complex fault. Three principal stress orientations are involved: (1) The orientation for the spatially uniform, $\boldsymbol{\sigma}^B$, (2) the orientation of the far-field tectonic stressing rate, $\dot{\boldsymbol{\sigma}}^T$, and (3) the orientation of the spatial mean of the static stress change defined in a region, $\Delta\bar{\boldsymbol{\sigma}}^F(\mathbf{x})$, from the main shock. For the case of a spatially uniform static stress change, $\Delta\boldsymbol{\sigma}^F$ (see Figs. 3, 4, and 5), $\dot{\boldsymbol{\sigma}}^T$ is aligned with $\boldsymbol{\sigma}^B$, but $\Delta\boldsymbol{\sigma}^F$ is misaligned. The misalignment of $\Delta\boldsymbol{\sigma}^F$ is used to test for possible biasing effects in the rotation of the inferred maximum horizontal compressive stress, σ_H, from stress inversions. All stress inversions are done using a bootstrapping technique with Andy Michael's program "slick" (MICHAEL, 1984, 1987). When slip on a finite fault is

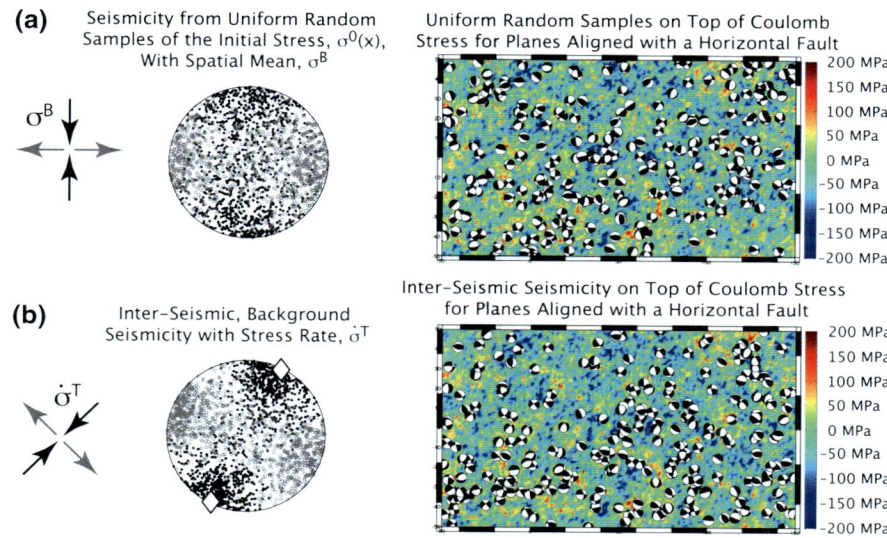

Figure 2

In **a**, a uniform random sampling of the heterogeneous stress field, $\sigma^0(\mathbf{x})$, with its associated optimally oriented failures, produces the synthetic seismicity. In **b**, a spatially homogeneous stress rate $\dot{\sigma}^T$ is applied at 45° relative to σ^B. Note that the stereographic projections of $\dot{\sigma}^T = 0.02$ MPa/year and $\sigma^B = 50$ MPa are not to scale. They simply show the orientation of the maximum and minimum compressive principal stresses. Seismicity is generated as a random Poissonian process, where the seismicity rate at each point in the grid is governed by the resolved Coulomb stressing history on heterogeneous failure planes/slip directions through the rate-state seismicity equations. The resultant inferred σ_H from a stress inversion of focal mechanism is rotated approximately 20° relative to the same quantities in **a**. This bias toward the stressing rate reproduces an effect described by SMITH (2006) and SMITH and HEATON (2010). Our calculation, however, uses rate-state seismicity equations and Coulomb stress, as opposed to a plastic failure criterion

used to create the mainshock static stress change, the misalignment of $\Delta\sigma^F(\mathbf{x})$ is spatially variable.

6. Background Seismicity at Constant Stressing Rate

The first model we examine is that of steady-state seismicity at a constant stressing rate (Fig. 2). In this model, the heterogeneous stress field, $\sigma^0(\mathbf{x})$, has a spatial mean with the most compressive stress, σ_1, oriented N–S and the least compressive stress, σ_3, oriented E–W. The heterogeneity parameters used are ($\alpha = 0.7$, $HR = 2.5$), and the deviatoric stress amplitude is 50 MPa. The heterogeneous population of faults, optimally oriented for initial stress, $\sigma^0(\mathbf{x})$, and coefficient of friction, $\mu = 0.4$, is subjected to a homogeneous stressing rate, $\dot{\sigma}^T$, of amplitude 0.02 MPa/year. $\dot{\sigma}^T$ has a maximum compressive principal stressing rate, $\dot{\sigma}_1$, aligned with ($Az. = N45°E$, $\delta = 0°$) and a least compressive principal stressing rate, $\dot{\sigma}_3$, aligned with ($Az. = N45°W$, $\delta = 0°$).

Figure 2a illustrates focal mechanisms that would arise from a spatially uniform sample of the failure planes/slip directions in the 3-D grid. The sampled failure planes/slip directions reflect the spatially heterogeneous initial stress, $\sigma^0(\mathbf{x})$, which has a spatial mean $\approx \sigma^B$. Since we allow for both right-lateral and left-lateral failures with $\mu = 0.4$, we have clusters of P-T axes on either side of the σ^B orientation; however, the orientation heterogeneity is large enough to smear together the two clusters so it appears that the average P axis is aligned with most compressive principal stress, σ_1, for σ^B and the average T axis is aligned with the least compressive principal stress, σ_3, for σ^B.

Figure 2b shows the seismicity and focal mechanisms generated by the model in response to a stressing rate, $\dot{\sigma}^T$, resolved onto the failure planes/slip directions from $\sigma^0(\mathbf{x})$ with spatial mean $\approx \sigma^B$; namely, $\dot{\sigma}^T$ is resolved onto failure planes/slip directions defined by $(\mathbf{n}^{RL}, \mathbf{l}^{RL})$ and $(\mathbf{n}^{LL}, \mathbf{l}^{LL})$ to calculate the background Coulomb stressing rate, $\dot{S}_r$, on the two possible failure planes/slip directions at each grid point. From Eq. 10, the relative seismicity rates are $\propto \dot{S}_r$, producing spatial variability in the background seismicity rate. Last, events are assumed to be random Poissonian processes non-stationary seismicity rates; hence, each potential failure plane/slip direction, with positive $\dot{S}_r$, provides a possible source of seismicity governed by its associated seismicity rate. Using an exponential random number generator to produce failure times for each potential seismicity source, we plot ~the first 1,000 events. This creates focal mechanism P–T axes in Fig. 2b rotated approximately 20° away from $\boldsymbol{\sigma}^B$, toward the stressing rate, $\dot{\boldsymbol{\sigma}}^T$. The rotation is purely a result of biased sampling of the failure planes that are oriented toward the optimal direction for the stressing rate, $\dot{\boldsymbol{\sigma}}^T$, rather than initial stress, $\boldsymbol{\sigma}^0(\mathbf{x})$. We employ Coulomb stress and rate-state seismicity equations to generate seismicity and reproduce the interseismic biasing effect found by SMITH (2006) and SMITH and HEATON (2010) who used a plastic failure criterion.

7. Spatially Uniform Static Stress Change, $\Delta\boldsymbol{\sigma}^F$

We next examine a simple model with a spatially uniform static stress, $\Delta\boldsymbol{\sigma}^F$, of deviatoric amplitude, 2 MPa. Again, the stress heterogeneity parameters are ($\alpha = 0.7$, $HR = 2.5$) for $\boldsymbol{\sigma}^0(\mathbf{x})$. $\boldsymbol{\sigma}^B$ has a 50 MPa deviatoric stress amplitude, and $\dot{\boldsymbol{\sigma}}^T$ has a 0.02 MPa/year deviatoric stress amplitude. The principal axes of the stress parameters $\boldsymbol{\sigma}^B$ and $\dot{\boldsymbol{\sigma}}^T$ are co-axially aligned with a σ_1 direction ($Az. = N45°E$, $\delta = 0°$) and a σ_3 direction ($Az. = N45°W$, $\delta = 0°$); however, the orientation of $\Delta\boldsymbol{\sigma}^F$ is varied with respect to the other stresses, which permits explicit tests for rotation of σ_H from stress inversions of focal mechanisms.

In Fig. 3, we simulate a series of models, using various differential angles between $\Delta\boldsymbol{\sigma}^F$ and $\boldsymbol{\sigma}^B$. Using Eqs. 10, 11, and 12, aftershock seismicity rates are evaluated at the same time shortly following the stress step ($10^{-3}\, t_a$), which would be a few days to a week for a typical aftershock sequence. Events arise when we randomly generate a set of failure times based on the spatially varying seismicity rates, extract events with failure times $\leq 10^{-3}\, t_a$, and plot P–T axes for 1,000 of these events with times $\leq 10^{-3}\, t_a$. A stress inversion is then applied to this aftershock seismicity for each differential angle between $\Delta\boldsymbol{\sigma}^F$ and $\boldsymbol{\sigma}^B$ to compute the inferred orientation of the maximum horizontal compressive stress, σ_H. The P–T plots show samples of this synthetic seismicity for varying differential angles, where the open diamonds are the inferred σ_H orientations for the background seismicity given the $\boldsymbol{\sigma}^B$ and $\dot{\boldsymbol{\sigma}}^T$ orientations, and the black circles are the inferred σ_H orientations one would obtain from aftershock focal mechanisms at $t = 10^{-3}\, t_a$; hence, any angular difference between the black circles and open diamonds indicates a rotation of the inferred σ_H. Below the P–T plots are two lines, a solid line representing the rotation of inferred σ_H from stress inversions of aftershock seismicity, which we call an "apparent" rotation, and a dashed line that represents the "true" rotation one would expect from the summation of stress, $\boldsymbol{\sigma}^B + \Delta\boldsymbol{\sigma}^F$.

We find major differences between the true stress rotation and the apparent stress rotation from focal mechanism inversions. While the maximum true stress rotation due to the stress step is <2°, the maximum apparent rotation from focal mechanisms is >30°. This large apparent rotation occurs because the change in seismicity rate depends exponentially on the change in stress from Eq. 11. Consequently, planes that are toward the optimal orientation for $\Delta\boldsymbol{\sigma}^F$ experience a much greater increase in seismicity than unfavorably oriented planes.

SMITH (2006) and SMITH and HEATON (2010) showed that biasing of stress orientations, as determined from stress inversions of focal mechanisms, depends on the value of HR up to some limit, $HR \approx 10$. If $HR = 0.0$, there is no biasing due to stress heterogeneity, and as HR increases, the biasing of inferred stress orientations also increases. Therefore, if $HR = 0.0$ in our aftershock simulations, there should be no biasing, and the maximum apparent rotation should be close to zero. (Remember that in our end-member simulations, all changes in σ_H and β are due entirely to changes in the biased sampling of pre-set failure planes/slip directions plus minimal measurement error. There is no updating of the pre-set failure planes/slip directions due to true stress changes.) Then as HR increases, we would expect the maximum apparent rotation to also increase.

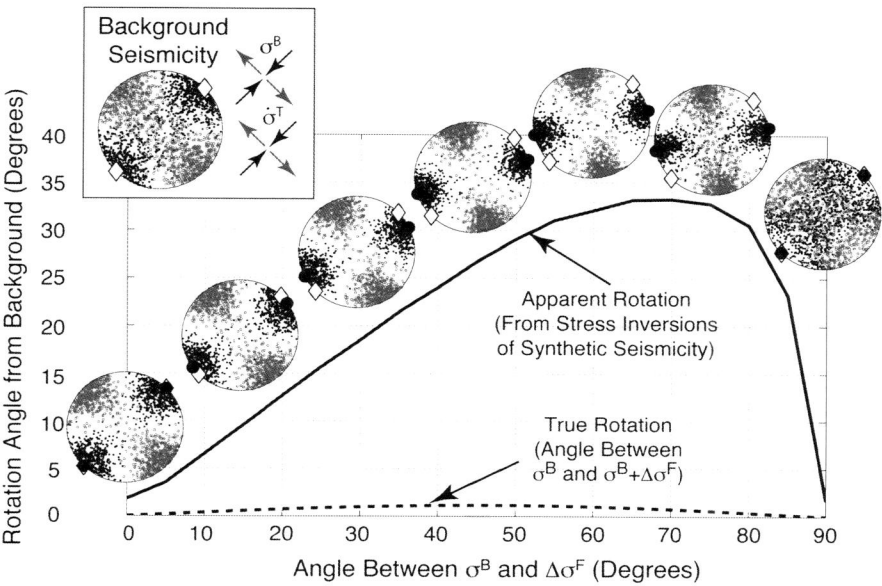

Figure 3
Plot of "apparent" rotation of the maximum horizontal compressive stress, σ_H, from inversions of synthetic aftershock seismicity versus expected "true" rotation from the static stress change, $\Delta\boldsymbol{\sigma}^F$. $\boldsymbol{\sigma}^B$ represents the approximate spatial mean of the initial stress field, and $\Delta\boldsymbol{\sigma}^F$ represents the static stress change. The principal axes of the stressing rate, $\dot{\boldsymbol{\sigma}}^T$, are aligned with those of $\boldsymbol{\sigma}^B$. In this example, $\Delta\boldsymbol{\sigma}^F$ is spatially uniform. Using the stress parameters described in the text, aftershock seismicity is evaluated at the same time, $10^{-3}\, t_a$, for various $\boldsymbol{\sigma}^B$ and $\Delta\boldsymbol{\sigma}^F$ angular differences. The "true" rotation of the stress field is plotted with the *dashed line*, and the "apparent" rotation of the maximum horizontal compressive stress, σ_H, from stress inversions of aftershock focal mechanisms is drawn with the *solid line*. Plots of synthetic aftershock P–T axes are plotted above the *solid line* where the black circles show the orientation of inferred σ_H for this data, and the open diamonds show the background seismicity σ_H orientation; hence, the angular difference between the circles and diamonds also show the "apparent" rotation of σ_H. The "apparent" rotation is considerably larger than the "true" rotation at every point

In Figs. 4 and 5, we now explore the time evolution of aftershock seismicity for our model with a spatially uniform stress step, $\Delta\boldsymbol{\sigma}^F$, by setting the angle between $\Delta\boldsymbol{\sigma}^F$ and $\boldsymbol{\sigma}^B$ to 45°, and using rate-state seismicity equations to evaluate seismicity rates at different times. The seismicity rate, normalized by the background seismicity rate is plotted as a function of time in Fig. 4. It shows approximately Omori law behavior, with a slope of $1/t^p$, where $p \approx 0.9$. Above the seismicity rate are plots of P–T axes from synthetic mechanisms and inferred σ_H orientations as a function of time. The rotated focal mechanism solutions produce an initial jump in the inferred σ_H orientation immediately after the applied static stress change, $\Delta\boldsymbol{\sigma}^F$, as seen by the angular difference between the open diamonds and black circles. Again the open diamonds represent the inferred σ_H orientation of background seismicity, and the black circles represent the inferred σ_H orientations from stress inversions of aftershocks. The angular difference between the open diamonds and black circles visually demonstrates the "apparent" rotation of σ_H. With time, the σ_H "apparent" rotation decays as σ_H returns to the reference orientation seen for background seismicity. The "apparent" rotation of σ_H, with a decay back to its original value, is explicitly plotted in Fig. 5 along with temporal changes in the mean misfit angle, β. In Fig. 5, β initially decreases as biasing effects kick in and then increases in time.

8. Spatially Variable Static Stress Change, $\Delta\boldsymbol{\sigma}^F(\boldsymbol{x})$, Through Slip on Finite Faults

We model aftershock patterns that might be expected from 10 m uniform slip on finite faults and their associated spatially nonuniform static stress changes, $\Delta\boldsymbol{\sigma}^F(\mathbf{x})$. The finite faults run 100 km long in the x direction and 20 km deep in the z direction, where the dimensions of the simulation space is

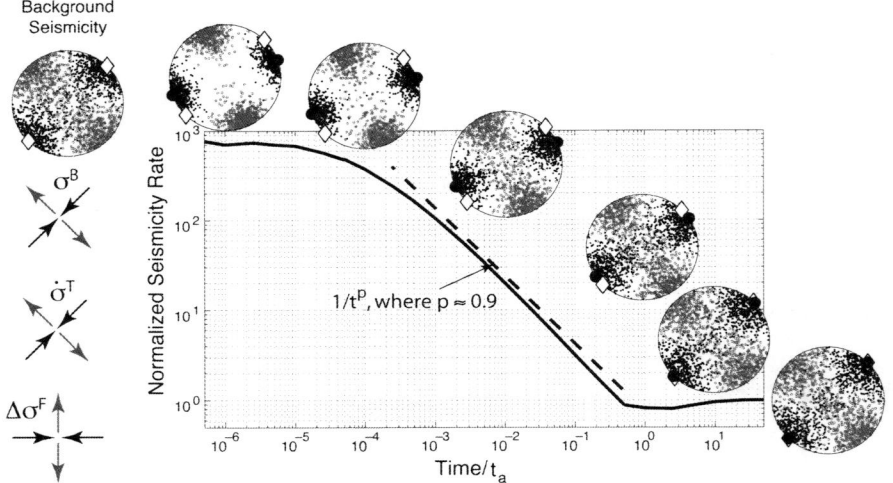

Figure 4
Evolution of seismicity and focal mechanisms with time following a stress step. Same experimental set-up as in Fig. 3, only the angular difference between σ^B and $\Delta\sigma^F$ is fixed to 45°. Plots of the focal mechanism P–T axes show snapshots of the aftershocks at different times. *Open diamonds* show the inferred σ_H orientation for the background seismicity, and the *black circles* show the inferred σ_H orientation from a stress inversion of the aftershocks at each time. There is an initial step rotation of σ_H at the onset of the spatially uniform stress step and then a decay as time progress. The seismicity rate, normalized by the background seismicity, has approximately Omori law-like behavior one would expect from rate-state controlled processes. Note that the stereographic projections of $\dot{\sigma}^T = 0.02$ MPa/year, $\sigma^B = 50$ MPa, and $\Delta\sigma^F = 2$ MPa are not to scale

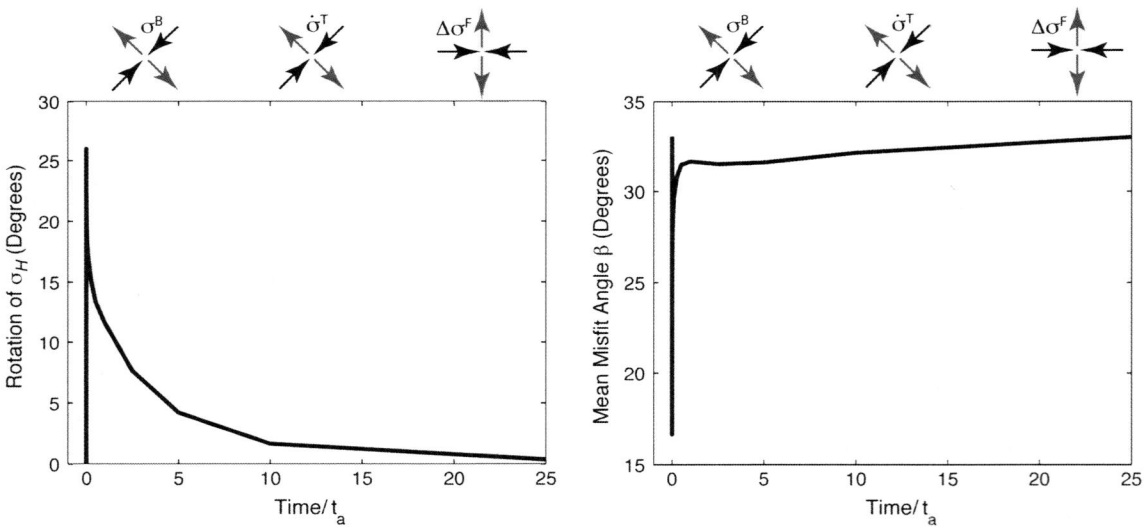

Figure 5
Rotation of σ_H and change in the mean misfit angle, β, from focal mechanism solutions using the synthetic seismicity from Fig. 4. The rotation of σ_H decays rapidly at first; hence, estimates of σ_H from stress inversions might only measure a 10°–15° rotation at the onset of the step stress change rather than the 27° rotation shown. The mean misfit angle, β, decreases at first, then increases, the opposite of what is seen in real data; however, the stress change applied for these figures is spatially uniform

200 km × 100 km × 50 km, with 1 km resolution. We use both a planar fault and a single geometrically complex fault with fractal-like topography. The geometrically complex fault used in the simulations has surface roughness amplitude that goes as *Amplitude* $\propto Bl^H$, as used in DIETERICH and SMITH (2009).

In this case, the exponent H is set to 1.0, which gives a self-similar roughness, and the rms slope has been set to $B = 0.07$. The initial stress, $\boldsymbol{\sigma}^0(\mathbf{x})$, again has stress heterogeneity parameters, ($\alpha = 0.7$, $HR = 2.5$), and a spatial mean deviatoric amplitude of 50 MPa. The stressing rate $\dot{\boldsymbol{\sigma}}^T$ has a 0.02 MPa/year deviatoric stress amplitude.

The orientations of $\boldsymbol{\sigma}^B$ and $\dot{\boldsymbol{\sigma}}^T$ with respect to the fault and each other significantly affect the results; therefore, we carefully choose these orientations for the simulations. For the planar fault, which serves as our "Reference" model, $\boldsymbol{\sigma}^B$ and $\dot{\boldsymbol{\sigma}}^T$ have principal stress axes ($Az. = N45°E$, $\delta = 0°$) for the σ_1 direction and ($Az. = N45°W$, $\delta = 0°$) for the σ_3 direction. For the geometrically complex or "Rough" fault, we examine three different scenarios. In "Rough" fault model #1, the principal axes of $\boldsymbol{\sigma}^B$ and $\dot{\boldsymbol{\sigma}}^T$ are the same as the planar fault, "Reference" model, where the most compressive principal stress axes for $\boldsymbol{\sigma}^B$ and $\dot{\boldsymbol{\sigma}}^T$ are at 45° with respect to the overall trend of the fault. In "Rough" fault model #2, $\dot{\boldsymbol{\sigma}}^T$ has its maximum compressive principal direction, $\dot{\sigma}_1$, $\perp$ to the fault trend so that $\dot{\sigma}_1$ is aligned with ($Az. = N0°E$, $\delta = 0°$) and $\dot{\sigma}_3$ is aligned with ($Az. = N90°E$, $\delta = 0°$). Last, for "Rough" fault model #3, $\boldsymbol{\sigma}^B$ has its maximum compressive principal direction, σ_1, $\perp$ to the fault trend so that ($Az. = N0°E$, $\delta = 0°$) for its maximum compressive principal stress, σ_1, and ($Az. = N90°E$, $\delta = 0°$) for its minimum compressive principal stress direction, σ_3.

Figures 6 and 7 show aftershock distributions for all four finite fault simulations. The top three rows, a, b, and c, show the instantaneous aftershock spatial distributions based on seismicity rates at a given instant in time. Specifically, we use rate-state friction equations to evaluate the seismicity rates at each point in the 3-D model region for the specified time. Then using these instantaneous rates, a random Poissonian process generates 2,000 events. The bottom row, d, for both Figs. 6 and 7, shows a normalized cumulative aftershock spatial distribution at $t = 0.1\ t_a$. This is a summation of all the aftershock seismicity that has occurred up until $t = 0.1\ t_a$, normalized by the background seismicity rate. In a sense, rows a–c in Figs. 6 and 7 plot the un-normalized probability density functions (PDFs) for seismicity at different time slices as a function of space, and row d plots the normalized time integration of the spatial pdfs until time, $t = 0.1\ t_a$.

Aftershocks for slip on the planar fault "Reference" model versus slip on the "Rough" fault model #1 are compared in Fig. 6. Again, $\boldsymbol{\sigma}^B$ and $\dot{\boldsymbol{\sigma}}^T$ have their most compressive principal stress at 45° with respect to the overall fault trend for both the "Reference" model and "Rough" fault model #1. The instantaneous aftershock seismicity concentrates initially on the Coulomb stress increase areas then migrates with time to an approximately spatially uniform distribution, which is the background seismicity spatial distribution in these models. Interestingly, even for the "Reference" model that has uniform slip on a planar fault, a few events occur in the stress shadow zone. This occurs because the pre-existing spatially heterogeneous stress field, $\boldsymbol{\sigma}^0(\mathbf{x})$, provides sufficient potential failure orientation heterogeneity that a few planes will be activated. Induced aftershock seismicity in the traditional stress shadow zone is even more pronounced for "Rough" fault model #1, especially near or on the fault trace. Slip on the geometrically complex fault produces small-scale stress asperities close to the fault trace, including zones of Coulomb stress increases that can especially generate aftershock seismicity.

Figure 7 illustrates the two examples where either $\boldsymbol{\sigma}^B$ or $\dot{\boldsymbol{\sigma}}^T$ have their most compressive principal stress axis $\perp$ with respect to the overall trend of the fault. "Rough" fault model #2 is shown on the left, where $\dot{\boldsymbol{\sigma}}^T$ has its most compressive principal stress rate oriented $\perp$ to the fault. "Rough" fault model #3 is shown on the right, where $\boldsymbol{\sigma}^B$ has its most compressive principal stress oriented $\perp$ to the fault. A significant percentage of the initial aftershock seismicity for model #2 occurs in the stress shadow zone, demonstrating a distinctly different aftershock pattern from model #1 in Fig. 6 when both $\boldsymbol{\sigma}^B$ and $\dot{\boldsymbol{\sigma}}^T$ are aligned 45° with respect to the fault. The aftershock distribution for model #3 in Fig. 7, however, looks very similar to the spatial distribution seen for model #1 in Fig. 6. Of interest, model #3, which has aftershock seismicity more realistic than that seen in model #2, is similar to some models of the Southern San Andreas (TOWNEND and ZOBACK, 2004), where the maximum compressive principal stress direction of

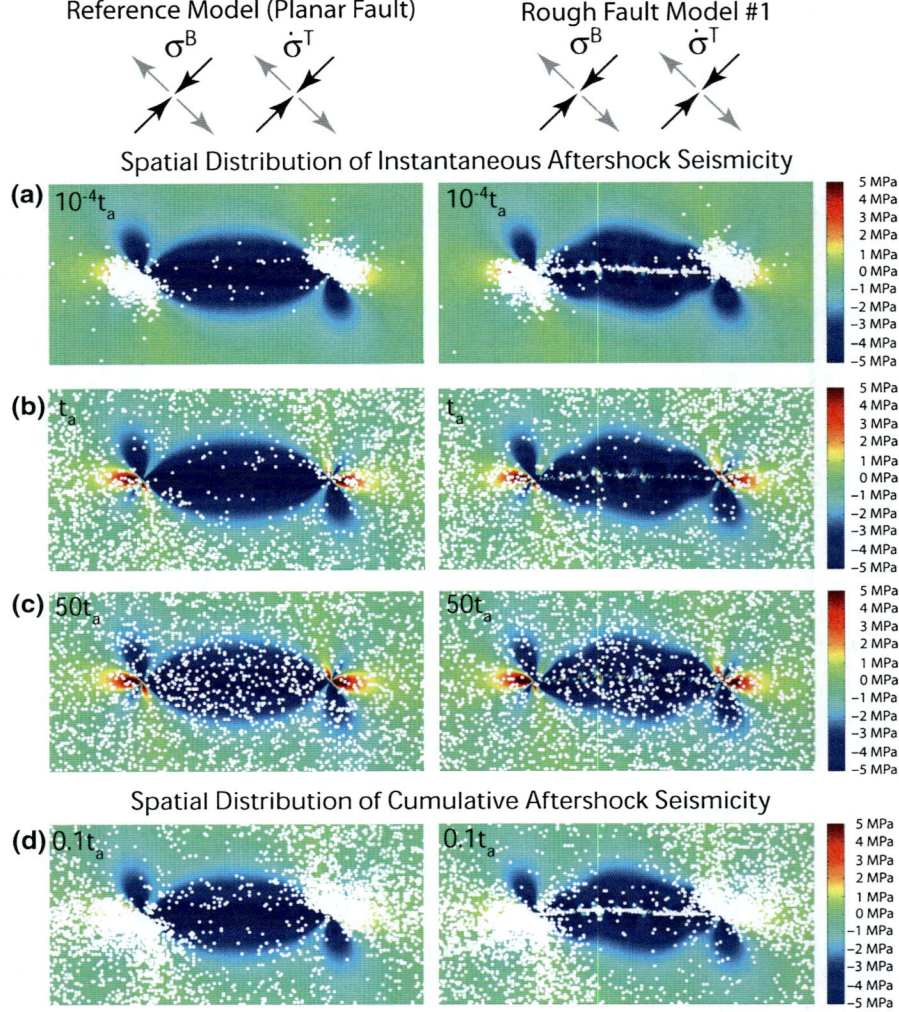

Figure 6

Aftershock seismicity for 10 m uniform slip on a planar fault and 10 m uniform slip on a geometrically complex fault. $\dot{\sigma}^T$ and σ^B orientated 45° with respect to the overall fault trend in both models. Note that the stereographic projections of $\dot{\sigma}^T = 0.02$ MPa/year and $\sigma^B = 50$ MPa are not to scale. The color scale goes from ±5 MPa, and the Coulomb stress change is calculated for planes parallel to the planar fault. For each panel in **a**, **b**, and **c**, seismicity rates are evaluated at the specified time. Then 2,000 random events are generated using a non-stationary random Poissonian process with the instantaneous seismicity rates. The panels in **d** show a normalized cumulative aftershock seismicity for $t = 0.1\ t_a$. The heterogeneous failure plane population enables the "Reference" model, with uniform slip on a planar fault, to experience a few failures in the stress shadow zone. Stress asperities from slip on the geometrically complex fault, in "Rough" model #1, create aftershock seismicity directly on or near the fault trace. Last, seismicity initially concentrates near the Coulomb stress increase areas and eventually becomes spatially uniform as the system transitions to the background seismicity state

σ^B is inferred to be perpendicular to the fault. Again, as in Fig. 6, there is a migration with time to an approximately spatially uniform seismicity distribution, which is the background seismicity distribution for our models.

Figures 9 and 10 present seismicity rates, rotations of the inferred maximum horizontal compressive stress, σ_H, from stress inversions, and changes in the stress inversion mean misfit angle, β, for "Rough" fault models #1–#3. To employ the synthetic data in a way that is similar to what is done in stress rotation studies, these quantities are plotted for the entire upper 15 km of the modeled region and for a subsection close to the fault trace (see Fig. 8).

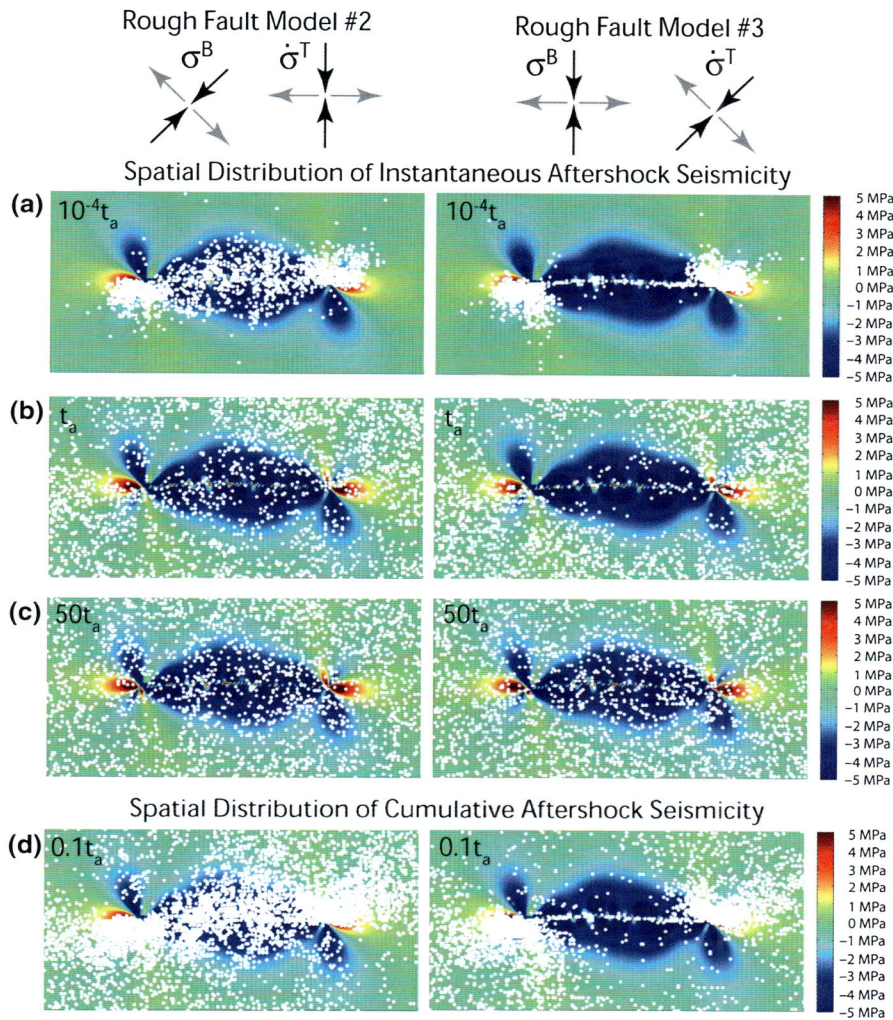

Figure 7
Similar to Fig. 6, only this time either σ^B or $\dot{\sigma}^T$ have their maximum compressive principal stress ⊥ with respect to the major fault trend. For "Rough" fault model #2, the principal compressive axis of $\dot{\sigma}^T$ is ⊥ to the overall fault trend, and for "Rough" fault model #3, the principal compressive axis of σ^B is ⊥ to the overall trend of the fault. The aftershock seismicity distribution for model #2 has a large percentage of its seismicity in the stress shadow region; whereas, the aftershock seismicity for model #3 looks fairly similar to model #1 in Fig. 6, where both σ^B and $\dot{\sigma}^T$ are aligned at 45° with respect to the fault

Seismicity rates and the behavior of aftershock seismicity as a function of time are shown in Fig. 9. The seismicity rates, normalized by the background rate for the upper 15 km of the modeled region, approximately follow Omori law, $1/t^p$, where $p \approx 0.87$ for the upper 15 km (dashed line) and $p \approx 0.87$ for the subsection (solid line). For the subsection, especially models #1 and #3, the seismicity rate bottoms out at t_a with a value significantly below its normalized background rate of ≈ 0.09. (Note that the background rate for the subsection will be less than 1.0 since the subsection represents a fraction of the upper 15 km.) Eventually, the seismicity rate for the subsection climbs back up for large times, at approximately $t = \frac{\Delta S^F}{S_r}$. This effect has been seen before with models that use rate-state equations when the static stress change is in the opposite direction of the stressing rate (SCHAFF et al., 1998); hence, the static stress change temporarily suppresses the seismicity rate.

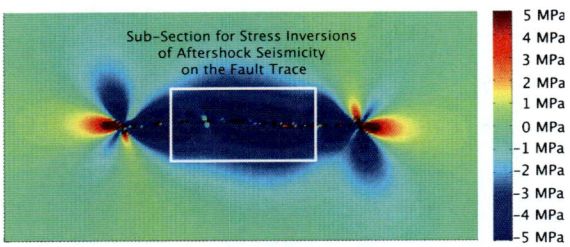

Figure 8
The subsection of the model region used for creating the P–T plots in Fig. 9 and the *solid lines* in Figs. 9 and 10. It is intended to capture seismicity close to or on top of the fault trace similar to aftershock studies

The P–T plots in Fig. 9 represent instantaneous aftershock seismicity from the subsection at different snapshots in time. The open diamonds represent the inferred σ_H orientation from stress inversions of background seismicity, and the solid circles represent the inferred σ_H orientation from stress inversions of aftershock seismicity; therefore, the angular differences between the diamonds and circles represent rotations of the inferred σ_H for the subsection. When $\boldsymbol{\sigma}^B$ and $\dot{\boldsymbol{\sigma}}^T$ have their most compressive principal stress oriented 45° with respect to the fault, as in "Rough" fault model #1, there is little to no rotation of σ_H. Any misalignment between the open diamonds and black circles may be simply due to random processes such as the random sampling of the failure planes to create the seismicity or the statistical noise that is added to the failure orientations. When $\boldsymbol{\sigma}^B$ or $\dot{\boldsymbol{\sigma}}^T$ have their maximum compressive principal stress axis $\perp$ with respect to the major fault trend, as in model #2 and model #3, there is a greater rotation of σ_H from stress inversions of the aftershock seismicity. Model #2, which had a larger percentage of the aftershock seismicity in the stress shadow zone, especially experiences a rotation of σ_H.

In Fig. 10, the σ_H rotations and changes of β from stress inversions of aftershock seismicity are plotted for our three "Rough" fault models. The rotation of σ_H for the subsection (solid line) can range from <5° for $\boldsymbol{\sigma}^B$ and $\dot{\boldsymbol{\sigma}}^T$ oriented 45° with respect to the fault (model #1) to almost 35° when $\dot{\boldsymbol{\sigma}}^T$ has its maximum compressive principal stress direction $\perp$ to the fault (model #2). Increases in the mean misfit angle, β, for the subsection (solid line) can range from 5° to over 17°, depending on the relative orientations of the background stress, $\boldsymbol{\sigma}^B$, and the tectonic stressing rate, $\dot{\boldsymbol{\sigma}}^T$. Rotations of σ_H and increases of β are usually smaller and have shorter decay times when calculated for the entire upper 15 km of the model region as shown by the dashed lines in Fig. 10.

9. Conclusions

A version of DIETERICH (1994) rate-state formulation for seismicity rates is combined with models of 3-D spatially heterogeneous stress to create a modeling environment for studying aftershock sequences. We assume that faults in a region represent fixed sources of seismicity, oriented favorably with respect to the local stresses. A spatially uniform tectonic stressing rate, $\dot{\boldsymbol{\sigma}}^T$, and a 3-D static stress change, $\Delta\boldsymbol{\sigma}^F$, are resolved onto the right-lateral and left-lateral "potential" failure planes/slip directions at every grid point to define a reference Coulomb stressing rate, $\dot{S}_r$, and Coulomb stress change, ΔS^F. The Coulomb stressing history, $\dot{S}_r$ and ΔS^F, drives the seismicity rate as a function of time at each point through rate-state seismicity equations (DIETERICH, 1994; DIETERICH et al., 2003). Each "potential" failure plane/slip direction, with its associated seismicity rate, is assumed to be a Poissonion source of seismicity with non-stationary rate; hence, there is some random probability that each "potential" failure plane/slip direction, with positive $\dot{S}_r$, will fail within a prescribed time and produce a synthetic focal mechanism for the catalog.

This model captures in a unified manner several aftershock features. For two of the three rough fault simulations, there is initial clustering of aftershocks in the Coulomb stress increase areas with a temporal migration back to a spatially uniform seismicity. Seismicity rates for all three models decay with approximately Omori law behavior. Aftershocks also occur in the traditionally Coulomb stress shadow regions. This occurs for two reasons: (1) The heterogeneous "potential" failure planes/slip directions, defined from the initial stress, engender a sufficient variation in resolved Coulomb stress for a few points to fail in the traditional Coulomb stress shadow zone. (2) Slip on geometrically complex faults produces small Coulomb stress increase asperities within the

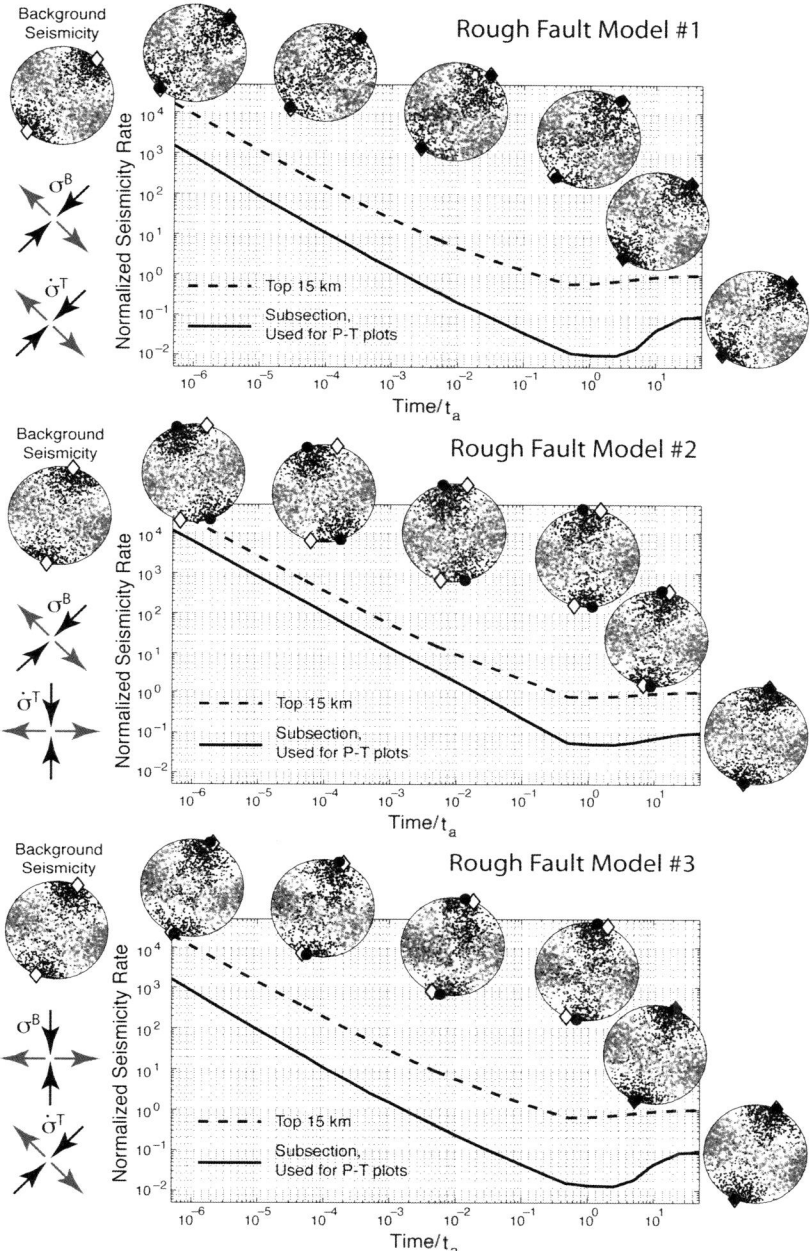

Figure 9
Plots of normalized seismicity rates, P–T axes, and inferred σ_H orientations for aftershock seismicity. Note that the stereographic projections of $\dot{\sigma}^T = 0.02$ MPa/year and $\sigma^B = 50$ MPa are not to scale. In model #1, σ^B and $\dot{\sigma}^T$ are both oriented at 45° with respect to the overall fault trend. In model #2, $\dot{\sigma}^T$ has its maximum compressive axis $\perp$ to the fault trace, and in model #3, σ^B has its maximum compressive axis $\perp$ to the fault trace. *Dashed lines* represent seismicity rates calculated for the entire upper 15 km of the model region, and *solid lines* represent seismicity rates calculated for the near fault subsection shown in Fig. 8. P–T plots show snapshots of focal mechanisms generated by aftershock seismicity in the subsection, and the angular difference between the open diamonds and the black circles shows the rotation of inferred σ_H from stress inversions. Seismicity rates for both the subsection and entire model region for models #1 through #3 show Omori law-like, $1/t^p$ behavior with $p \approx 0.87$; however, the rate for the subsection overshoots its background rate then climbs back up at long times. The smallest rotation of inferred σ_H occurs when σ^B and $\dot{\sigma}^T$ are both oriented 45° with respect to the fault in model #1, and the largest inferred rotation occurs when $\dot{\sigma}^T$ has its maximum compressive stress axis $\perp$ to the fault as in model #2

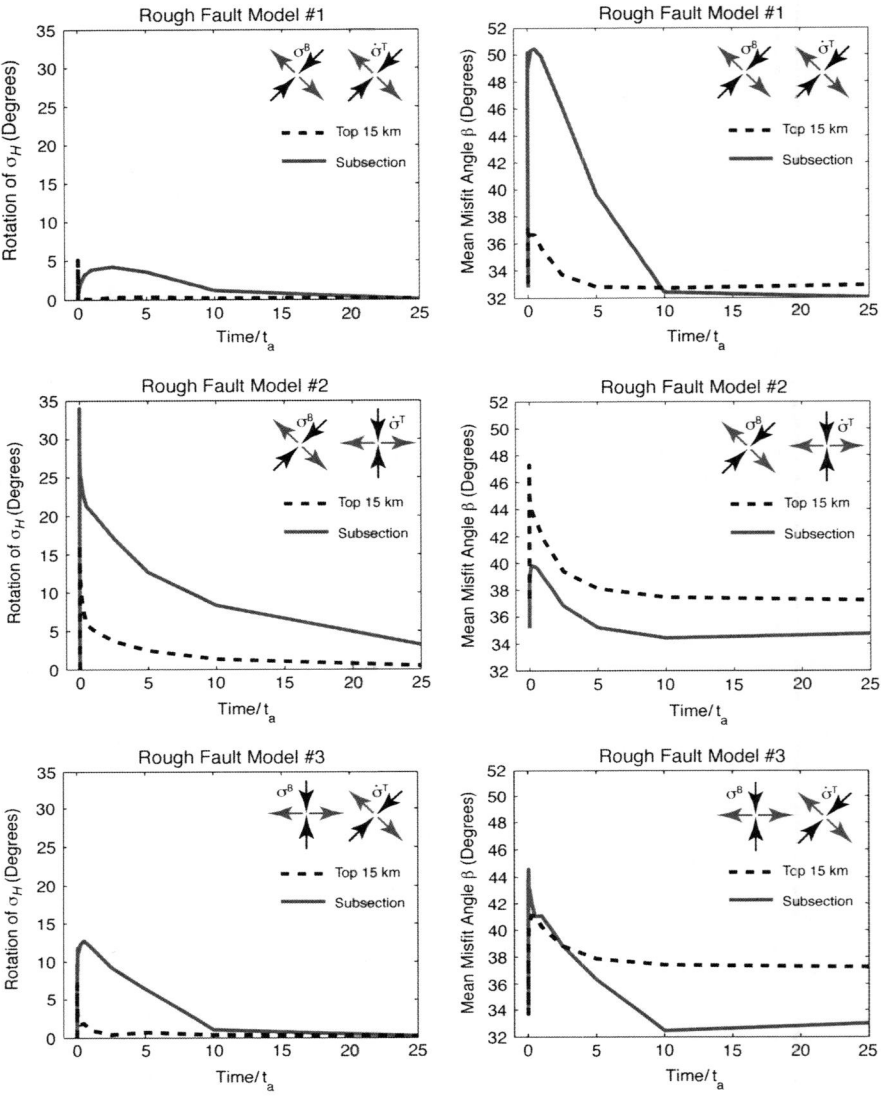

Figure 10
Rotations of inferred σ_H on the left as a function of time and evolution of β as a function of time on the right for the "Rough" fault models. Results are based on stress inversions of the synthetic aftershock focal mechanisms for different specified times. The *solid lines* represent seismicity from the subsection, and the *dashed lines* represent seismicity for the upper 15 km of the model region. Seismicity for the entire upper 15 km tends to have smaller changes in σ_H and β and much shorter decay times. Rotations of σ_H can range from less than 5° to almost 35°. Increases in β can range from 5° to over 17° for the three scenarios shown

overall Coulomb stress shadow zone. These asperities occur close to and on the fault; hence, they concentrate aftershock seismicity along the fault trace. Both of these mechanisms may affect real aftershock sequences, and they may help explain why Coulomb static stress change analysis only partially correlates with aftershock seismicity.

This model also shows that synthetic focal mechanisms can produce large "apparent" rotations of the maximum horizontal compressive stress, σ_H, when a static stress change, $\Delta\boldsymbol{\sigma}^F$, is applied to a spatially heterogeneous stress field. For a 2 MPa spatially uniform stress change, $\Delta\boldsymbol{\sigma}^F$, and an initial stress field, $\boldsymbol{\sigma}^0(\mathbf{x})$, with a 50 MPa spatial mean and stress heterogeneity parameters ($\alpha = 0.7$, $HR = 2.5$), the model can produce an "apparent" rotation of σ_H anywhere from 4°–33°, depending on the relative angle between $\Delta\sigma^F$ and σ^B. The expected "true" rotation of σ_H from the summation, $\sigma^B + \Delta\sigma^F$, is less than 2°, much smaller than the "apparent" rotation of σ^B calculated from stress inversions of synthetic aftershocks. Models of uniform slip on geometrically complex faults can also produce significant "apparent" rotations of inferred σ_H from inversions of synthetic aftershock focal mechanisms. At the same time, slip on these "rough" faults can create aftershock focal mechanisms that boost the stress inversion mean misfit angle parameter, β, anywhere from 5° to over 17°, yielding an "apparent" increase in the stress heterogeneity. These effects, rotations of inferred σ_H and increases in β, arise from the same highly nonlinear response of seismicity to a stress step that generates bursts of seismicity following stress perturbations that follow Omori's aftershock decay law. In a heterogeneous system with different fault plane orientations (reflecting heterogeneity of the initial stress), the nonlinear response of seismicity means that failure orientations favorably aligned toward the stress change will have a greater increase of seismicity than less favorably aligned orientations. Consequently, the seismicity following a stress change provides a sample of the fault planes and their associated slip directions, where the sample is biased in favor of failures aligned toward the optimal orientation for the stress perturbation.

These results indicate one cannot directly use rotations of σ_H from stress inversions of aftershock seismicity to estimate the magnitude and orientations of stress in the Earth's crust. Additionally, these results indicate one must be careful when interpreting temporal changes in β during aftershock sequences to study the time variation of stress heterogeneity. In our model of aftershocks, we can create a significant increase and subsequent decay of the mean misfit angle parameter, β, by updating as a function of time the ensemble of seismicity rates on temporally stationary failure orientations, rather than through "true" changes in stress; in other words, we can modify β as a function of time through biasing effects alone. For several aftershock sequences, an increase in the parameter β immediately after the mainshock has been observed, followed by a temporal decay (WOESSNER, 2005). While similar to our synthetic results, the aftershock data typically demonstrates β variations with an amplitude at least double what we produce for the synthetic aftershock sequences in this paper. Undoubtedly, stress heterogeneity evolves due to the mainshock and during the aftershock sequence; hence, the safest conclusion is that changes in β may need to be interpreted as a combination of both "true" changes in stress heterogeneity and biasing effects.

Understanding to what degree rotations of observed σ_H from stress inversions are due to "apparent" versus "true" rotations could be important in resolving conflicting observations of crustal stress. Studies of aftershock seismicity have assumed that rotations of inferred σ_H from aftershock stress inversions reflect a "true" rotation of the spatially homogeneous component of the total stress field and can be used to estimate the crustal stress (HARDEBECK, 2001; HARDEBECK and HAUKSSON, 2000, 2001; HAUKSSON, 1994; PROVOST and HOUSTON, 2003; RATCHKOVSKI, 2003; WOESSNER, 2005); therefore, if the static stress change due to the main shock is relatively small and changes in inferred σ_H are "true" rotations, then a low average crustal stress over the region, sometimes <10 MPa, is necessary. Yet, other measurements of crustal strength, such as borehole breakouts, can estimate considerably larger crustal stress of the order ≥80 MPa (HICKMAN and ZOBACK, 2004; TOWNEND and ZOBACK, 2000, 2004; ZOBACK and TOWNEND, 2001; ZOBACK et al., 1993).

In this paper, we demonstrate one potential solution to the reported crustal stress discrepancy by examining "apparent" rotations of σ_H that naturally arise from stress inversions in a spatially heterogeneous stress field. Our simulations show that significant "apparent" rotations of inferred σ_H can be created using moderate crustal strengths of 50 MPa; hence, one cannot definitively conclude weak crustal strengths of <10 MPa from rotations of σ_H, where σ_H is inferred from stress inversions of aftershock seismicity.

Acknowledgments

This research was supported by the Southern California Earthquake Center. SCEC is funded by NSF

Cooperative Agreement EAR-0106924 and USGS Cooperative Agreement 02HQAG0008. The SCEC contribution number for this paper is 1301. This material is based upon work supported by the National Science Foundation under Grant No. EAR-0636064. Deborah would also like to thank her former graduate advisor, Thomas Heaton, for his encouragement and for their discussions at the California Institute of Technology, which led to her extrapolating the idea of biasing in a spatially heterogeneous stress field to aftershock sequences.

Open Access This article is distributed under the terms of the Creative Commons Attribution Noncommercial License which permits any noncommercial use, distribution, and reproduction in any medium, provided the original author(s) and source are credited.

References

ANDREWS, D. J. (1980), *A stochastic fault model: 1) Static case*, J. Geophys. Res., 85, 3867–3877.

ANDREWS, D. J. (1981), *A stochastic fault model: 2) Time-dependent case*, J. Geophys. Res., 86, 821–834.

ARNOLD, R. and TOWNEND, J. (2007), A Bayesian approach to estimating tectonic stress from seismological data, Geophys. J. Int. doi:10.1111/j.1365-246X.2007.03485.x.

BARTON, C. A. and ZOBACK, M. D. (1994), *Stress perturbations associated with active faults penetrated by boreholes: Possible evidence for near-complete stress drop and a new technique for stress magnitude measurement*, J. Geophys. Res. 99, 9373–9390.

BEN-ZION, Y. and SAMMIS, C. G. (2003), *Characterization of fault zones*, Pure Appl. Geophys. 160, 677–715.

DENG, J. and SYKES, L. R. (1997a), *Evolution of the stress field in southern California and triggering of moderate-size earthquakes: A 200-year perspective*, J. Geophys. Res. 102, 9859–9886.

DENG, J. and SYKES, L. R. (1997b), *Stress evolution in southern California and triggering of moderate-, small-, and micro-sized earthquakes*, J. Geophys. Res. 102, 24411–24435.

DIETERICH, J. H. (1992), *Earthquake nucleation on faults with rate-dependent and state-dependent strength*, Tectonophysics 211, 115–134.

DIETERICH, J. H. (1994), *A constitutive law for rate of earthquake production and its application to earthquake clustering*, J. Geophys. Res.-Solid Earth 99, 2601–2618.

DIETERICH, J. H. (ed.), Applications of Rate- and State-Dependent Friction to Models of Fault Slip and Earthquake Occurrence, 107–129 pp. (Elsevier B. V. 2007).

DIETERICH, J. H. et al. (2003), Stress changes before and during the Puu Oo-Kupaianaha eruption. In: *The Puu Oo-Kupaianaha Eruption of Kilauea Volcano, Hawaii: The First 20 Years*, U.S. Geolog. Survey Prof. Paper, 1676, 187–202.

DIETERICH, J. H. and SMITH, D. E. (2009), *Non-planar faults: Mechanics of slip and off-fault damage*, Pure Appl. Geophys. 166, 1799–1815.

ELBANNA, A. E. and HEATON, T. H. (2010, in preparation), *Scale dependence of strength in systems with strong velocity weakening friction failing at multiple length scales*.

HARDEBECK, J. L. (2001), *The crustal stress field in Southern California and its implications for fault mechanics*, 148 pp., California Institute of Technology, Pasadena, California.

HARDEBECK, J. L. (2006), *Homogeneity of small-scale earthquake faulting, stress and fault strength*, Bull. Seismol. Soc. Am. 96, 1675–1688.

HARDEBECK, J. L. and HAUKSSON, E. (2000), *The San Andreas Fault in Southern California: A weak fault in a weak crust*, 3rd Conf. Tectonic Problems of the San Andreas Fault System, Stanford, CA.

HARDEBECK, J. L. and HAUKSSON, E. (2001), *Crustal stress field in southern California and its implications for fault mechanics*, J. Geophys. Res.-Solid Earth 106, 21859–21882.

HARDEBECK, J. L. et al. (1998), *The static stress change triggering model: Constraints from two southern California aftershock sequences*, J. Geophys. Res. 103, 24427–24437.

HARDEBECK, J. L. and SHEARER, P. M. (2003), *Using S/P amplitude ratios to constrain the focal mechanisms of small earthquakes*, Bull. Seismol. Soc. Am. 93, 2434–2444.

HARRIS, R. A. and SIMPSON, R. W. (1996), *In the shadow of 1857-the effect of the great Ft. Tejon earthquake on subsequent earthquakes in southern California*, Geophys. Res. Lett. 23, 229–232.

HARRIS, R. A. et al. (1995), *Influence of static stress changes on earthquake locations in southern California*, Nature 375, 221–224.

HAUKSSON, E. (1994), *State of stress from focal mechanisms before and after the 1992 Landers earthquake sequence*, Bull. Seismol. Soc. Am. 84, 917–934.

HELMSTETTER, A. and SHAW, B. E. (2006), *Relation between stress heterogeneity and aftershock rate in the rate-and-state model*, J. Geophys. Res. 111.

HELMSTETTER, A. and SHAW, B. E. (2009), *Afterslip and aftershocks in the rate-and-state friction law*, J. Geophys. Res. 114.

HERRERO, A. and BERNARD, P. (1994), *A kinematic self-similar rupture process for earthquakes*, Bull. Seismol. Soc. Am. 84, 1216–1228.

HICKMAN, S. and ZOBACK, M. (2004), *Stress orientations and magnitudes in the SAFOD pilot hole*, Geophys. Res. Lett. 31, doi: 10.1029/2004GL020043.

KING, G. C. P. et al. (1994), *Static stress changes and triggering of earthquakes*, Bull. Seismol. Soc. Am. 84, 935–953.

LAVALLEE, D. and ARCHULETA, R. J. (2003), *Stochastic modeling of slip spatial complexities of the 1979 Imperial Valley, California, earthquake*, Geophys. Res. Lett. 30, Art. No. 1245.

LUND, B. and TOWNEND, J. (2007), *Calculating horizontal stress orientations with full or partial knowledge of the tectonic stress tensor*, Geophys. J. Int. doi:10.1111/j.1365-246X.2007.03468.x.

MAI, P. M. and BEROZA, G. C. (2002), *A spatial random field model to characterize complexity in earthquake slip*, J. Geophys. Res.-Solid Earth 107, Art. No. 2308.

MANIGHETTI, I. et al. (2005), *Evidence for self-similar, triangular slip distributions on earthquakes: Implications for earthquake and fault mechanics*, J. Geophys. Res.-Solid Earth, 110, Art. No. B05302.

MANIGHETTI, I. et al. (2001), *Slip accumulation and lateral propagation of active normal faults in Afar*, J. Geophys. Res. 106, 13667–13696.

McGill, S. F. and Rubin, C. M. (1999), *Surficial slip distribution on the central Emerson fault during the June 28, 1992 Landers earthquake, California*, J. Geophys. Res.-Solid Earth *104*, 4811–4833.

Michael, A. J. (1984), *Determination of stress from slip data: Faults and folds*, J. Geophys. Res.-Solid Earth *89*, 11517–11526.

Michael, A. J. (1987), *Use of focal mechanisms to determine stress: A control study*, J. Geophys. Res.-Solid Earth *92*, 357–368.

Okada, Y. (1992), *Internal deformation due to shear and tensile faults in a half-space*, Bull. Seismol. Soc. Am. *82*, 1018–1040.

Oppenheimer, D. H. et al. (1988), *Fault plane solutions for the 1984 Morgan Hill, California earthquake sequence: Evidence for the state of stress on the Calaveras fault*, J. Geophys. Res. *93*, 9007–9026.

Pollitz, F. F. and Sacks, I. S. (2002), *Stress triggering of the 1999 Hector Mine earthquake by transient deformation following the 1992 Landers earthquake*, Bull. Seismol. Soc. Am. *92*, 1487–1496.

Power, W. L. and Tullis, T. E. (1991), *Euclidean and fractal models for the description of rock surface roughness*, J. Geophys. Res. *96*, 415–424.

Provost, A.-S. and Houston, H. (2003), *Investigation of temporal variations in stress orientations before and after four major earthquakes in California*, Phys. Earth Planet. Inter. *139*, 255–267.

Ratchkovski, N. A. (2003), *Change in stress directions along the central Denali fault, Alaska after the 2002 earthquake sequence*, Geophys. Res. Lett. *30*, 2017, doi:2010.1029/2003GL017905.

Reasenberg, P. A. and Simpson, R. W. (1992), *Response of regional seismicity to the static stress change produced by the Loma Prieta earthquake*, Science *255*, 1687–1690.

Schaff, D. P. et al. (1998), *Postseismic response of repeating aftershocks*, Geophys. Res. Lett. *25*, 4549–4552.

Scholz, C. H., The Mechanics of Earthquakes and Faulting, 2nd ed., 471 pp. (Cambridge University Press 2002).

Scholz, C. H. and Aviles, C. A. (eds.) (1986), The fractal geometry of faults and faulting, 147–156 pp., Am. Geophys. Union, Washington D.C.

Smith, D. E. (2006), *A new paradigm for interpreting stress inversions from focal mechanisms: How 3-D stress heterogeneity biases the inversions toward the stress rate*, California Institute of Technology, Pasadena.

Smith, D. E. and Heaton, T. H. (2010), *Simulations of 3D spatially heterogeneous stress–Potential biasing of stress orientation estimates derived from focal mechanism inversions*, Bull. Seismol. Soc. Am., (submitted).

Stein, R. S. (1999), *The role of stress transfer in earthquake occurrence*, Science *402*, 605–609.

Stein, R. S. et al. (1997), *Progressive failure on the North Anatolian fault since 1939 by earthquake stress triggering*, Geophys. J. Int. *128*, 594–604.

Stein, R. S. et al. (1994), *Stress triggering of the 1994 M = 6.7 Northridge, California, earthquake by its predecessors*, Science *265*, 1432–1435.

Toda, S. and Stein, R. (2003), *Toggling of seismicity by the 1997 Kagoshima earthquake couplet: A demonstration of time-dependent stress transfer*, J. Geophys. Res. *108*, 2567.

Toda, S. et al. (1998), *Stress transferred by the 1995 M-w = 6.9 Kobe, Japan, shock: Effect on aftershocks and future earthquake probabilities*, J. Geophys. Res.-Solid Earth and Planets *103*, 24543–24565.

Townend, J., What do faults feel? Observational constraints on the stresses acting on seismogenic faults. In Earthquakes: Radiated Energy and the Physics of Faulting, pp. 313–327 (American Geophysical Union 2006).

Townend, J. and Zoback, M. (2000), *How faulting keeps the crust strong*, Geology *28*, 399–402.

Townend, J. and Zoback, M. D. (2001), Implications of earthquake focal mechanisms for the frictional strength of the San Andreas fault system. In The Nature and Tectonic Significance of Fault Zone Weakening, eds R. E. Holdsworth et al., pp. 13–21, Special Publication of the Geological Society of London.

Townend, J. and Zoback, M. D. (2004), *Regional tectonic stress near the San Andreas fault in central and southern California*, Geophys. Res. Lett. *31*, 1–5.

Walsh, D. et al. (2008), *A Bayesian approach to determining and parameterizing earthquake focal mechanisms*, Geophys. J. Int. doi:10.1111/j.1365-246X.2008.03979.x.

Wilde, M. and Stock, J. (1997), *Compression directions in southern California (from Santa Barbara to Los Angeles Basin) obtained from borehole breakouts*, J. Geophys. Res. *102*, 4969–4983.

Woessner, J. (2005), *Correlating statistical properties of aftershock sequences to earthquake physics*, Swiss Federal Institute of Technology, Zürich.

Zoback, M. and Townend, J. (2001), *Implications of hydrostatic pore pressures and high crustal strength for deformation of the intraplate lithosphere*, Tectonophysics *336*, 19–30.

Zoback, M. D. et al. (1993), *Upper-crustal strength inferred from stress measurements to 6 km depth in the KTB borehole*, Nature *365*, 633–635.

Zoback, M. D. and Beroza, G. C. (1993), *Evidence for near-frictionless faulting in the 1989 (M 6.9) Loma Prieta, California, earthquake and its aftershocks*, Geology *21*, 181–185.

(Received August 21, 2008, revised May 21, 2009, accepted July 15, 2009, Published online March 23, 2010)

Earthquake Recurrence in Simulated Fault Systems

James H. Dieterich[1] and Keith B. Richards-Dinger[1]

Abstract—We employ a computationally efficient fault system earthquake simulator, RSQSim, to explore effects of earthquake nucleation and fault system geometry on earthquake occurrence. The simulations incorporate rate- and state-dependent friction, high-resolution representations of fault systems, and quasi-dynamic rupture propagation. Faults are represented as continuous planar surfaces, surfaces with a random fractal roughness, and discontinuous fractally segmented faults. Simulated earthquake catalogs have up to 10^6 earthquakes that span a magnitude range from ~M4.5 to M8. The seismicity has strong temporal and spatial clustering in the form of foreshocks and aftershocks and occasional large-earthquake pairs. Fault system geometry plays the primary role in establishing the characteristics of stress evolution that control earthquake recurrence statistics. Empirical density distributions of earthquake recurrence times at a specific point on a fault depend strongly on magnitude and take a variety of complex forms that change with position within the fault system. Because fault system geometry is an observable that greatly impacts recurrence statistics, we propose using fault system earthquake simulators to define the empirical probability density distributions for use in regional assessments of earthquake probabilities.

Key words: Seismicity, earthquake simulations, earthquake recurrence, fault roughness.

1. Introduction

Many processes and interactions undoubtedly affect earthquake occurrence, and each may imprint its own signature on earthquake statistics. Heterogeneities in fault strength and stress conditions have a primary impact on the size/frequency distributions of earthquake ruptures (Rundle and Klein, 1993; Stirling et al., 1996; Ben-Zion and Rice, 1997; Steacy and McCloskey, 1999). Heterogeneities may develop as a remnant of dynamical complexity during earthquake rupture, from interactions during slip of geometrically complex fault systems, from heterogeneous material properties, and through external processes such as spatially non-uniform pore fluid pressure changes or off-fault yielding. Also, earthquake nucleation, because it determines both the time of occurrence and place of origin of earthquake ruptures, can strongly affect the space-time patterns of seismicity, particularly following stress perturbations. This study employs a fault system earthquake simulator to explore earthquake recurrence statistics. Our focus is on the possible imprinting of earthquake nucleation processes and fault system geometry on earthquake recurrence statistics.

The simulations incorporate time- and stress-dependent earthquake nucleation as required by rate- and state-dependent fault properties. The rate- and state-dependent constitutive formulation quantifies observed characteristic dependencies of sliding resistance on slip, sliding speed and contact time; and it provides a framework to unify observations of dynamic/static friction, displacement weakening at the onset of macroscopic slip, time-dependent healing, slip history dependence, and slip speed dependence (Dieterich, 1979, 1981; Ruina, 1983; Tullis, 1988; Marone, 1998). Laboratory studies of earthquake nucleation processes (Dieterich and Kilgore, 1996) and studies of earthquake nucleation with rate- and state-dependent constitutive properties (Dieterich, 1992, 1994; Rubin and Ampuero, 2005) indicate that nucleation processes are highly time- and stress-dependent. Seismicity models that incorporate nucleation with rate- and state-dependent friction reproduce a variety of characteristics observed in seismicity data including foreshocks and aftershocks with Omori-type

---

[1] Department of Earth Sciences, University of California, Riverside, CA 92521, USA. E-mail: Dieterichj@ucr.edu; keithrd@ucr.edu

temporal clustering (DIETERICH, 1987, 2007; GOMBERG et al., 1997, 1998, 2000; BELARDINELLI et al., 2003; ZIV and RUBIN, 2003).

Fault system geometry is an obvious system-level structural heterogeneity that is both observable and persistent. Faults in nature are not geometrically flat surfaces, and they do not exist in isolation, but form branching structures and networks. These structural features are evident over a wide range of length scales. Individual faults exhibit roughness at all length scales that can be modeled as mated surfaces with random fractal topography (SCHOLZ and AVILES, 1986; POWER and TULLIS, 1991; SAGY et al., 2007). Fault step-overs (OKUBO and AKI, 1987) and fault system geometry (BONNET et al., 2001; BEN-ZION and SAMMIS, 2003) also have fractal characteristics. Slip of faults with these features results in strong geometric incompatibilities and interactions that do not occur in planar fault models. For example, fault step-overs may break a fault into weakly connected segments that serve as persistent barriers that inhibit rupture propagation. Also, non-planarity of faults and fault branches gives rise to geometric incompatibilities that may similarly inhibit rupture growth. The fractal characteristics of faults and fault system geometry mean that these interactions operate over a wide range of length scales. Indeed WESNOUSKY (1994) proposes that individual faults making up a regional fault system have a strong tendency to generate characteristic earthquakes that essentially rupture an entire fault and that the characteristic Gutenberg–Richter earthquake magnitude–frequency distribution reflects the size distribution of faults in a region. This view is supported by idealized model studies (RUNDLE and KLEIN, 1993; STIRLING et al., 1996; BEN-ZION and RICE, 1997; STEACY and MCCLOSKEY, 1999) but the issue remains an open question.

Previous modeling studies of earthquakes and slip in geometrically complex faults include investigation of slip of wavy faults (SAUCIER et al., 1992; CHESTER and CHESTER, 2000), slip through idealized fault bends (NIELSEN and KNOPOFF, 1998), rupture propagation into fault branches (OGLESBY et al., 2003; FLISS et al., 2005), and rupture jumps across gaps (HARRIS et al., 1991; DUAN and OGLESBY, 2006; SHAW and DIETERICH, 2007). Seismicity simulations that implement region-specific models of fault systems (WARD, 1996, 2000; RUNDLE et al., 2004; ROBINSON and BENITES, 1995) have demonstrated that plausible seismicity models can be implemented that replicate basic characteristics of regional seismicity. In this work we investigate the individual and combined effects of several of these forms of complexity on the recurrence statistics of earthquakes.

2. Simulations

This study employs synthetic catalogs with up to 10^6 earthquakes that are generated using an efficient simulation procedure developed by DIETERICH (1995). The current model, RSQSim, uses 3-D boundary elements based on the solutions of either OKADA (1992) or MEADE (2007), and it accepts different modes of fault slip (normal, reverse, strike-slip) as well as mixed slip modes. In this study we examine only strike slip faults. With the current single processor version of the computer code, up to 30,000 elements are used to represent fault surfaces. This permits quite detailed 3-D representations of fault system geometry and fault interaction effects. In this study the simulations generally employ 1 km × 1 km or 1.5 km × 1.5 km elements, and seismicity catalogs span a magnitude range from roughly M 4–M 8. Although the simulations employ large-scale approximations and simplifications to achieve computational efficiency, comparisons with fully dynamic 3-D finite-element models described below indicate the calculations are quite accurate. Details of the computations together with an overview of the dynamic characteristics of individual events and characteristics of the synthetic catalogs are given by RICHARDS-DINGER and DIETERICH (in preparation). In the following we briefly describe the model computations and outline some important characteristics of the model.

RSQSim is based on a boundary element formulation whereby interactions among the fault elements are represented by an array of 3-D elastic dislocations, and stresses acting on the centers of the elements are

$$\tau_i = K_{ij}^\tau \delta_j + \tau_i^{\text{tect}} \qquad (1)$$

$$\sigma_i = K_{ij}^\sigma \delta_j + \sigma_i^{\text{tect}}, \qquad (2)$$

where i and j run from 1 to N, the total number of fault elements; τ_i and σ_i are the shear stress in the directions of slip and fault-normal stress on the ith element, respectively; the two K_{ij} are interaction matrices derived from elastic dislocation solutions; δ_j is slip of fault element j; τ_i^{tect} and σ_i^{tect} represent stresses applied to the ith element by sources external to the fault system (such as far field tectonic motions); and the summation convention applies to repeated indices. The code uses full 3-D boundary element representations and can employ rectangular (OKADA, 1992) or triangular (MEADE, 2007) fault elements.

The model employs a rate- and state-dependent formulation for sliding resistance (DIETERICH, 1979, 1981; RUINA, 1983; RICE, 1983):

$$\tau = \sigma\left[\mu_0 + a\ln\left(\frac{\dot\delta}{\dot\delta^*}\right) + b\ln\left(\frac{\theta}{\theta^*}\right)\right], \qquad (3)$$

where μ_0, a, and b are experimentally determined constants; $\dot\delta$ is sliding speed; θ is a state variable that evolves with time, slip, and normal stress history; and $\dot\delta^*$ and θ^* are normalizing constants. In the simulations fault strength is fully coupled to normal stress changes through the coefficient of friction and through θ, which evolves with changes of normal stress as given by LINKER and DIETERICH (1992):

$$\dot\theta = 1 - \frac{\dot\delta\theta}{D_c} - \frac{\alpha\theta\dot\sigma}{b\sigma}. \qquad (4)$$

At constant normal stress, the evolution of θ takes place over a characteristic sliding distance D_c, and for a constant sliding speed $\dot\delta$ will approach a steady-state of $\theta_{ss} = D_c/\dot\delta$. See MARONE (1998) and DIETERICH (2007) for detailed reviews of rate- and state-dependent friction and a discussion of applications.

A central feature of the method is the use of event-driven computational steps as opposed to time-stepping at closely spaced intervals (DIETERICH, 1995). The cycle of stress accumulation and earthquake slip at each fault segment is separated into three distinct phases designated as sliding states 0, 1,

and 2 that are based on more detailed models with rate- and state-dependent fault constitutive properties. Previously DIETERICH (1995) and ZIV and RUBIN (2003) employed this three-state approach to model foreshock and aftershock processes. A fault element is at state 0 if stress is below the steady-state friction, as defined by rate- and state-dependent friction. In the model this condition is approximated as a fully locked element in which the fault strengthens as the frictional state-variable θ increases with time, e.g., $\theta = \theta_0 + t$ at constant normal stress, but modified by effects arising from normal stress changes using the LINKER and DIETERICH (1992) formulation.

The transition to sliding state 1 occurs when the stress exceeds the steady-state friction. During state 1, conditions have not yet been met for unstable slip, although the fault progressively weakens as described by rate- and state-dependent fault constitutive properties. Analytic solutions for nucleation of unstable slip (DIETERICH, 1992) generalized for varying normal stress (DIETERICH, 2007; RICHARDS-DINGER and DIETERICH, in preparation), together with stressing rate determine the transition time to state 2, which is earthquake slip. At tectonic stressing rates, earthquake nucleation typically requires a year or more, however during earthquake slip the high stressing rates at the rupture front compress the duration of state 1 to a fraction of a second. Hence, during an earthquake rupture, state 1 in effect forms a process zone at the rupture front, where time-dependent breakdown of fault strength occurs. The slip during nucleation is negligible compared to coseismic earthquake slip and is therefore ignored for purposes of computing stress changes on other elements.

During earthquake slip (state 2), the model employs a quasi-dynamical representation of the gross dynamics of the earthquake source based on the relationship for elastic shear impedance together with the local dynamic driving stress. From the shear impedance relation (BRUNE, 1970) the fault slip rate is

$$\dot\delta_j^{EQ} = \frac{2\beta\Delta\tau_j}{G}, \qquad (5)$$

where the driving stress $\Delta\tau_j$ is the difference between the stress at the initiation of slip and the sliding friction at element j; β is the shear-wave speed; and G

is the shear modulus. This provides a first-order representation of dynamical time scales and slip rates for the coseismic portion of the earthquake simulations. In the simulations described here a single rupture slip speed was used that is based on average values of $\Delta\tau_j$. An element ceases to slip and reverts to state 0 when the stress decreases to some specified stress determined by the sliding friction (with inertial overshoot of stress to levels less than the sliding friction as an adjustable model parameter).

The computational efficiency of the model is obtained from the use of event-driven computational steps, use of analytic nucleation solutions, and specification of earthquake slip speed from the shear impedance relation. Determination of the sliding state changes requires computation of the stress state as a function of time at each fault element. Note that stressing rates are constant between state changes, and the change of stressing rate at any element i resulting from the initiation or termination of earthquake slip at element j is given by

$$\dot{\tau}_i = \dot{\tau}_i \pm K_{ij}^\tau \dot{\delta}_j^{EQ} \tag{6}$$

$$\dot{\sigma}_i = \dot{\sigma}_i \pm K_{ij}^\sigma \dot{\delta}_j^{EQ} \text{ (no summation)}, \tag{7}$$

where the $+$ and $-$ refer to $1 \rightarrow 2$ and $2 \rightarrow 0$ transitions on element j, respectively. Hence, these state transition events require only one multiply and add operation at each element to update stressing rates everywhere in the model (no system-scale updates are required for the $0 \rightarrow 1$ transition). These changes to the stressing rates are applied instantaneously to all patches in the model (but note that the stresses themselves do not change discontinuously). A possible improvement to the model, with which we plan to experiment in the future, would be to delay the changes by a suitable wave propagation speed. Because the transition times depend only on initial conditions and stressing rates, computation proceeds in steps that mark the transition from one sliding state to the next without calculation of intermediate steps. This approach completely avoids computationally intensive solutions of systems of equations at closely spaced time intervals. Computation time for an earthquake event of some fixed size, embedded in a model with N fault elements, scales approximately by N^1.

For this study, stressing-rate boundary conditions drive fault slip and are set using the back-slip method (SAVAGE, 1983; KING and BOWMAN, 2003). With this method, the stressing rates acting on individual fault elements are found through a one-time calculation in which all fault elements slide backwards at specified long-term geologic rates. This insures that long-term stressing rates are consistent with observed slip rates. The method provides a lumped representation of all stressing sources, including tectonic stressing and stress transfer from off-fault yielding, consistent with prescribed/observed long-term fault slip rates. A characteristic of backslip stressing is that regions of uniform long-term slip rate require non-uniform stressing rates—stressing rates vary most strongly at the ends and bottom of the fault.

3. Model Characteristics

Except as noted, the simulations employ fault models with uniform initial normal stresses of 150 MPa and uniform constitutive properties of $a = 0.012$, $b = 0.015$, $\mu_0 = 0.6$, and $D_c = 10^{-5}$ m; these are typical laboratory values (DIETERICH, 2007). Three fault surface geometries are employed in isolation or as components of fault systems: (1) continuous planar surfaces, (2) continuous surfaces with random fractal roughness, and (3) discontinuous fractally segmented faults in which segment boundaries are delineated by fault step-overs.

The fractally rough surfaces are generated using the method of random mid-point displacement (FOURNIER et al., 1982) whereby the fault surface is repeatedly divided and the midpoints of the new divisions are randomly displaced by a normally distributed random variable with a standard deviation given by

$$y = \beta l^H, \tag{8}$$

where l is the current subdivision length; the factor β is the rms slope at a reference division length $l = 1$; and the exponent H has values 0–1. In the following we use $H = 1$, which generates self-similar profiles. At large scales (wavelengths > 1 km) real faults have discernible roughness indicating values of β approximately in the range 0.01–0.05 (DIETERICH and SMITH, 2009).

For fractally segmented faults we again employ the random mid-point displacement method used to generate the fractally rough faults but with two modifications. First, during the subdivision process every segment is not necessarily subdivided; instead there is some probability for a segment to be subdivided (the probability is 0.85 for the models used in this study). Second, the resulting points are taken as the centers of planar segments, all of which are parallel to the overall fault (rather than as the vertices of a continuous triangulated surface). This leads to a fractal (power-law) distribution of segment sizes and offsets between them. Any segments larger than the desired patch size are subdivided down to the desired patch size but with all these patches being coplanar and continuous.

Examples of isolated faults with fractal roughness and fractal segmentation are shown in Fig. 1. The slip events that are shown in Fig. 1 are taken from simulations of 500,000 earthquake events on those faults. Compared to planar faults, which tend to have smooth displacement profiles along the rupture, the somewhat patchy slip for the events in Fig. 1 appears to be characteristic of the fractal faults. Larger earthquake ruptures on faults with fractal roughness break through both releasing and constraining bends, however smaller earthquake ruptures tend to occur preferentially along constraining fault bends.

The simulations produce a range of rupture characteristics that are comparable to those obtained in detailed fully dynamical calculations. Rupture speeds for large earthquakes in these simulations generally range 2.0–2.4 km/s, which is reasonable given the implied shear-wave speed of 3.0 km/s used to set slip speed. Rupture growth and slip can be crack-like, or consist of a narrow slip-pulse (HEATON, 1990). Factors favoring crack-like behavior in the simulations are relatively smooth initial stresses and weak healing (re-strengthening of the fault) following termination of slip, while slip-pulse behavior arises with heterogeneous initial stresses and strong fault healing following rupture termination. This behavior is consistent with fully dynamical rupture simulations (BEROZA and MIKUMO, 1996; ZHENG and RICE, 1998). In our simulations, healing is set by the rate-state

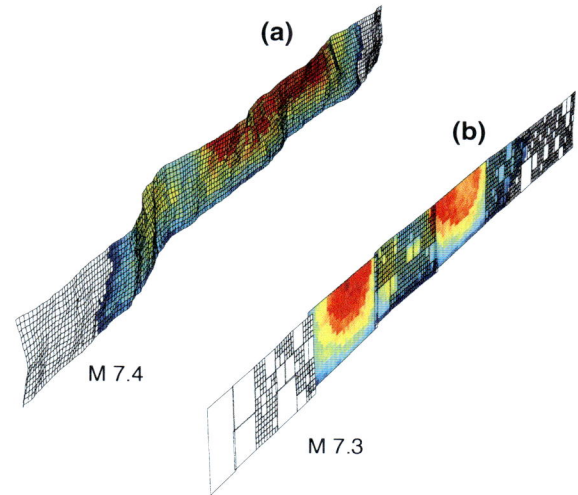

Figure 1
Coseismic slip on isolated strike-slip faults with **a** fractal roughness and **b** fractal segmentation. The *color scale* indicates slip in a single large earthquake that occurred in simulations with 500,000 events. The rough fault uses an exceptionally large amplitude factor ($\beta = 0.10$) to illustrate the character of the fractal roughness. With the segmented fault only the segment boundaries are shown—individual segments are made up of 1 km × 1 km elements. The amplitude factor for the segmented fault is $\beta = 0.04$. Both fault models have 3,015 elements.

frictional properties and by a dynamic stress overshoot parameter that determines the shear stress at the termination of slip relative to the sliding friction. During an earthquake, if sliding stops at stresses that are sufficiently below the sliding friction, then healing outpaces re-stressing from continuing slip on adjacent regions of the fault. This inhibits renewed or continuing slip and leads to pulse-like ruptures. Conversely, if sliding stops at or only slightly below the sliding friction, then continuing slip on adjacent regions of the fault can immediately trigger renewed sliding before healing can occur. This effect favors on- and off-switching of slip, which approximates continuous slip over broad regions at slower slip speeds, which is characteristic of crack-like ruptures.

Although the simulations employ approximations of the earthquake rupture processes to achieve computational efficiency, we believe those approximations do not seriously distort the model results. The key performance measure for earthquake rupture calculations in seismicity simulations is the accuracy with which the calculations predict (a) the size of

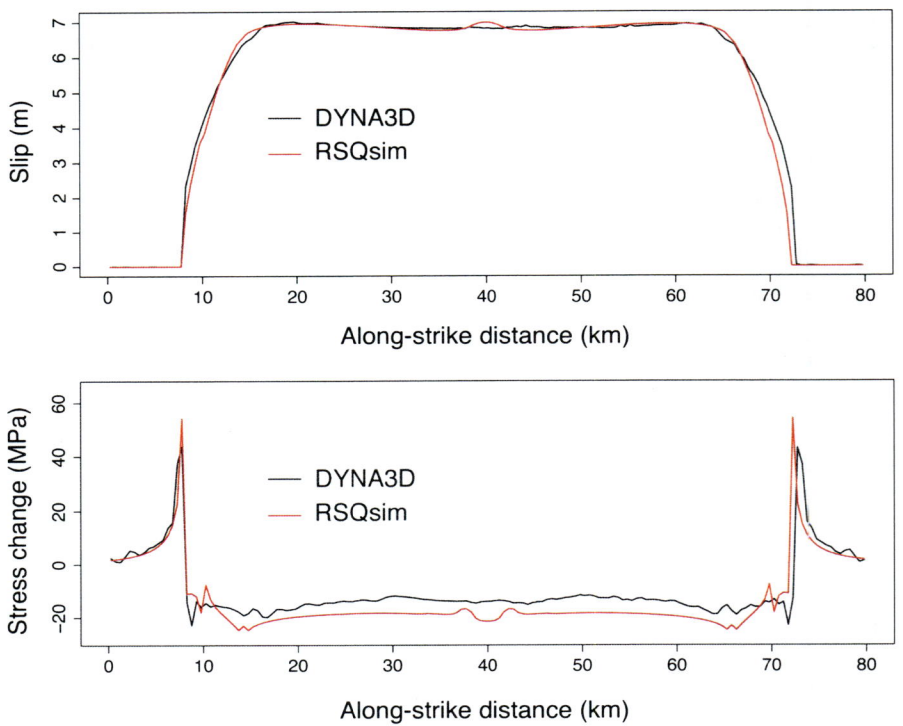

Figure 2
Comparison of slip and shear stress change from 3-D bilateral rupture simulations on a planar strike slip fault with RSQSim and DYNA3D (RICHARDS-DINGER et al., in preparation). The total rupture length is 64 km and slip extends from the surface to a depth of 8 km. The computations employ 500 m × 500 m fault elements.

earthquake rupture given a stress state at the initiation of an earthquake, and (b) the slip distribution in that rupture, which determines the details of the stress state in the model following an earthquake (and therefore subsequent earthquake history). In collaboration with our colleague David Oglesby and with the assistance of student interns Christine Burrill and Jennifer Stevens we have undertaken a program of tests that compare single-event RSQSim simulations with detailed fully dynamic finite element calculations (RICHARDS-DINGER et al., in preparation). Figure 2 shows one in a series of comparisons of RSQSim with DYNA3D, a fully dynamic 3-D finite-element model. DYNA3D employs slip-weakening friction at the onset of earthquake slip with specified static and sliding friction. Hence, it was necessary to match the rate-state friction parameters and initial conditions as closely as possible to the friction, stress, and slip-weakening conditions in DYNA3D. The example in Fig. 2 is for a bilateral rupture on a strike-slip fault with uniform initial stress and sliding resistance during earthquake slip. Other comparisons of simple bilateral and unilateral ruptures under conditions of uniform initial stress give similar results.

Similarly, models with heterogeneous stresses are in good agreement. This includes models with heterogeneous normal stress that produce highly complex rupture histories with heterogeneous earthquake stress drop. The principal mismatch between the simulation methods occurred in a case in which initial shear stress was smoothly tapered over a distance of 20 km to progressively impede rupture propagation. Both models produced very similar slip and stress patterns, however the fully dynamic rupture penetrated about 3 km farther into the low stress region than the quasi-dynamic rupture, resulting in final rupture lengths of 57 and 60 km for RSQSim and DYNA3D, respectively. The somewhat longer rupture obtained with the dynamic finite-element model may arise from dynamic stress effects, which are not represented in RSQSim. Alternatively it may be caused by differences in the failure laws that

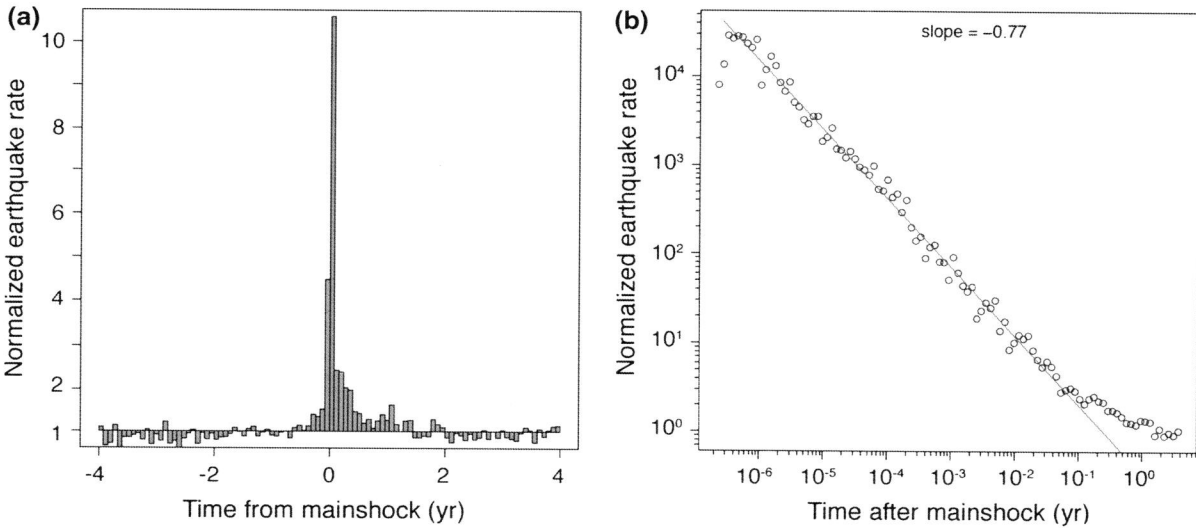

Figure 3
Foreshocks and aftershocks from a simulation of 500,000 earthquakes spanning 16,370 years. The simulations use the smooth fault version of the idealized fault system model described below with 6760 1 km × 1 km elements. These records are composite plots formed by stacking the rate of seismic activity relative to main-shock times. Main shocks are $6 < M < 7$ separated by at least 4 years from any other events $M > 6$. Earthquake rates are normalized by the average background rate. The same data set is used in **a, b**. **a** Composite plot of showing foreshocks and aftershocks relative to main-shock time. **b** Characteristic decay of aftershocks by t^{-p}, with $p = 0.77$. Foreshocks (not shown) have a similar power-law decay by time before the main shock with $p = 0.92$.

control rupture growth used in the two codes. The direct rate strengthening effect with rate- and state-dependent friction law used in RSQSim results in transient rate-strengthening at the rupture front that tends to impede rupture growth relative to the rate-independent slip weakening law in DYNA3D. Additional tests are underway.

The simulations produce clustered seismicity that includes foreshocks, aftershocks and occasional large earthquake clusters. Composite seismic histories constructed by stacking seismic activity relative to main-shock occurrence times (Fig. 3) replicate the Omori aftershock decay of aftershock rates by t^{-p} with $p \sim 0.8$, and foreshocks have Omori-like dependence of foreshock rates by time before the main shock with $p \sim 0.9$. Because clusters of large events sometimes occur that produce overlapping aftershock sequences, the stacking procedure used to construct the record in Fig. 3 employed an added constraint that rejected sequences if more than one earthquake $M \geq M_{\min}$ occurred in a ±4 year interval. The example presented in Fig. 3 was obtained with the smooth fault section version of an idealized fault system described below (e.g., Fig. 5) that consisted of 13 fault sections of various lengths. Clustering in systems with fractal roughness and fractal segmentation is somewhat greater than simulations with planar faults. Previously, Dieterich (1995) showed that productivity of foreshocks and aftershocks (i.e., the ratios of the numbers of foreshocks and aftershocks to main shocks in a magnitude interval) is controlled by the product $a\sigma$, where a is the rate-state parameter appearing in Eq. 3 and σ is normal stress.

The magnitude frequency distributions of simulated earthquakes for isolated planar faults consistently follow a power law up to about $M6.0$, together with a pronounced peak that marks characteristic earthquakes that rupture the entire fault. There is a very pronounced deficiency of events between $M6.0$ and $M7.4$. The upper limit of earthquake magnitude for the power-law portion of the distributions corresponds to rupture dimensions of about 10 km (compared to a vertical fault dimension of 15 km). The characteristic earthquake behavior reflects a strong tendency of earthquake ruptures that reach dimensions greater than about 10 km to continue propagating to the limits of the model. Following such end-to-end ruptures the stress conditions are reset to a similar average value, which in turn results in highly periodic recurrence of the largest earthquakes.

The principal difference seen in simulations with fractally rough faults is slight enrichment of earthquakes in the magnitude interval between the power-law region and the characteristic earthquakes. However, there remains a strong deficiency of events in this range, even using an extreme roughness with $\beta = 0.10$. The use of fractal segmentation has a significantly stronger impact on filling the deficiency between the power-law region and the characteristic earthquakes. Also, as the amplitude of fractal offsets increases, the frequency of characteristic earthquakes decreases—and their recurrence becomes less periodic. At the extreme of fractal segmentation that we studied ($\beta = 0.04$) no end-to-end ruptures occurred in a simulation of 10^6 events.

4. Recurrence Distributions

We have assembled distributions of the time intervals separating earthquakes above a minimum magnitude M_{min} that affect the same point on a fault. Distributions of this type reflect local characteristics of fault stressing and failure processes and form the basis for estimating conditional time-dependent probabilities along a section of a fault given the time of the previous earthquake. The density distributions are constructed by first binning the recurrence intervals for individual fault elements then summing the binned data with the data from other elements that make up a designated fault section. Fault sections consist of many elements (320 to >1,000) and represent distinct structural components such as an isolated fault or the portion of a fault that lies between two branching points in a fault system. To construct the distributions, sequences of 5×10^5 to 10^6 earthquakes were simulated representing records of extending to about 35,000 years at fault slip rates of 25 mm/year.

Models of seismicity on single isolated strike-slip faults employ a planar fault surface, a fault with random fractal roughness, and a fault with random fractal segmentation. In each case the fault dimensions are 201 km long, 15 km deep and consist of 3,015 elements with nominal dimensions of 1 km^2. The long-term slip rate is 25 mm/year. Because the effect of fractal roughness on the recurrence statistic is rather weak, results are shown only for the extreme case with $\beta = 0.10$. The amplitude parameters for faults with fractal segmentation use $\beta = 0.02, 0.03$, and 0.04. In all non-planar models $H = 1.0$.

The recurrence distributions for each of the single fault models (Fig. 4) differ in minor details, nonetheless all share several common characteristics. (1) The distributions change with earthquake magnitude. (2) There is a very narrow peak at the shortest intervals (0–12 years). This peak is strongest for the smallest magnitude threshold $M_{min} \geq 5$ and decreases as M_{min} increases. When examined in detail, the earthquakes in this interval are found to represent foreshocks, aftershocks, and regions of overlapping slip for earthquake pairs. Within this 0–12 year interval the recurrence rates have the characteristic Omori decay by t^{-p} shown in Fig. 3. (3) There is a pronounced peak of recurrence times around 150–200 years, indicating a strong periodic component to recurrence. This peak appears in all the distributions using different M_{min}, but it results from periodicity of large characteristic earthquakes that rupture the most or all of the fault. (4) The distributions that employ smaller magnitude thresholds $M_{min} \geq 5.0$ are somewhat complex with a more-or-less uniform density of recurrence times prior to the characteristic earthquake peak. The close similarity of the distributions $M_{min} \geq 6.0$ and $M_{min} \geq 7.0$ reflect the relative dearth of earthquakes $6 \leq M < 7$ compared to characteristic earthquakes $M \geq 7.0$ that rupture most or all of the fault.

The distributions obtained with the isolated planar fault and with faults which have fractal roughness are quite similar. The principal difference in the density distributions is a progressive shifting of the characteristic earthquake peak to shorter times as roughness increases. The peak in the distributions of recurrence time for the planar fault is ≈ 190 years compared to ≈ 150 years for a very rough fault with $\beta = 0.10$. Also, the longest recurrence interval for the planar fault is approximately 220 years, while the simulations with fractal roughness have a continuing low incidence of recurrence exceeding 400 years. These differences arise because fractal roughness introduces

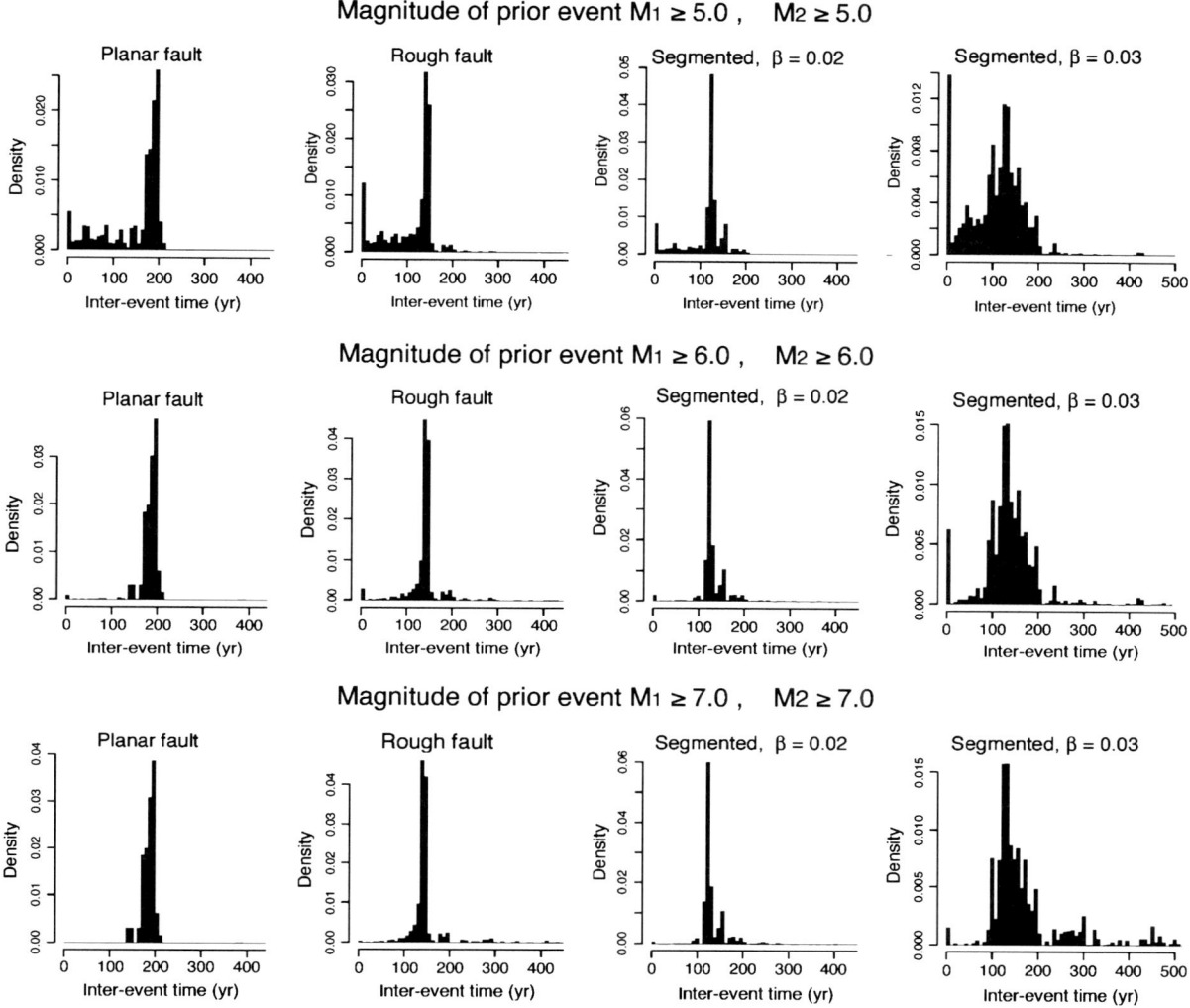

Figure 4
Density distributions of recurrence times for single isolated faults. The distributions give the density of inter-event times between successive earthquake pairs above the minimum magnitudes M 1 and M 2 for the first and second events, respectively, that define a pair.

weak barriers that inhibit slip and sometimes interrupt full growth of large earthquakes over the entire fault. This results in less slip and shorter recurrence time (on average) for large characteristic earthquakes, and occasionally skipped recurrence cycles.

Segmentation more strongly alters the distributions than fault roughness. Segmented faults with $\beta \leq 0.02$ produce distributions that are nearly identical to the rough fault with $\beta = 0.10$. However, at $\beta \geq 0.03$ fractal segmentation significantly broadens the quasi-periodic peaks in the recurrence distributions. For example using $M_{min} \geq 7.0$ the standard deviations for recurrence with a planar fault and a fault with fractal roughness ($\beta = 0.10$) are 14 and 45 year, respectively; compared to standard deviations of 28, 111, and 296 year for segmented faults with $\beta = 0.02$, 0.03, and 0.04, respectively. With increasing separation across segmentation boundaries (increasing β) the rate of end-to-end earthquake ruptures decreases. At $\beta \geq 0.04$, no end-to-end earthquake ruptures occurred in a simulation with 200,000 events, although a broad peak in the distributions persists.

Simulations with a more complex but highly idealized fault system model (Fig. 5) were conducted to examine the effects of the geometric component of fault interactions on recurrence statistics. In the following we use the term fault section to indicate the portion of an individual fault that lies between branch points in the fault system. The model consists of parallel and branching faults and incorporates a variety of configurations, fault section lengths, and slip rates (see Fig. 5). The model consists of 6,760 elements with nominal dimensions of 1.5 km × 1.5 km. In addition to the model with smooth fault sections, several versions were implemented with fractal roughness and fractal segmentation using a range of values of β. To test for possible model resolution effects, the smooth fault version also used 1 km × 1 km and 3 km × 3 km patches. One simulation was carried out with a different set of rate-state friction parameters ($a = 0.007$ and $b = 0.010$).

Representative density distributions for earthquake recurrence on the smooth fault version of the fault system model version are shown in Fig. 5. The characteristic features of distributions for single isolated faults described above are also seen in the density distributions for the fault system (peak at short times, magnitude dependence, quasi-periodicity, complexity at small M_{min}). Also it is very evident that the density distributions change significantly with position within the fault system and have a greater variety of forms than the isolated fault simulations. For example, at $M_{min} \geq 5$ the forms include an approximately monotonic decay of density with time (Fig. 5, section 7), long interval of constant density followed by comparably long tail with decaying density (Fig. 5, section 4), and multi-peaked distributions (Fig. 5, sections 6, 9 and 10).

At larger magnitudes ($M_{min} \geq 6$, $M_{min} \geq 7$, and $M_{min} \geq 7.5$) the distributions maintain strong positional dependencies, but generally take somewhat simpler forms. Of the 13 fault sections in the model, all but three density distributions (including sections 4 and 6 in Fig. 5) have a single well-defined peak indicating quasi-periodicity of recurrence times. However, there are large differences in the shapes and widths of the peaks. Compared to the isolated fault models, the distributions generally have much larger spreads of recurrence times than the isolated fault models, which is expected given the increased complexity of interactions that determine the stressing history of the faults.

Figure 5
Idealized strike-slip fault system and density distributions for recurrence for representative fault sections (section numbers are given in the top panel as circled numbers, e.g., ④). The faults extend from the surface to a depth of 15 km. Motion on the fault is right lateral and the slip rates for each fault section are indicated in the top panel. The probability density distributions are for the smooth fault version of the model. See Fig. 6 for comparisons with models that employ fractal roughness and fractal segmentation of the fault sections.

We attempted to fit a variety of analytic probability distributions (e.g., Weibull, log-normal, and Brownian passage time) to these recurrence distributions. None of these analytic forms fit any of the entire (i.e., including the short-time power-law behavior) empirical distributions. If the short-time part of the empirical distributions is removed (or, equivalently, we attempt to fit the empirical distributions with the sum of a power law and one of the aforementioned analytic distributions) then a few of the distributions can be fit reasonably well by one or the other of the analytic forms, however most cannot.

Comparisons of distributions of interevent times for smooth fault sections with those using fractal roughness and fractal segmentation are summarized in Fig. 6. To facilitate comparisons we use cumulative distributions, which permits results to be plotted together. The surfaces with fractal roughness (with β up to 0.10) closely follow those with smooth surfaces. Indeed, the differences between the rough and smooth surfaces are smaller with the fault system model than with the isolated fault models. This perhaps indicates that stress interactions that are linked to system geometry override local fault geometry in setting recurrence characteristics. Similarly, weak to moderately segmented fault surfaces ($\beta \leq 0.03$) produce distributions that are very similar to the distributions with rough surfaces and are not plotted. The distributions with strongly segmented faults ($\beta = 0.04$), which are shown in Fig. 6, diverge somewhat from the other distributions, but generally retain the shapes of the other distributions. The single exception to this is at fault section 4, which is a short section with low slip rate that branches from longer fault sections with higher slip rates. Because large earthquakes have longer rupture lengths than the length section 4, of necessity such earthquakes on

Earthquake Recurrence in Simulations

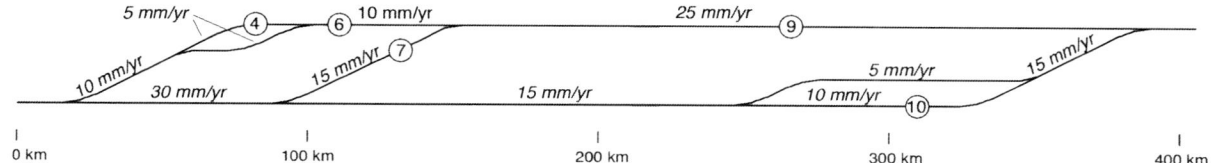

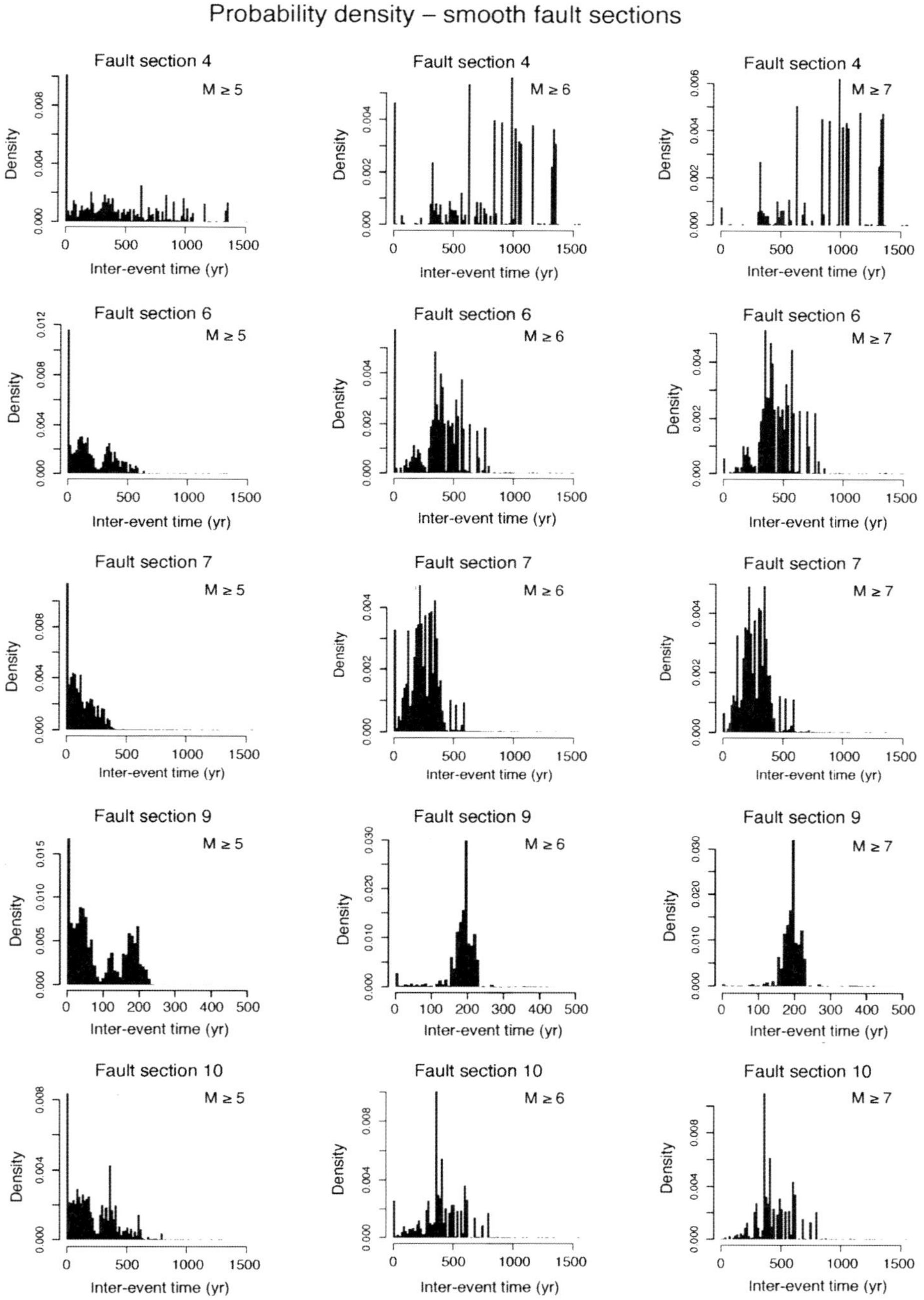

Probability density – smooth fault sections

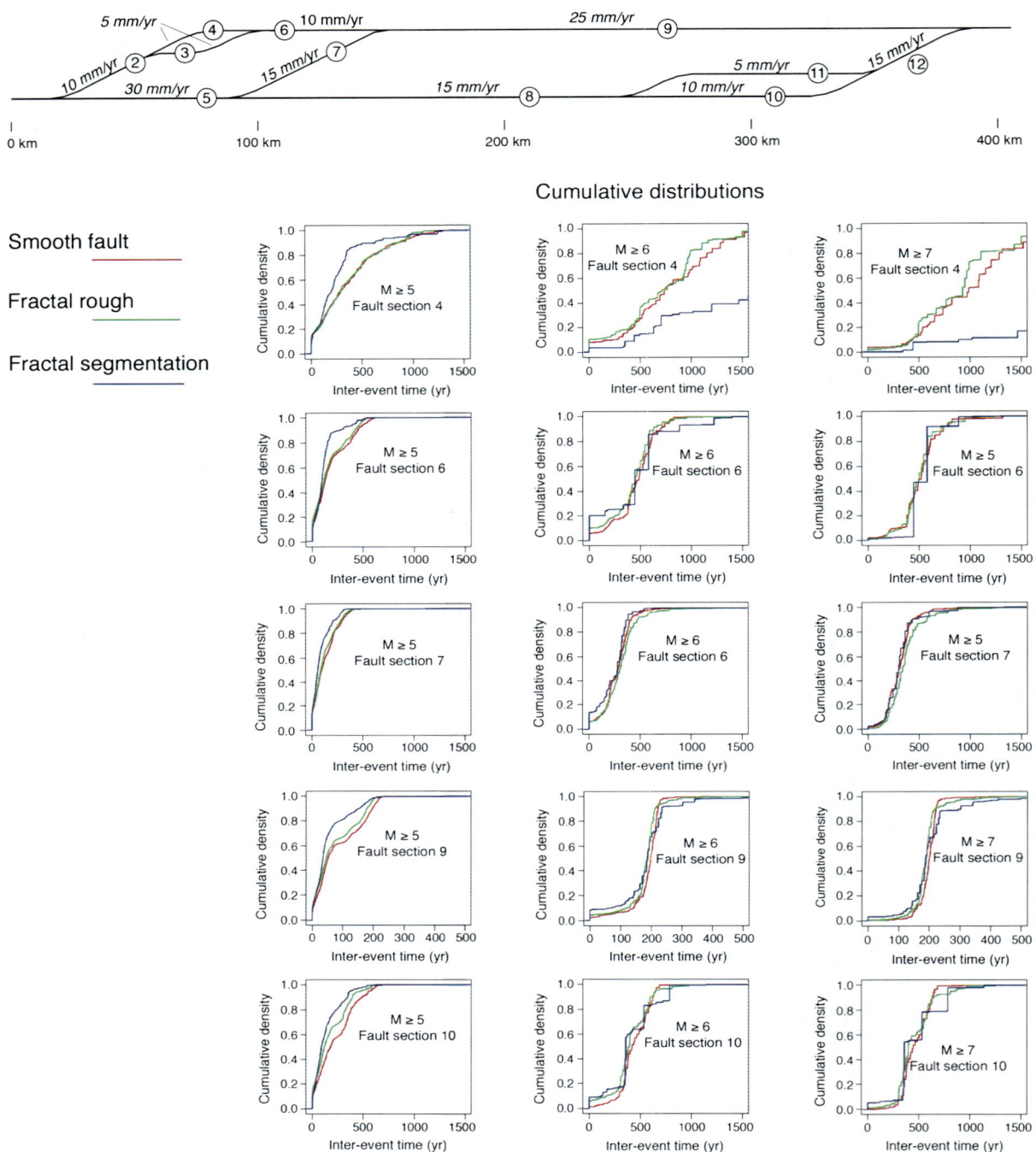

Figure 6
Cumulative distributions of recurrence times for earthquakes of various magnitudes on selected sections of an idealized fault system (*upper panel*) with three different forms of small-scale geometry: Smooth fault sections (*red*), fractal roughness with $\beta = 0.1$ (*green*), and fractal segmentation with $\beta = 0.04$ (*blue*).

Table 1
Clustering of earthquakes $M \geq 7$ in fault system simulations

Model	Total number of events	Number ($M \geq 7$)	Single events	Double events	Triple events
Planar faults	299,000	196	130	27.5	3.6
Fractal roughness $\beta = 0.1$	377,000	237	152	35.8	4.6
Fractal segmentation $\beta = 0.02$	394,000	221	144	36.1	1.8
Fractal segmentation $\beta = 0.04$	607,000	274	58.4	32.1	38.0

All numbers are per 10,000 years of simulated time

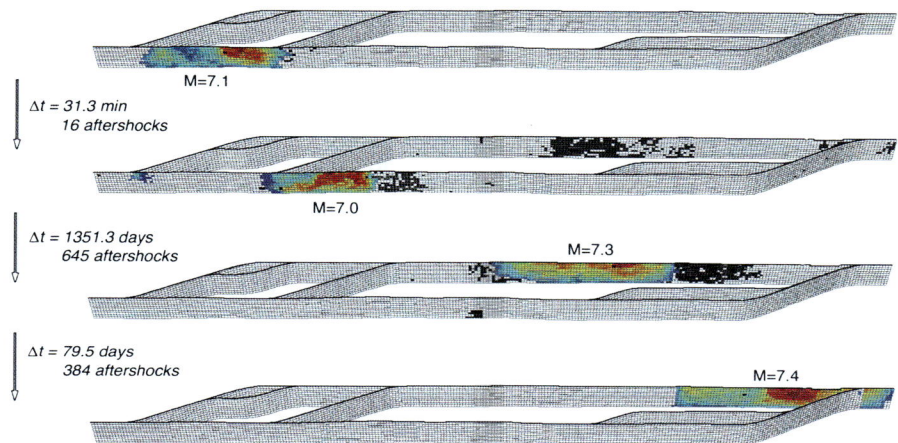

Figure 7

An example of a cluster of four large earthquakes occurring within a 4-year period. In each panel the *colors* indicate the amount of slip in one of the large earthquakes; the hypocenter of the large earthquake is marked in *black*; and, in addition, the hypocenters of all events taking place after the given large event (but before the subsequent large event) are also shown in *black*. The *colorscale* for slip runs from cool to hot colors for small to large values of slip, respectively. The maximum slip in the four large events is 4.3, 3.3, 4.9, and 5.4 m, in chronologic order.

section 4 must also involve a neighboring segment. Apparently, with increasing β there is a progressive decoupling of slip across section boundaries that reduces frequency of large earthquakes on section 4.

The regional differences in the distributions appear to be quite stable and independent of model details. Simulations with different combinations of model parameters were tested. These include reversing the sense of slip in the fault system from right lateral to left lateral, use of different element dimensions (1, 1.5, 3 km) and different combinations of constitutive parameters. The distributions must be sensitive to earthquake stress drop, because larger stress drops will require greater elapsed time to recover stress—consequently the alternative model using a different set of constitutive parameters was designed to give the same average earthquake stress drop. The only one of these variations that produced substantial differences in the recurrence statistics was the most coarsely resolved model (patches with side-lengths of 3 km). This model produced considerably longer average recurrence intervals for the largest ($M \geq 7$) events. This dearth of large events is presumably due to the diffculties in propagating ruptures in such a model. As models with patches of side-length ≤ 1.5 km agree with one another, we interpret the 3-km patch size model to be too coarsely parameterized for our current purposes. Other than this one exception, we find that these various changes have only minor effects on the distributions that are comparable to the variations seen in Fig. 6.

An interesting feature of the fault system simulations is occasional clustering of large earthquake events. Clusters of large events, though relatively uncommon, are certainly a well-established characteristic of earthquake occurrence (KAGAN and JACKSON, 1991, 1999). With the idealized fault system, earthquakes $M \geq 7$ occur somewhere in the system at an average frequency of one every 36–51 years, and most are isolated by four years or more from other large earthquakes. However, some large events occur as pairs, and even more rarely as triples (Table 1). Figure 7 shows an unusual set of four large events that propagate across much of the fault system. The intervals between large earthquakes in clusters vary from several seconds to 4 years, which is an arbitrary maximum interval used here in defining event clusters. The distribution of intervals between large events in clusters decays by Omori's law (with $p \sim 0.9$). In some cases the regions of slip in a cluster very slightly overlap. As shown in the example in Fig. 7, the subsequent large earthquake ruptures during particularly strong aftershock sequences, and the point of nucleation falls within this aftershock region.

5. Summary and Discussion

Earthquake nucleation with rate- and state-dependent friction strongly affects the statistics of earthquake recurrence in the simulations, particularly at short time intervals and at smaller earthquake magnitudes. Density distributions of recurrence intervals have very narrow peaks at the shortest times (0–12 years) that consist of foreshocks, aftershocks, and earthquake clusters. Rates of recurrence within this peak decay by $t^{-0.8}$. Clustering in the form of large-earthquake pairs (and more rarely triples) is a consistent feature of the fault system simulations, but at low rates ($\sim 20\%$ of $M \geq 7$ events are followed within 4 years by another such event). Intervals between large earthquake pairs vary from a few seconds to 4 years (our arbitrary cutoff to define large event clusters) and also follow an Omori decay, which is consistent with earthquake pairs in nature (KAGAN and JACKSON, 1991, 1999). From a regional earthquake hazard perspective the clusters represent a continuing interval of significantly increased hazard following large earthquakes. The follow-on events in large earthquake clusters initiate in the aftershock regions of the prior events and their occurrence correlates with especially high aftershock rates. There is little or no overlap of the areas of slip in the clusters.

The shapes of the recurrence distributions with isolated faults change with earthquake magnitude threshold $M_{\min}$ and form a narrow characteristic earthquake peak at high magnitudes. The characteristic earthquake peak occurs because earthquake ruptures that reach a critical size (about 10 km for faults that extend from the surface to 15 km) have a strong tendency to continue to propagate to the limits of the model. The resulting end-to-end ruptures are highly periodic because the stress after the earthquakes is reset to a similar average state following each end-to-end rupture. Strong segmentation of faults reduces the periodicity and in the extreme eliminates end-to-end ruptures.

Recurrence distributions for individual fault sections within a fault system depend on magnitude and take great variety of forms that change with position within the fault system. In addition, the recurrence intervals have considerably wider distributions than isolated faults. The distributions appear to be quite insensitive to local details such as the addition of fault roughness. Limited tests that vary element dimensions and use different combinations of constitutive parameters reveal that the results are quite stable. These characteristics indicate that gross fault system geometry plays a primary role in establishing the characteristics of stress evolution that control earthquake recurrence. Above some limiting separation, fault step-overs form effective impediments to the propagation of earthquake ruptures and have a significant though lesser impact on the recurrence distributions.

One reason for undertaking this study was to begin to explore possible applications of earthquake simulations to assessments of regional earthquake probabilities. Current standard methodologies for assessing time-dependent earthquake probabilities employ models of regional seismicity that include information of past earthquakes (such as time and extent of earthquake slip) together with idealized probability density functions (PDFs) for the recurrence of earthquake slip. However, major sources of uncertainty in such assessments relate to both the choice of an appropriate functional form for a PDF and in specifying parameters for implementing the idealized PDF.

Questions surrounding current usage of generic PDFs in assessment of earthquake probabilities arise for a number of reasons. First, fundamentally different classes of PDFs based on Omori-type clustering, Poisson statistics, and quasi-periodicity individually capture well-established aspects of earthquake recurrence statistics, however no single distribution fully represents the range of observed behavior. For example, recent assessments of earthquake probabilities in California (e.g., WORKING GROUP ON CALIFORNIA EARTHQUAKE PROBABILITIES (WGCEP), 2007) used weighted estimates based on quasi-periodic and Poisson (exponential distribution) models of earthquake occurrence, which individually yield very different probabilities. Also, a number of uncertainties arise in implementing the generic PDFs because largely ad hoc assumptions must be made regarding relationships between stress accumulation and failure, characteristic earthquakes, probabilities of multi-segment earthquakes, and magnitude–frequency statistics of large earthquakes on specific faults. Finally, the results of this study indicate that the distributions have significant magnitude and positional dependencies that are not considered in current approaches.

In place of idealized PDFs the use of empirical density distributions for probabilistic assessments could potentially address these shortcomings. An advantage of such an approach is that one would not be restricted to simple functional forms that cannot describe intrinsically complicated statistics, and most implementation and scaling issues relating to the use of PDFs are completely avoided. Also, magnitude dependencies and strong local variations in the recurrence distributions that are tied to fault system geometry (an observable) could be incorporated into probabilistic assessments. Ideally, one would like to use earthquake data for this purpose, however, long earthquake histories covering many average recurrence times of the largest events of interest are required to define local empirical distributions—clearly historic and paleoseismic data are inherently inadequate for this purpose.

Of necessity and design the simulations in this exploratory study are quite idealized. Certainly the practical use of fault system simulators in the assessment of time-dependent earthquake probabilities will require additional study. These include detailed region-specific simulations, and proper quantification of the effects of uncertain model parameters on the distributions. Our results demonstrate that gross fault system geometry strongly affects the shape of probability distributions for the recurrence of earthquake slip, and as a general rule the distributions are quite insensitive to small-scale geometric details. A possible exception may be sensitivity of the distributions to segmentation beyond some threshold in step-over distance. Because such features may be difficult to characterize at seismogenic depths, this effect may represent a significant source of uncertainty and merits close attention. In addition to time-dependent earthquake nucleation and the effect of fault system geometry in recurrence statistics investigated here, other model parameters will impact earthquake recurrence statistics. These include fault constitutive parameters, earthquake stress drop, and processes that produce stressing transients. A first-order dependence of mean recurrence time on fault slip rate and stress drop has been previously explored and characterized by WARD (1996) and RUNDLE et al. (2004). In our simulations, stress drop is controlled by fault normal stress and fault constitutive properties. Fault creep and viscoelastic relaxation following large earthquakes are widely documented and produce stressing rate transients that may impact recurrence statistics. Similarly, effective stress transients due to pore-fluid pressure changes could possibly affect recurrence statistics as well, though such effects have proven difficult to document. Though meriting further investigation, the effect of stress transients on earthquake occurrence appears to be at least partially mitigated by the rate-state nucleation process which is strongly self-driven, making nucleation times relatively insensitive to transient changes of stressing rates (DIETERICH, 1994).

Finally we note that with current standard methods, based on PDFs for earthquake recurrence intervals, the calculation of time-dependent probabilities using paleoseismic data and historical records of past earthquakes requires a number of interpretive and modeling steps that substantially increase uncertainties in ways that are difficult to quantify. Essentially, these steps convert very limited data on timing of an earthquake, and information on magnitude or amount of slip at a point on a fault, to a spatial

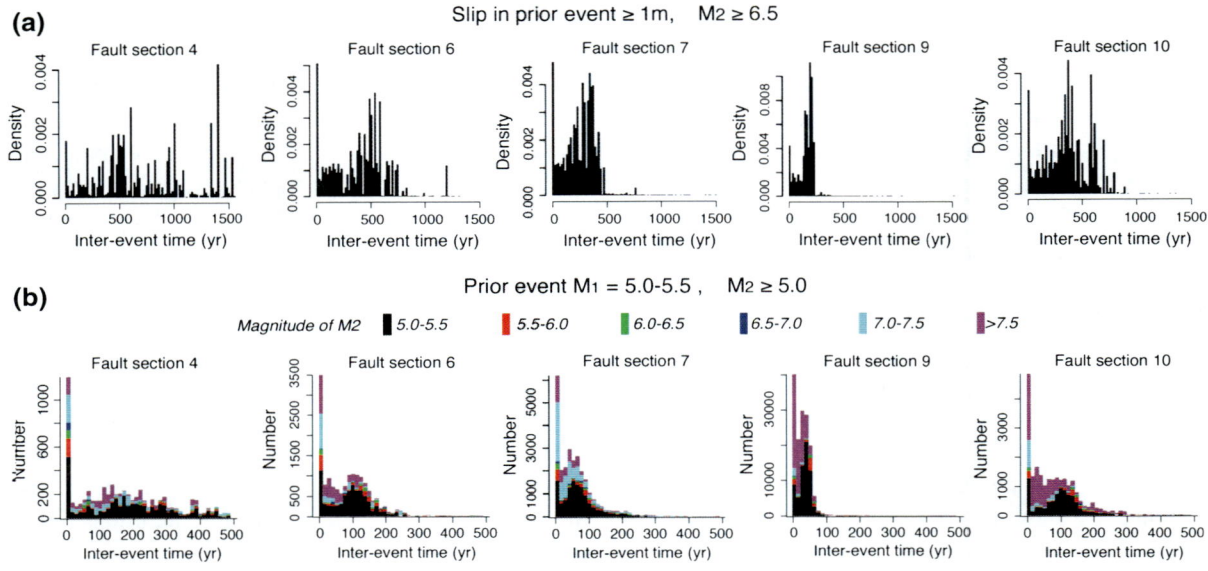

Figure 8
Examples of alternative parameterizations of density distributions of recurrence times. Data are from the smooth fault version of the fault system model of Figs. 5 and 6.

distribution of slip over an assigned section of fault. Simulations provide the capability to define specialized empirical density distributions that directly utilize primary observational data without the modeling steps and assumptions of current methods. Figure 8 illustrates two examples of alternative distributions. The first distribution (Fig. 8a) is defined in terms of magnitude of slip at an observation point in the prior earthquake. It is intended to directly utilize paleoseismic data on the amount of slip in the prior earthquake at some point on a fault, with no other direct information on earthquake magnitude or extent of slip. The second distribution (Fig. 8b) is intended to represent a case in which the time and magnitude (with some uncertainty) of the prior earthquake are both known. The distribution provides information about both the time of the following event and also its magnitude. Both distributions relax the assumptions of characteristic earthquakes and allow for earthquakes of varying sizes.

The results in Fig. 8b are rather interesting. Broad quasi-periodic peaks for earthquakes M5–M5.5 following an earthquake M5–M5.5 are quite evident in these distributions, but the sub-distributions for $M \geq 7.5$ following a M5.5 earthquake decay monotonically and roughly follow an exponential distribution indicating a constant Poisson rate of occurrence following a M5.5 event. Some other examples of specialized density distributions that might be assembled directly from the synthetic catalogs include (a) situations in which historical records indicate the prior earthquake may lie within a region although causative fault is uncertain, (b) recurrence of slip exceeding some amount at a specific site, in some time interval (of possible interest for lifelines that cross faults), and (c) probability of future earthquake by time and distance from a site.

Acknowledgements

We thank Euan Smith and two anonymous reviewers for comments which improved the manuscript. Funding for this work was provided by grants from the USGS (#G09AP00009), and the Southern California Earthquake Center (SCEC#08092). SCEC is funded by NSF Cooperative Agreement EAR-0529922 and USGS Cooperative Agreement 07HQAG0008. The SCEC contribution number for this paper is 1273.

Open Access This article is distributed under the terms of the Creative Commons Attribution Noncommercial License which permits any noncommercial use, distribution, and reproduction in any medium, provided the original author(s) and source are credited.

REFERENCES

BELARDINELLI, M. E., BIZZARRI A., and COCCO, M. (2003), *Earthquake triggering by static and dynamic stress changes.* J. Geophys. Res. (Solid Earth) *108*, 2135±, doi:10.1029/2002JB001779.

BEN-ZION, Y. and RICE, J. R. (1997), *Dynamic simulations of slip on a smooth fault in an elastic solid.* J. Geophys. Res. *102*, 17771–17784, doi:10.1029/97JB01341.

BEN-ZION, Y., and SAMMIS, C. G. (2003), *Characterization of fault zones*, Pure App. Geophy. *160*, 677–715

BEROZA, G. C., and MIKUMO, T. (1996), *Short slip duration in dynamic rupture in the presence of heterogeneous fault properties*, J. Geophys. Res. *101*, 22449–22460, doi:10.1029/96JB02291.

BONNET, E., BOUR, O., ODLING, N. E., DAVY, P., MAIN, I., COWIE, P., and BERKOWITZ, B. (2001), *Scaling of fracture systems in geological media.* Rev. Geophys. *39*, 347–384, doi:10.1029/1999RG000074.

BRUNE, J. (1970), *Tectonic stress and the spectra of seismic shear waves from earthquakes.* J. Geophys. Res. 75(26), 4997–5009.

CHESTER, F. M., and CHESTER, J. S. (2000), *Stress and deformation along wavy frictional faults.* J. Geophys. Res. *105*, 23,421–23,430, doi:10.1029/2000JB900241.

DIETERICH, J. (1981), *Constitutive properties of faults with simulated gouge.* In CARTER, N. L., FRIEDMAN, M., LOGAN, J.M., and STERNS, D. W. (eds), *Monograph 24, Mechanical behavior of crustal rocks*, Am. Geophys. Union, Washington, D.C., pp. 103–120.

DIETERICH, J. (1987) *Nucleation and triggering of earthquake slip: effect of periodic stresses.* Tectonophysics *144*, 127–139, doi:10.1016/0040-1951(87)90012-6.

DIETERICH, J., *Applications of rate-and-state-dependent friction to models of fault slip and earthquake occurrence.* In SCHUBERT, G. (ed.) *Treatise on Geophysics*, Vol. 4 (Elsevier, Oxford 2007).

DIETERICH, J., and SMITH, D. (2009), *Non-planar faults: mechanics of slip and off-fault damage*, Pure Appl. Geophys., *166*, 1799–1815.

DIETERICH, J.H. (1979), *Modeling of rock friction 1. Experimental results and constitutive equations.* J. Geophys. Res. *84*, 2161–2168.

DIETERICH, J. H. (1992), *Earthquake nucleation on faults with rate- and state-dependent strength.* Tectonophysics *211*, 115–134.

DIETERICH, J. H. (1994), *A constitutive law for rate of earthquake production and its application to earthquake clustering*, J. Geophys. Res. *99*, 2601–2618.

DIETERICH, J. H. (1995), *Earthquake simulations with time-dependent nucleation and long-range interactions.* J. Nonlinear Proc. Geophys. *2*, 109–120.

DIETERICH, J. H., and KILGORE, B. (1996) *Implications of Fault Constitutive Properties for Earthquake Prediction.* Proc. Natl. Acad Sci. USA *93*, 3787–3794.

DUAN, B. and OGLESBY, D.D. (2006), *Heterogeneous fault stresses from previous earthquakes and the effect on dynamics of parallel strike-slip faults*, J. Geophys. Res. *111*(B10), 5309±, doi:10.1029/2005JB004138.

FLISS, S., BHAT, H. S., DMOWSKA, R., RICE, J. R. (2005), *Fault branching and rupture directivity.* J. Geophys. Res. *110*(B9), 6312±, doi:10.1029/2004JB003368.

FOURNIER, A., FUSSELL, D., and CARPENTER, L. (1982), *Computer rendering of stochastic models*, Commun. ACM *25*(6), 371–384, doi:10.1145/358523.358553.

GOMBERG, J., BLANPIED, M. L., and BEELER, N. M. (1997), *Transient triggering of near and distant earthquakes*, Bull. Seismol. Soc. Am. *87*(2), 294–309, http://www.bssaonline.org/cgi/content/abstract/87/2/294, http://www.bssaonline.org/cgi/reprint/87/2/294.pd.

GOMBERG, J., BEELER, N.M., BLANPIED, M.L., and BODIN, P. (1998), *Earthquake triggering by transient and static deformations*, J. Geophys. Res. *103*, 24411–24426, doi:10.1029/98JB01125.

GOMBERG, J., BEELER, N., and BLANPIED, M. (2000), *On rate-state and Coulomb failure models*, J. Geophys. Res. *105*, 7857–7872, doi:10.1029/1999JB900438.

HARRIS, R. A., ARCHULETA, R. J., DAY, S. M. (1991) *Fault steps and the dynamic rupture process: 2-D numerical simulations of a spontaneously propagating shear fracture*, Geophys. Res. Lett. *18*, 893–896.

HEATON, T.H. (1990), *Evidence for and implications of self-healing pulses of slip in earthquake rupture*, Phys. Earth Planet. Inter. *64*, 1–20, doi:10.1016/0031-9201(90)90002-F.

KAGAN, Y. Y., and JACKSON, D. D. (1991), *Long-term earthquake clustering*, Geophys. J. Int. *104*(1), 117–134, doi:10.1111/j.1365-246X.1991.tb02498.x.

KAGAN, Y. Y., and JACKSON, D. D. (1999), *Worldwide doublets of large shallow earthquakes*, Bull. Seismol. Soc. Am. *89*(5), 1147–1155.

KING, G. C. P., and BOWMAN, D. D. (2003), *The evolution of regional seismicity between large earthquakes*, J. Geophys. Res. *108*, 2096±, doi:10.1029/2001JB000783.

LINKER, M. F., and DIETERICH, J. H. (1992), *Effects of variable normal stress on rock friction—observations and constitutive equations*, J. Geophys. Res. *97*, 4923–4940.

MARONE, C. (1998), *Laboratory-derived friction laws and their application to seismic faulting.* Annual Rev. Earth Planet. Sci. *26*, 643–696, doi:10.1146/annurev.earth.26.1.643.

MEADE, B. J. (2007), *Algorithms for the calculation of exact displacements, strains, and stresses for triangular dislocation elements in a uniform elastic half space*, Comp. Geosci. *33*, 1064–1075, doi:10.1016/j.cageo.2006.12.003.

NIELSEN, S. B., and KNOPOFF, L. (1998), *The equivalent strength of geometrical barriers to earthquakes*, J. Geophys. Res. *103*, 9953–9966, doi:10.1029/97JB03293.

OGLESBY, D. D., DAY, S. M., LI, Y. G., VIDALE, J. E. (2003), *The 1999 Hector Mine earthquake: the dynamics of a branched fault system*, Bull. Seismol. Soc. Am. *93*, 2459–2476

OKADA, Y. (1992), *Internal deformation due to shear and tensile faults in a half-space*, Bull. Seismol. Soc. Am. *82*, 1018–1040.

OKUBO, P. G., and AKI, K. (1987), *Fractal geometry in the San Andreas fault system*, J. Geophys. Res. *92*, 345–356.

POWER, W. L., and TULLIS, T. E. (1991), *Euclidean and fractal models for the description of rock surface roughness*, J. Geophys. Res. *96*, 415–424.

RICE, J. R. (1983), *Constitutive relations for fault slip and earthquake instabilities*, Pure Appl. Geophys. *121*, 443–475, doi:10.1007/BF02590151.

ROBINSON, R., and BENITES, R. (1995) *Synthetic seismicity models of multiple interacting faults*, J. Geophys. Res. *100*, 18229–18238, doi:10.1029/95JB01569.

RUBIN, A. M., and AMPUERO, J. P. (2005), *Earthquake nucleation on (aging) rate and state faults*. J. Geophys. Res. (Solid Earth) *110*(B9), 11312±, doi:10.1029/2005JB003686.

RUINA, A. (1983), *Slip instability and state variable friction laws*, J. Geophys. Res. *88*, 10359–10370.

RUNDLE, J. B., and KLEIN, W. (1993), *Scaling and critical phenomena in a cellular automaton slider-block model for earthquakes*. J. Statist. Phys. *72*, 405–412, doi:10.1007/BF01048056.

RUNDLE, J. B., RUNDLE, P. B., DONNELLAN, A., and FOX, G. (2004), *Gutenberg–Richter statistics in topologically realistic system-level earthquake stress-evolution simulations*, Earth, Planets, and Space *56*, 761–771.

SAGY, A., BRODSKY, E. E., and AXEN, G. J. (2007), *Evolution of fault-surface roughness with slip*, Geology *35*, 283±, doi:10.1130/G23235A.1.

SAUCIER, F., HUMPHREYS, E., and WELDON, R. I. (1992), *Stress near geometrically complex strike-slip faults - Application to the San Andreas fault at Cajon Pass, southern California*, J. Geophys. Res. *97*, 5081–5094.

SAVAGE, J. C. (1983), *A dislocation model of strain accumulation and release at a subduction zone*. J. Geophys. Res. *88*, 4984–4996.

SCHOLZ, C. H. and AVILES, C. A. (1986), *The fractal geometry of faults and faulting*. In Das S., Boatwright J., Scholz C. H. (eds), *Earthquake source mechanics* (Maurice Ewing Volume 6), Am. Geophys. Union, Washington, D.C., pp. 147–155.

SHAW, B. E. and DIETERICH, J. H. (2007), *Probabilities for jumping fault segment stepovers*, Geophys. Res. Lett. *34*:L01,307, doi:10.1029/2006GL027980.

STEACY, S. J. and MCCLOSKEY, J. (1999), *Heterogeneity and the earthquake magnitude–frequency distribution*. Geophys. Res. Lett. *26*, 899–902, doi:10.1029/1999GL900135.

STIRLING, M. W., WESNOUSKY, S. G., and SHIMAZAKI, K. (1996), *Fault trace complexity, cumulative slip, and the shape of the magnitude–frequency distribution for strike-slip faults: a global survey*, Geophys. J. Internatl. *124*, 833–868, doi:10.1111/j.1365-246X.1996.tb05641.x.

TULLIS, T. E. (1988), *Rock friction constitutive behavior from laboratory experiments and its implications for an earthquake prediction field monitoring program*, Pure Appl. Geophys. *126*, 555–588, doi:10.1007/BF00879010.

WARD, S. N. (1996), *A synthetic seismicity model for southern California: Cycles, probabilities, and hazard*, J. Geophys. Res. *101*, 22393–22418, doi:10.1029/96JB02116.

WARD, S. N. (2000), *San Francisco Bay Area earthquake simulations: A step toward a standard physical earthquake model*, Bull. Seismol. Soc. Am. *90*, 370–386, doi:10.1785/0119990026.

WESNOUSKY, S. G. (1994), *The Gutenberg–Richter or characteristic earthquake distribution, which is it?* Bull. Seismol. Soc. Am. *84*(6), 1940–1959, http://www.bssaonline.org/cgi/content/abstract/84/6/1940, http://www.bssaonline.org/cgi/reprint/84/6/1940.pd.

WORKING GROUP ON CALIFORNIA EARTHQUAKE PROBABILITIES (WGCEP) (2007), *The Uniform California Earthquake Rupture Forecast*, version 2 (UCERF 2). USGS Prof. Pap. 2007-1437, http://pubs.usgs.gov/of/2007/143.

ZHENG, G. and RICE, J. R. (1998), *Conditions under which velocity-weakening friction allows a self-healing versus a cracklike mode of rupture*, Bull. Seismol. Soc. Am. *86*, 1466–1483.

ZIV, A. and RUBIN, A. M. (2003), *Implications of rate-and-state friction for properties of aftershock sequence: Quasi-static inherently discrete simulations*, J. Geophys. Res. *108*, 2051, doi:10.1029/2001JB001219.

(Received September 25, 2008, revised February 5, 2009, accepted August 6, 2009, Published online April 9, 2010)

Continuous Observation of Groundwater and Crustal Deformation for Forecasting Tonankai and Nankai Earthquakes in Japan

SATOSHI ITABA,[1] NAOJI KOIZUMI,[1] NORIO MATSUMOTO,[1] and RYU OHTANI[1]

Abstract—In 2006, we started construction of an observation network of 12 stations in and around Shikoku and the Kii Peninsula to conduct research for forecasting Tonankai and Nankai earthquakes. The purpose of the network is to clarify the mechanism of past preseismic groundwater changes and crustal deformation related to Tonankai and Nankai earthquakes. Construction of the network of 12 stations was completed in January 2009. Work on two stations, Hongu-Mikoshi (HGM) and Ichiura (ICU), was finished earlier and they began observations in 2007. These two stations detected strain changes caused by the slow-slip events on the plate boundary in June 2008, although related changes in groundwater levels were not clearly recognized.

Key words: Groundwater, strain, tremor, slow-slip event, Nankai earthquake, Tonankai earthquake.

1. Introduction

The Geological Survey of Japan, AIST has a network of about 40 groundwater observation stations in and around the Tokai and Kinki areas in Japan (Fig. 1). It is one of the best equipped groundwater observation networks for earthquake-prediction research in the world. Based on the pre-slip model of the impending Tokai Earthquake in the Suruga Trough, and the assumption that groundwater level changes are proportional to volumetric strain changes, it has been found that our network has the ability to detect preseismic groundwater level changes (MATSUMOTO *et al.*, 2007). A pre-slip is an aseismic slow slip, in and around the focal region, expected to start a few days before the main shock. These groundwater data can be accessed from http://riodb02.ibase.aist.go.jp/gxwell/GSJ_E/index.shtml.

We have been monitoring groundwater in the Tokai area for earthquake prediction since the 1970s. However, the possibility of the occurrence of Tonankai and Nankai earthquakes, which have occurred in the Nankai Trough next to the Suruga Trough at intervals of 100–200 years, has been increasing recently. In addition, hydrological anomalies related to past Nankai earthquakes were often reported in Shikoku and the Kii Peninsula in historical documents (USAMI, 2003). SATO *et al.* (2005) pointed out that there might have been a large drop in the discharge of the Yunomine hot spring just after the 1944 Tonankai Earthquake, although the Yunomine hot spring has shown coseismic and postseismic drops in discharge related to past Nankai earthquakes (USAMI, 2003). Therefore, in 2006 we started construction of an observation network of 12 stations in and around Shikoku and the Kii Peninsula for research into groundwater changes and crustal deformation related to Tonankai and Nankai earthquakes. Construction of the network of 12 stations was completed in January 2009.

In 2007, we finished construction of two stations and started monitoring groundwater changes and crustal deformation at Hongu-Mikoshi (HGM) and Ichiura (ICU) in the southern part of the Kii Peninsula, which is near the epicenters of the 1944 Tonankai and 1946 Nankai earthquakes (Fig. 1). Figure 1 shows the location of the other 10 observation stations in and around Shikoku and the Kii Peninsula.

---

[1] National Institute of Advanced Industrial Science and Technology, Geological Survey of Japan, Active Fault and Earthquake Research Center, Site C-7, 1-1-1, Tsukuba, Ibaraki 305-8567, Japan. E-mail: itaba-s@aist.go.jp

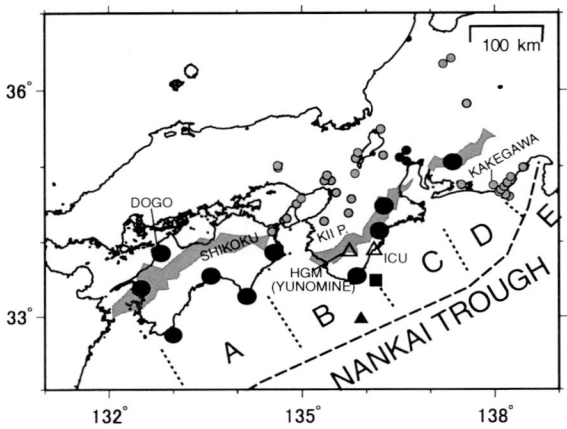

Figure 1
Map of Southwest Japan, showing the Nankai and Suruga Troughs. The Nankai Trough is divided into four sections (A–D; ANDO, 1975); Section E is the Suruga Trough. The location of the Dogo and Yunomine hot springs is also shown. Kii P. stands for the Kii Peninsula. Two *open triangles* and ten *black circles* show new observation stations, which have been under construction since 2006. Observations at HGM and ICU stations, which are shown as the two *open circles*, started in 2007. Small *gray circles* indicate existing groundwater observation stations. The *shadow zones* show the areas where non-volcanic tremors occur. The *black square* and *triangle* denote the epicenters of the 1944 Tonankai Earthquake and 1946 Nankai Earthquake, respectively

In this paper, we will introduce our new observation stations and show the preliminary results of observations at HGM and ICU.

2. Observation

2.1. Nankai, Tonankai and Tokai Earthquakes

Nankai, Tonankai and Tokai earthquakes are large interplate earthquakes that have occurred repeatedly in the Nankai and Suruga Troughs at intervals of around 100–200 years since A.D. 684 (ANDO, 1975; SANGAWA, 1992). The Nankai Trough is divided into four sections (A–D; ANDO, 1975), with large earthquakes that occur in Sections A and B being referred to as Nankai earthquakes. Earthquakes that occurred in Sections C–D or C–E were referred to as Tokai earthquakes prior to the 1944 Tonankai Earthquake, which occurred in Sections C and D. As the earthquake that is expected to occur in the Suruga Trough (E in Fig. 1) is actually called the Tokai Earthquake, and as earthquakes occurring in Sections C and D are currently referred to as Tonankai earthquakes, the accepted nomenclature for Tonankai and Tokai earthquakes will be adopted in this paper. Historical records spanning the past 1,300 years indicate that Nankai earthquakes have occurred eight or nine times since A.D. 684, making Nankai earthquakes one of the most well-known large interplate earthquakes in the world. Tonankai earthquakes have occurred six times since 1096. The groundwater level or discharge at the Dogo and Yunomine hot springs, which are old, well-known hot springs in Japan, has coseismically and postseismically decreased several times during past Nankai and Tonankai earthquakes (USAMI, 2003), although it is not clear whether those decreases began prior to the earthquakes.

The M 7.9 1944 Tonankai Earthquake on December 7, 1944, followed a preseismic crustal deformation in Kakegawa in the Shizuoka Prefecture (MOGI, 1982, his Fig. 1), which can be explained by the pre-slip or the preseismic aseismic slip on the plate boundary. The groundwater level at the Yunomine hot spring coseismically and postseismically dropped more than 1 m at the time of the 1944 Tonankai Earthquake (SATO et al., 2005). The M 8.0 1946 Nankai Earthquake on December 21, 1946, followed 11 preseismic drops in well-water levels and one decrease in discharge from a hot spring near the coastal regions of Shikoku and the Kii Peninsula (Fig. 2, HYDROGRAPHIC BUREAU, 1948; DISASTER PREVENTION RESEARCH INSTITUTE, KYOTO UNIVERSITY, 2003b). Similar drops in well-water levels also occurred before the 1854 event (SHIGETOMI et al., 2005). Those groundwater changes can be qualitatively explained by the pre-slip (DISASTER PREVENTION RESEARCH INSTITUTE, KYOTO UNIVERSITY, 2003a). The groundwater level at the Dogo hot spring dropped more than 10 m at the time of the 1946 Nankai Earthquake (RIKITAKE, 1947). The coseismic large groundwater level drop can be quantitatively explained by static volumetric strain changes at the Dogo hot spring calculated from the fault model of the 1946 Nankai Earthquake (ITABA and KOIZUMI, 2007).

These findings suggest that continuous observation of groundwater and crustal deformation near the expected source region of the Tonankai and Nankai

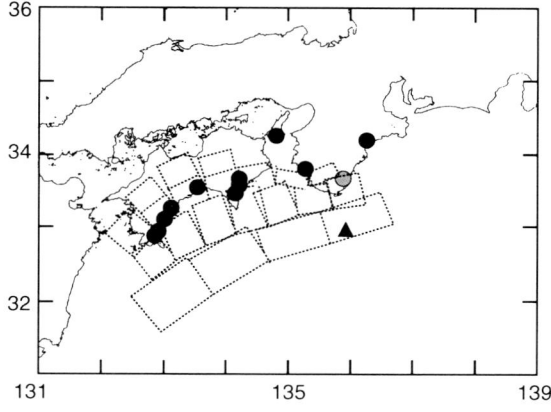

Figure 2
Distribution of groundwater changes prior to the 1946 Nankai Earthquake. The 11 *black circles* represent areas where unconfined groundwater levels fell 1–10 days prior to the 1946 Nankai Earthquake, the *gray circle* indicates an area where there was a decrease in hot-spring discharge 6 h prior to the 1946 Nankai Earthquake (HYDROGRAPHIC BUREAU, 1948). The *solid triangle* indicates the epicenter of the 1946 Nankai Earthquake. The *dotted rectangles* show the western parts of the fault models of SAGIYA and THATCHER (1999), which are considered to correspond to the 1946 Nankai Earthquake

earthquakes should enable us to detect preseismic changes related to pre-slip.

2.2. Details of Past Groundwater Changes related to Past Nankai Earthquakes

If a reverse slip occurs on the plate boundary in the Nankai Trough, irrespective of whether the reverse slip is a pre-slip or coseismic slip, the southern coasts of Shikoku and the Kii Peninsula can rise, and large areas in them will be in extension (DISASTER PREVENTION RESEARCH INSTITUTE, KYOTO UNIVERSITY, 2003a). Therefore unconfined groundwater levels in the coastal region, which are related to sea level, can decrease relative to the surface and pressure of the confined groundwater, or hot springs can drop (LINDE and SACKS, 2002). In other words, reported groundwater level drops or discharge decreases can be explained by a reverse slip on the plate boundary in the Nankai Trough.

However, the rise in land level is at most several centimeters, which was calculated from the pre-slip model of the DISASTER PREVENTION RESEARCH INSTITUTE, KYOTO UNIVERSITY (2003a), which is assumed to be 10% of the slip on the deepest parts of the fault models for the 1946 Nankai Earthquake by SAGIYA and THATCHER (1999). This means that drops in groundwater level resulting from the assumed land upheaval alone could only be several centimeters at most, which is considerably smaller than reported groundwater level drops, said to be larger than several tens of centimeters (HYDROGRAPHIC BUREAU, 1948). On the other hand, as the extension (dilatation) can be 5×10^{-7} due to the same pre-slip model, and the strain sensitivity of some confined groundwater can be $10 \text{ cm}/10^{-7}$ strain (e.g., ROELOFFS, 1996), the level of confined groundwater or hot springs can drop by more than several tens of centimeters. However, most of the reported drops in groundwater levels have occurred not in confined groundwater or hot springs, but in shallow unconfined groundwater in coastal regions. This suggests there are unknown mechanisms by which a small crustal deformation translates into large changes in the levels of unconfined groundwater. Movement between unconfined and confined groundwater is one possible mechanism. Another plausible mechanism is that if tensile cracks open in rock near the surface, then fluid will flow into them and water levels will fall.

The 11 drops in groundwater level and one hot spring discharge decrease were found in a wide area around the source region of the 1946 Nankai Earthquake as the result of a survey by the HYDROGRAPHIC BUREAU (1948). However, more than 160 places had been surveyed in the coastal regions of Shikoku and the Kii Peninsula. In addition, even in the areas where preseismic groundwater level drops were reported, some well-water levels dropped, but other well-water levels did not (SHIGETOMI *et al.* 2005). In other words, a very few drops in groundwater levels, whose amplitudes are large, occurred in a wide area around the source region of the 1946 Nankai Earthquake. This suggests that there were common small preseismic crustal deformations related to the 1946 Nankai Earthquake, and a local special mechanism by which the small crustal deformation transformed into detectable drops in groundwater levels.

In any event, a small crustal deformation caused by a pre-slip is not enough for a quantitative evaluation, and observations that are more precise are needed to improve any such evaluation.

2.3. Observation Systems

Based on the above considerations, we planned the following observation system (Fig. 3). Each of the new observation stations has three wells, 30-, 200- and 600-m deep. The groundwater levels and temperatures are observed at each well to monitor any changes in shallow unconfined groundwater levels and deep-confined groundwater pressure, and to observe groundwater movement among the three wells. Crustal strain and tilt are also observed by means of a multicomponent borehole strainmeter and borehole tiltmeter installed at the bottom of the 600-m-deep well or the 200-m-deep well, because reported groundwater changes are considered to be responses to seismic crustal deformation. GPS observation also is carried out if there are no Geographical Survey Institute GPS stations in the vicinity. A borehole seismometer is also positioned in each of the wells. The sampling rate is relatively high for observations of groundwater, crustal strain and tilt (Table 1). The data obtained are sent to the Geological Survey of Japan, AIST in real time.

The first two stations (HGM and ICU in Fig. 1) were constructed in the southern part of the Kii Peninsula. This is because the rupture of both the 1944 Tonankai and Nankai earthquakes started off the south coast of the Kii Peninsula (Fig. 1). It is also because the next Tonankai and Nankai earthquakes will most likely occur in the same place (HORI, 2006). HGM is also located in the neighborhood of the Yunomine hot spring, where groundwater changes

Table 1

Sampling rate for each observation

Observation	Sampling rate (Hz)
Groundwater level	1
Groundwater temperature	1
Crustal strain	20
Crustal tilt	10
Seismometer	100–1,000

related to the Tonankai and Nankai earthquakes are expected to occur, as described above.

We expect to observe groundwater changes related to crustal deformation associated not only with the pre-slips and main shocks for the Tonankai and Nankai earthquakes, but also with the episodic slow-slip events (SSEs) accompanied by non-volcanic tremors on the plate boundary (OBARA et al., 2004; OBARA and HIROSE, 2006, their Fig. 1). As those episodic slow slips, which are similar to the expected pre-slip, occur repeatedly within our observation network several times per year, observation data during the slow slips can be easily obtained, and should yield information useful to an understanding of the mechanism of past preseismic groundwater changes related to Tonankai and Nankai earthquakes. Therefore, the purpose of our current observations is to gain an understanding of usual crustal deformation and groundwater level changes, and evaluate groundwater changes related to the 1944 Tonankai and 1946 Nankai earthquakes, as well as to monitor groundwater changes and crustal deformation related to the episodic slow slips precisely in order to research their mechanism. This research will enable us to improve the model of the process for the Tonankai and Nankai earthquakes. This will prove useful in forecasting earthquakes.

3. Results

3.1. Long-term Changes from May 2007 to June 2008

Observations at ICU and HGM started in May 2007 and June 2007, respectively (Table 2). Figure 4a and b shows observations at ICU and HGM from July 2007 to June 2008. All of the strain and groundwater levels and pressure readings at HGM1 and HGM2 show

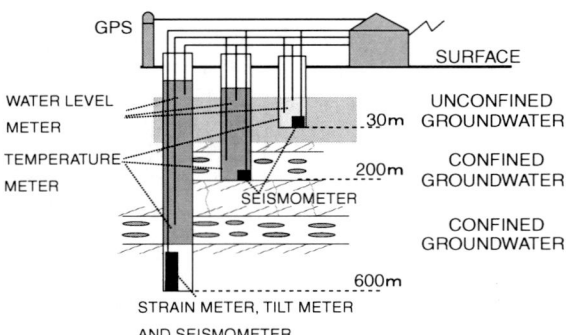

Figure 3
Schematic diagram of a system at a new observation station. The scale is arbitrary. At some of the 12 new stations shown in Figure 1, the strainmeter is positioned in a well not at a depth of 600 m, but at 200 m

Table 2

Depths of screens and strainmeters and tiltmeters at HGM and ICU

Station	Latitude (°)	Longitude (°)	Altitude (m)	Well	Screen depth (m)	Depth of the strain and tilt meters (m)
HGM	33.87	135.73	123	HGM1	320.4–331.3	368.2–375.0
				HGM2	180.9–191.8	–
				HGM3	24.3–29.8	–
ICU	33.90	136.14	27	ICU1	522.4–533.4	583.9–590.8
				ICU2	95.7–106.6	–
				ICU3	13.4–18.8	–

Screen casing pipe with slots through which groundwater flows in and out

exponential changes, probably due to the construction of the boreholes. The exponential strain changes at ICU1 are smoother than those at HGM1. Groundwater levels at HGM3 and ICU3 are affected by rainfall because HGM3 and ICU3 are shallow wells (Table 2). The data of Strain-2 (N67E component) at HGM1 have been fluctuating since April 2008.

There were four SSEs with non-volcanic tremors on the plate boundary during the observation period (Fig. 4a, b). During the SSEs, there were no large changes in groundwater level such as those that occurred prior to the 1946 Nankai Earthquake. During the fourth SSE in June 2008, when the exponential decay part of the strain changes was

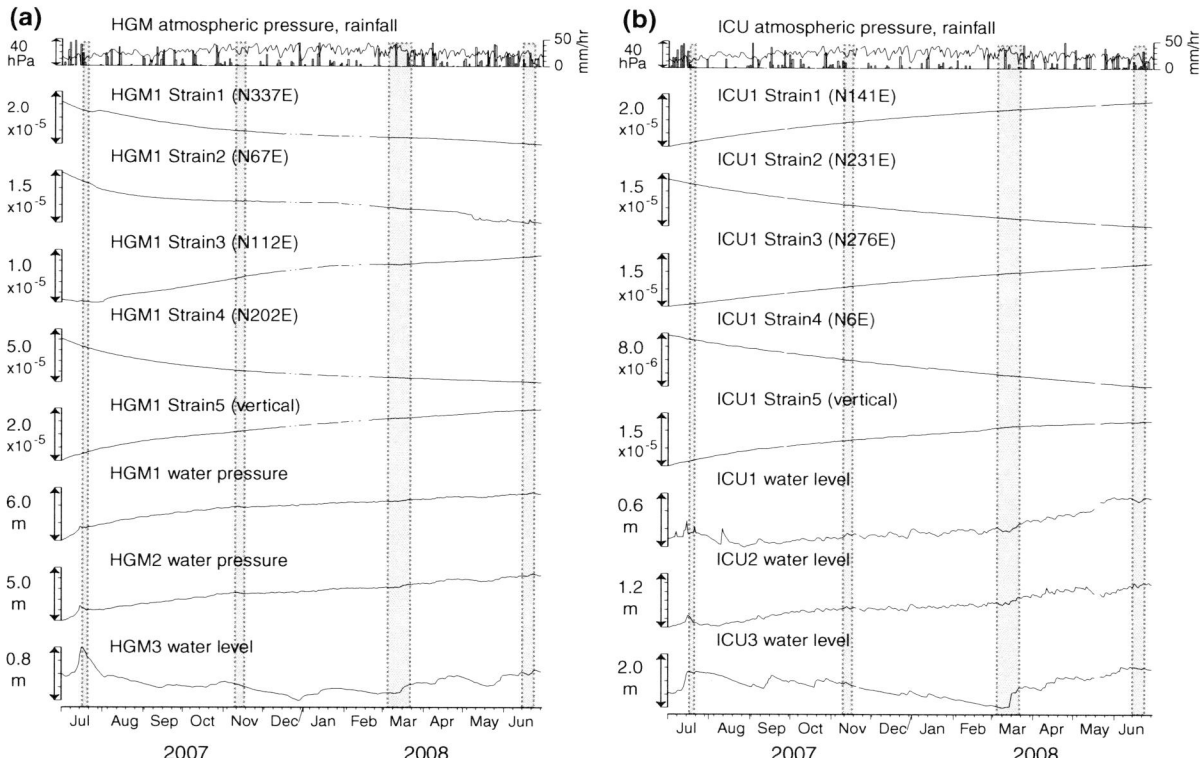

Figure 4

Observed results of strain and groundwater level or pressure at HGM (**a**) and ICU (**b**) from July 2007 to June 2008. Daily data are shown. The wells of HGM1 and HGM2 are sealed and the other wells are open. The water pressures are observed at HGM1 and HGM2, although they are expressed as head levels. The four *shadow zones* show the periods when SSEs occurred with the tremors

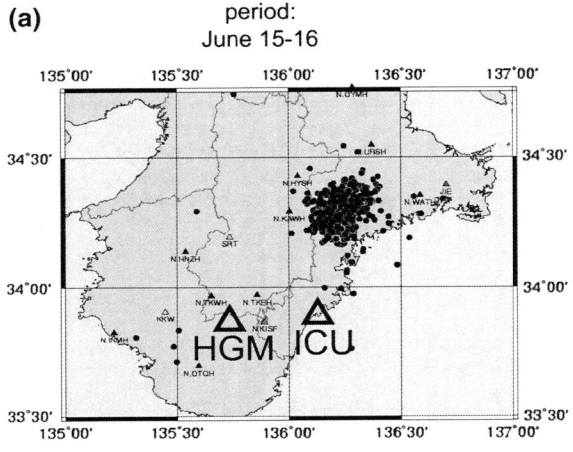

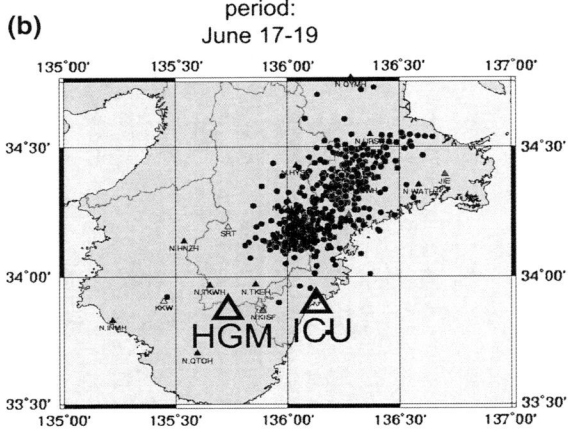

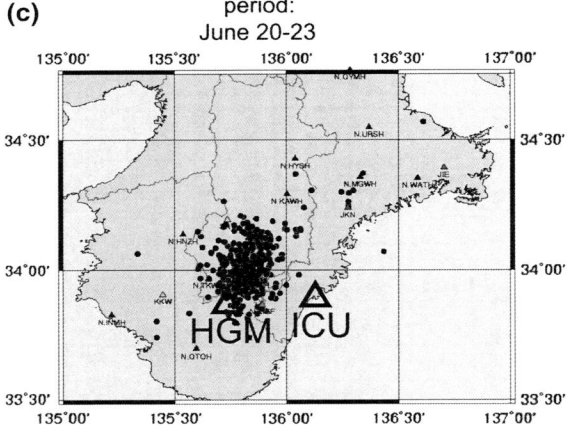

Figure 5
Distribution of epicenters of tremors (*small circles*) detected from June 12–25, 2008 by the Automatic Tremor Monitoring System (ATMOS) of Hiroshima University. *Small triangles* show the observation points used for the hypocenter determination

smaller (Fig. 4), strain changes at HGM1 and ICU1 were clearly recognized (Fig. 6a, b).

3.2. Strain and Groundwater Changes in June 2008

According to the Automatic Tremor Monitoring System of Hiroshima University (ATMOS: http://tremor.geol.sci.hiroshima-u.ac.jp/), during the period from June 15 to June 23, 2008, active tremors were generated (Fig. 5). According to the distribution of the tremors, the period was divided into three: A-period: June 15–16, B-period: June 17–19, and C-period: June 20–23 (Fig. 5).

Figure 6a and b shows the observation results at ICU and HGM during the period May 13 to June 27, 2008. In Figure 6a and b, the linear trend of each strain data series during the same period is removed. Tidal components and the effect of atmospheric loading on the strain and groundwater level as well as pressure are also removed by BAYTAP-G, a program for tidal analysis (TAMURA *et al.*, 1991). On June 12, maintenance was carried out on the strainmeters at ICU and HGM and the changes caused by the maintenance were recognized in the strain data at ICU and HGM. Therefore, we consider that the strain changes during the period from June 12 to June 15 are related to that maintenance, and that they should be neglected (Fig. 6a, b).

During the A-period, the tremors occurred a little further from ICU and HGM, and there were small strain changes in the horizontal components. During the B-period, the area for tremors became larger and the southern part of the area was approaching ICU and HGM. The strain changes were larger, especially in the N276E and N6E components at ICU. During the C-period, the tremor area was the nearest to HGM and ICU, and could be regarded as in the vicinity of HGM. At HGM, the strain changes were clearly recognized as Strain-3 (N112E). However, there were step-like changes at Strain-1 (N337E) and Strain-2 (N67E) on June 20. Similar changes are often recognized at Strain-1 and Strain-2, regardless of tremors and rainfall. We think that they are caused by local deformation in the vicinity of the sensor of Strain-1 or Strain-2 and that they have no tectonic

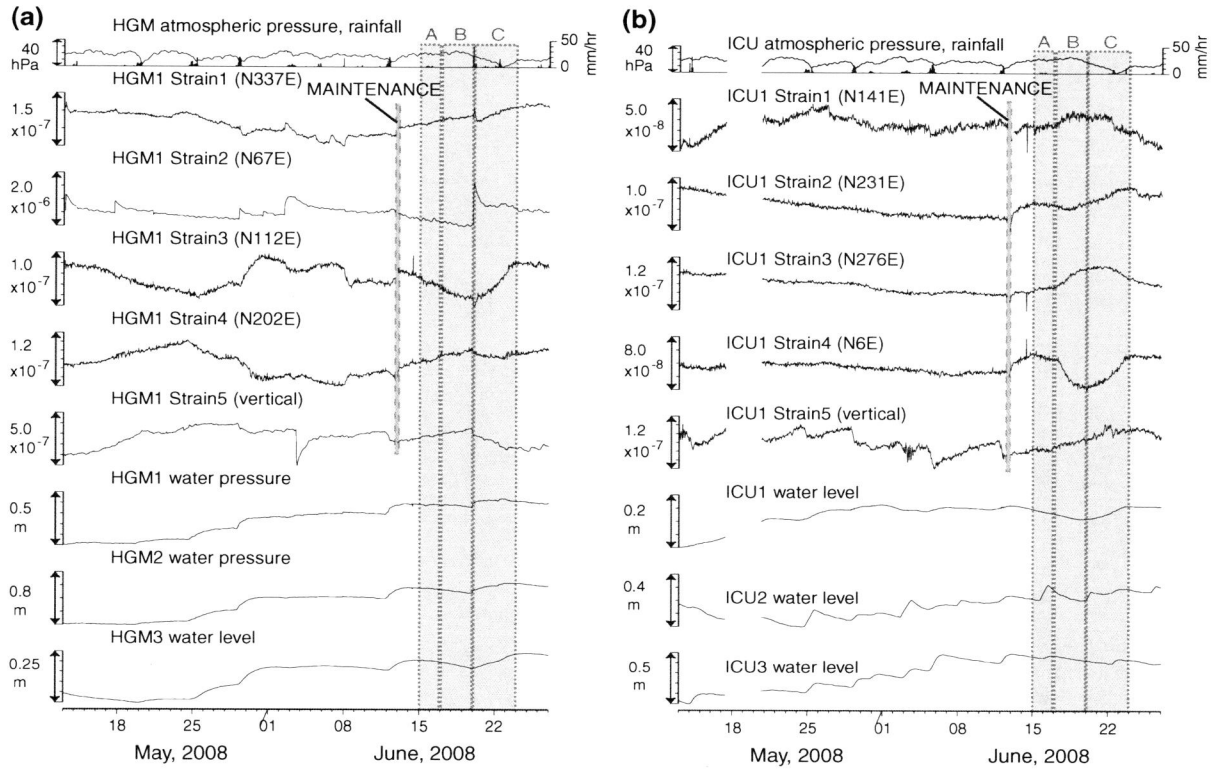

Figure 6
Hourly strain and groundwater level changes at HGM (**a**) and ICU (**b**) accompanying the active tremors in June 2008. Tidal changes and the effect of atmospheric loading were removed by BAYTAP-G (TAMURA *et al.*, 1991). The linear trend also is removed from the strains. The *shadow zones* indicate maintenance work and SSEs

meaning. The step-like changes at Strain-2 are much larger than those at Strain-1. Therefore, we do not use the data at Strain-2 in the following analysis. As to Strain-1 at HGM, the step-like changes were compensated in the following modeling. At ICU, there were also clear strain changes (Fig. 6a, b). The vertical strains and groundwater levels or pressures showed no clear changes during periods A–C.

4. Discussion

The strainmeter is set in homogeneous granite porphyry or crystal tuff at ICU. However, the strainmeter at HGM is set in tilted fractured sedimentary rocks, i.e., accretionary prism. In addition the water pressure at HGM1 is very high, more than 60 m H_2O. It is possible that this high water pressure could cause microfractures in the vicinity of the strain sensors, which in turn could cause extremely localized deformation. These are possibly the main reasons the S/N (signal-to-noise) ratio of the strain data at ICU is higher than at HGM.

SSEs in Japan are too small to be detected by GPS monitoring. Neither are the SSEs in the southern part of the Kii Peninsula detectable by tilt observations made by the National Research Institute for Earth Science and Disaster Prevention, Japan (OBARA and HIROSE, 2006). In order to check whether the horizontal strain changes at HGM and ICU during the active tremors in June 2008 can be explained by SSEs on the plate boundary, we constructed forward rectangular fault models for SSEs as follows. First, we decided the fault shape on the plate boundary of SATAKE (1993) in consideration of the epicenters of the tremors (Figs. 5, 7). Second, a reverse slip was assumed and a slip direction was chosen to be consistent with the plate motion. Finally, the amplitude

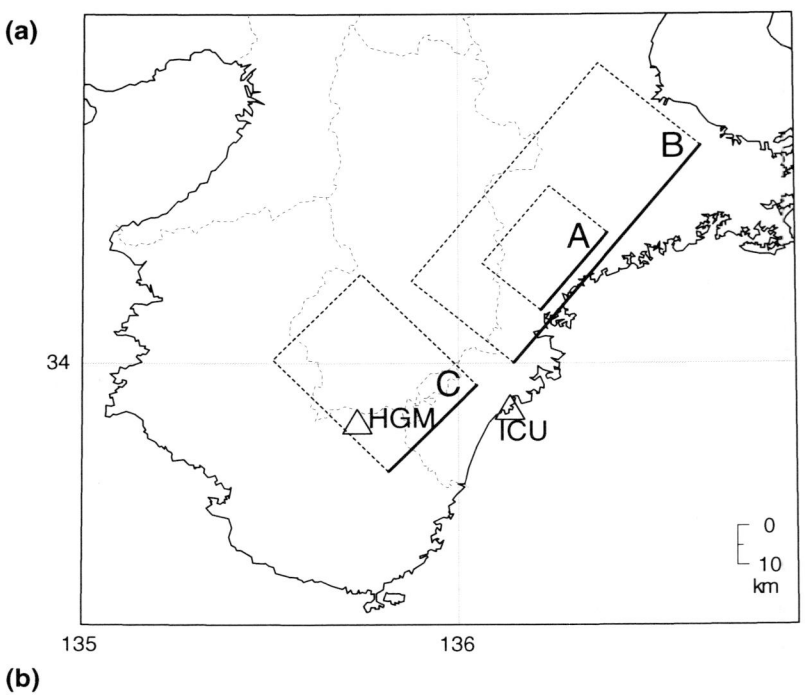

Figure 7

a Fault models for SSEs in June 2008 and **b** principal strains estimated from fault models and observed strain changes at HGM and ICU. For A, B and C in (**a**) refer Table 3

Table 3

Fault parameters for SSEs

Period	Latitude (°)	Longitude (°)	Depth (km)	Strike (°)	Length (km)	Width (km)	Dip (°)	Rake (°)	Slip (mm)	M_w
A:6/15–16	34.30	136.40	30	220	25	20	25	85	20	5.6
B:6/17–19	34.50	136.65	30	220	70	35	25	85	40	6.3
C:6/20–23	33.95	136.05	35	235	30	45	30	90	40	6.1

M_w moment magnitude

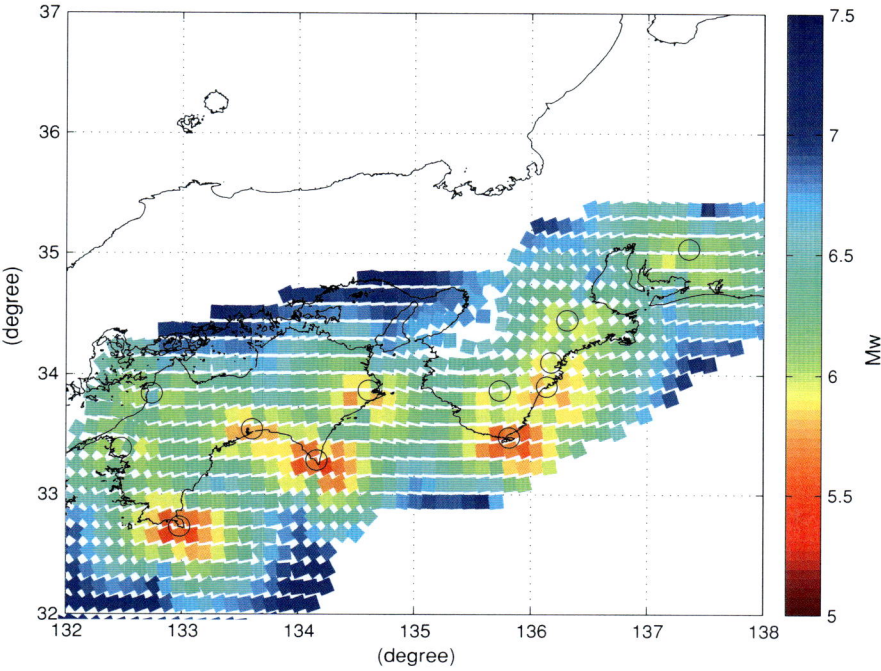

Figure 8
Detectability of episodic slow slip on the plate boundary. This *figure* shows the smallest SSE that can be detected by the 12 new observation stations, which are shown by *open circles*. The depth of the plate boundary is after HIROSE *et al.* (2007). M_w moment magnitude

of the slip was decided to match the observed strain changes at ICU. In calculating the strain changes caused by the assumed fault slip, we used the program devised by NAITO and YOSHIKAWA (1999), which was modified from the program of OKADA (1992). The parameters are shown in Figure 7 and Table 3.

As there are four components of horizontal strain at ICU, and three of the four components can decide the principal strain, four patterns of the principal strain were estimated for each of the periods (Fig. 7). The four patterns at ICU during each of the periods are very similar. This means that the estimated strain changes are reliable. At HGM, there are also four components in the horizontal strain. However, Strain-2 (N67E) was not usable as previously mentioned, so we only used the other three components, and determined only one principal strain set at HGM. The theoretical principal strain estimated from the fault model (Fig. 7 and Table 3) explains well the observation results at ICU (Fig. 7). This suggests that SSEs could cause the strain changes observed during the periods A–C at ICU. However, at HGM there are rather large differences between the theoretical and observed principal strains, except for period-B.

OHTANI *et al.* (2009) estimated the detectability of SSEs for our 12 new observation stations. They assumed that the noise level of the strain observation is 2.0×10^{-8}, the value of which is obtained from previous statistics at other Geological Survey of Japan, AIST observation sites. Actually, this noise level is almost achieved at ICU, but it is slightly smaller than the current noise level at HGM (Fig. 6a, b). According to OHTANI *et al.* (2009), the 12 new observation stations can detect SSEs of M_w 6.0–6.5 in and around the source region of the Tonankai and Nankai earthquakes (Fig. 8).

5. Conclusion

The Geological Survey of Japan, AIST has been constructing a new observation network of 12 stations for research on the Tonankai and Nankai earthquakes since 2006. Construction of the network of 12 stations was completed in January 2009. They are designed to clarify the mechanism of past preseimic groundwater changes and crustal deformation related to Tonankai and Nankai earthquakes. The first two stations of HGM and ICU started observations in 2007. Strain changes caused by SSEs in June 2008 were detected at HGM and ICU, although related changes in groundwater levels were not clearly recognized. We believe that the network will make a contribution to forecasting future Tonankai and Nankai earthquakes and help reduce the hazards they will present.

Acknowledgments

We wish to thank the members of the Tectono-Hydrology Team, Geological Survey of Japan, AIST for their cooperation and advice. We are grateful to the many people who have assisted us with our groundwater observations. We also wish to thank Dr. Townend and one anonymous reviewer for their helpful advice.

REFERENCES

ANDO, M. (1975), *Source mechanisms and tectonic significance of historical earthquakes along the Nankai trough, Japan*, Tectonophysics *27*, 119–140.

DISASTER PREVENTION RESEARCH INSTITUTE, KYOTO UNIVERSITY (2003a), *A fault model for the precursory slip that might have induced groundwater changes*, Rep. Coord. Com. Earthq. Pred. *70*, 402–403 (in Japanese).

DISASTER PREVENTION RESEARCH INSTITUTE, KYOTO UNIVERSITY (2003b), *On the decrease of the well water prior to the Nankai earthquake — A mechanism of the amplification*, Rep. Coord. Com. Earthq. Pred. *70*, 423–428 (in Japanese).

HIROSE, F., NAKAJIMA, J., and HASEGAWA, A. (2007), *Three-dimensional velocity structure in Southwestern Japan and configuration of the Philippine sea slab estimated by double-difference tomography*, J. Seismol. Soc. Jpn. *60*, 1–20 (in Japanese).

HYDROGRAPHIC BUREAU, *Report on the Nankai earthquake in 1946* (Change of the land surface and damage), (Hydrographic Bulletin, special number, 1948) (in Japanese).

HORI, T. (2006), *Mechanisms of separation of rupture area and variation in time interval and size of great earthquakes along the Nankai Trough, southwest Japan*, J. Earth Simulator *5*, 8–19.

ITABA, S. and KOIZUMI, N. (2007), *Earthquake-related changes in groundwater levels at the Dogo hot spring, Japan*, Pure Appl. Geophys. *164*, 2397–2410.

LINDE, A.T. and SACKS, I.S. (2002), *Slow earthquakes and great earthquakes along the Nankai trough*, Earth. and Planet. Sci. Lett. *203*, 265–275.

MATSUMOTO, N., KITAGAWA, Y., and KOIZUMI, N. (2007), *Groundwater-level anomalies associated with a hypothetical perslip prior to the anticipated Tokai earthquake: Detectability using the groundwater observation network of the Geological Survey of Japan*, AIST, Pure Appl. Geophys. *164*, 2377–2396.

MOGI, K. (1982), *Temporal variation of the precursory crustal deformation just prior to the 1944 Tonankai earthquake*, J. Seismol. Soc. Jpn. *35*, 145–148 (in Japanese).

NAITO, H. and YOSHIKAWA, S. (1999), *A program to assist crustal deformation analysis*, J. Seismol. Soc. Jpn. *52*, 101–103 (in Japanese).

OBARA, K., HIROSE, H., YAMAMIZU, F., and KASAHARA, K. (2004), *Episodic slow-slip events accompanied by non-volcanic tremors in southwest Japan subduction zone*, Geophys. Res. Lett. *31*, L23602, doi:10.1029/2004GL020848.

OBARA, K. and HIROSE, H. (2006), *Non-volcanic deep low-frequency tremors accompanying slow slips in the southwest Japan subduction zone*, Tectonophysics *417*, 33–51.

OHTANI, R., KOIZUMI, N., TAKAHASHI, M., MATSUMOTO, N., SATO, T., KITAGAWA, Y., and ITABA, S. (2009), *Appraisal of the detectivity of hypothetical preslip of the Tonankai and Nankai Great Earthquakes using the integrated groundwater observatories of the Geological Survey of Japan, AIST*, Bull. Geol. Surv. Japan *60(11/12)*, 511–525 (in Japanese).

OKADA, Y. (1992), *Internal deformation due to shear and tensile faults in a half-space*, Bull. Seismol. Soc. Am. *82*, 1018–1040.

RIKITAKE, T. (1947), *Groundwater changes at the Dogo hot spring related to the Nankai earthquake*, Bull. Earthq. Res. Inst. *5*,189–194 (in Japanese).

ROELOFFS, E.A. (1996), *Poroelastic techniques in the study of earthquake-related hydrologic phenomena*, Adv. Geophys. *37*, 135–195.

SAGIYA, T. and THATCHER, W. (1999), *Coseismic slip resolution along a plate boundary megathrust: The Nankai trough, southwest Japan*, J. Geophys. Res. *104*, 1111–1129.

SANGAWA, A., Jishin-Kokogaku (Earthquake Archaeology), (Chuokoron-sha, 1992) (in Japanese).

SATAKE, K. (1993), *Depth distribution of coseismic slip along the Nankai Trough, Japan, from joint inversion of geodetic and tsunami data*, J. Geophys. Res. *98*, 4553–4565.

SATO, T., KOIZUMI, N., and NAKABAYASHI, K. (2005), *Did the groundwater flow stop in the Yunomine Spa associated with the 1946 Nankai earthquake?*, Chishitsu News *609*, 31–42 (in Japanese).

SHIGETOMI, K., UMEDA, Y., ONOUE, K., ASADA, T., HOSO, Y., KONDO, K. and TAKUMI, T. (2005), *Well water decrease and ocean tide abnormality before Nankai earthquake found out by the literature and hearing investigation*, Annuals, Disaster Prevention Research Institute, Kyoto University *48B*, 191–195 (in Japanese).

TAMURA, Y., SATO, T., OOE M. and ISHIGURO, M. (1991), *A procedure for tidal analysis with a Bayesian information criterion*, Geophys. J. Int. *104*, 507–516.

USAMI, T., *Materials for comprehensive list of destructive earthquakes in Japan*, (416)-2001, (University Tokyo Press 2003) (in Japanese).

(Received August 19, 2008, revised February 4, 2009, accepted July 21, 2009, Published online March 11, 2010)

Anomalies of Seismic Activity and Transient Crustal Deformations Preceding the 2005 M 7.0 Earthquake West of Fukuoka

YOSIHIKO OGATA[1]

Abstract—If aseismic slip occurs on a fault or its deeper extension, both seismicity and crustal deformation around the source would be affected. Anomalous phenomena of this kind are revealed from earthquake occurrence data and geodetic records during a period of 10 years leading up to the March 2005 M 7.0 earthquake west of Fukuoka that occurred off the northern coast of Kyushu, Japan. Seismicity rate anomalies (quiescence and activation) took place relative to the rates expected by the ETAS model in a number of seismic zones in and around the Kyushu District. The seismic zone of the relative quiescence and activation consistently corresponds to the zone of the negative and positive ΔCFS (Coulomb failure stress change), respectively, assuming the precursory aseismic slips on the M 7.0 source fault. In addition, we consider the time series of geodetic baseline distances between permanent GPS stations in the Kyushu District for the same period, which also supports the possible precursory slips rather than the known slow slips beneath the Bungo Straight, off the eastern coast of Kyushu.

Key words: Change-point analysis, Coulomb stress changes, ETAS model, GPS geodetic time series, precursory slip, seismic activation and quiescence.

1. Introduction

The decreasing and increasing rates of earthquake occurrence (seismicity, quiescence and activation, respectively) have attracted much attention as intermediate-term precursors to large earthquakes, possibly providing useful information on their location, time and size (INOUYE, 1965; UTSU, 1968; OHTAKE *et al.*, 1977; WYSS and BURFORD, 1987; KISSLINGER, 1988; KEILIS-BOROK and MALINOVSKAYA, 1964; SEKIYA, 1976; EVISON, 1977; SYKES and JAUME, 1990). However, in most cases, these anomalies are not clearly visible solely from cumulative number and magnitude of earthquakes against time because of complex earthquake clusters.

To detect such anomalies taking the clustering effect into consideration, it is useful to apply the epidemic-type aftershock sequence (ETAS) model (see Appendix 1) that captures the normal clustering effect caused by triggering interaction among the contiguous complex faults within a closed geophysical region. Namely, we fit the ETAS model to the occurrence data of earthquake sequences from a considered region to detect the anomalous changes of seismicity rates relative to the expected rates by the ETAS model (OGATA, 2001, 2005a, b, 2006a). Then, assuming a slip somewhere, we link such seismicity changes to the ΔCFS (see Appendix 2), which is an important indicator to explain seismicity rate changes in terms of the rate/state-dependent friction law of DIETERICH (1994). We examine whether the seismicity rate changes relative to the ETAS model are in qualitative agreement with the ΔCFS.

To date, silent earthquakes or slow-slip events have been reported worldwide on many plate boundaries and elsewhere, either in post-seismic observations (KAWASAKI *et al.*, 1995; HEKI *et al.*, 1997; GSI, 2005b, 2007), pre-seismically (SEGALL *et al.*, 2006; OGATA, 2007), or independently (WYSS *et al.*, 1990b; HIROSE *et al.*, 1999; DRAGERT *et al.*, 2001; OZAWA *et al.*, 2002; GSI, 2004). This paper intends to provide further evidence of pre-seismic slip and suggests that silent slip can take place on a shallow intra-plate fault or down-dip extension.

We will investigate whether the precursory slow slip on the 2005 earthquake west of the Fukuoka Prefecture can explain the observed seismicity

---

[1] Institute of Statistical Mathematics and Graduate University of Advanced Studies, 10-3 Midori-cho, Tachikawa-city, Tokyo 190-8562, Japan. E-mail: ogata@ism.ac.jp

anomalies and transient changes of the crustal deformations in and around the Kyushu District. However, there were other known slow slips beneath the Bungo Straight, which is located between the Kyushu and Shikoku Islands. Therefore, we will also discuss whether or not the seismicity and geodetic anomalies are entirely caused by the precursory slow slip. Throughout, we use the earthquake catalogue assembled and compiled by the Japan Meteorological Agency (JMA) in addition to the global positioning system (GPS) displacement data compiled by the Geographical Survey Institute (GSI) from the GPS Earth Observation Network (GEONET).

2. Seismic Activity in and around Kyushu Preceding the 2005 Earthquake of M 7.0 West of Fukuoka

The earthquake of M 7.0 west of Fukuoka struck the Fukuoka Prefecture, Japan at 10:53 a.m. JST on 20 March 2005, and was followed by on- and off-fault aftershocks (see OGATA, 2006a for the aftershock study). The Fukuoka area is not as seismically active as many other parts of Japan and was known prior to the earthquake as one of Japan's safest locations in terms of natural disasters. The strike-slip rupture fault model of the main shock described in Fig. 1a is obtained by the inversion of displacements of the GPS stations (GSI, 2005b).

To make an ETAS analysis, we selected seismicity zones not only to include enough earthquakes within a rectangular region in which the seismic activity is less interactive with that of the outside areas, but also to include many earthquakes of similar mechanisms so that we can reasonably assume a set of predominant orientations of earthquake faults. Such seismic zones (see Table 1, Fig. 1) include: (A) the plate boundary off the coast of the Miyazaki Prefecture, (B) the Southern Nagasaki Prefecture, (C) the aftershock region of the 2001 Geiyo earthquake, (D) the Suo-Nada region, (E) the Kego Fault region in the Fukuoka Prefecture, (F) the Beppu region of the Oita Prefecture, (G) the region off the coast of the Oita Prefecture, (H) the fault zone in the Yamaguchi-Shimane Prefectures, (I) the Kumamoto Plain and (J) the aftershock area of the 1997 Northwestern Kagoshima Prefecture earthquakes.

The predominant fault angles of each region are listed in Table 1, taking into consideration the stress field, the tectonic environment, orientations of known active faults, alignment of the epicenters and fault mechanisms of the past large earthquakes of the region. Fault mechanisms are taken from the Harvard global catalogue (DZIEWONSKI et al., 1981; DZIEWONSKI and WOODHOUSE, 1983) and the full-range seismograph network (F-net) catalogue of the National Research Institute for Earth Science and Disaster Prevention (NIED, 2007). To remove ambiguity in the fault mechanism, strike, dip and rake angles were inferred from the alignment of the hypocentres and active fault orientations, and also from the stress field, which is mainly the east–west compression except for the following zone: The crust of central Kyushu (the east–west zone around 33°N in latitude), is gradually spreading in the north–south direction from the Beppu-Shimabara Graben, which is the spreading axis (TADA, 1993). Most of the shallow earthquakes in the Beppu-Shimabara Graben have fault-plane solutions that are normal or strike-slip faults along a north–south extensional axis, as shown for the regions B, F, G and I in Table 1. Furthermore, we assume reverse faulting for the interplate earthquakes on the plate boundary over the subducting Philippine Sea Plate (regions A and G). Mechanisms of earthquakes in the regions dominated by aftershocks (C, H and J) are taken as those of the main shocks.

In each region, the seismic activity during the 10-year period until the Fukuoka earthquake, from 1995 to 23 May 2005, is analysed by fitting the ETAS model (see Appendix 1 for the procedure). In particular, the change-point analysis is necessary for the significance of the seismicity rate change, and the analysed result of the earthquake sequence from each zone is indicated in Table 1; the parameter estimates are printed in each panel in Fig. 2. The figure indicates that the times of the pre-seismic seismicity-rate changes vary from place to place due to the estimate of the change-point. The theoretical cumulative function of the ETAS model is calculated using the estimated parameters obtained by fitting to the sequence of earthquakes in either the whole period or the first part of the period before the change-point. In the latter case, the cumulative function is

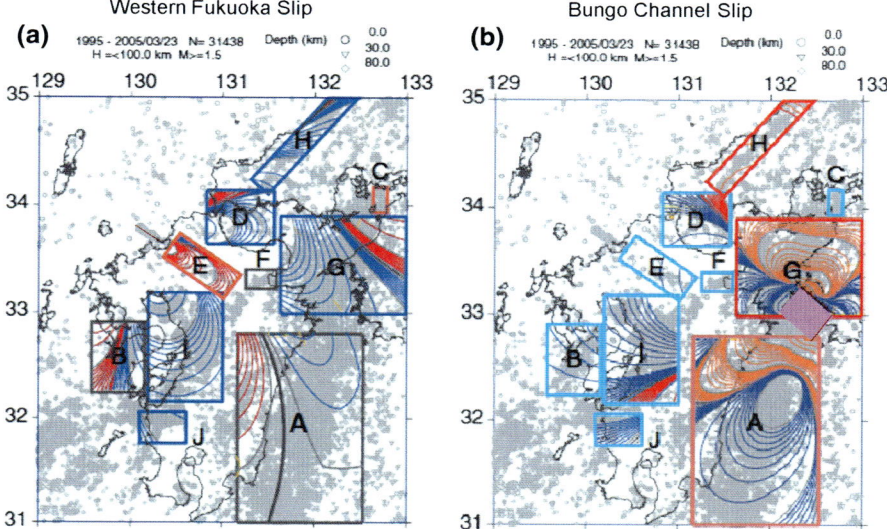

Figure 1

Epicentres of earthquakes with M ≥1.5 and depth ≤100 km during the period of 1995 until the main shock, 23 March 2005. The selection of the regions *A–J* is described in the text, and ΔCFS pattern of the most frequent angles of receiver faults for respective depths of the considered regions (cf. Table 1) due to the assumed rupture on the fault models of **a** the earthquake at the Fukuoka Prefecture offshore (*brown segment near zone E*; latitude = 33.68, longitude = 130.30, depth = 0.0 km, length = 24.9 km, width = 14.9 km, strike = 301°, dip = 85°, rake = −3° and slip size = 0.75 m) and **b** slow slip of the Bungo Straight (*pink rectangle*; longitude = 132.37, latitude = 33.41, depth = 40.0 km, length = 44 km, width = 60 km, strike = 227°, dip = 10°, rake = 86° and slip size = 0.11 m), which are due to GSI (2005b) and GSI (2004), respectively. The region of *red and blue contours* in logarithmic scale shows positive and negative ΔCFS values, respectively. Here, the ΔCFS values in Fig. 1a and Table 1 are calculated by assuming 10% of the slip size of the Fukuoka main shock. The *red and blue colour* of the region boundary indicates that the predominant ΔCFS is positive or negative, respectively, while *grey* indicates a neutral value

Table 1

Assumed receiver fault configurations and ΔCFS values

Zone	Strike (°)	Dip (°)	Rake (°)	Depth (km)	ΔCFS[a] (millibars)		ΔAIC[b]	Seismicity change
					(a)	(b)		
A	210	30	90	20–45	+1 to +2	−50 to +150	+	Normal
B	45	90	180	10	0	−2	+	Normal
	90	45	−90		0	−1		
C	179	55	−82	45	4	−8	+	Normal
D	135	90	0	10	−90 to −0	−7	−8.6	Quiet
E	135	90	0	10	+50 to +500	−4	−1.1	Activate
F	90	90	180	10	−10	−20	−2.2	Quiet
	90	45	−90		10	−8		
G	170	75	−90	40–90	−4 to −1	−1 to 1 bars	−75.2	Quiet
	330	35	−110		−4 to −1	−1 to 1 bars		
H	45	90	180	10	−1 to −10	8	−65.6	Quiet
I	225	45	180	10	−40	−2	−29.2	Quiet
	45	90	180		−40	−4		
	90	80	−50		−40	−2		
	90	45	−90		−30	−1		
J	280	90	0	10	−10	−0.3	−194.2	Quiet

[a] The parameters of the source fault models (a) the west of Fukuoka earthquake and (b) the Bungo-Channel slow slip are given in Fig. 1

[b] For the significance, the modified ΔAIC taken a change-point into consideration is defined in Appendix 2

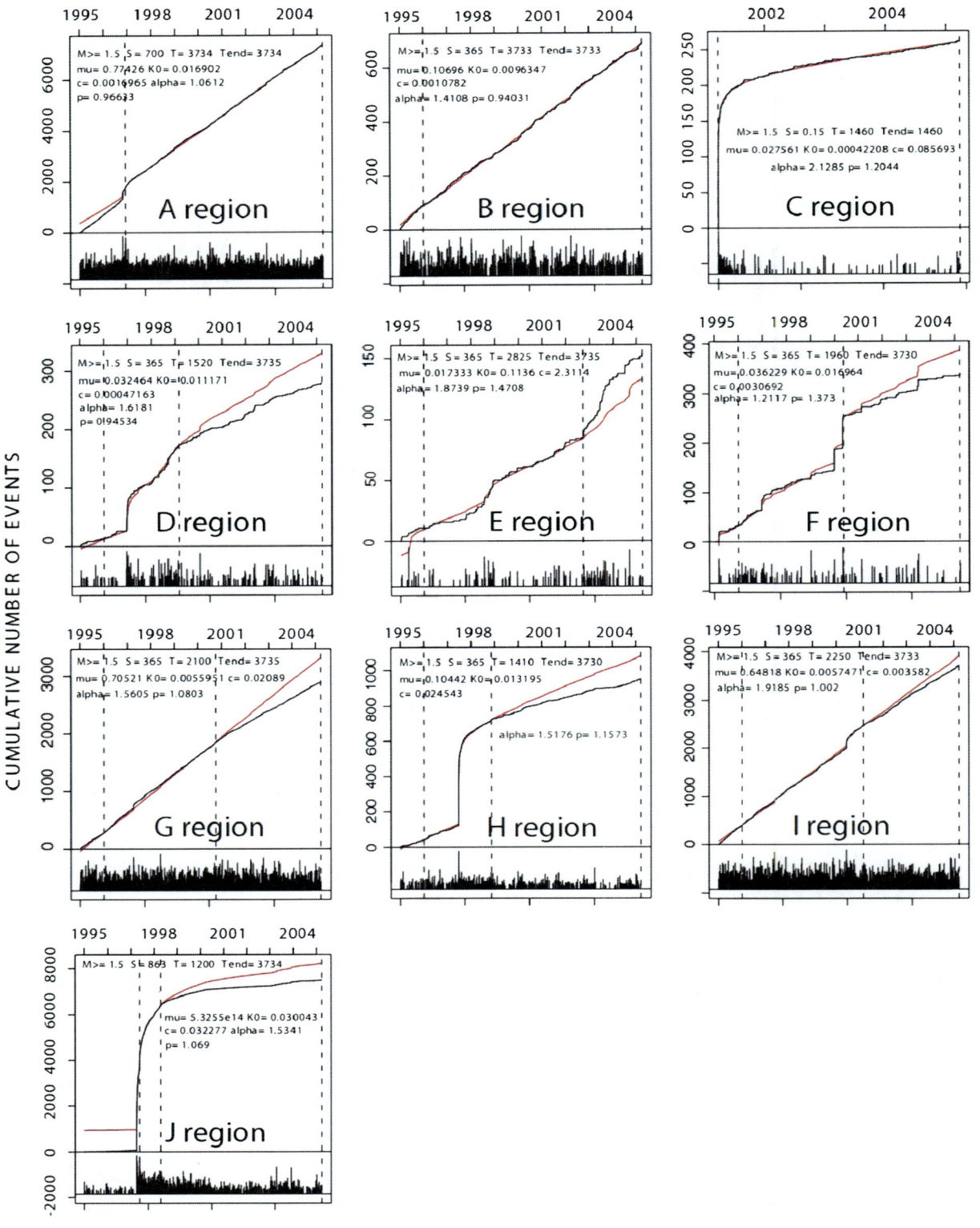

Figure 2

The ETAS model is applied to the sequence of events during the period from 1995 to 23 March 2005 in the respective regions A–J indicated in Fig. 1, where the cumulative number and magnitude of aftershocks are plotted against ordinary time. The empirical cumulative curve (*black*) is superimposed by the theoretical cumulative curve (*red*) calculated by the ETAS during the earlier period before the change-point, and it is extrapolated for the later period. The activity is fit well for the whole period in regions A–C. For the other regions, the two-states ETAS model with a change-point shows a better fit than the single ETAS model for the whole region even if the model complexity including the change point parameter is taken into account (see Appendix 1; Table 1). The activation relative to that predicted by ETAS model is indicated in region E, and the relative quiescence in regions D and F–J are significant

Table 2

Considered GEONET stations

Station	Name	ID	Lon (°)	Lat. (°)	Height (m)
A	Kamitsushima	950456	129.4821	34.6556	67.8712
B	Mitsushima	950457	129.3115	34.2682	43.4302
C	Koga	940087	130.4768	33.7307	49.0861
D	Noogata	960685	130.7496	33.7455	59.9349
E	Yukuhashi	960686	131.0165	33.6974	55.3950
F	Chikushino	950451	130.5220	33.5004	87.7073
G	Yamaguchitoyota	960670	131.0657	34.1798	70.0864
H	Mitou	950411	131.3461	34.1895	135.8400
I	Ikata	940086	132.2812	33.4690	191.7279
J	Aki	940088	131.6914	33.4615	51.3467
K	Saganoseki	950473	131.7981	33.2394	54.4193
L	Ooita	960709	131.5795	33.2284	106.9117
M	Hiji	960706	131.5884	33.3499	39.3855
N	HonyabakeI	950471	131.1689	33.4962	129.1613
O	Ishida	950458	129.7347	33.7427	127.9762
P	Genkai	940091	129.8503	33.4761	39.0842
Q	Hirado	950459	129.5370	33.3622	172.3720
R	Uku	960691	129.1255	33.2558	92.0530
S	Wakamatsu	960692	129.0263	32.8856	60.6681
T	Fukue	950462	128.8431	32.6694	150.7818
U	Tagawa	960687	130.8239	33.6405	87.4691
V	Koishiwara	950452	130.8288	33.4655	527.4753

extrapolated for the remaining period using the integrated rate of the estimated ETAS model of the first interval. The actual cumulative number of events (black) is compared with the theoretical cumulative curve (red) of the ETAS model to determine whether it deviates upward or downward from the expected value, which we call either by the relative quiescence or relative activation, respectively. If any change-point is not significant, then we call it the normal activity throughout the period. Thus, the result of the ETAS fitting is summarised in Table 1.

At the same time, assuming the precursory slip source on the fault of the forthcoming Fukuoka earthquake (see Fig. 1a for the source parameters), we calculated ΔCFS values at the centroid co-ordinate of the epicentres in each region using the receiver fault angles in Table 1 unless the region is wide, in which case we put the ranges of ΔCFS values in Table 1. Then, all the zones of the negative ΔCFS (D, G, H, I and J) show relative quiescence in Fig. 2. The zones of positive ΔCFS (C and E) and the neutral (A, B) show either relative activation (E) or the normal seismic activity as expected by the ETAS model (A, B and C). This agreement suggests that slow slip on the fault of the earthquake west of Fukuoka is likely to have taken place during the latter part of the 10 years before the earthquake.

The seismicity in the F zone shows relative quiescence, though it has two different types of predominant mechanisms (cf. Table 1). This is not an exception, because according to the rate/state-dependent friction law of DIETERICH (1994), a negative ΔCFS has significantly greater inhibitory effects than a commensurate positive ΔCFS has activation effects (see also OGATA, 2004). The stress changes due to aseismic slip can be of small values in the order of a few tens of millibars or less, as shown in Table 1, which are comparable to or smaller than fluctuations in daily earth tides, but the number of faults of small sizes to be either triggered or inhibited in a seismic zone can be substantial if the faults' orientations are similar.

We should note here that the report GSI (2004) shows that there were two periods of slow-slip events in 2001 and 2003 beneath the Bungo Straight, which is located between the Kyushu and Shikoku Islands (see Fig. 1b) in southwestern Japan. Therefore, one may be concerned with the possibility of whether the seismicity changes are affected by these slow-slip events. Therefore, assuming this slow-slip event, similar calculations of ΔCFS were implemented for the receiver zones A–J. These are also listed in Table 1 and shown in Fig. 1b. In this case, the agreements between the seismicity changes and the corresponding ΔCFS signs are observed only in the zones proximal to the Bungo Channel (D, F and G), besides the regions I and J. Thus, the available evidence gives less support to this explanation compared to that of a slow slip on the fault of the Fukuoka earthquake. We therefore still maintain the hypothesis that the seismicity changes are caused by the stress changes due to the precursory slip on the fault of the Fukuoka earthquake.

3. *Geodetic Time Series of Baseline Distances*

The sources of crustal deformation in northern Kyushu do not appear simple. The stress there is characterized by the gradient of horizontal stresses both in the east–west and north–south directions, which can be explained not only by simple plate interactions or by crust/plate structural variation but also by the viscous drag exerted by the flow spreading laterally from the mantle upwelling plume in the East China Sea west of Kyushu (SENO, 1999).

The permanent GPS network throughout Japan is called the GPS Earth Observation Network (GEONET). The development of the GEONET was established in October 1994 by the GSI and has since been expanded more densely. The accuracy of the data catalogue has been reported by MIYAZAKI *et al.* (1998) and HATANAKA *et al.* (2001a, b, 2003), which also describes the data processing to avoid various biases due to incorrect modelling of different antenna phase characteristics, etc. It is expected that the baseline distance between the stations, in comparison with the displacement of the station locations relative to a station set as origin, can cancel the various effects from displacements at other locations such as removal of the reference-frame errors (cf., HATANAKA *et al.*, 2001a, b, 2003). In particular, the F2 solutions in the GEONET catalogue are based on unified data taken from sub-networks of different types of antennas, and compensated for the artificial changes due to the antenna replacement of the station or its environmental changes. For the F2 data, it has been demonstrated that the error variability is within 2 mm in horizontal directions (HATANAKA *et al.*, 2005; HATANAKA, 2006).

In this study, we use the time series of the baseline distance calculated from the F2 catalogue (GSI, 2005a). The baseline distances are simply Euclidean distances between the three-dimensional (x, y, z) coordinates of the GPS stations in the catalogue. Table 2 lists the considered GEONET stations A–V in Figs. 3a and 4a around the rupture source that were installed prior to 1997.

Figures 3a and 4a show the baselines between GPS stations in and around northern Kyushu. The red and blue vectors on the lattice locations, which are the same in both Figs. 3a and 4a, show the expected sizes and directions contributed by the slow strike-slip on the fault of the earthquake of Fukuoka and the reverse-faulting slow slip on plate boundary (dark red rectangle) beneath the Bungo Channel, respectively (cf., Table 1 for the source parameters).

The most clear transient variation in 2003 in the baseline distances shown in Fig. 3b owes to the slow slip beneath the Bungo Channel, which agrees with the expected displacement by the blue vectors. It is also observed during the period of 1996–1997 when a similar slow-slip event took place. The similar but less clear changes due to the Bungo slip are seen as the slope changes of the time-series trend of other baselines in Figs. 3c, 4b and c (see light blue shaded periods). These are useful to discriminate changes due to transient stress changes from other sources, including the suspected slow slip on the fault of the earthquake of Fukuoka.

The slips on the fault of the earthquake of Fukuoka should cause stations C, D, E and F in Fig. 3a to move westward relative to the stations A and B according to the red vectors, thereby accelerating the contraction of the baselines between the two groups of stations. In fact, the trend of the time series in Fig. 3c becomes steeper in slope as shown by red lines during the first yellow shaded period, and these again become steeper in the last yellow shaded period, in which the time series are ended by co-seismic downward jumps, as shown by the arrows.

According to the red vectors in Fig. 4a, stations A, B and C are expected to move eastward relative to stations C, D, E and F on the other side of the baseline as a result of the slip on the fault of the earthquake of Fukuoka. This time the movement corresponds to a deceleration of the baseline contraction due to the east–west compressional stress field. In fact, the time series in Fig. 4b have gentler slopes during the first and second yellow shaded

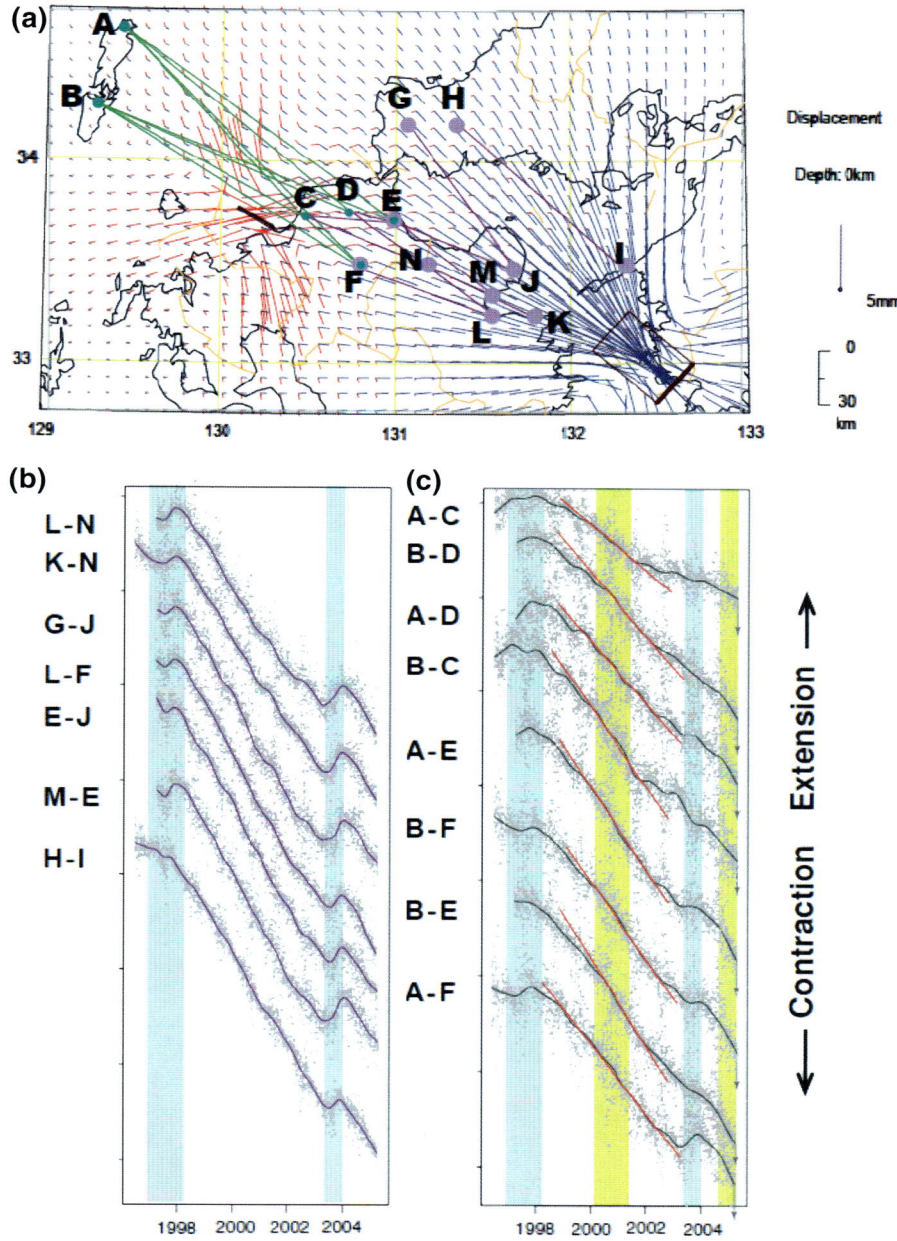

Figure 3

The baselines between the GEONET stations around the rupture source (*dark red thick segment* around 33.90°N, 130.20°E; see Fig. 1 and the caption for the source parameters) of the earthquake west of Fukuoka and Bungo Channel slip (*dark red rectangle in the bottom left corner*; see Fig. 1). The *red and blue vectors* on the lattice locations show calculated cumulative displacements due to the slip on the fault of the earthquake of Fukuoka and the Bungo Channel slip, respectively, where 10% of the main shock slip size (cf. Table 1) is assumed for the former precursory slip. The daily time series records of each coloured baseline distance between the stations are indicated on the *left side of panels* **b–c**. The *smooth blue curve* shows the average within a 365-day moving window. The *arrow* at the time end, 23 March 2005, shows the direction of displacement (jump) due to the rupture. The *shaded light blue colour* shows periods of reported Bungo Channel slip, and the *shaded yellow colour* shows the suspected precursory slip

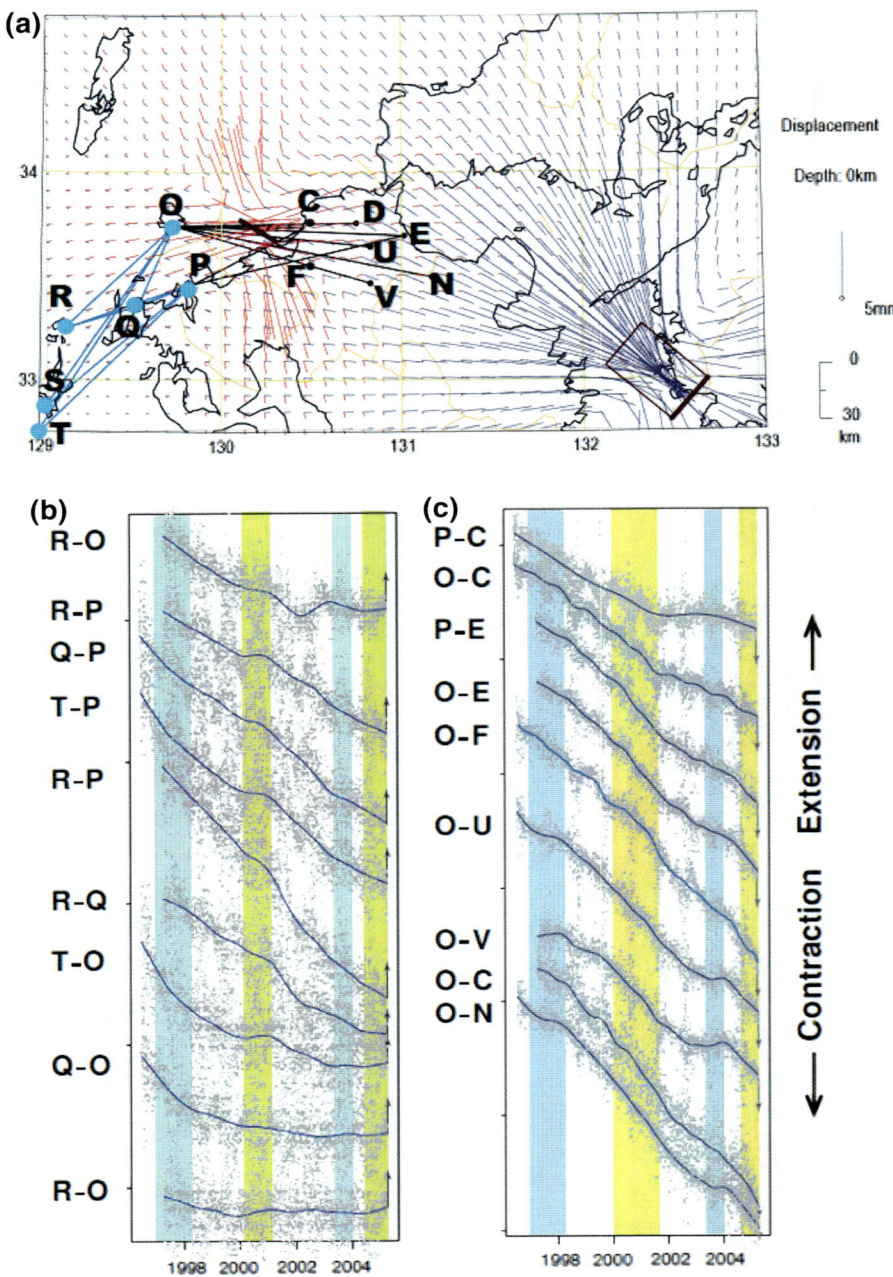

Figure 4
The baselines between the GEONET stations around the rupture source (*dark red thick segment* around 33.90°N, 130.20°E; see Fig. 1 and the caption for the source parameters) of the earthquake west of Fukuoka and Bungo Channel slip (*dark red rectangle in the bottom left corner*; see Fig. 1). The *red and blue vectors* on the lattice locations show calculated cumulative displacements due to the slip on the fault of the earthquake of Fukuoka and the Bungo Channel slip, respectively, where 10% of the main shock slip size (cf. Table 1) is assumed for the former precursory slip. The daily time series records of each coloured baseline distance between the stations are indicated on the *left side of panels* **b–c**. The *smooth blue curve* shows the average within a 365-day moving window. The *arrow* at the time end, 23 March 2005, shows the direction of displacement (jump) due to the rupture. The *shaded light blue colour* shows periods of reported Bungo Channel slip, and the *shaded yellow colour* shows the suspected precursory slip except for the different baselines and the corresponding daily time series from GEONET

period in which the time series are ended by co-seismic upward jumps, as shown by the arrows.

Finally, from the red vectors in Fig. 4a, the baselines across the fault, between stations A, B and H–M, are expected to shrink by the slow slip, which corresponds to an acceleration of the contraction of the baselines. In fact, the time series trends in Fig. 4c during the first and second yellow-shaded period becomes steeper, where the time series are ended by co-seismic downward jumps.

To summarise this section, we have observed geodetic anomalies (i.e., non-constant changes) that are not due to the slow slip beneath the Bungo Straight, but are consistent with changes of crust deformation under the assumed precursory slow slips on the fault of the Fukuoka earthquake.

4. Discussion

This case study is retrospective and entirely based on the fault model of the focal earthquake that already occurred. We have discussed the seismicity anomalies on the selected zones, where the predominant mechanism is known as the receiver faults, to check the possible slips on the fault of rupture. However, in general, inference of such subtle slow slips based on seismicity anomalies is not easy. Indeed, the slip location, the angles of the slipping fault, and its imminence to a major rupture are difficult to identify based only on seismicity anomalies. Most of these details are unknown to us unless any other relevant data or constraints are available.

For example, OGATA (2005c) observed a significant relative quiescence in the aftershock sequence of the same M 7.0 earthquake before the occurrence of the largest aftershock, and considered several likely and/or unlikely speculative scenarios of stress transfers from some possible aseismic slip of known faults around the source. The concern was whether these slow slips could have been promoted or inhibited by the main M 7.0 rupture; moreover, whether the stress shadows due to the triggered slip should have, in turn, covered the majority of the focal aftershock region. Eventually, the largest aftershock of M 5.8 occurred in the southern end of the first aftershock volume, from which we know that the strike angle of the rupture fault is slightly different from that of the main fault, but has a significantly different ΔCFS configuration covering the aftershock volume. Knowing such information for the fault mechanism, we were able to develop a more probable scenario of the precursory slip that explains the detailed space–time changes in both activities of the aftershocks and the off-fault events preceding the M 5.8 event (OGATA, 2006a).

It is even more difficult to identify whether the suspected slip is an imminent precursor to a rupture and to assess its size. Currently, few empirical relationships have been found to forecast this. Incidentally, TSUBOKAWA (1969) suggested a statistical relationship between the magnitude of the expected earthquakes and the duration of geodetic anomalies. OHTAKE (1993) observed a similar correlation between the magnitude of the major earthquakes and length of the period during which seismic activity is quiet relative to the ETAS model over a wide region (OGATA, 1992). Nevertheless, through the current case studies, no helpful clues on these aspects were found.

It should be also noted from the prediction viewpoint that, in some regions, aseismic slips are not necessarily the immediate precursors to a large event. For example, the seismic quiescence and geodetic anomaly in the Parkfield zone (WYSS et al., 1990a, b) were not followed by the main rupture until 2004. Moreover, a number of aseismic slips, or silent earthquakes, have been observed in the same region with no subsequent large events (HIROSE et al., 1999). Also, we have some reported and several unreported empirical results that (relative) quiescence is not always followed by a large event. Therefore, the identification of an aseismic slip leading to the rupture of an asperity remains an even more difficult research topic in earthquake prediction. At present, this issue can only be described in terms of probabilistic forecasting (e.g., OGATA, 2001), and the efficiency of forecasting performance could be enhanced by further investigations of stress changes to discriminate precursory slip.

5. Conclusion

In this paper we have used the ETAS model to examine seismicity rate changes in the selected seismic zones in and around Kyushu. The increase or decrease of the seismicity rates from those predicted by the ETAS model (the relative activation and relative quiescence, respectively) were found in some sub-regions during the latter period preceding the 2005 earthquake of M 7.0 west of Fukuoka. Assuming precursory slow slips on the fault of the main shock, the seismic zones of negative and positive increments of the CFS consistently correspond to those of relative quiescence and activation, respectively. The hypothesis of the precursory slips can also explain the transient crustal movements around the source, namely, the time series of baseline distances between the permanent GPS stations have velocity changes at common time points that are basically consistent with the horizontal displacements of the stations due to the assumed slip.

Acknowledgment

Comments by David Vere-Jones, Mark Bebbington, and Massimo Cocco helped me with clarification of the paper. I have used the TSEIS visualisation program package (TSURUOKA, 1996) for the study of hypocentre data, and the PC program MICAP-G (NAITO and YOSHIKAWA, 1996; OKADA, 1992), and CFFVS from Manabu Hashimoto, DPRI, Kyoto University for spatial visualization of Coulomb stress changes. I used the earthquake hypocenter catalogue of the JMA (which contains the original data from the Universities and the NIED), the GEONET-GPS data of the GSI, and the F-net catalogue of the NIED. This study is partially supported by the Japan Society for the Promotion of Science under grant-in-aid for Scientific Research no. 17200021 and 20240027; and by the 2007 and 2008 projects of the Institute of Statistical Mathematics and the Transdisciplinary Research Integration Centre of the Research Organization of Information and Systems, Inter-University Research Institute Corporation.

Appendix 1: ETAS Model and Change-point Analysis for Seismicity

The epidemic-type aftershock sequence (ETAS) model is a standard model to predict a short-term seismic activity in a closed geophysical region such that the occurrence rate of an earthquake at time t is given by

$$\lambda(t|H_t) = \mu + \sum_i^{t_i<t} K_0 (t - t_i + c)^{-P} e^{\alpha(M_i - M_c)}, \quad (1)$$

where $H_t = \{(t_i, M_i), t_i < t\}$ is the history of the occurrence times and magnitudes of earthquakes before time t; μ, K_0, c, α, p are empirical parameters, and M_i and M_c represent the magnitude of ith earthquakes and the cut-off magnitude, respectively. To estimate the parameters (μ, K_0, c, α, p), OGATA (1988) proposed a method that maximizes the log-likelihood function

$$\ln L(\theta) = \sum_{\{i; S < t_i < T\}} \ln \lambda_\theta(t_i|H_{t_i}) - \int_S^T \lambda_\theta(t) dt \quad (2)$$

with respect to the parameters $\theta = (\mu, K_0, c, \alpha, p)$, where $\{(t_i, M_i), i = 1, 2, \ldots\}$ is the data consisting of occurrence times and magnitude of earthquakes in the time interval (S, T). Here, we also use the data in the precursory period $(0, S)$ for history prior to the target period (S, T). Then we can see how well or how poorly the model fits an earthquake sequence by comparing the cumulative number $N(S, t)$ of earthquakes with the rate predicted by the model

$$\Lambda(t|H_t) = \int_0^t \lambda(s|H_s) ds \quad (3)$$

in the time interval $S < t < T$. Furthermore, from a given series of magnitudes of earthquakes, we can calculate the predicted cumulative curve for the extrapolated period (T, T_{end}) to compare this with the empirical cumulative function. The readers are referred to OGATA (2006b) for the FORTRAN programs of these methods and their manuals.

The Akaike information criterion, AIC = $(-2) \max_\theta \log L(\theta) + 2 \dim\{\theta\}$, where $\dim\{\theta\}$ means the number of adjusted parameters, is useful to compare the goodness-of-fit of the competing models to a given data set (AKAIKE, 1974). The model with a smaller AIC value shows a better fit to the data. To examine whether or not the temporal seismicity pattern changed at a suspected time t_c on a time interval

(S, T) in a given data set, we consider an ETAS model with a change point for the parameter set (two-states ETAS model), applied to the occurrence data sets on the separated sub-intervals (S, t_c) and (t_c, T) to calculate the corresponding AICs, AIC1 = $-2 \max_{\theta_1} \ln L(\theta_1; S, t_c) + 2 \dim\{\theta_1\}$ and AIC2 = $-2 \max_{\theta_2} \ln L(\theta_2; t_c, T) + 2 \dim\{\theta_2\}$, respectively. Then, we compare AIC1 + AIC2 with AIC0 = $(-2) \max_\theta \ln L(\theta; S, T) + 2 \dim\{\theta\}$ of the single ETAS model for the fit to the whole data from the period (S, T). To validate the significance of the seismicity change, we compare AIC0 with AIC1 + AIC2 + $2q(N)$ where $q(N)$ is the penalty value for the change-point time t_c, since we actually search for t_c that minimizes AIC1 + AIC2. Here, the penalty $q(N)$ varies monotonically from 2 to 3 depending on the total number N of events in the interval (S, T), while the penalty in AIC for the other parameter is unity [see OGATA (1999)] for the function form of $q(N)$. The criteria for this comparison takes into account the over-fitting bias due to the greater complexity of the two-state model, and also the freedom in searching for a change-point (OGATA, 1992). In Table 1, we list ΔAIC = AIC1 + AIC2 + $2q(N)$ − AIC0. If ΔAIC takes a positive value, this indicates that the single ETAS model is selected for the seismicity in the whole period. If ΔAIC takes a negative value, the seismicity change is regarded to be significant.

Theoretical cumulative function of the ETAS model is calculated using the estimated parameters obtained by fitting to the sequence of earthquakes in either the entire period or the first part of the period before the change-point. In the latter case, the cumulative function is extrapolated for the remaining period using the function (3) that integrates the rate of the estimated ETAS model of the first interval.

Appendix 2: Coulomb Stress Change

Our concern is the relationship to the change pattern of the Coulomb failure stress (CFS) transferred from a rupture or silent slip elsewhere. Changes in seismic activity rate are often reported (REASENBERG and SIMPSON, 1992; TODA et al., 1998) to correlate with the calculated Coulomb failure stress change

$$\Delta\text{CFS} = \Delta(\text{shear-stress}) - \mu' \Delta(\text{normal-stress}),$$

where $\mu' = 0.4$ is assumed for the apparent coefficient of friction and positive normal stress means the compression. The Coulomb stress change in an elastic half-space (OKADA, 1992) is calculated by assuming a shear modulus 3.2×10^{11} dyn cm^{-2} and a Poisson ratio of 0.25. Positive values of ΔCFS promote failure while negative ones inhibit failure. The region with negative ΔCFS is called a stress shadow (HARRIS and SIMPSON, 1996). We calculate ΔCFS assuming that the size of the precursory slip is tentatively taken 10% as large as the main rupture unless any relevant information is available.

This paper assumes that there is no threshold value of ΔCFS capable of affecting seismic changes. The stress changes due to aseismic slip can be small values, of the order of a few tens of millibars or less. The small stress change may have minor effect for each receiver fault, but the number of faults of small sizes to be triggered or inhibited in a seismic zone can be substantial if the fault orientations are similar.

REFERENCES

AKAIKE, H. (1974), *A new look at the statistical model identification*, IEEE Trans. Autom. Control *19*, 716–723.

DIETERICH, J. (1994), *A constitutive law for rate of earthquake production and its application to earthquake clustering*, J. Geophys. Res. *99*, 2601–2618.

DRAGERT, H. K., WANG, K., and JAMES, T. S. (2001), *A silent slip event on the deeper Cascadia subduction interface*, Science *292*, 1525–1527.

DZIEWONSKI, A. M. and WOODHOUSE, J. H., (1983), *An experiment in systematic study of global seismicity: Centroid-moment tensor solutions for 201 moderate and large earthquakes of 1981*, J. Geophys. Res. *88*, 3247–3271.

DZIEWONSKI, A. M., CHOU, T.-A., and WOODHOUSE, J. H. (1981), *Determination of earthquake source parameters from waveform data for studies of global and regional seismicity*, J. Geophys. Res. *86*, 2825–2852.

EVISON, F. F. (1977), *The precursory earthquake swarm*, Phys. Earth Planet. Inter. *15*, 19–23.

GSI (2004), *Crustal Movements in the Chugoku, Shikoku and Kyushu Districts*, Report of the Coordinating Committee for Earthquake Prediction *71*, 680–694.

GSI (2005a), *Crustal movement in Japan detected by GPS-based control station*, http://mekira.gsi.go.jp/ENGLISH/.

GSI (2005b), *Crustal movement in the Hokkaido District*, Report of the Coordinating Committee for Earthquake Prediction *74*, 53–68.

GSI (2007), *Crustal movement in the Hokkaido District*, Report of the Coordinating Committee for Earthquake Prediction *77*, 44–64.

HARRIS, R. A. and SIMPSON, R. W. (1996), *In the shadow of 1857—The effect of the great Ft. Tejon earthquake on subsequent earthquakes in southern California*, Geophys. Res. Lett. *23*, 229–232.

HATANAKA, Y., SAWADA, M., HORITA, A., and KUSAKA, M. (2001a), *Calibration of antenna-radome and monument-multipath effect of GEONET—Part 1: Measurement of phase characteristics*, Earth Planets Space *53*, 3–21.

HATANAKA, Y., SAWADA, M., HORITA, A., KUSAKA, M., JOHNSON, J.M., and ROCKEN, C. (2001b), *Calibration of antenna-radome and monument-multipath effect of GEONET—Part 2: Evaluation of the phase map by GEONET data*, Earth Planets Space *53*, 23–30.

HATANAKA, Y., IIZUKA, T., SAWADA, M., YAMAGIWA, A., KIKUTA, Y., JOHNSON, J.M., and ROCKEN, C. (2003), *Improvement of the Analysis Strategy of GEONET*, Bull. Geograph. Surv. Inst. *49*, 11–34.

HATANAKA, Y., YAMAGIWA, A., YUTSUDO, T., and MIYAHARA, B. (2005), *Evaluation of routine solutions of GEONET*, J. Geograph. Surv. Inst. *108*, 49–56 (in Japanese).

HATANAKA, Y. (2006), *Enhancement of continuous GPS observation networks as geoscience sensors—resolving signal/noise of GPS observable*, J. Geodetic Soc. Jpn. *52*, 1–19 (in Japanese).

HEKI, K., MIYAZAKI, S., and TSUJI, H. (1997), *Silent fault slip following an interpolate thrust earthquake at the Japan Trench*, Nature *386*, 595–597.

HIROSE, H., HIRAHARA, K., KIMATA, F., FUJII, N., and MIYAZAKI, S. (1999), *A slow thrust slip event following the two 1996 Hyuga-nada earthquakes beneath the Bungo Channel, southwest Japan*, Geophys. Res. Lett. *26*, 3237–3240.

INOUYE, W. (1965), *On the seismicity in the epicentral region and its neighborhood before the Niigata earthquake*, Kenshin-jiho (Q. J. Seismol.) *29*, 139–144 (in Japanese).

KAWASAKI, I., ASAI, Y., TAMURA, Y., SAGIYA, T., MIKAMI, N., OKADA, Y., SAKATA, M., and KASAHARA, M. (1995), *The 1992 Sanriku-oki, Japan, ultra-slow earthquake*, J. Phys. Earth *43*, 105–116.

KISSLINGER, C. (1988), *An experiment in earthquake prediction and the 7th May 1986 Andreanof Islands Earthquake*, Bull. Seismol. Soc. Am. *78*, 218–229.

KEILIS-BOROK, V. I. and MALINOVSKAYA, L. N. (1964), *One regularity in the occurrence of strong earthquakes*, J. Geophys. Res. *70*, 3019–3024.

MIYAZAKI, S., HATANAKA, Y., SAGIYA, T., and TADA, T. (1998), *The nationwide GPS array as an Earth observation system*, Bull. Geograph. Surv. Inst. *44*, 11–22.

NAITO, H. and YOSHIKAWA, S. (1996), *A program to assist crustal deformation analysis*, Zisin II (J. Seismol. Soc. Japan) *52*, 101–103 (in Japanese).

NIED (2007), *Earthquake Mechanism Search in F-net Broadband Seismograph Network data*, http://www.fnet.bosai.go.jp/freesia/event/search/search.html.

OGATA, Y. (1988). *Statistical models for earthquake occurrences and residual analysis for point processes*, J. Am. Stat. Assoc. *83*, 9–27.

OGATA, Y. (1992), *Detection of precursory relative quiescence before great earthquakes through a statistical model*, J. Geophys. Res. *97*, 19845–19871.

OGATA, Y. (1999), *Seismicity analyses through point-process modelling: a review*, Pure Appl. Geophys. *155*, 471–507.

OGATA, Y. (2001), *Increased probability of large earthquakes near aftershock regions with relative quiescence*, J. Geophys. Res. *106*, B5, 8729–8744.

OGATA, Y. (2004), *Static triggering and statistical modeling*, Report of the Coordinating Committee for Earthquake Prediction *72*, 631–637 (in Japanese).

OGATA, Y. (2005a), *Detection of anomalous seismicity as a stress change sensor*, J. Geophys. Res. *110*, B5, B05S06, doi:10.1029/2004JB003245.

OGATA, Y. (2005b), *Synchronous seismicity changes in and around the northern Japan preceding the 2003 Tokachi-oki earthquake of M 8.0*, J. Geophys. Res. *110*, B5, B08305, doi:10.1029/2004JB003323.

OGATA, Y. (2005c), *Relative quiescence reported before the occurrence of the largest aftershock (M5.8) in the aftershocks of the 2005 earthquake of M 7.0 at the western Fukuoka, Kyushu, and possible scenarios of precursory slips considered for the stress-shadow covering the aftershock area*, Report of the Coordinating Committee for Earthquake Prediction *74*, 529–535 (in Japanese).

OGATA, Y. (2006a), *Monitoring of anomaly in the aftershock sequence of the 2005 earthquake of M 7.0 off coast of the western Fukuoka, Japan, by the ETAS model*, Geophys. Res. Lett. *33*, L01303, doi:10.1029/2005GL024405.

OGATA, Y. (2006b), *Statistical Analysis of Seismicity—SASeis 2006*, Computer Science Monographs, No. 33, The Institute of Statistical Mathematics, 4-6-7 Minami-Azabu, Minato-Ku, Tokyo, Japan. http://www.ism.ac.jp/editsec/csm/index.html.

OGATA, Y. (2007), *Seismicity and geodetic anomalies in a wide area preceding the Niigata-Ken-Chuetsu Earthquake of October 23, 2004, central Japan*, J. Geophys. Res. *112*, B10, B10301. doi:10.1029/2006JB004697.

OHTAKE, M., MATUMOTO, T., and LATHAM, G. V. (1977), *Seismicity gap near Oaxaca, southern Mexico as a probable precursor to a large earthquake*, Pure Appl. Geophys. *115*, 375–385.

OHTAKE, M. (1993), *Does earthquake occur in seismicity gap?* Zisin J. *15*, 12–19, Association for the Development of Earthquake Prediction, Tokyo (in Japanese).

OKADA, Y. (1992), *Internal deformation due to shear and tensile faults in a half-space*, Bull. Seism. Soc. Am. *82*, 1018–1040.

OZAWA, S., MURAKAMI, M., KAIZU, M., TADA, T., SAGIYA, T., HATANAKA, Y., YARAI, H., and NISHIMURA, T. (2002), *Detection and monitoring of ongoing aseismic slip in the Tokai region, central Japan*, Science *298*, 1009–1012.

REASENBERG, P.A. and SIMPSON, R.W. (1992), *Response of regional seismicity to the static stress change produced by the Loma Prieta earthquake*, Science *255*, 1687–1690.

SEGALL, P., DESMARAIS, E. K., SHELLY, D., MIKLIUS, A., and CERVELLI, P. (2006), *Earthquakes triggered by silent slip events on Kilauea volcano, Hawaii*, Nature *442*, 71–74, doi:10.1038/nature04938.

SEKIYA, H. (1976), *The seismicity preceding earthquakes and its significance in earthquake prediction*, Zisin II (J. Seismol. Soc. Japan) *29*, 299–312 (in Japanese).

SENO, T. (1999). *Syntheses of the regional stress fields of the Japanese Islands*, The Island Arc *8*, 66–79.

SYKES, L. R. and JAUME, S. C. (1990), *Seismic activity on neighboring faults as a long-term precursor to large earthquakes in the san francisco Bay area*, Nature *348*, 595–599.

TADA, T. (1993), *Crustal deformation in central Kyusyu, Japan and its tectonic implication : Rifting and spreading of the Beppu-Shimabara Graben*, The memoirs of the Geological Society of Japan *41*, 1–12.

TODA, S., STEIN, R. S., REASENBERG, P., DIETERICH, J. H., and YOSHIDA, A. (1998), *Stress transferred by the $M_w = 6.9$ Kobe, Japan, shock: Effect on aftershocks and future earthquake probabilities*, J. Geophys. Res. *103*, 24543–24545.

TSUBOKAWA, I. (1969), *On relation between duration of crustal movement and magnitude of earthquake expected*, J. Geod. Soc. Japan *15*, 75–88.

TSURUOKA, H. (1996), *Development of seismicity analysis software on workstation*, Tech. Res. Rep. *2*, 34–42, Earthquake Res. Inst., Univ. Tokyo (in Japanese).

UTSU, T. (1968), *Seismic activity in Hokkaido and its vicinity*, Geophys. Bull. Hokkaido Univ. *13*, 99–103 (in Japanese).

WYSS, M. and BURFORD, R. O. (1987), *Occurrence of a predicted earthquake on the San Andreas fault*, Nature *329*, 323–325.

WYSS, M., BODIN, P., and HABERMANN, R. E. (1990a), *Seismic quiescence at Parkfield: an independent indication of an imminent earthquake*, Nature *345*, 426–428.

WYSS, M., and BURFORD, R. O. (1987), *Occurrence of a predicted earthquake on the San Andreas fault*, Nature *329*, 323–325.

WYSS, M., SLATER, L., and BURFORD, R. O. (1990b), *Decrease in deformation rate as a possible precursor to the next Parkfield earthquake*, Nature *345*, 428–431.

(Received August 1, 2008, revised February 3, 2009, accepted June 9, 2009, Published online March 31, 2010)